A Student's Companion in Molecular Cell Biology

A Student's Companion in Molecular Cell Biology

David Rintoul and **Ruth Welti**

Kansas State University

Muriel Lederman and **Brian Storrie**

Virginia Polytechnic Institute and State University

Robert Van Buskirk

State University of New York at Binghamton

SCIENTIFIC
AMERICAN
BOOKS

Distributed by W. H. Freeman and Company, New York

Cover illustration by Tomo Narashima

ISBN 0-7167-2110-4

Printed in the United States of America

Scientific American Books is a subsidiary of Scientific American, Inc.
Distributed by W. H. Freeman and Company, 41 Madison Avenue, New
York, New York 10010 and 20 Beaumont Street, Oxford OX1 2NQ
England

3 4 5 6 7 8 9 0 VB 9 9 8 7 6 5 4 3 2

Preface

Claude Bernard once wrote "Man can learn nothing unless he proceeds from the known to the unknown." In *A Student's Companion in Molecular Cell Biology* we have attempted to adhere to this insightful adage, which is particularly relevant to the teaching of molecular cell biology today. Our principal goal in preparing this *Companion* was to expose students to the concepts, problems, and open questions of the field in a question and answer format. Thus, true to Bernard, the questions and problem sets for each chapter are organized in a sequence that leads from the known (Parts A and B), to extensions of what is known (Parts C and D), to the relatively unknown (Part E). Many of the questions are designed to introduce some of the flavor of the laboratory experience into the learning process.

Although inspired by and closely parallel to *Molecular Cell Biology*, Second Edition, by James Darnell, Harvey Lodish, and David Baltimore, the value of *A Student's Companion* as an independent text is enhanced by a complete answer key, a glossary, and an index. Its uses will be as varied as are the students of molecular cell biology. It can be used as a required supplement, as a self-study guide for undergraduate and graduate students taking a molecular cell biology course, or as an aid in preparation for tests that include cell biological material (e.g., MCAT, GRE, or Ph.D. preliminary examinations). Students of all levels will find this volume a welcome learning and review tool.

A student using *Molecular Cell Biology* should be able to read the relevant chapter in the text and easily answer the objective questions in Parts A and B. Thus prepared, the student can readily move to the Part C questions that attempt to further elucidate important concepts from the chapter. Answering these questions will require the student to synthesize and re-express the material in his or her own words, to confirm that it is understood. Problem sets requiring application of these facts and concepts are found in Part D, whereas problem sets that require additional information, concepts, or logic are to be found in Part E.

Many of the problems in this *Companion* are based on published experimental results. Sometimes the transformation from experimental result to teaching tool required simplification or deletion of some data; we hope that researchers whose results are the bases for questions will understand the pedagogical utility of this strategy. In all cases we have attempted to retain the logic and conclusions of important experimental results; we sincerely hope that our respect for the work of other cell biologists is apparent to the reader.

In some cases the answers to problems in Parts D and E may seem incomplete to readers familiar with the research areas. This is due in part to the adaptation of experimental results to a question and answer format. More important, however, it is because interpretations and conclusions at the cutting edge of knowledge are always incomplete. We believe that this experience of the experimental method — identifying the limits of our information and formulating the next questions to be answered and the experiments that might answer them — will promote much of the learning to be gained from *A Student's Companion in Molecular Cell Biology*.

Many of the questions and problems have been reviewed by students; indeed, many of these students were unwitting reviewers during our years of teaching cell biology using *Molecular Cell Biology*. All the questions have been reviewed by colleagues, who have contributed much of their valuable time toward improving the clarity and quality of this book. These colleagues include James Darnell, Harvey Lodish,

David Baltimore, Thomas M. Roberts, Lon S. Kaufman, Larry G. Williams, Susan E. Conrad, Patrick Bender, Brian Spooner, Donald Roufa, Daniel Raben, Joseph Baldassare, S. Keith Chapes, and G. William Fortner. We, the authors, however, are ultimately responsible for any mistakes, misinterpretations, omissions, or oversights, and we would greatly appreciate any comments from either teachers or students who use the book. Your comments should be directed to our Publisher, W. H. Freeman and Company/Scientific American Books, Inc. at 41 Madison Ave., New York, NY 10010. Our appreciation goes to the staff of W. H. Freeman and Company/Scientific American Books, Inc., particularly Kay Ueno, Patrick Fitzgerald, Margot Getman, Philip McCaffrey, Ruth Steyn, Howard Johnson, Andrew Kudlacik, Mara Kasler, and Ellen Cash for their role in initiation, organization, and assembly of this volume, and for helping to sustain our efforts during its preparation.

David Rintoul
Ruth Welti
Muriel Lederman
Brian Storrie
Robert Van Buskirk

Contents

A Student's Companion in Molecular Cell Biology

C H A P T E R *1*

Chemical Foundations

PART A: *Reviewing Basic Concepts*

Fill in the blanks in statements 1–17 using the most appropriate terms from the following list:

acceptor	exothermic
activation energy	free energy
anabolism	glucose
antibodies	hydrogen
ATP	hydrogen bonds
base	hydrophobic
catabolism	interactions
catalyst	hydroxyl
covalent bonds	ionic bonds
dipolar	inordinate
donor	oxidation
endergonic	reduction
endothermic	sodium
enthalpy	van der Waals
entropy	interactions
enzymes	water
exergonic	zwitterion

1. _____ hold the atoms in a molecule together; these bonds have strengths ranging from 50–200 kcal/mol.

2. A covalent bond that is slightly negatively charged at one end and slightly positively charged at the other end is said to be _____.

3. A hydrogen bond is a relatively weak association between an electronegative atom, also known as the _____ atom, and a hydrogen atom covalently bonded to another atom, also known as the _____ atom.

4. _____ are the noncovalent bonds responsible for the high melting and boiling points of water.

5. In aqueous solutions, simple ions of biological significance, such as Na^+ and Cl^-, are surrounded by a stable, tightly bound shell of _____.

6. Weak attractive forces between noncovalently bonded atoms, resulting from the formation of transient dipoles are called _____.

7. Water dissociates into hydronium and *hydroxyl* ions.

8. Any molecule or ion that can combine with a hydrogen ion is called a(n) *base*.

9. A molecule with both positively and negatively charged

atoms, such as an amino acid at pH 7, is called a(an) _____.

10. In a(n) _exothermic_ reaction, heat is given off and ΔH is negative.

11. _entropy_ is a measure of the degree of randomness or disorder.

12. The loss of electrons from an atom is called _oxidation_; the gain of electrons from an atom is called _reduction_.

13. _ATP_ is an important cellular molecule for capturing and transferring free energy.

14. The process by which organic molecules are degraded to provide energy for living organisms is known as _____.

15. A reaction that absorbs energy is known as a(n) _endergonic_ reaction.

16. The input of energy required to initiate a reaction is known as the _activation energy_

17. A(n) _catalyst_ is any substance that increases the rate of a reaction without being permanently changed; proteins that perform this function are known as _enzymes_.

PART B: _Linking Concepts and Facts_

Circle the letters corresponding to the most appropriate terms/phrases that complete items 18–27; more than one of the choices provided may be correct in some cases.

18. Forms of energy that can be quantitated in calories include
   a. heat.
   b. kinetic energy.
   c. concentration gradients.
   d. chemical bond energies.
   e. entropy.

19. If the pH of a solution is 9, then
   a. the solution is said to be acidic.
   b. the hydrogen ion concentration is $10^{-9}\ M$.
   c. the solution has three times more hydrogen ions than a solution of pH 12.
   d. the hydroxyl ion concentration is $10^{-5}\ M$.
   e. the solution has 100 times more hydroxyl ions than a solution of pH 7.

20. If the pH of a solution of HA, which dissociates to $H^+ + A^-$, equals the pK_a, then
   a. the pH of the solution is 7.
   b. the concentration of the acid form of the compound (HA) equals the concentration of the dissociated form of the compound (A^-).
   c. the solution has a greater capacity for buffering than at any other pH.

   d. the hydrogen ion concentration is equal to the equilibrium constant for the reaction describing the dissociation of the acid, $HA \rightleftharpoons H^+ + A^-$.

   e. if acid is added to the solution, protons released by the added acid will be taken up by the HA form of the compound.

21. If the equilibrium constant for the reaction $A \rightleftharpoons B$ is 0.5 and the initial concentration of B is 10 mM and of A is 20 mM, then

   a. the reaction will proceed in the direction it is written, producing a net increase in the concentration of B.

   b. ΔG equals 0.

   c. the reaction will produce energy, which can be used to drive ATP synthesis.

   d. the reaction is at equilibrium.

   e. the reaction will proceed in the reverse direction, producing a net increase in the concentration of A, if a catalyst is added to the reaction mixture.

22. An increase in entropy

   a. is equivalent to an increase in the total bond energies of the reactants.

   b. is an increase in order.

   c. occurs when a NaCl solution is diluted.

   d. occurs when a hydrocarbon molecule is removed from an aqueous environment.

   e. occurs in the system when amino acids are linked to form a protein.

23. If, for a biochemical reaction, $\Delta H < 0$ and $\Delta S > 0$, then

    a. the reaction is spontaneous.

    b. the disorder in the system will increase if the reaction proceeds.

    c. the reaction is endothermic.

    d. the reaction is endergonic.

    e. ΔG is positive.

24. The conditions that apply to the standard free-energy change $\Delta G^{\circ\prime}$ for a reaction include

    a. pH 7.0.

    b. 1 atm.

    c. 1 M initial concentrations of all reactants and products except protons and water.

    d. 1 liter reaction volume.

    e. 298 K (25°C).

25. The standard change in reduction potential for an oxidation-reduction reaction $\Delta E_0'$,

    a. is positive when $\Delta G^{\circ\prime}$ is negative.

    b. is the sum of the standard reduction potentials of the individual oxidation or reduction steps in the reaction.

    c. is negative when an oxidation-reduction reaction, under standard conditions, is spontaneous.

    d. is defined as the change in reduction potential at 25°C and 1 atm with 1 M reactants.

    e. is positive when electrons move toward atoms or molecules with more positive reduction potentials.

26. A cellular process coupled to the hydrolysis of the phosphoanhydride bonds of ATP

    a. can often proceed even if the free energy change of the uncoupled process (excluding the ATP hydrolysis) is positive.

    b. produces glucose from CO_2 and O_2 in photosynthetic plants and bacteria.

    c. generates ion gradients across cell membranes.

    d. is used to connect simple sugars to form oligosaccharides.

    e. usually is catalyzed by an enzyme.

27. The presence of a catalyst in a reaction mixture may affect

    a. the rate of the reaction.

    b. the equilibrium constant.

    c. the standard free energy change.

    d. the activation energy.

    e. the structure of the catalyst permanently.

28. In the space provided, write the letter(s) corresponding to the types of chemical(s) or interaction(s) to which each of the following phrases applies. C = covalent bonds; H = hydrogen bonds; I = ionic bonds; V = van der Waals interactions; and Hy = hydrophobic interactions.

    a. Have a strength of 1–5 kcal/mol ____

    b. Dissociate in aqueous solution ____

    c. Form when two hydrocarbon molecules interact in water ____

    d. Play a role in the structure of biological macromolecules, such as proteins ____

    e. Are oriented at precise angles to one another ____

In items 29–34, write the requested word or symbol on each line marked with an asterisk in the accompanying figure.

29. In Figure 1-1, label the partial negative or partial positive charge on each atom in the water molecule.

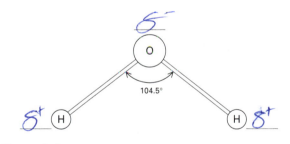

▲ **Figure 1-1**

30. The oxygen molecules shown in Figure 1-2 are in van der Waals contact. Name each of the distances shown.

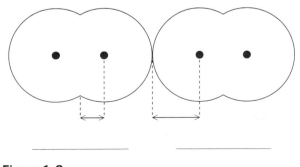

▲ **Figure 1-2**

31. Indicate the pK_a of the acid HA whose titration curve is shown in Figure 1-3.

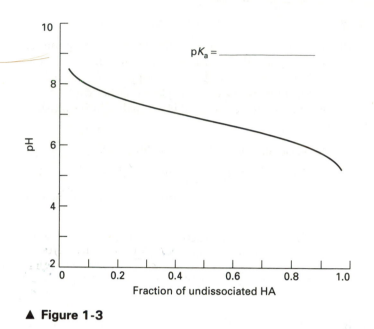

pK_a = _____

▲ **Figure 1-3**

32. In Figure 1-4, label the indicated bonds.

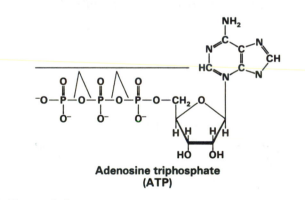

Adenosine triphosphate (ATP)

▲ **Figure 1-4**

33. In Figure 1-5, label the indicated differences in potential energies.

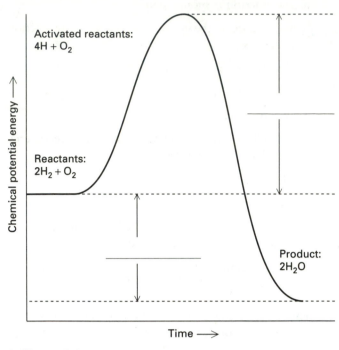

Activated reactants: 4H + O$_2$

Reactants: 2H$_2$ + O$_2$

Product: 2H$_2$O

▲ **Figure 1-5**

34. In Figure 1-6, one curve represents a catalyzed reaction and the other an uncatalyzed reaction. Indicate which curve is which.

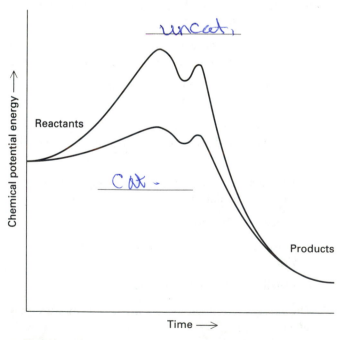

▲ **Figure 1-6**

PART C: *Putting Concepts to Work*

35. Triacylglycerols or fats form water-excluding "droplets" within the cytoplasm of adipocytes or fat storage cells. In contrast, more polar molecules such as ATP dissolve in aqueous solutions such as cytoplasm.

    a. What is the interaction between the fat molecules called?

    b. What is the effect on the entropy S when ATP is dissolved in water?

    c. What would be the effect on S if triacylglycerols were dissolved in water? Why?

    d. What is the effect of forming fat droplets on the enthalpy H? Why?

    e. What do your answers to parts (c) and (d) imply about the change in free energy G for fat droplet formation in an aqueous system?

36. A solution of 8 M urea is sometimes used in the isolation of protein molecules. When this solution is prepared by dissolving urea in water at room temperature, it becomes cold.

    a. What can you infer about the ΔH, ΔS, and ΔG values for the dissolution of urea in water?

    b. How would you expect the ΔG for this process to change if you tried to dissolve urea in the cold room rather than at room temperature?

37. Catabolic reactions are involved in the breakdown of cellular fuels, whereas anabolic reactions are involved in the biosynthesis of cellular materials. Which type of reaction do you think would generally produce ATP? Which type of reactions would consume it?

PART D: *Developing Problem-solving Skills*

38a. If 1 ml of a solution of 0.01 M HCl is diluted to 100 ml at 25°C, what is the pH of the resulting solution?

    b. If 1 ml of a solution of 1 M NaOH is diluted to 100 L at 25°C, what is the pH of the resulting solution?

39. The pK_a for the dissociation of acetic acid is 4.74.

    a. Calculate the amounts of acetic acid (HAc) and sodium acetate (Ac^-) which you would need to use to make a buffer that is 0.1 M in total acetate concentration and at pH 5.

    b. Calculate the pH of this buffer after the addition of 0.01 mol/L HCl.

40. A titration curve, such as that shown in Figure 1-3, indicates how the fraction of molecules in the undissociated form of an acid (HA) depends on the pH.

    a. At one pH unit below the pK_a of an acid, 91 percent of the molecules are in the HA form, and at one pH unit above the pK_a, 91 percent are in the A^- form. How is this determined?

    b. What fraction is in the HA form at 0.5 units above the pK_a?

    c. What fraction is in the HA form at 2 units below the pK_a?

41. For the reaction $A + B \rightleftharpoons C + D$, $\Delta G°' = -2.4$ kcal/mol.

    a. What would be the net direction of the reaction under standard conditions (1 M reactants and products, 298 K, 1 atm)?

    b. Calculate ΔG and predict the net direction of the reaction if the initial concentrations were [A] = 0.01 mM, [B] = 0.01 mM, [C] = 5 mM, and [D] = 10 mM and the temperature is 37°C.

42. The following reaction is catalyzed by the enzyme phosphoglycerate kinase:

    1,3-bisphosphoglycerate + ADP $\rightleftharpoons$

                    3-phosphoglycerate + ATP

    $\Delta G°'$ for this reaction is -4.5 kcal/mol. Assuming

that the ratio of [ATP] to [ADP] is 10 and the temperature is 25°C, calculate the ratio of 3-phosphoglycerate to 1,3-bisphosphoglycerate at equilibrium.

43. Hydrolysis of ATP to produce ADP under standard conditions has a ΔG of -7.3 kcal/mol. The hydrolysis of ATP is often used as a source of energy to "pump" substances against a concentration gradient into or out of a cell. Assume that the hydrolysis of one molecule of ATP is coupled to the transport of one molecule of substance A from the inside to the outside of a cell and that ΔG for ATP hydrolysis under cellular conditions is -7.3 kcal/mol at 25°C. At 25°C, if the concentration of substance A inside the cell is 100 μM, what is the maximum concentration of substance A outside the cell against which the pump can export it?

44. Ubiquinone, also called coenzyme Q or CoQ, is found in inner mitochondrial membranes, where it serves as an "electron carrier." In this capacity, CoQ undergoes an oxidation-reduction reaction:

$$CoQ + 2e^- + 2H^+ \rightleftarrows CoQH_2 \qquad E_0' = 0.10$$

In another oxidation-reduction reaction, O_2 acts as the ultimate electron acceptor in the mitochondria:

$$\tfrac{1}{2}O_2 + 2e^- + 2H^+ \rightleftarrows H_2O \qquad E_0' = 0.82$$

Calculate the electric-potential change ΔE and the free-energy change ΔG when $CoQH_2$ is oxidized by O_2 under standard conditions in the following reaction:

$$CoQH_2 + \tfrac{1}{2}O_2 \rightleftarrows CoQ + H_2O$$

CHAPTER 2

Molecules in Cells

PART A: Reviewing Basic Concepts

Fill in the blanks in statements 1–18 using the most appropriate terms from the following list:

active site	K_M
allosteric site	L
amphipathic	negatively
condensation	nucleation
conservative	peptide
D	phosphorylation
denatured	positively
disulfide bridge(s)	primary
fibrous	prosthetic group
globular	secondary
glycine	substrates
hydrocarbons	tyrosine
hydrogen	V_{max}
hydrophilic	x-ray crystallography
hydrophobic	zymogens
hydroxyl groups	

1. At neutral pH, the amino acids arginine and lysine are _____ charged.

2. The side chains of the amino acids alanine, isoleucine, leucine, phenylalanine, and valine consist only of _hydrocarbons_; thus these amino acids are _hydrophobic_.

3. All amino acids except _glycine_ have two stereoisomeric forms; the _L_ forms of amino acids are generally incorporated into <u>proteins</u>.

4. The reaction called _condensation_ connects two amino acids via a(n) _peptide_ bond.

5. Some polypeptides can fold into a specific three-dimensional structure that resembles a long rod, as in the _fibrous_ proteins that give tissues their rigidity. Other polypeptides can fold into a compact ball structure, as in a(n) _globular_ protein.

6. _Secondary_ structure refers to the folding of parts of polypeptides into regular structures such as α helices and β pleated sheets.

7. Phospholipids and some α helices are _amphipathic_, they contain both hydrophobic and hydrophilic parts.

8. The three-dimensional structure of many proteins has been determined using a process called _x-ray crystallography_

9. α Helices and β pleated sheets are primarily stabilized by noncovalent _hydrogen_ bonds.

10. When an amino acid in a protein is replaced by a chemically similar amino acid, the change is said to be _conservative_

11. A small molecule that binds to a protein and plays a crucial role in its function is termed a(n) _prosthetic group_.

12. Important determinants of the shape of many proteins are the covalent _____ between one or more pairs of cysteine residues in the same polypeptide chain or different chains.

13. The activity of many enzymes is regulated by their state of _____.

14. When the native structure of a protein is altered by treatments that disrupt its weak bonds and cause unfolding, the protein is said to be _denatured_.

15. The chemicals that undergo a change in a reaction catalyzed by an enzyme are the _unfolding_ of that enzyme.

16. The collective name for the regions of an enzyme that bind the substrate and catalyze the reaction is the _active site_

17. Inactive precursors of proteolytic enzymes are called _____.

18. The _Km_ of an enzyme is a measure of the affinity of an enzyme for its substrate.

Fill in the blanks in statements 19–35 using the most appropriate terms from the following list:

active site
allosteric site
antibody
antiparallel
asymmetric
cellulose
gangliosides
glycogen
glycolipids
glycoproteins
glycosidic
hapten
ligand
liposomes

nicked
nucleoside
nucleotide
parallel
peptide
phosphodiester
phosphoglyceride
purines
pyrimidines
renatured
saturated
triacylglycerol
unsaturated

19. Effectors bind to an enzyme at a(n) _____.

20. _____ is a general term for a molecule other than an enzyme substrate that binds specifically to a macromolecule.

21. A protein that is produced by an animal in response to a foreign substance and that binds the substance specifically is called a(n) _____.

22. A(n) _____ is a small chemical group that is capable of eliciting antibody production.

23. A(n) _nucleotide_ has three parts: a phosphate, a pentose, and an organic base.

24. Adenine and guanine are _purines_.

25. The bonds joining nucleotides are called _____ bonds.

26. In a DNA double helix, the orientation of the two strands is _antiparallel_

27. If a phosphodiester bond in one strand of double-stranded circular DNA is broken, the DNA is said to be _____.

28. Fatty acids with no double bonds are said to be _saturated_; those with at least one double bond are _unsat._.

29. A(n) _____ contains a phosphate group, a glycerol, and two fatty acyl chains.

30. _____ are spherical bilayer structures with an aqueous interior; they can be produced from phospholipids in the laboratory.

31. Carbon atoms 2, 3, 4, and 5 in the linear form of glucose are _____.

32. Two monosaccharides are linked by a(n) _____ bond.

33. The most common storage carbohydrate in animal cells is _glycogen_, a very long, branched polymer of glucose.

34. Glycolipids that contain N-acetylneuraminic acid are called _____.

35. The classes of molecules that constitute the human blood group antigens are _____ and _____.

PART B: *Linking Concepts and Facts*

Circle the letters corresponding to the most appropriate terms/phrases that complete or answer items 36–50; more than one of the choices provided may be correct in some cases.

36. Which amino acid(s) could substitute for tyrosine in a polypeptide without changing the overall charge of the polypeptide at neutral pH?

    a. serine

    b. glutamic acid

    c. asparagine

    d. lysine

    e. tryptophan

37. Which element(s) of protein structure depend(s) on the existence of noncovalent bonds?

    a. primary structure

    b. β pleated sheet

    c. quaternary structure

    d. tertiary structure

    e. α helix

38. Disulfide bridges

    a. are covalent.

    b. are formed by a reduction reaction.

    c. are generally found in intracellular proteins.

    d. form before proinsulin is cleaved to form insulin.

    e. form in the presence of high concentrations of glutathione.

39. Denaturation of a protein

    a. can involve disruption of hydrogen bonds.

    b. can involve disruption of hydrophobic interaction.

    c. can be caused by acidification of the protein's environment.

    d. can cause precipitation of the protein from solution.

    e. is sometimes reversible.

40. Enzyme active sites

    a. usually consist of amino acids that are contiguous in the primary structure of the protein.

    b. may contain an amino acid residue that forms a covalent bond with the substrate.

    c. are generally preserved when the protein is denatured.

    d. may have similar structures in enzymes with similar functions.

    e. of zymogens will hydrolyze substrate rapidly.

41. Coenzymes

    a. may be derived from vitamins.

    b. are not essential to the activity of the enzymes that bind them.

    c. may be prosthetic groups.

    d. are generally proteins.

    e. increase the activation energy of an enzymatic reaction.

42. The K_M of an enzyme-catalyzed reaction

    a. is equal to the catalytic rate when all substrate sites are full.

    b. describes the affinity of an enzyme for its substrate.

    c. is dependent on the enzyme concentration.

    d. is higher when the enzyme binds its substrate more tightly.

    e. is equal to the substrate concentration when the rate of the reaction is maximal.

43. Which base pair(s) occur(s) in double-stranded DNA?

    a. G-C

    b. G-A

    c. A-U

    d. G-T

    e. A-T

44. Denaturation of double-stranded DNA

    a. involves its separation into single strands.

    b. results from the destabilization of hydrogen bonds.

    c. tends to occur more readily as the temperature of the DNA solution is decreased.

    d. results in an increase in the absorbance of light at 260 nm by the DNA solution.

    e. can be reversed if the salt concentration and temperature of the solution are adjusted appropriately.

45. RNA

    a. is usually double-stranded.

    b. is a polymer of nucleotides.

    c. can hybridize with other RNA molecules but not with DNA.

d. has fewer hydroxyl groups than DNA.

e. can act as a catalyst.

46. Amphipathic lipids

  a. include phospholipids.

  b. are the major components of cytoplasmic fat droplets.

  c. spontaneously associate with each other in a noncovalent manner to form lipid bilayers.

  d. have polar head groups, which can interact with water molecules.

  e. include glycolipids.

47. Components of cellular membranes include

  a. triacylglycerols.

  b. proteins.

  c. cholesterol.

  d. oligosaccharide chains.

  e. phosphatidylcholine.

48. In cells, monosaccharides are a component of *or* can be found covalently bound to

  a. RNA.

  b. proteins.

  c. lipids.

  d. other monosaccharides.

  e. hyaluronic acid.

49. Monosaccharides that have the same chemical formula as glucose include

  a. valine.

  b. mannose.

c. lactose.

d. ribose.

e. galactose.

50. The oligosaccharide chains of a glycoprotein

  a. are noncovalently associated with the protein.

  b. may contain branches.

  c. may be linked to tyrosine residues.

  d. may contain mannose.

  e. generally play a role in the catalytic function of the protein.

Classify each substance listed in items 51–64 by writing in the corresponding letter: C = carbohydrate; L = lipid; N = nucleic acid. P = protein; and O = none of these.

51. Antibody ____

52. Actin ____

53. Fructose ____

54. Tristearin ____

55. Cellulose ____

56. RNA ____

57. β-Galactosidase ____

58. Heme ____

59. Phospholipase C ____

60. Sucrose ____

61. Phenylalanine ____

63. Glycogen ____

63. Cholesterol ____

64. Hemoglobin ____

PART C: *Putting Concepts to Work*

65. Because of the partial double bond character of the C-N peptide bond, the peptide group is planar. There are two possible configurations about the C-N bond: one in which the two α carbons are *cis* and one in which the two α carbons are *trans*:

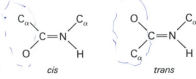

cis   trans

Which configuration do you think is energetically favored? Why?

66. A polypeptide has the following sequence:

Leu-Leu-Asp-Met-Val-Ala-Leu-Gln-His-Ser-Val-Val-Val-Leu-
  1   2   3   4   5   6   7   8   9 10 11 12 13 14

Gly-Pro-Tyr-Gly-Ala-Met-Val-Thr-His-Leu-Phe-Ala-Glu-Met
 15 16 17 18 19 20 21 22 23 24 25 26 27 28

It is known that this peptide has two α-helical segments. Where in the primary sequence would you predict the break between the two α helices to occur?

67. Lipoproteins are solid, spherical complexes of proteins and lipids. They are found in the blood of animals. Their function is to transport *nonpolar* lipids such as triacyl-glycerols and cholesterol esters (cholesterol esterified to a fatty acid) from one place to another in the body. In addition to the nonpolar lipids, some phospholipids are found in lipoproteins. The proteins found in lipoproteins often have α-helical structure. Based on your knowledge of the chemical properties of nonpolar lipids, phospholipids, and α-helical proteins, how do you think these components are arranged in a lipoprotein particle? Which components do you think would be on the outside in contact with the blood plasma? Which components do you think would be on the inside?

68. Would a *short* peptide be more likely to form a parallel or an antiparallel *β*-pleated sheet? Assume you want to maximize hydrogen bonding.

69. Why do some proteins that are denatured under reducing conditions renature spontaneously to their native state when removed from the denaturing conditions, whereas others do not?

70. You have a protein with a molecular weight of 45,000. You determine the amino acid composition and find that the protein has about three tyrosine residues, about five phenylalanine residues, and one tryptophan residue. Since chymotrypsin is specific for the peptide bond on the C-terminal side of these residues, you attempt to degrade the native protein with chymotrypsin but are unable to obtain any hydrolysis. Why might this be the case?

71. Consider the following metabolic pathway:

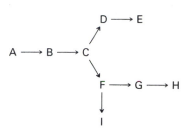

a. Which enzymatic step is likely to be inhibited by the accumulation of product E?

b. Which step is likely to be inhibited by the accumulation of product H?

c. What is the likely mechanism by which E and H effect inhibition?

72. You have purified a small amount of a protein that represents a small fraction of the total protein from a complex mixture of proteins. The purification procedure you used has many steps and requires over a month to complete. Now you find that your experiments will require a continuous supply of the purified protein that you have obtained. How might you reduce the time and effort required in further purification procedures?

73. Under the same ionic conditions, which double-stranded DNA fragment (1 or 2) would denature at the lower temperature?

5'A-A-G-T-A-G-T-T-T-C-A 3'	5'G-C-A-G-G-T-C-G-C-C-T 3'
3'T-T-C-A-T-C-A-A-A-G-T 5'	3'C-G-T-C-C-A-G-C-G-G-A 5'
1	2

PART D: *Developing Problem-solving Skills*

74. You wish to learn as much as possible about the sequence of a peptide you have isolated. Although you have only limited laboratory equipment, you are able to carry out enzymatic reactions and have a gel filtration chromatography column, which allows you to determine the molecular weight of peptides. You hydrolyze a small quantity of the peptide and send it to a friend who performs an amino acid analysis and provides you with the data shown at the right:

% composition		% composition	
Ala	8.1	Leu	17.0
Arg	0	Lys	8.0
Asn	4.2	Met	0
Asp	3.9	Phe	3.8
Cys	3.8	Pro	0
Gln	3.9	Ser	3.4
Glu	7.8	Thr	3.5
Gly	8.4	Trp	0
His	4.0	Tyr	3.4
Ile	4.1	Val	12.7

You determine that the intact peptide has a molecular weight of about 3000. Extensive treatment of the peptide with trypsin yields two fragments with molecular weights of 1000 and 2000. Treatment of the peptide with chymotrypsin yields three fragments with molecular weights of 1300, 500, and 1200.

a. Approximately how many amino acid residues are present in the peptide?

b. Based on the data provided, which amino acid may occur at the carboxyl end of the peptide?

c. Can you suggest an experiment that would help you map the location of the chymotrypsin sites with respect to the trypsin site?

75. Both diisopropylfluorophosphate (DFP) and L-1-(p-toluenesulfonyl)-amido-2-phenylethylchloromethyl ketone (TCPK) inhibit the activity of chymotrypsin; DFP also inhibits trypsin, whereas TCPK does not. Based on the structure of TCPK shown in Figure 2-1, propose a mechanism by which TCPK might act to inhibit chymotrypsin and explain why it does not inhibit trypsin.

76. When rats are given high dosages of antibiotics that kill their intestinal bacteria, their health declines (e.g., they lose their hair, develop dermatitis, and lose muscular coordination. In addition, the activity of acetyl CoA carboxylase, the rate-limiting enzyme in fatty acid synthesis, is very low in the tissues of treated rats. Develop a hypothesis that would explain the effects of antibiotics on *both* the health of the rats and on the activity of acetyl CoA carboxylase.

77. HMGCoA reductase, a critical enzyme in cholesterol biosynthesis, converts HMGCoA to mevalonic acid. When cells from a mammalian cell line are grown in the presence of the compound compactin, many more molecules of the enzyme HMGCoA reductase are produced than in the same number of cells grown in the absence of compactin.

a. Cells grown in the presence and absence of compactin are ground up and the activity of HMGCoA reductase in the homogenate (the ground-up cells) is measured as a function of HMGCoA concentration (in the absence of compactin) with the enzyme rate expressed as mevalonic acid formed/(min)(mg cell protein). How would the K_M and V_{max} compare for the cells grown with and without compactin?

b. The HMGCoA reductase is purified from the cells grown with and without compactin and the enzyme activity is again determined with the rate expressed as mevalonic acid formed/(min)(mg purified protein). How would the K_M and V_{max} compare for these two enzyme preparations?

78. Enzyme M acts on substrate A to produce product B. Compound C is an allosteric activator of enzyme M. With genetic techniques to be discussed in later chapters, the valine at position 57 in enzyme M was altered. The altered and unaltered proteins were purified, and their activities were measured in a standard assay system, which include $10\ \mu mol$ A/ml of reaction mixture (a saturating level of A), in the presence and absence of C, the allosteric effector. The assay results are presented in Table 2–1.

a. What is the probable functional location of valine 57 in enzyme M?

b. Speculate as to why the substitution of serine, glutamine, and alanine for valine had different effects on the activity of enzyme M.

79. You want to synthesize 1-stearoyl, 2-oleoylphosphatidylcholine. (Stearoyl refers to a saturated fatty acyl chain with 18 carbons; oleoyl refers to a chain with 18 carbons and one double bond between the 9th and 10th carbon atoms. The numerical prefixes refer to the carbon on the glycerol that is esterified to each acyl chain, with 2- referring to the middle carbon.) You read that it is possible to acylate glycerol phosphorylcholine (Figure 2–2) using fatty acid anhydrides as fatty acyl donors. You synthesize oleic anhydride and stearic anhydride, mix them in equal proportions, and prepare to acylate glycerol phosphorylcholine. What is the problem with this procedure for synthesizing 1-stearoyl, 2-oleoyl-phosphatidylcholine?

Table 2-1

	Enzyme activity (nmol B produced/(min)(mg enzyme M))	
Alteration	No C added	1 mM C added
None	10.3	51.4
Val 57 → Ser 57	10.5	30.2
Val 57 → Glu 57	10.2	11.1
Val 57 → Ala 57	10.1	49.5

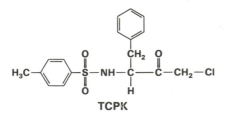

TCPK

▲ **Figure 2-1**

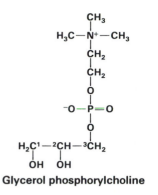

Glycerol phosphorylcholine

▲ Figure 2-2

80. Storage polysaccharides such as glycogen are highly branched, whereas most structural polysaccharides are linear. What might be the functional advantages of a branched structure for storage polysaccharides?

81. You isolate a bacterial toxin (a protein) that binds to the surface of mammalian intestinal cells, causing serious gastrointestinal distress in whole animals. You are interested in determining what type of cell-surface molecule in membranes binds the toxin. In order to remove most of the lipids from the membranes, you extract the membranes with organic solvents and keep the aqueous phase. You then add detergents to keep the membrane proteins "solubilized" and dialyze the aqueous phase to remove traces of organic solvents. The mixture is then passed over a column to which the bacterial toxin has been covalently attached. Chemical analysis indicates that some material did bind to the column, but to your surprise, there is no bound protein. Further analysis indicates that N-acetylneuraminic acid is present in the bound material.

    a. What might the material be?

    b. How would you test your hypothesis as to the identity of the material?

C H A P T E R 3

Synthesis of Proteins and Nucleic Acids

PART A: Reviewing Basic Concepts

Fill in the blanks in statements 1–20 using the most appropriate terms from the following list:

amino	N-formylmethionine
aminoacyl-tRNA synthetase	Okazaki fragments
	polymerases
anticodon	primers
carbohydrates	protein
carboxyl	ribosome
codon	rRNA
cysteine	spliceosome
elongation	suppressor
exons	template
gobble	termination
initiation	topoisomerase
introns	transcription
methionine	translation
missense	tRNA
mRNA	wobble
nonsense	

1. A coupling enzyme called _____ links an amino acid to the 3′ hydroxyl of the ribose of the terminal adenosine of a tRNA.

2. Short segments of DNA, synthesized during DNA replication, are called _____.

3. _____ is the name of the process whereby DNA is used as a template for the synthesis of RNA; _____ is the name of the process whereby RNA serves as a coded message for the synthesis of protein.

4. Enzymes that catalyze the synthesis of long, correctly ordered chains of subunits are generally classified as _____.

5. A nucleotide triplet on a tRNA that forms a stable, hydrogen-bonded complex with a complementary triplet in mRNA is called a(n) _____; its complementary triplet on the mRNA is called a(n) _____.

6. A protein that assists in binding mRNA to a ribosome is called a(n) _____ factor.

7. AUG is the _____ codon in mRNA.

8. Some chain-terminating mutations in bacteria are corrected if the gene containing the mutation is transferred to a(n) _____ strain.

9. Nonstandard base-pairing between nucleotides in codon-anticodon pairs is called _____.

10. The cellular structure that recognizes both tRNA and mRNA is called a(n) _____.

11. _____ is the initial amino acid added to the peptide chain during protein synthesis in eukaryotes.

12. A _____ mutation generates a new UAG codon in the coding region of a gene.

13. During RNA processing in eukaryotes, sequences called _____ are removed from the primary transcript.

14. One DNA strand serves as a(n) _____ during DNA replication; short segments of RNA serve as _____ during this same process.

15. Nucleic acid bases in _____ are highly modified after synthesis; these modifications often involve the addition of a methyl group to specific bases.

16. Cellular protein synthesis begins at the _____ end and stops at the _____ end.

17. The three generally recognized stages of protein synthesis in both prokaryotes and eukaryotes are _____, _____, and _____.

18. The stop codons UAG, UAA, and UGA are recognized by proteins called _____ factors.

19. An enzyme that nicks DNA to provide a swivel point during replication is called a(n) _____.

20. The macromolecules that compose a ribosome are _____ and _____.

PART B: *Linking Concepts and Facts*

Circle the letter(s) corresponding to the most appropriate terms/phrases that complete or answer items 21–30; more than one of the choices provided may be correct in some cases.

21. Cellular DNA replication

    a. is known as transcription.

    b. requires the DNA double helix to be unwound.

    c. employs an enzyme called DNA ligase.

    d. occurs at a structure called a "growing fork."

    e. often involves the synthesis of small pieces of RNA.

22. Transfer RNA

    a. is synthesized by a process known as translation.

    b. binds an amino acid covalently.

    c. has many modified bases.

    d. contains a sequence of nucleotides known as a "codon."

    e. is present only in eukaryotic cells.

23. A ribosome

    a. consists of a large and two small subunits.

    b. is composed of protein and carbohydrate.

    c. contains identical components in prokaryotes and eukaryotes.

    d. has two sites to which tRNA can bind.

    e. is the site of DNA replication.

24. RNA

    a. is more susceptible than DNA to degradation at high pH.

    b. contains thymine.

    c. is found in all cells.

    d. has three primary roles in protein synthesis.

    e. contains subunits added one at a time during synthesis.

25. The genetic code

    a. is a triplet code.

    b. is degenerate.

    c. is vastly different in bacteria and plants.

    d. is "commaless."

    e. was deciphered in the 1950s and 1960s.

26. Which of the following proteins are involved in translation?

    a. topoisomerase

b. ribosomal RNA

c. initiation factors

d. aminoacyl-tRNA synthetase

e. elongation factors

27. In which of these polymers are the monomers added one at a time?

a. DNA   d. tRNA

b. mRNA   e. rRNA

c. protein

28. The wobble position of a codon

a. is the first (5′) base.

b. is the third (3′)base.

c. may form a nonstandard base pair with the anticodon.

d. is the second base.

e. often contains inosine.

29. Operons

a. are found in eukaryotes only.

b. can be controlled by repressors.

c. contain a DNA sequence called a promoter.

d. are normally active at all times.

e. contain genes.

30. Primary transcripts of eukaryotic mRNA

a. are found in the cytoplasm.

b. contain only introns.

c. are usually larger than cytoplasmic mRNA.

d. usually contain introns.

e. are translated immediately.

For each component or process listed in items 31–45, write in the letter indicating if it is found only in eukaryotes (E), only in prokaryotes (P), or in both (EP).

31. tRNA ____

32. Nuclear RNA ____

33. Small ribosomal subunit ____

34. Operons ____

35. DNA polymerase ____

36. DNA ligase ____

37. Reading frames ____

38. Elongation factors ____

39. Amber suppressor ____

40. Spliceosome ____

41. Modified bases in tRNA ____

42. AUG used as a start codon ____

43. Stem-loop structures in rRNA ____

44. Shine-Dalgorno sequence ____

45. Okazaki fragments ____

PART C: *Putting Concepts to Work*

46. What purpose is served by having mRNA, tRNAs, and various enzymes associated with a large, complicated structure (the ribosome) during protein synthesis?

47. Explain the reasoning behind the conclusion that any amber-suppressible mutation is in a gene that encodes a protein, rather than in a gene that encodes a tRNA or an rRNA.

48. What is one possible reason why nonstandard base-pairing (wobble) is allowed during protein synthesis?

49. What are four general similarities in the polymeric structure and synthesis of proteins and nucleic acids?

50. What are two differences between RNA and DNA that help explain the greater stability of DNA? What implications does this have for the function of DNA?

51. What is one conclusion that can be drawn from the observation that the genetic code is nearly identical in all cells on earth?

52. What are the major differences between prokaryotic and eukaryotic messenger RNAs?

53. How was the genetic code broken (deciphered)?

54. Why is DNA synthesis discontinuous; that is, why is DNA ligase needed to join fragments of one strand of DNA?

55. What is the experimental evidence that aminoacyl-tRNA synthetases recognize the anticodon regions of at least some tRNA species?

PART D: *Developing Problem-solving Skills*

56. Scientists have discovered extraterrestrial bacteria, which arrived on earth *via* a meteorite. Nucleic acids and proteins were found as components of this organism, but the nitrogenous base composition of the nucleic acids was different than that of earth-based forms. Specifically, RNA from the bacteria consisted of four different bases called W, X, Y, and Z; DNA contained one additional base called Q. Proteins, however, were found to contain only the 20 amino acids regularly found as components of terrestrial proteins. You are assigned to decipher the genetic code of this creature and decide to emulate the Nobel Prize–winning approach of Khorana, Nirenberg, and Ochoa by synthesizing polynucleotides and determining the structure of translation products made from these sequences. You obtain the following data:

Synthetic nucleotide	Peptide sequence obtained
(5′)XYXYXYXYXYXY(3′) etc.	(Pro-Leu)$_n$
(5′)XXYXXYXXYXXY(3′) etc.	(Pro)$_n$, (Ala)$_n$, and (Thr)$_n$
(5′)XXXYXXXYXXXY(3′) etc.	(Trp-Thr-Pro-Ala)$_n$

If the genetic code of this organism is a commaless triplet code, what are the codons for proline (Pro), leucine (Leu), threonine (Thr), alanine (Ala), and tryptophan (Trp)?

57. If the adenine content of DNA from an organism is 36 percent, what is the guanine content?

58. In Table 3-1, reproduced from MCB, most of the codons for an individual amino acid are found in the same "box" defined by the 5′ nucleotide. What is the explanation for this observation?

59. In order to function as a proper defense system, mammalian immune cells must rapidly respond to the presence of bacteria. Based on what you have learned in this chapter about the similarities and differences between prokaryotic and eukaryotic synthetic pathways, what specific macromolecular characteristic, usually found only in prokaryotes, might be a good candidate for an "early warning" signal of bacterial infection?

60. Although it seems clear that most of the triplet code could be deciphered by the synthetic polynucleotide approach pioneered by Nirenberg and his coworkers (see Figure 3-8 in MCB), it is not clear that this approach could yield the identity of the stop codons. Based on the knowledge and technology available in the early 1960s, when scientists could not sequence naturally occurring mRNA or DNA but could sequence proteins and obtain genetic information, how do you think that these triplets were identified? (*Hint:* Amber mutations were also known at the time.)

61. As shown in Table 3-1, methionine and tryptophan, which are relatively rare amino acids in most proteins, each have only one codon. In addition, the codon for methionine is the start codon, and the codon for tryptophan is in the same box as the stop codons. Both of these codons contain G in the third position and thus are not subject to wobble (see problem 58). In contrast, leucine and serine, which are quite prevalent in many proteins, each have six codons. What do these observations suggest about the evolution of the genetic code?

62. In the experiments that led to the deciphering of the genetic code, synthetic mRNAs such as polyuridylate were incubated with a cell-free *E. coli* translation system. Although translated slowly (relative to the rates observed for biological mRNAs), these synthetic polynucleotides were translated and peptides were produced in sufficient quantity to be analyzed. You are probably wondering (if not, you should be) why these synthetic mRNAs were ever translated, since they do not contain start codons. The answer lies in the relatively high concentrations of Mg^{2+} (0.02 *M*) used by Nirenberg and his coworkers in these experiments.

    The effects of Mg^{2+} were demonstrated by incubating bacterial ribosomes, a synthetic polyribonucleotide, initiation and elongation factors, tRNAs, and nucleotide triphosphates with 0.005 *M* Mg^{2+} or 0.02 *M* Mg^{2+}

Table 3-1 The genetic code (RNA to amino acids)*

First position (5' end)	Second position				Third position (3' end)
	U	C	A	G	
U	Phe	Ser	Tyr	Cys	U
	Phe	Ser	Tyr	Cys	C
	Leu	Ser	Stop (och)	Stop	A
	Leu	Ser	Stop (amb)	Trp	G
C	Leu	Pro	His	Arg	U
	Leu	Pro	His	Arg	C
	Leu	Pro	Gln	Arg	A
	Leu	Pro	Gln	Arg	G
A	Ile	Thr	Asn	Ser	U
	Ile	Thr	Asn	Ser	C
	Ile	Thr	Lys	Arg	A
	Met (start)	Thr	Lys	Arg	G
G	Val	Ala	Asp	Gly	U
	Val	Ala	Asp	Gly	C
	Val	Ala	Glu	Gly	A
	Val (Met)	Ala	Glu	Gly	G

* Stop (och) stands for the ochre termination triplet, and Stop (amb) for the amber, named after the bacterial strains in which they were identified. AUG is the most common initiator codon; GUG usually codes for valine, but it can also code for methionine to initiate an mRNA chain.

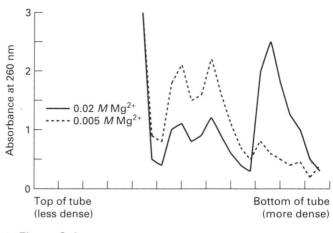

▲ **Figure 3-1**

for 2 min. These mixtures were then centrifuged for 2 h at 100,000g on a 15–40 percent sucrose density gradient. This procedure separates macromolecules on the basis of mass, or S value (see Chapter 5 of MCB for more information about centrifugation methods). After centrifugation, the centrifuge tubes were punctured and the contents allowed to drip slowly into a series of other tubes. These *fractions* were then assayed for RNA content by measuring the absorbance at 260 nm, a wavelength at which RNA absorbs quite strongly (the bulk of the RNA in these preparations is rRNA). Results of such an analysis are shown in Figure 3-1; open symbols are data from the mixture with 0.005 M Mg^{2+} and closed symbols are data from the mixture containing 0.02 M Mg^{2+}. Translation assays also were conducted by adding amino acids to each incubation mixture and measuring the amount of protein formed. Protein synthesis was observed at the higher magnesium concentration; no protein synthesis could be detected at the lower magnesium concentration.

a. How do you explain the effect of Mg^{2+} on translation of synthetic polyribonucleotides in these extracts?

b. What would be the predicted profile of a similar fractionation performed on a mixture of bacterial ribosomes, initiation and elongation factors, tRNAs, nucleotide triphosphates, and a biologically synthesized mRNA in 0.005 M Mg^{2+}? Why?

63. Ionizing radiation (such as that obtained from radioisotope decay near Chernobyl or from UV lights at your local tanning salon) has many effects on DNA structure. Two of the most damaging effects are induction of breaks in the DNA (both single-strand and double-strand breaks) and the formation of cross-linked thymine bases *(thymine dimers)* when two thymines are side-by-side in one strand of DNA. Much of this damage is mediated by the ability of radiation to generate reactive compounds such as peroxides (from molecular oxygen); these reactive compounds interact with sites in the DNA and cause the damage. Reactive compounds similar to these are used in normal cellular metabolism, but the concentration of these normal intermediates is lower and more tightly controlled than is the concentration of the UV-induced compounds. It is possible to isolate mutants of bacterial or animal cells that exhibit altered sensitivity to ionizing radiation. The response of two of these mutants to radiation is shown in Figure 3-2. Based on your knowledge of DNA synthetic mechanisms, and taking into account the information presented above, answer the following questions.

a. What can you conclude about the number of radiation-dependent events needed to kill a bacterial cell and about the nature of the mutations from the shapes of the survival curves for the wild-type, resistant, and sensitive bacteria?

b. Which enzymes or pathways are likely to be different in the radiation-resistant mutant?

c. Which enzymes or pathways are likely to be different in the radiation-sensitive mutant?

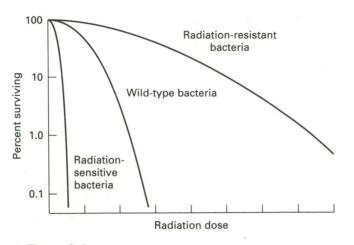

▲ **Figure 3-2**

d. Would either of these mutants be more sensitive than wild-type cells to the effects of drugs that modify bases in double-stranded DNA? If so, which mutant and why?

64. The compound known as AZT (3′-azido-2′,3′-dideoxythymidine), shown in Figure 3-3, is used to treat patients with acquired immunodeficiency syndrome (AIDS). The effects of AZT treatment vary considerably in different patients, but AZT therapy can result in longer survival times for many AIDS patients. This disease is thought to be caused by the human immunodeficiency virus (HIV), which is a member of the class of viruses known as retroviruses. Retroviruses contain RNA as their genetic material; a DNA copy of the viral RNA is made during infection by a viral enzyme called

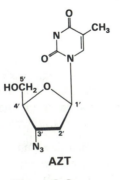

AZT

▲ **Figure 3-3**

reverse transcriptase. (See Chapter 6 of MCB for additional information about virus classes and their characteristics.)

a. AZT treatment reduces the amount of HIV present in some patients. What do you think is the mode of action of this drug?

b. Is AZT the active form of the drug; that is, is AZT or some metabolite of AZT responsible for the biological effects in these patients?

c. Recent evidence has indicated that long-term treatment with AZT is associated with the appearance of HIV strains that are resistant to the actions of the drug. What is a likely biochemical or molecular explanation of this observation?

65. Many antibiotics, most of which have been isolated from fungi, act by inhibiting initiation, elongation, or termination of peptides during protein synthesis. These compounds have been very useful in the study of protein synthesis, mainly because an individual antibiotic usually interferes with protein synthesis at a single well-defined step in the complex process of translation. In general, a researcher who uses such defined antibiotics to study protein synthesis can obtain three different kinds of information, depending on the experimental design and the antibiotics used. *Sequential* information —i.e., does a particular step or process occur before or after another—can be easily obtained using two different antibiotics and comparing the combined effects with the effect of each used separately. *Chemical* information about the translation pathway can also be obtained because blockage of a process often results in accumulation of intermediates that would be very short-lived during uninhibited protein synthesis. Purification and analysis of these intermediates can give valuable information about the mechanism(s) of the reactions involved or the metabolic sources of the chemical groups found in the intermediate. A final type of information that can be obtained is *structural* information, i.e., which component of the translational machinery is affected by the antibiotic.

Assume that you have isolated a novel antibiotic from the fungus *Pilobolus*. This antibiotic inhibits bacterial growth and bacterial protein synthesis but does not affect growth or protein synthesis in eukaryotic cells. Describe an approach that you would take to determine the nature and identification of the macromolecule that is affected by this novel antibiotic.

PART E: *Working with Research Data*

66. The replication of DNA is quite complicated and requires the participation of many different enzymatic activities. These include an enzyme (DNA primase) that synthesizes short segments of base-paired RNA, whose 3'-OH ends serve as initiation sites for the actual DNA polymerase activity. Obviously DNA primase can synthesize polynucleotides without the benefit of a 3'-OH primer; indeed, it can catalyze the hydrolysis and subsequent linkage of two nucleoside triphosphates without any complementary strand whatsoever. Why do you think that this enzyme synthesizes short segments of RNA, which must be eliminated and replaced with DNA before replication can be completed?

67. Most mRNAs are translated almost immediately after they are processed and transported to the cytoplasm. A notable exception to this generalization occurs in the eggs of the sea urchin, *Strongylocentrotus purpuratus.* Maternal mRNAs, coding predominantly for histones, tubulin, and actin, are deposited in the cytoplasm of the mature sea urchin egg and serve eventually as the template for almost all protein synthesis for the first 6 h after fertilization. Until fertilization, however, these mRNAs are not translated but remain in the cytoplasm in structures called *maternal ribonucleoprotein particles* (RNPs). Two general types of hypotheses have been generated to explain these observations: (1) maternal mRNA in these particles is "masked" and cannot be translated until something is removed, or (2) some component of the translational machinery is missing until after fertilization.

Experiments designed to determine the parameter(s) that limit protein synthesis in the egg have been performed, with the following results:

(i) Comparison of cell-free extracts of eggs and 30-min zygotes (fertilized eggs 30 min after fertilization) with regard to protein synthetic capability. Jagus and coworkers have shown that rates of protein synthesis are 10- to 15-fold higher in cell-free extracts of zygotes than in similar extracts of eggs.

(ii) Addition of foreign mRNA to cell-free extracts of sea urchin eggs. Several laboratories have shown that foreign mRNA (e.g., viral mRNA or globin mRNA) is translated in such extracts; however, addition of this mRNA does not cause an increase in total protein synthesis.

(iii) Comparison of the effects of adding sea urchin egg cell-free lysate to an active translational system such as that obtained from the rabbit reticulocyte. (Despite the evolutionary distance between the rabbit and the sea urchin, translation of sea urchin mRNA proceeds normally in the rabbit system.) Data obtained by Hershey's laboratory are shown in Figure 3-4. Note that the scale of the ordinate in panel C is different from the scale in panels A and B.

In Hershey's work, cell-free extracts of unfertilized eggs (A), 30-min zygotes (B), or swimming sea urchin larvae (pluteus stage, C) were mixed in varying ratios with rabbit reticulocyte lysate; the amount of RNA in each urchin extract was standardized such that equal amounts were added in the experiments shown. The rabbit reticulocyte lysate was first treated with a nuclease to destroy any rabbit mRNA, but both lysates

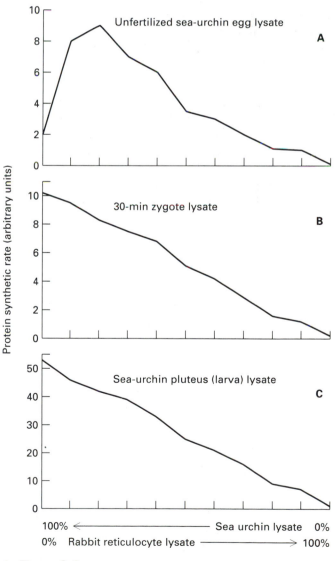

▲ **Figure 3-4**

should be capable of translating any added mRNA efficiently.

a. Are data from these three experimental approaches (i–iii) consistent with each other? If so, which hypothesis do these data support? If not, how would you explain the inconsistencies?

b. What additional experiments, using the same materials described in experimental approach (iii), would you design to determine the identity of the components involved in limiting protein synthesis in the sea urchin egg?

c. What result would you expect if you performed a "shift assay" (see problem 62) using RNA from eggs and RNA from zygotes in the rabbit reticulocyte lysate system?

68. High-fidelity replication of DNA is obviously important to all living organisms. Indeed, the error rate for enzymatic synthesis of DNA in eukaryotes approaches one mistake in 10^{10} bases; obviously cells have developed very accurate synthetic and error-correcting mechanisms in order to achieve such precision. Yet these mechanisms are to no avail if the DNA is damaged after synthesis; such damage can occur if an organism is exposed to radiation, chemicals, or ultraviolet light. Ultraviolet light, which is strongly absorbed by DNA, is in fact lethal to bacteria (Figure 3-5, panel A). However, bacteria can recover from the effects of ultraviolet light in some cases. In the experiment depicted in Figure 3-5, panel B, bacteria were first irradiated with UV light and then either irradiated with visible light or left in the dark. It is apparent that irradiation with visible light increased the survival of UV-irradiated bacteria. This process is called *photoreactivation;* cells treated in this way are said to be *photoreactivated.* In another experiment, bacteria were irradiated with UV light as before and then held in the dark in a non-nutritive medium (in which no cell growth occurs) for several hours before the surviving fraction was determined. As shown in Figure 3-5, panel C, survival of UV-irradiated cells was also enhanced by holding the cells in a non-nutritive liquid medium for several hours immediately after irradiation.

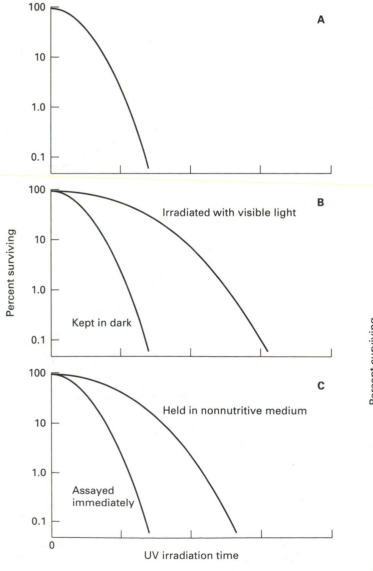

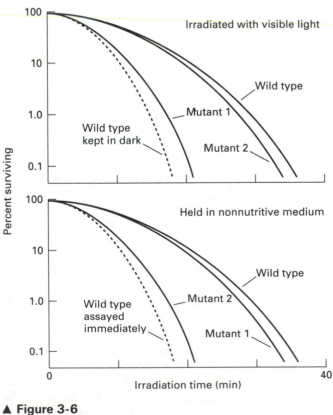

▲ **Figure 3-5**

▲ **Figure 3-6**

a. What do you think happens to the DNA in photo-reactivated cells (panel B)? What is the basis for this opinion?

b. What do you think happens to the DNA in the cells held in the non-nutritive medium (panel C)? Is this the same process that occurs in the photoreactivated cells?

c. The results of irradiation experiments with two bacterial mutants are shown in Figure 3-6. How do these additional data affect your answer to part (b) above?

d. Why do you think that bacteria that normally live "where the sun doesn't shine," have evolved these systems for recovering from UV damage?

69. Protein synthesis is dependent not only on the various types of RNA but also on a large number of ribosomal proteins. In yeast, for example, there are approximately 73 proteins per ribosome. These proteins are transcribed from widely scattered genes in the yeast genome, and the messenger RNAs for these proteins are translated just like any other mRNA on cytoplasmic ribosomes. The regulatory apparatus controlling ribosomal protein synthesis is complex and incompletely understood at present. It is clear, however, that ribosomal proteins and ribosomal RNAs are synthesized in stoichiometric quantities; that is, cells never accumulate excess ribosomal proteins or excess ribosomal RNA.

Attempts to understand the regulation of this crucial class of proteins have included a series of elegant experiments by Warner's group at Albert Einstein College of Medicine. These workers have taken advantage of the observation that baker's yeast (*Saccharomyces cerevisiae*) can be grown on different media with different growth rates. Yeast grown oxidatively with ethanol as a carbon source have a generation time of approximately 6 h. Yeast grown fermentatively with glucose as a carbon source have a generation time of approximately 2 h and contain more than twice the number of ribosomes per cell than do yeast grown on ethanol. When yeast cultures are shifted from the ethanol medium to the glucose medium ("shift-up"), they exhibit a rapid acceleration of both ribosomal RNA and ribosomal protein synthesis. This increased synthesis leads to accumulation of ribosomes until a steady-state level, characteristic of yeast grown on a rich medium, is achieved.

Yeast were grown in ethanolic medium containing radioactive [^{14}C]leucine for 34 h. Under these conditions all cellular proteins were labeled to steady-state conditions. At intervals before and after shift-up to the glucose-containing medium, samples of this culture were incubated for 5 min with [^{3}H]leucine; the radioactive amino acid was then removed. This is called a *pulse-chase* experiment (consult Chapter 6 of MCB for further details about this type of experiment). Cells were then harvested, ribosomal and nonribosomal proteins were isolated, and the ratio of ^{3}H/^{14}C in these proteins was measured. These data were used to calculate the relative rate of synthesis of the proteins as follows:

$$A_i = \text{relative rate of synthesis of protein}_i = \frac{(^{3}\text{H}/^{14}\text{C})_i}{(^{3}\text{H}/^{14}\text{C})_{total\ protein}}$$

Results for several proteins are shown in Table 3-2. Data pertaining to the rates of rRNA synthesis in cells at various times after shift-up compared with the rate before shift-up are shown in Figure 3-7.

a. What do these data demonstrate about the coordination or lack of coordination of synthesis of different ribosomal proteins?

Table 3-2

Protein	A_i values at time (min) after shift-up					
	0	5	15	30	60	90
Ribosomal proteins:						
2	0.97	1.6	2.03	2.36	2.5	2.65
11	1.09	1.7	2.88	2.62	4.14	3.55
27	0.76	1.41	2.06	2.71	3.54	2.59
61	0.84	1.31	5.53	2.66	3.42	2.68
Average for 13 proteins	.89	1.65	2.29	2.7	3.38	3.01
Nonribosomal proteins:						
A	1.36	1.13	1.33	1.03	0.93	0.88
B	1.16	0.86	0.29	0.59	0.50	0.42
C	1.07	1.09	0.69	1.08	0.94	1.15
D	1.08	0.85	2.05	2.24	2.30	2.25
Average for 4 proteins	1.27	1.11	1.06	1.19	1.32	1.15

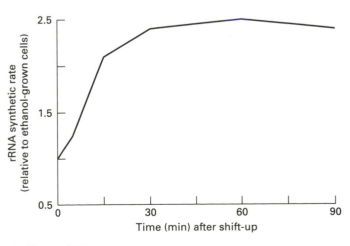

▲ **Figure 3-7**

b. What do these data demonstrate about the coordination or lack of coordination of synthesis of ribosomal proteins and ribosomal RNA?

c. What changes (if any) would you predict in the relative rates of synthesis of tRNAs before and after a shift-up such as described above?

4

The Study of Cell Organization and Subcellular Structure

PART A: Reviewing Basic Concepts

Fill in the blanks in statements 1 – 20 using the most appropriate terms from the following list:

brush border

chloroplasts

chromosomes

cilia

coated vesicles

connective tissue

cristae

cytoskeleton

endocytosis

exocytosis

glyoxysomes

Golgi complex

Gram-negative

histone

hypertonic

light

lysosomes

microvilli

mitochondria

nucleolar organizer

nucleoplasm

nucleus

organelle

peroxisomes

photosynthesis

plasma membrane

receptor

resolving power

rough endoplasmic reticulum

scanning electron

smooth endoplasmic reticulum

stroma

transmission electron

turgor

vacuole

1. The _____ contains many copies of the DNA that codes for ribosomal RNA.

2. The nuclear DNA of all diploid eukaryotic organisms is divided between two or more *chromosomes*.

3. The *resolving power* of a microscope is a measure of its ability to distinguish between two very closely positioned objects.

4. A general term for membrane-limited structures in eukaryotic cells is *organelle*.

5. The *plasma membrane* is the only type of membrane found in prokaryotic cells.

6. _____ bacteria have a semipermeable inner membrane and a more permeable outer membrane.

7. The array of fibrous proteins present in the cytoplasm of most eukaryotic cells is called the *cytoskeleton*.

8. Animal cell organelles that are bounded by a double membrane include the *nucleus* and the *mitochondria*.

9. In prokaryotes, membrane-bound ribosomes are found on the *plasma mem*; in eukaryotes, they are found on the *rough ER*.

10. The _____ is the major site of lipid synthesis in animal cells.

11. The fusion of an intracellular vesicle with the plasma membrane is called _____.

12. Enzymes called acid hydrolases are found in _____.

13. Hydrostatic pressure caused by the entry of water into the vacuole of a plant cell is called _____.

14. Organelles called _____ in animal cells and _____ in plant cells produce hydrogen peroxide as a by-produce of fatty acid and amino acid metabolism.

15. Nonnuclear organelles that contain DNA include _____ and _____.

16. A major function of the _plasma membrane_ in eukaryotes is to communicate and interact with other cells.

17. Specimens for visualization in the _____ microscope must be only 50–100 nm in thickness.

18. Enzymes in the organelle known as the _____ act to modify and sort proteins destined for secretion to the extracellular space.

19. Small structures known as _____ act to shuttle membrane constituents and lumen contents between organelles.

20. Oxygen is generated during the chemical process called _photosynthesis_; this occurs in plant cell organelles called _chloroplasts_.

PART B: *Linking Concepts and Facts*

Circle the letters corresponding to the most appropriate term/phrases that complete or answer items 21–30; more than one of the choices provided may be correct in some cases.

21. Compared with prokaryotes, eukaryotes
    a. usually have more DNA.
    b. usually have less DNA.
    c. have a smaller average cell size.
    d. have organelles.
    e. grow in harsher environments.

22. Which of the following organelles or structures are found in both plant and animal cells?
    a. nucleus
    b. mitochondria
    c. endoplasmic reticulum
    d. chloroplasts
    e. Golgi complex

23. Which of the following types of organisms can be photosynthetic?
    a. fungi
    b. plants
    c. protozoa
    d. animals
    e. bacteria

24. Mammalian cells that synthesize and export large quantities of protein for use by other cells in the body usually contain prominent

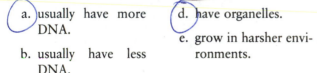

    a. mitochondria.
    b. lysosomes.
    c. Golgi complexes.
    d. endoplasmic reticulum.
    e. contractile vacuoles.

25. The conventional light microscope can give detailed, high-resolution images of structures such as
    a. whole cells.
    b. nucleoli.
    c. coated vesicles.
    d. mitochondria
    e. chloroplasts.

26. Elements of the cytoskeleton of eukaryotic cells include
    a. microfilaments.
    b. mitochondria.
    c. intermediate filaments.
    d. microtubules.
    e. lysosomes.

27. In order to visualize cell constituents using immunofluorescence microscopy, you would need a
    a. specific antibody.
    b. scanning electron microscope.
    c. light microscope equipped with specific filters.
    d. fluorescent dye.
    e. transmission electron microscope.

28. Which of the following structures are organelles?
    a. nucleus
    b. mitochondria

c. microtubules

e. lysosomes

d. inclusion bodies

29. Macromolecules that can be degraded by enzymes found in lysosomes are

a. ribonucleic acid.

d. protein.

b. deoxyribonucleic acid.

e. phospholipids.

c. carbohydrates.

30. Functions of the plasma membrane of prokaryotes include

a. endocytosis.

d. amino acid transport.

b. carbohydrate transport.

e. photosynthesis.

c. ion transport.

31. For each of the following phrases, write the corresponding letter indicating whether the phrase is true of some or all prokaryotes (P), some or all eukaryotes (E), or both (PE).

a. Synthesize DNA PE

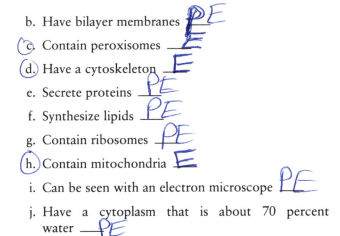

b. Have bilayer membranes PE

c. Contain peroxisomes E

d. Have a cytoskeleton E

e. Secrete proteins PE

f. Synthesize lipids PE

g. Contain ribosomes PE

h. Contain mitochondria E

i. Can be seen with an electron microscope PE

j. Have a cytoplasm that is about 70 percent water PE

32. In the table below, indicate all the cellular structures or compartments in which each process occurs.

Process	Eukaryotes	Prokaryotes
Photosynthesis		
DNA synthesis		
Amino acid transport		
Protein synthesis		

PART C: *Putting Concepts to Work*

33. What are the structural differences between gram-negative and gram-positive bacteria?

34. Based on what you know about the differences between light microscopy and transmission electron microscopy, indicate which technique would be best for visualizing each structure or phenomenon listed below and why. If light microsopy is best, which particular technique (dark-field, phase-contrast, etc.) would be most useful?

a. Motion of chloroplasts in plant cells

b. Viral particles

c. Motion of bacterial cells

d. Cells containing a specific protein in a tissue such as brain

35. Give two reasons why lysosomal enzymes do not degrade macromolecules located in the cytosol or nucleus of intact cells.

36. What is the function of the contractile vacuole present in many protozoans? How does it differ in function from the vacuole present in plant cells?

37. The cells lining the cavity of the small intestine are characterized by substantial folding of the plasma membrane to form structures called microvilli. What is the purpose of this structural modification?

38. The protein concentration of the cytosol of many cells (both prokaryotic and eukaryotic) is 20–30 percent. What problem does this high protein concentration pose for biochemists interested in cytosolic protein-protein interactions? What problem does this pose for an electron microscopist?

39. A major difference between prokaryotes and eukaryotes is the presence of organelles in the latter. Since we think that eukaryotes are more advanced than prokaryotes, a logical conclusion is that the evolution of organelles must confer some advantage(s) to eukaryotes. Describe at least two advantages of organellar structures.

40. What are the functions of the cytoskeleton?

41. What are the functions of the cell walls of plants and bacteria?

42. In terms of the basis for separation of particles, what is the major difference between differential-velocity centrifugation and equilibrium density-gradient centrifugation?

PART D: *Developing Problem-solving Skills*

43. It is estimated that the human genome contains sufficient DNA to code for about 50,000 different proteins. It is also estimated that human lymphocytes can make between 10^7 and 10^9 different specific proteins (antibodies). Explain this discrepancy.

44. It is difficult to appreciate the relative sizes of cellular structures from their dimensions because they are all much smaller than familiar everyday objects. In order to obtain a better grasp of the relative sizes of various structures, indicate the "magnified" size of each structure listed in the table below based on a scale in which the diameter of a ribosome (∼ 25 nm) is magnified to the diameter of a BB (0.25 cm). For each calculated size, suggest a familiar object with approximately the same dimensions.

Cellular structure	Actual size	"Magnified" size	Familiar object
Bacterial cell	1 μm diameter	_____	_____
Mitochondrion	$1 \times 1 \times 2$ μm	_____	_____
Muscle actin filament	0.007 μm thick $\times$ 1 μm long	_____	_____
Nucleus	5 μm diameter	_____	_____
Intestinal epithelial cell	$7.5 \times 7.5 \times 15$ μm	_____	_____
Human egg cell	70 μm diameter	_____	_____

45. The ability of a microscope to discriminate between two objects separated by a distance D is dependent upon the wavelength of the radiation (λ), the numerical aperture of the optical apparatus (α), and the refractive index of the medium between the specimen and the objective lens (N), according to the following equation:

$$D = \frac{0.61\,\lambda}{N\,\sin\alpha} \qquad \text{(see MCB, p. 119)}$$

a. If you are viewing an object with visible light at a wavelength of 600 nm, in an instrument with a numerical aperture of 70°, what would be the resolution if air (refractive index = 1) was the medium between the specimen and the objective?

b. What would be the resolution if immersion oil with a refractive index of 1.5 were placed between the specimen and the objective?

c. What would be the resolution if you used the same immersion oil and viewed the specimen in blue light (wavelength of 450 nm)?

d. Under any of the conditions specified above, could you see a mitochondrion (average size 1×2 μm)?

46. Many biologists think that mitochondria represent the descendants of oxidative bacteria, which either parasitized early cells or were engulfed by early cells. Based on the structural and functional characteristics of bacteria and mitochondria, what evidence can you offer in support of this hypothesis?

47. Mouse liver cells are homogenized and the homogenate is subjected to equilibrium density-gradient centrifugation, using sucrose gradients. Fractions obtained from these gradients are assayed for *marker molecules* (i.e., molecules that are limited to specific organelles). Results of these assays are shown in Figure 4-1. The marker molecules have the following functions: cytochrome oxidase catalyzes electron transfer during oxidative phosphorylation; ribosomal RNA forms part of the protein-synthesizing ribosomes; catalase catalyzes decomposition of hydrogen peroxide; acid phosphatase

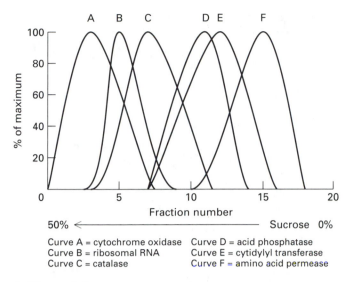

Curve A = cytochrome oxidase Curve D = acid phosphatase
Curve B = ribosomal RNA   Curve E = cytidylyl transferase
Curve C = catalase   Curve F = amino acid permease

▲ **Figure 4-1**

hydrolyzes monophosphoric esters at acid pH; cytidylyl transferase is involved in phospholipid biosynthesis; and amino acid permease aids in transport of amino acids across membranes.

a. Indicate the marker molecule for each organelle listed in the table below and the number of the fraction that is *most* enriched for each organelle.

Organelle	Marker molecule	Enriched fraction (no.)
Lysosomes		
Peroxisomes		
Mitochondria		
Plasma membrane		
Rough endoplasmic reticulum		
Smooth endoplasmic reticulum		

b. Is the smooth endoplasmic reticulum denser or lighter than the rough endoplasmic reticulum?

48. Just the evolution of organelles allows separation of diverse biochemical processes into specific compartments, the evolution of multicellular organsims allows particular cell types to specialize in different activities. This cellular specialization is often accompanied by organelle-specific specialization; that is, cells optimized for a particular role often have an abundance of the particular organelle or organelles involved in that role. Based on what you know about organellar functions, which organelle(s) would you predict would be overrepresented in each of the following cell types?

    a. Osteoclast (involved in degradation of bone tissue)

    b. Anterior pituitary cell (involved in secretion of peptide hormones)

    c. Palisade cell of leaf (involved in photosynthesis)

    d. Brown adipocyte (involved in lipid storage and metabolism, also thermogenesis)

    e. Ceruminous gland cell (involved in secreting earwax, which is mostly lipid)

    f. Schwann cell (involved in making myelin, a membranous structure that envelops nerve axons)

    g. Intestinal brush border cell (involved in absorption of food materials from gut)

    h. Leydig cell of testis (involved in production of male sex steroids, which are oxygenated derivatives of cholesterol)

PART E: *Working with Research Data*

49. A fluorescence-activated cell sorter (FACS) can be used to identify and purify cells that have a specific fluorescent antibody bound to them. In addition, this instrument can also identify and purify cells with varying amounts of DNA. For example, cells that have just divided and have X amount of DNA can be separated from cells that have duplicated their DNA ($2X$) and are preparing to divide again. This is done by incubating the cells in a fluorescent dye called chromomycin A_3, which binds strongly to DNA. Because the fluorescence of this dye is directly proportional to the DNA content, cells containing $2X$ DNA are twice as fluorescent as those containing only $1X$ DNA. Data from two such analyses are shown in Figure 4-2. The cells used were mouse erythroleukemia cells, grown as a single cell suspension, fixed in methanol, and stained with chromomycin A_3

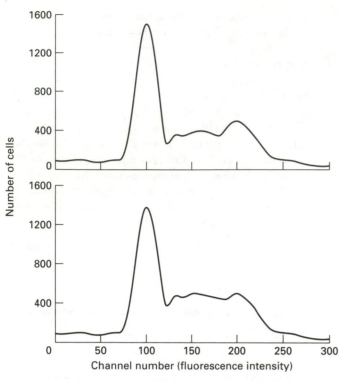

Number of cells

Channel number (fluorescence intensity)

▲ **Figure 4-2**

before analysis. The x-axis (channel number) indicates the level of fluorescence of a given cell; a higher channel number means that more fluorescence was measured. The y-axis indicates the number of cells with a given fluorescence level.

a. In panels A and B of Figure 4-2, indicate the portions of the graph corresponding to the following cells: G_1 cells, which have not replicated their DNA (1X DNA); G_2 cells, which have replicated their DNA (2X DNA); and S-phase cells that are in the process of replicating.

b. Were the cells used in analysis A (panel A) dividing more or less rapidly than those used in analysis B (panel B)? Explain your answer.

c. How would the pattern shown in panel A differ if the culture was contaminated with yeast cells? Why?

50. During cell division, contents of other organelles besides the nucleus need to be evenly divided between the daughter cells. Cells usually have sufficient numbers of each organelle so that organelles are evenly distributed after cell division. However, some algal cells have only one Golgi apparatus and only one chloroplast.

a. How do you think that these algal cells equally apportion the Golgi apparatus and the chloroplast during cytokinesis? Is the process likely to be the same for both organelles. Why or why not?

b. What do you think happens to the nuclear membrane during cell division; that is, how is it distributed equally between the daughter cells? Remember that at the light microscopic level the nuclear membrane disappears before cell division and reappears in the daughter cells.

51. Mammalian cells grown in culture usually contain a representative population of organelles such as mitochondria, Golgi vesicles, lysosomes, etc. These cells can be used as a model system in which to study the formation and dynamics of organelles. In one such study, it was found that cultured hamster cells, when grown in the presence of 0.03 *M* sucrose, accumulated numerous refractile (very bright in the phase-contrast microscope), sucrose-containing vacuolar structures (conveniently called *sucrosomes*), as shown diagrammatically in Figure 4-3. These vacuoles were found to be derived from lysosomes, since cytochemical staining techniques indicated the presence of the enzyme acid phosphatase in these structures. The structures persisted for many days in the cells but apparently had no ill effects on cellular metabolism. Hamster cells grown in the presence of sucrose and thus containing many sucrosomes were then incubated in the presence of the enzyme yeast *invertase*, which catalyzes the cleavage of sucrose into its monosaccharide components. The number of sucrosomes per cell was monitored during this incubation; the data are shown in Figure 4-4 (upper line).

a. Based on this experiment, can you conclude that invertase is internalized by these cells? If so, what can you conclude about its subcellular localization? What additional experiments could you perform in order to test this hypothesis?

b. In a second type of experiment, hamster cells grown in the presence of sucrose were mixed with hamster cells grown in the presence of invertase. The two cell types were fused together by the addition of a fuso-

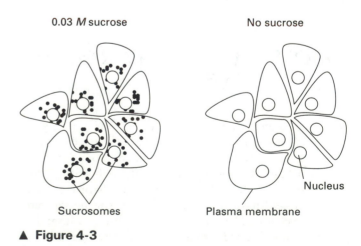

0.03 *M* sucrose   No sucrose

Nucleus

Sucrosomes   Plasma membrane

▲ **Figure 4-3**

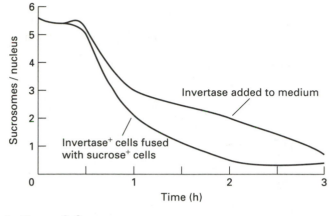

▲ **Figure 4-4**

genic agent, in this case a virus called vesicular stomatitis virus (see Chapter 5 of MCB for additional information about this technique). During the course of cell fusion the number of sucrosomes per cell (actually sucrosomes per nucleus) was monitored by phase-contrast microscopy, as shown by the lower line in Figure 4-4. From these data, what can you conclude about the subcellular localization of invertase in the fused cells? Are these data consistent with your hypothesis regarding the fate of invertase in unfused cells? Based on these experiments, is there a dynamic exchange of molecules between subcellular organelles?

52. Peroxisomes and glyoxysomes are organelles that have only been recently recognized and appreciated, and their functions are still being identified. One of these functions might be related to the fact that peroxisomes, like mitochondria, can utilize oxygen. However, unlike reactions in mitochondria, peroxisomal reactions that utilize oxygen do not produce energy in the form of ATP. An additional fact about oxygen utilization by organelles is that mitochondrial oxygen consumption is saturated at about 2 percent O_2 (atmospheric O_2 is about 20 percent), whereas peroxisomal oxygen consumption increases linearly with oxygen concentration up to 100 percent O_2.

   a. Which organelle do you think first appeared during the evolution of eukaryotes, the mitochondrion or the peroxisome? Why? (*Hint:* O_2 is very reactive and is toxic to some organisms, such as anaerobic eubacteria and archaebacteria.)

   b. What additional, modern role for peroxisomes is suggested by these observations?

5

Growing and Manipulating Cells and Viruses

PART A: *Reviewing Basic Concepts*

Fill in the blanks in statements 1–20 using the most appropriate terms from the following list:

amino acid	monoclonal antibody
Arabidopsis	myoblasts
auxotroph	nucleic acid
bacteriophages	plaque
blastocyst	plasmids
cell cycle	retroviruses
clone	*Saccharomyces cerevisiae*
de novo	
differentiated cell	salvage
Drosophila melanogaster	single cell
	somatic cells
envelope	togaviruses
erythroleukemia	transduction
fibroblasts	transformation
heterokaryons	trophectoderm
integration	undifferentiated cell
meiosis	virion
mitosis	

1. _Plasmids_ are small circles of DNA that are capable of independent replication in bacterial cells.

2. Phases called G_1, S, G_2, and M make up the eukaryotic _Cell Cycle_

3. A culture of cells isolated from a single cell is called a(n) _Clone_.

4. A mutant cell strain that requires a nutrient (e.g., an amino acid) not required by the parental cell is called a(n) _____.

5. _Transformation_ is the term that originally referred to the genetic change in a bacterium after incubation with purified DNA but that now is also used to describe the process by which eukaryotic cells become capable of indefinite growth.

6. Cells called _____ are found in mammalian connective tissue and are thought to be primarily involved in wound healing.

7. The _blastocyst_ stage of mammalian development is the source from which researchers isolate embryonic stem cells.

8. Cultured _____ cells can be induced to produce hemoglobin.

9. Cells from an organism's body are called _somatic_; these are distinct from the germ-line cells involved in generation of gametes.

10. Fused eukaryotic cells in which more than one nucleus is found are called _____.

11. Folic acid antagonists interfere with early stages in the de novo synthesis of *nucleic acid* precursors.

12. Thymidine kinase and hypoxanthine-guanine phosphoribosyl transferase are examples of enzymes in the so-called nucleotide _____ pathway.

13. A _____ is produced in quantity by a cultured clone of the specialized cell type called a B-lymphocyte.

14. Viruses that infect prokaryotes are called *bacterio- phages*.

15. Viruses in which the RNA strand directs the synthesis of a DNA copy are called _____.

16. An infectious viral particle is called a(n) *virion*.

17. HeLa cells, which cannot synthesize specialized cellular products, are an example of a(n) *undifferentiated cell*.

18. *Meiosis* is the process that generates haploid cells.

19. The yeast known as _____ has been widely used in molecular cell biology.

20. A *plaque* is a visible clear area, signifying the death of a group of virus-infected cells on a sheet of uninfected cells.

PART B: *Linking Concepts and Facts*

Circle the letters corresponding to the most appropriate terms/phrases that complete or answer items 21–30; more than one of the choices provided may be correct in some cases.

21. DNA synthesis occurs in a precisely limited portion of the cell cycle in

    a. cultured human cells.

    b. bacterial cells.

    c. *Saccharomyces cerevisiae*.

    d. cultured mouse cells.

    e. cultured *Drosophila* cells.

22. Which of the following phrases apply to *Saccharomyces cerevisiae*?

    a. requires a more complex medium than *E. coli*

    b. does not possess a true nucleus

    c. grows faster than most types of eukaryotic cells

    d. can exist in both haploid and diploid forms

    e. has a cell wall

23. Transformed eukaryotic cells

    a. usually do not perform specialized functions.

    b. may exhibit altered growth patterns.

    c. require the same media as yeast cells.

    d. require vitamins for growth.

    e. cease to grow after 50–100 generations in culture.

24. Hybridoma cells

    a. cannot divide in culture.

    b. can produce monoclonal antibodies.

    c. are products of the fusion of lymphocytes and myeloma cells.

    d. cannot be cloned.   e. are immortal.

★ 25. Which of the following statements are true of viruses?

    a. All contain nucleic acid.

    b. All contain lipid.

    c. All can reproduce outside of living cells.

    d. Some can infect plants.

    e. They always lyse the cells that they infect.

26. Which of the cultured mammalian cells listed below are examples of specialized or differentiated cell types?

    a. erythroleukemia cells

    b. myoblasts

    c. teratocarcinoma cells

    d. HeLa cells

    e. fibroblasts

27. Techniques for introducing foreign DNA into eukaryotic cells include

    a. somatic-cell hybridization.

    b. electroporation.

    c. addition of yeast cells to mammalian cell cultures.

    d. DNA injection into fertilized eggs.

    e. addition of calcium phosphate and precipitated DNA.

28. The Ti plasmid will introduce genetic material into

    a. monocots.   d. yeast.

    b. dicots.   e. HeLa cells.

    c. fruit flies.

29. *E. coli* is particularly suitable for molecular cell biology research because it

    a. has a short generation time.

    b. contains single-copy DNA.

    c. is susceptible to genetic manipulation by many techniques.

    d. requires only a simple growth medium.

    e. contains multiple organelles.

30. Steps in the lytic cycle of viral infection include

    a. replication.

    b. penetration.

    c. release of infectious particles.

    d. absorption.

    e. integration into the host chromosome.

31. Fill in the missing information in the following table.

Cell type	Type of organism (e.g., bird, fungus, yeast)	Generation time (approx.)	Advantages for studies in molecular cell biology
Escherichia coli	_____	_____	(1) _____
			(2) _____
			(3) _____
Saccharomyces cerevisiae	_____	_____	(1) _____
			(2) _____
			(3) _____
Cultured human cells	_____	_____	(1) _____
			(2) _____
			(3) _____

32. In Figure 5-1, the diagrams labeled a–e represent the stages in mitosis. Indicate the order in which these stages occur by writing numbers 1 through 8 (starting with interphase = 1) on the lines above the diagrams. Write in the name of each stage also.

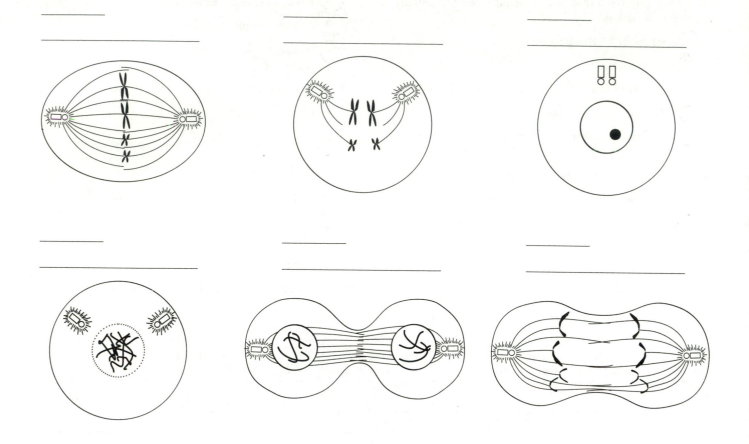

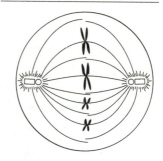

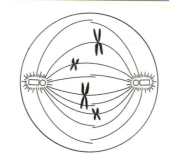

▲ **Figure 5-1**

PART C: *Putting Concepts to Work*

33. How would you isolate a strain of bacteria that requires the amino acid leucine?

34. What is the probability of a gene that you inherited from your father ending up in any one of your gametes? Which meiotic division determines this probability? What is the name for the process of allocating chromosomes to different daughter cells?

35. Much biological research has concentrated on the molecular biology of the events accompanying viral infection of prokaryotic and eukaryotic cells. List three features of viral infections that explain this emphasis.

36. What are two techniques commonly used to minimize genetic heterogeneity in commonly used strains of bacterial cells?

37. Why do cultured human fibroblasts require the amino acid glutamine, whereas the human organism does not?

38. How do prokaryotes and eukaryotes differ in the timing of DNA synthesis before cell division?

39. What are the advantages and disadvantages of using cultured cells rather than an a whole organ, such as the liver, in molecular cell biology research?

40. How does integration of DNA into the host DNA differ in yeast and higher eukaryotes? What are two characteristics of the yeast integration process that are advantageous for researchers?

41. Animal viruses have been categorized into six major classes (I–VI) depending on the genetic strategy they employ. What are the specific characteristics used to classify viruses according to this scheme?

42. Foreign DNA can be introduced into cultured mammalian cells in many ways, including somatic-cell hybridization and transfection with naked DNA. What are three advantages of somatic-cell hybridization over transfection methods?

PART D: *Developing Problem-solving Skills*

43. A mouse cell that does not contain the enzyme thymidine kinase is fused with a primary human cell containing this enzyme. The cells are then grown in the absence of thymidine. After several generations of growth, several cell clones are isolated and assayed for thymidine kinase. In addition, the human chromosomes retained by these hybrids are identified, using chromosome-banding techniques (see Chapter 9 of MCB). The data are shown in Table 5-1. Which human chromosome contains the gene for thymidine kinase?

44. Assume that a_1 and a_2 are two alleles of the same gene; b_1 and b_2 are two alleles of another gene. A female mouse who is heterozygous for both genes mates with a male mouse who is homozygous for both genes, having only the a_2 and b_2 alleles. The offspring have two genotypes: $a_1 a_2/b_2 b_2$ and $a_2 a_2/b_1 b_2$. What can you infer about the location of these genes?

45. Investigators want to order two events in a yeast synthetic pathway. A⁻ cells accumulate product a. A⁺ cells have normal function. B⁻ cells accumulate produce b. B⁺ cells have normal function. The genes A and B are not linked. Haploid A⁻B⁺ cells are mated with haploid A⁺B⁻ cells to produce diploid cells, which are then allowed to undergo meiosis. Each of the haploid progeny cells is then mated with each of the parent cells. This mating allows the detection of the haploid progeny that show the same phenotype as each of the parents when a diploid is formed by combining with the parent. The phenotype of the haploid progeny meeting this criterion is that it accumulates product b.

  a. What is the genotype of this haploid progeny?

  b. Which event occurs first in this pathway?

46. Tobacco mosaic virus (TMV) can spontaneously assemble from protein subunits and TMV RNA in a test tube. Electron micrographs of these preparations, taken at early time points during self-assembly, show two strands of RNA emerging from rods of various lengths,

Table 5-1

Clone no.	Thymidine kinase*	Human chromosomes present
1	+	9, 11, 17, 20
2	+	3, 11, 17, 19
3	−	2, 3, 11, 20
4	+	16, 17, 21
5	+	9, 12, 17, 19
6	−	9, 16, 21
7	+	2, 17

* A (+) indicates presence and (−) indicates absence of thymidine kinase.

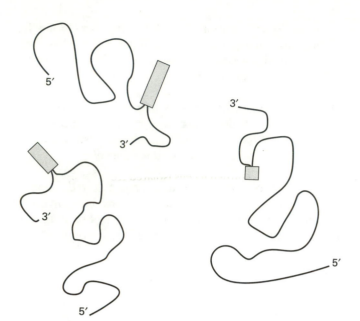

▲ **Figure 5-2**

as diagramed in Figure 5-2. Further analysis indicates that the 5′ tail (initially 5000 bases in length) becomes shorter as assembly progresses, whereas the length of the 3′ tail remains constant (about 1000 nucleotides). Deletion of 1200 nucleotides from the 5′ end of TMV RNA has no effect on the initiation of self-assembly, but deletion of 1200 nucleotides from the 3′ end completely inhibits the initiation of self-assembly.

a. Since the structure of TMV is essentially a hollow protein rod containing a coiled RNA, what do these observations indicate about the location of the assembly initiation site on the RNA?

b. Describe a model for self-assembly that accounts for these observations.

47. You wish to assay a solution of Newcastle disease virus (NDV) sent to you by another investigator. One way to

Table 5-2

Amount of viral solution added	Plaques/dish (3 replicates)
Undiluted solution	
1.0 ml	No cells remaining in any culture
0.1 ml	No cells remaining in any culture
Dilution 1 (1 part virus and 999 parts buffer)	
1.0 ml	No cells remaining in any culture
0.1 ml	Too many to count
Dilution 2 (1 part virus and 99,999 parts buffer)	
1.0 ml	343, 381, 364
0.1ml	33, 38, 41

quantitate this virus is to measure the activity of a viral coat enzyme called hemagglutinin, which is present on both infectious and noninfectious viral particles. Previous investigators have found that there are 1×10^6 NDV particles per unit of hemagglutinin. You perform a hemagglutinin assay on the NDV solution and find that there are 400 hemagglutinin units per ml.

Next you perform a plaque assay by inoculating petri dishes containing a complete sheet (confluent monolayer) of chicken fibroblasts with various amounts of viral solution; 3 days later you count the cleared areas in the cell monolayer. These data are shown in Table 5-2.

a. What is the total concentration of NDV particles in the original solution?

b. What is the concentration of infectious viral particles in the original solution?

c. What percentage of the NDV particles in the original solution are infectious?

48. Cell division cycle (*cdc*) mutants in yeast are blocked at specific points in the cell cycle, and most such mutants are temperature sensitive.

a. How could you enrich for *cdc* mutants in a yeast culture based on their temperature sensitivity? Would this procedure yield only *cdc* mutants?

b. How could you rapidly distinguish between *cdc* mutants and other types of mutants that might be selected by temperature enrichment?

49. What characteristics of the nucleotide salvage pathway make it more useful than other metabolic pathways (e.g., lipid or amino acid biosynthetic pathways) as the basis for selection procedures in cell-fusion experiments?

50. You have isolated a protein found in the yeast cell wall and determined part of its amino acid sequence. Based on this information, you have synthesized an oligonucleotide probe and used this probe to obtain the complete gene encoding this protein from a yeast genomic library (plasmids carried by bacterial cells). How would you determine if the gene product is essential for growth?

51. You are interested in fusing primary human kidney cells with mouse fibroblasts. In order to do this, you need to have a selective culture system that will allow the hybrids (heterokaryons) but not the two parental cell lines to grow. The mouse fibroblast is resistant to bromodeoxyuridine and is deficient in thymidine kinase (TK). What selective system would you use to enrich for the hybrid cells? Why?

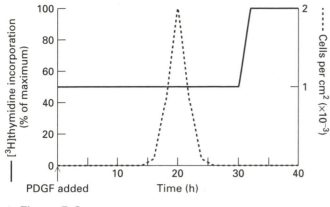

PDGF added

▲ **Figure 5-3**

a. What is the length of G₂ in these cells?

b. What is the approximate length of S?

c. In what portion of the cell cycle are the quiescent cells trapped?

d. How would you use this system to obtain 3T3 variants that were not dependent on PDGF for proliferation?

52. Mouse fibroblasts known as 3T3 cells are dependent upon a peptide hormone called platelet-derived growth factor (PDGF) for continuous growth. This hormone is usually supplied by the bovine serum used to supplement the culture medium. However, if 3T3 cells are grown in culture medium supplemented instead with calcified human plasma, which contains no PDGF, they cease to grow and become quiescent. After 3 days in this medium, very little macromolecular synthesis can be detected in these cells. If optimal amounts of PDGF are added at this time, the cells become activated to make DNA and divide, as indicated by the data in Figure 5-3 showing incorporation of [³H] thymidine into DNA and cell concentration.

53. Precancerous epithelial cells isolated from mouse mammary gland grow best in culture media supplemented with high levels of bovine serum. Culture in a defined medium, containing no serum, results in cessation of growth. However, supplementation of this minimal medium with medium in which mouse mammary fat pads have been cultured (adipocyte-conditioned medium) results in initiation of DNA synthesis in the mammary epithelial cells. (In the organism, the mammary fat pad acts as a source of lipid, which is used by the mammary epithelial cells as an energy source and as a source of lipid for milk production.) The factor in the conditioned medium that is responsible for the stimulation of DNA synthesis is stable after heating to 100°C and after treatment with trypsin at 37°C for 2 h.

a. Based on what you know about the functions of mammary fat pad cells, propose a hypothesis regarding the identity of the stimulatory factor.

b. What experiments would you perform to prove or disprove this hypothesis?

PART E: *Working with Research Data*

54. *E. coli* Hfr cells have a group of genes called an F element in their chromosome. The F element carries the genes necessary to establish conjugation with F⁻ cells. When an HFr cell establishes a conjugation tube with an F⁻ cell, a nick is formed in the F element in the chromosome. As the nicked strand of the chromosome of the Hfr cell starts to enter the F⁻ cell, the genes nearest the F element enter first. If conjugation is interrupted, only genes near the F element will be transferred to the F⁻ cell. In fact, the conjugation tubes are fragile and interruption of mating frequently occurs. If interruption were not to occur, it would take about 100 min for an entire strand of the Hfr *E. coli* chromosome to be transferred to the F⁻ cell. The transferred DNA recombines randomly with the chromosome of the F⁻ cell. By pur-

posefully interrupting mating at various times, one can determine the arrangement of genetic markers and the distance between these markers and the F element.

An ampicillin-sensitive, streptomycin-resistant *E. coli* Hfr strain with functioning genes x^+, y^+, and z^+ (which make products x, y, and z, all of which are required for cell growth), was mixed with an ampicillin-resistant, streptomycin-sensitive F⁻ strain with nonfunctioning genes, x^-, y^-, and z^-. Aliquots of cells were removed from the mixture at various times, and conjugation was interrupted by vigorously stirring the removed cells in a mechanical blender. The cells removed at each time were first grown on plates containing ampicillin (amp), streptomycin (str), x, y, and z. The colonies from these plates were replica plated onto three plates containing

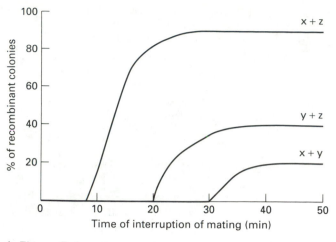

▲ **Figure 5-4**

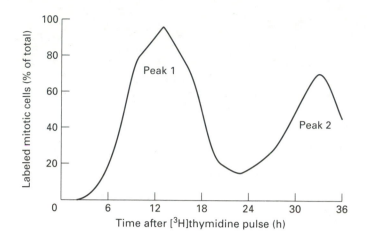

▲ **Figure 5-5**

y + z, x + z, and x + y. The percentages of recombinant colonies from the amp-str plates that grew on the replica plates are shown in Figure 5-4.

a. Which cells survived the amp-str plates?

b. For what genotype did each of the replica plates select?

c. From the data in Figure 5-4, what is the order of the genes *x*, *y*, and *z* as determined by the order in which they enter the F^- cell?

d. Only a fraction of the recombinant cells ever becomes x^+, y^+, or z^+. Why?

55. A rapidly growing culture of mammalian cells was incubated with [^{3}H]thymidine for 30 min, after which the radioactive thymidine was removed. The cells were sampled at various times thereafter by overlaying the cells with a photographic emulsion and preparing a autoradiogram. The proportion of *mitotic* cells (i.e., those in metaphase) that were radiolabeled was then determined and plotted as a function of time after the radioactive pulse, as shown in Figure 5-5. Note that the earliest samples have no labeled mitotic cells and that the proportion of labeled mitotic figures increases until it is nearly 100 percent in peak 1.

a. What do peaks 1 and 2 in Figure 5-5 represent?

b. Assuming that the mitotic division period M = 1 h, determine the following cell-cycle parameters from Figure 5-5: total cell-cycle time; length of G_1, length of G_2, and length of S.

c. How could the length of the M period be determined from these autoradiograms.

56. Certain strains of mouse fibroblasts (such as 3T3-L1) can differentiate into adipocytes when maintained in the proper hormonal and nutritional environment. This differentiation occurs only after the cells stop dividing and achieve confluence (completely cover the surface of the culture vessel). Over the course of several days, the cells assume morphological and enzymological characteristics typical of adipocytes in white adipose tissue. These cells are said to be terminally differentiated; they have undergone significant morphological, biochemical, and presumably genetic changes and will never divide again. However, the differentiation event does not occur in all the cells in the culture dishes; in all cases some cells that resemble fibroblasts remain. If these fibroblast cells are removed and placed in a new dish, they will divide normally and regenerate a confluent monolayer of cells. Again, after reaching confluence, a similar proportion of the cells will undergo differentiation into cells that resemble adipocytes. The experiment can be repeated several times with the same results: some fibroblast cells will remain which can be transferred to another dish where a certain (relatively constant) proportion will differentiate into adipocytes upon reaching confluency.

a. Do these observations imply the existence of two classes of fibroblasts, one of which is involved in wound healing and one which is an adipocyte precursor? If so, how would you test this hypothesis?

b. If not, what hypothesis could you propose to explain these observations? How would you test this hypothesis?

57. The Ti plasmid from *Agrobacterium tumefaciens* has been used as a source of foreign DNA to be introduced into plant cells. This plasmid has been modified, using recombinant DNA techniques, to eliminate its oncogenic properties (i.e., introduction of the modified plasmids does not cause growth of the crown gall tumor) and to make it capable of replication in other bacteria such as *E. coli*. You are interested in making a strain of

tomato plants that is resistant to the herbicide known as Roundup. The active ingredient of this herbicide is glyphosate, which acts as an inhibitor of an essential plant enzyme called 5-enolpyruvylshikimate-3-phosphate synthase (EPSPS). The basis of the inhibition is competitive; that is, glyphosate competes with the substrate for access to the enzyme active site. Based on this information, what strategies would you employ to develop a tomato strain resistant to Roundup? (*Hint:* Whole tomato plants can be grown up from single tomato cells in culture.)

C H A P T E R **6**

Manipulating Macromolecules

PART A: *Reviewing Basic Concepts*

Fill in the blanks in statements 1–20 using the most appropriate terms from the following list:

amino

antibodies

autoradiograph

carboxyl

carboxypeptidases

carrier

cDNA

clone

column chromatography

concentration

DNA polymerases

double-stranded

Edman degradation

equilibrium density-gradient

eukaryotic

exonucleases

genomic DNA

genomic library

heteroduplex

in situ hybridization

isoelectric focusing

isopycnic

isotonic

nucleic acids

photograph

plasmids

polymerase chain reaction

pool

prokaryotic

proteins

pulse-field

radioisotope

rate-zonal

restriction endonuclease(s)

restriction mapping

reverse transcriptase

SDS-polyacrylamide

single-stranded

site-directed mutagenesis

sodium dodecyl sulfate (SDS)

Southern blotting

specific activity

terminal transferase

toposiomerases

vector

1. A photographic emulsion that is exposed to radioactively labeled cells and developed to reveal the location of the radioactive components is called a(n) _____.

2. The most commonly used technique for determining the sequence of a polypeptide is called _____.

3. Enzymes that recognize short DNA sequences (from 4 to 8 base pairs in length) and cut the DNA at these sites are called _____.

4. A collection of bacteriophages containing inserted DNA sequences that are representative of the entire genome of another organism is called a(n) _____.

5. An electrophoretic technique that separates polypeptides primarily on the basis of their charge is called _____.

6. Pulse-chase experiments using radioactive precursors for cellular macromolecules are usually interpretable only if the size of the precursor _____ within the cells is small.

7. Rapid separation of cellular components on the basis of mass or size can be achieved by the technique known as _rate-zonal_ centrifugation.

8. The technique known as _____ allows a researcher to enzymatically synthesize any piece of DNA that lies between two known sequences.

9. DNA complementary to an mRNA sequence is called _____ and may be synthesized by the enzyme known as _____.

10. Large fragments of DNA, including whole yeast chromosomes, can be separated using _pulse-field_ electrophoresis.

11. A virus or plasmid used to transfer DNA from one organism to another is generally called a(n) _____.

12. A technique used to visualize the cellular location of specific mRNAs is called _____.

13. In the technique known as Western blotting, _____ are immobilized on a nylon or nitrocellulose membrane and reacted with specific _____.

14. The _____ of a radioisotope is the amount of radioactivity per unit of material.

15. The enzyme known as S1, isolated from *Aspergillus oryzae*, destroys _SS_ DNA but not _DS_ DNA.

16. Peptide chains that are chemically synthesized in the laboratory grow from the _carboxyl_ end to the _amino_ end.

17. Introduction of mutations, either deletions or point mutations, into a specific sequence of a cloned gene is called _____.

18. Enzymes called _____ remove nucleotides one at a time from the ends of DNA or RNA strands.

19. Proteins exposed to the detergent called _____ can be separated electrophoretically primarily on the basis of their size.

20. Centrifuged particles located in a solution that is equal to their own density are said to be _____ with the solution.

PART B: *Linking Concepts and Facts*

Circle the letters corresponding to the most appropriate terms/phrases that complete or answer items 21–30; more than one of the choices provided may be correct in some cases.

21. The technique known as fingerprinting can be used to compare samples of

    a. amino acids.
    b. lipid.
    c. protein.
    d. carbohydrate.
    e. RNA.

22. Which precursors listed below are impractical for pulse-chase experiments because of their large pool size?

    a. thymidine
    b. alanine
    c. tryptophan
    d. adenosine
    e. leucine

23. The most useful isotope in cellular autoradiographic studies is

    a. ^{14}C
    b. ^{35}S
    c. ^{15}N
    d. ^{32}P
    e. ^{3}H

24. Macromolecules that can be analyzed with the technique known as Southern blotting include

    a. proteins.
    b. oligosaccharides.
    c. antibodies.
    d. RNA.
    e. DNA.

25. Lambda phage vectors can be prepared containing DNA inserts from

    a. *Escherichia coli*
    b. *Saccharomyces cerevisiae*

c. *Caenorhabditis elegans*

e. *Homo sapiens*

d. *Drosophila melanogaster*

26. Advances in which of the following fields have contributed greatly to the development and increased use of recombinant DNA methodology?

a. enzymology

d. computer science

b. virology

e. electrophoresis

c. genetics

27. Parameters that control the motion of particles during rate-zonal centrifugation include

a. friction.

c. particle density.

b. peptide sequence (if the particle is a protein).

d. particle mass.

e. acceleration.

28. The Maxam-Gilbert method of determining a DNA sequence involves use of

a. restriction endonucleases.

d. end-labeling.

e. reverse transcriptase.

b. electrophoresis.

c. electron microscopy.

29. Which of the following isotopes are *not* radioactive?

a. ^{15}N

d. ^{35}S

b. ^{125}I

e. ^{16}O

c. ^{2}H

30. Parameters that greatly influence the extent of molecular hybridization that occurs in preparations of single-stranded eukaryotic DNA include

a. radioactivity.

d. time.

b. genome complexity.

e. nucleic acid concentration.

c. nucleic acid sequence.

31. In the table below, the sedimentation constants of several molecules and particles are listed. Predict the order in which these species will sediment during rate-zonal centrifugation by writing in the appropriate number from 1 (fastest sedimenting) to 5 (slowest sedimenting).

Molecule or particle	S constant	Order of sedimentation (fastest to slowest)
Human ribosomal RNA (large)	28	____
Cytochrome *c*	1.7	____
Bacterium	5000	____
Fibrinogen	7.6	____
Poliomyelitis virus	150	____

32. Prepare a table listing the source, the substrate, and the major uses in molecular cell biology of the following enzymes: *Eco*RI, reverse transcriptase, S1 nuclease, *Taq* polymerase, and trypsin.

PART C: *Putting Concepts to Work*

33. When is it advantageous to use radiolabeled compounds with the highest specific activity available?

34. Why are the effective densities of DNA and RNA greater in a $CsCl_2$ solution than in cells? Why is the effective density of RNA increased to a greater degree than that of DNA in a $CsCl_2$ solution?

35. What is the function of sodium dodecyl sulfate (SDS) during separation of proteins with polyacrylamide gel electrophoresis?

36. Name and briefly describe two different techniques for assessing the amount of a particular *functional* mRNA in mammalian cells.

37. Based on your knowledge of the differences between a cDNA clone of a complete transcript and a genomic clone of a complete gene, which kind of clone would be best for each of the following purposes?

a. A bacterial expression vector

b. A probe for in situ hybridization to detect mRNA content

c. DNA for packaging into a lambda phage vector

d. A probe for testing paternity by comparing restriction fragment patterns

38. The human β-globin gene, which has two introns, can be diagramed as follows:

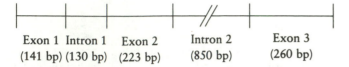

| Exon 1 | Intron 1 | Exon 2 | Intron 2 | Exon 3 |
| (141 bp) | (130 bp) | (223 bp) | (850 bp) | (260 bp) |

a. Draw a diagram of human β-globin mRNA annealing to a double-stranded DNA molecule containing the human β-globin gene. Use a dashed line for the RNA and a solid line for the DNA. Label the portions of the mRNA-DNA hybrid corresponding to the DNA introns and exons.

b. What are the intron portions of the hybrid called?

39. What would be the most efficient method for performing the following separations?

a. Separation of fragments of DNA ranging from 10 to 500 bp in length

b. Analytical separation of polypeptides differing in charge but with similar molecular weights

c. Separation of large DNA molecules containing 5×10^5 to 3×10^8 bp

d. Preparative separation of cellular organelles such as mitochondria and lysosomes

e. Analytical separation of polypeptides differing in mass but with similar charges

40. In selecting a protein sequence to be used as a template for construction of partial RNA sequences, many investigators look for sequences rich in tryptophan or methionine. Why?

41. What two aspects of basic microbiological research performed in the 1960s and 1970s were indispensable in the development of recombinant DNA technology? What does this tell you about the importance of basic (as opposed to applied) scientific research?

42. Why can pulses of radioactive amino acids be chased effectively in experiments using cultured mammalian cells? Why can pulses of thymidine be chased effectively?

PART D: *Developing Problem-solving Skills*

43. Nowadays, one can theoretically isolate, characterize, and prepare large quantities of any eukaryotic gene. It seems that this ability should make it possible to tackle virtually any problem in molecular cell biology. However, as you know, this is not quite true. What do you think is the basis for the present inability of medical scientists to cure diseases like retinoblastoma or cystic fibrosis, which are reasonably well characterized genetically?

44. Rapidly growing mammalian cells, in a medium with no added thymidine, were incubated for 20 min in 1×10^{-6} $M[^3H]$thymidine. After this time, the radioactive medium was removed and replaced with a culture medium containing no added thymidine. Samples of the cell culture were removed periodically, DNA was precipitated with acid, and the radioactivity in the DNA was measured by liquid scintillation counting. The results of this analysis are shown in Figure 6-1. This pulse-chase experiment was repeated with the changes described below. On Figure 6-1, sketch the curve that

would be obtained for each of these additional experiments.

a. Addition of 1×10^{-5} M nonradioactive thymidine after removal of the radioactive precursor at 20 min

b. Addition of an equal amount of 1×10^{-6} M nonradioactive thymidine in addition to the radioactive thymidine at time 0

c. Inhibition of thymidine phosphorylation at time 0

d. Inhibition of thymidine phosphorylation at 20 min

45. You have discovered a virus with a circular double-stranded DNA chromosome containing approximately 10,000 bp. You want to begin your characterization of the chromosome by making a map of the cleavage sites of some well-characterized restriction endonucleases. You digest the viral DNA under conditions that allow the endonuclease reactions to go to completion and then subject the digested DNA to electrophoresis on agarose to determine the lengths of the resulting DNA

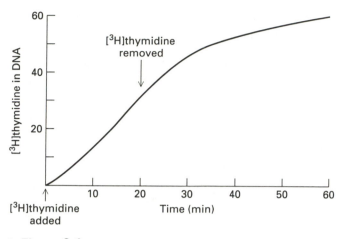

▲ **Figure 6-1**

fragments. The data are shown in Table 6-1. From the data given, draw a map of the viral chromosome indicating the relative positions of the cleavage sites for these restriction endonucleases.

46. Why are vertebrate genomic libraries made from embryonic or sperm DNA?

47. Analysis of the DNA content of *Drosophila melanogaster* reveals that a haploid cell contains about 1.5×10^8 bp.

   a. How many standard λ phage vectors would theoretically be required to constitute a complete *Drosophila* genomic library? How many λ vectors should you prepare in order to ensure that every sequence is included in the library?

   b. Recently it has been shown that DNA sequences can be cloned in the form of artificial yeast chromosomes; these fragments contain as much as 300 kb of DNA. How many yeast clones would theoretically be needed to contain all the genes of *D. melanogaster* using this cloning system? How many yeast clones should you prepare in order to ensure that every sequence is included in the library?

Table 6-1

Endonuclease	Length of fragments (kb)
*Eco*RI	6.9, 3.1
*Hind*III	5.1, 4.4, 0.5
*Bam*HI	10.0
*Eco*RI + *Hind*III	3.6, 3.3, 1.5, 1.1, 0.5
*Eco*RI + *Bam*HI	5.1, 3.1, 1.8
*Hind*III + *Bam*HI	4.4, 3.3, 1.8, 0.5
*Eco*RI + *Hind*III + *Bam*HI	3.3, 1.8, 1.5, 1.1, 0.5

48. You have chosen a sequence from the amino terminus of a protein in order to construct an oligonucleotide probe to search for the mRNA for this protein. The sequence is Met-Ala-Cys-His-Trp-Asn.

   a. How many possible oligonucleotide probes will you have to synthesize in order to account for all possible codon usages?

   b. How many would you have to synthesize if the tryptophan residue were replaced by a leucine?

49. In order to complete a grant proposal due in a week's time, you have to characterize a newly discovered bacterial plasmid (12 kb long). Unfortunately, you have only one restriction endonuclease available because your low-temperature freezer thawed over the weekend and all your other endonucleases were ruined. Since it would take too long to order a new supply of various endonucleases, you digest the plasmid sample for various lengths of time with the single endonuclease you have, so that at early times the sample is only partially digested; you then electrophorese the resulting fragment mixtures. A diagram of the electrophoretograms is shown in Figure 6-2.

   a. From this experiment, what can you conclude about the order of the fragments in the circular plasmid?

   b. Can you suggest another experiment, using only this one endonuclease, that could give more complete information about the order of the fragments?

50. You have isolated a protein with interesting properties from brain tissue. You wish to find out which cells in the brain are responsible for producing this protein. Which techniques, and in what sequence, would you use to identify these cells?

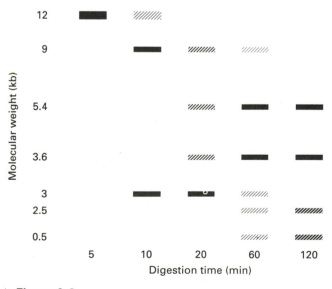

▲ **Figure 6-2**

51. A protein that you have isolated from brain contains sequences homologous to other proteins that are known to bind to DNA. Your experiments indicate that the presence of this protein in a cell is correlated with the presence of a specific neurotransmitter peptide, for which you obtain a genomic clone. How would you show that your protein binds to the gene for this neurotransmitter; that is, what characteristics would you predict for a DNA-protein complex? (*Hint:* You have an endonuclease that cleaves DNA at random sites and is capable of digesting it into single nucleotides if given enough time.)

52. You have two samples of purified DNA, one derived from a bacteriophage and one from *E. coli*. Each DNA sample has been cut into pieces of about 100 bp. Unfortunately, the labels fell off when you thawed the samples in a water bath, and you no longer know which sample is which. In order to quickly determine the identity of the samples, you decide to measure the rate of renaturation of each DNA. With both samples at the same concentration and in the same buffer, you heat them to 90°C in a thermostatted cuvette in a spectophotometer, then cool the samples to 60°C, and measure the absorbance at 260 nm at various time intervals. You obtain the curves shown in Figure 6-3. Based on these data, which sample is bacterial DNA and which is phage DNA?

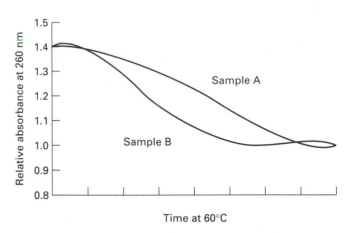

▲ **Figure 6-3**

PART E: *Working with Research Data*

53. Studies with a large number of restriction endonucleases have shown that most of these enzymes recognize and cut sequences, called palindromes, in which the two strands are the same when they are read in opposite directions. For example, the recognition sequence for *Eco*RI is as follows:

(5')GGATCC(3')

(3')CCTAGG(5')

Why do you think restriction endonucleases specifically recognize such palindromic sequences?

54. You want to obtain a genomic clone for a particular protein in order to study the characteristics of its gene. You have an antibody to the protein, a cell line that expresses the protein at a reasonable level, and access to standard molecular cell biological equipment and technology. How would you isolate a genomic clone for this protein?

55. Based on your knowledge of the genetic code and recombinant DNA technology, explain how would you recognize the *translation* initiation site in a cDNA with the sequence shown below, which contains the coding sequences for a 40-aa protein. Circle the start codon(s) and underline the coding sequence for the protein.

(3')CCCTTGTGGATCCACACCCTACCGGAGGACTATTAACTGTCCG

GCATACTTTGGCTGCGGTGTGGGGCAAGGTGAAGCTGGATGAA

GTTGGTGGTGAGGCCCTGGGGCAGACGTTGTATCAAGGTTTCA

AGACAGGTTTAAGGCAGACCAATAGAAACTGGGCGGCATTATT

GCATACATTGGCCCTCGGAGTGTCAGTTGCAATGCTAGCTAAG(5')

56. Gaucher's disease is a syndrome caused by lack of activity of a specific lysosomal enzyme called β-glucosidase. The lack of this enzyme activity leads to accumulation of a sphingolipid called glucocerebroside in tissue and in macrophages. The symptoms of this autosomal recessive disease appear only in homozygotes, and the enzyme activity appears to be near normal in known heterozygotes. Symptoms of the disease vary in severity among affected individuals, but most often consist of spleen and liver enlargement, neurologic disorders, and bone deterioration. On the basis of these symptoms and

age of onset, affected individuals have been grouped into three distinct types. Type 1 is found with high frequency among the Ashkenazic Jewish population; type 2 and type 3 are not associated with any particular ethnic group.

You have isolated a genomic clone containing the β-glucosidase gene from an Ashkenazic Jewish patient with type 1 Gaucher's disease; sequencing of this gene demonstrates that it contains a single-base mutation (adenosine to guanosine transition) in a single exon, resulting in the substitution of serine for asparagine at position 370 of the β-glucosidase polypeptide. Previously, other workers had discovered a single-base mutation (Leu 444 to Pro 444) in a type 2 individual; this allele was also found in heterozygous form in approximately 20 percent of the type 1 individuals tested.

a. The genotypes of individuals affected with Gaucher's disease and of normal controls were determined by a procedure that utilized radioactive oligomeric DNA probes encoding the sequence around position 370 in β-glucosidase. The number of individuals with each genotype are shown in Table 6-2: a +/+ denotes an individual with two wild-type genes (i.e., both encode asparagine at position 370); a +/− denotes a heterozygote (i.e., one normal gene and one encoding serine at position 370); and a −/− denotes an individual with two mutant genes (i.e., both encoding serine at position 370). Describe the experimental procedure used to determine the genotypes of these individuals?

b. Based on the data presented in this problem, how many genotypes of the β-glucosidase gene can be found in type 1 Gaucher's disease patients?

Table 6-2

Individuals	β-glucosidase genotype (No. of individuals)		
	+/+	+/−	−/−
Normal controls	12	0	0
Gaucher's disease			
Type 1	6	15	3
Type 2	6	0	0
Type 3	11	0	0

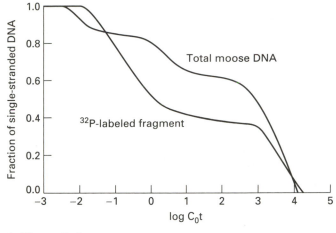

▲ **Figure 6-4**

c. Is type 2 disease, which is associated with neuropathologic symptoms, due to the same allele as type 1 disease?

d. How could you screen for heterozygotes among siblings of individuals with Gaucher's disease?

57. You have cloned a 2-kb fragment from an *Eco*RI genomic library of moose DNA. You think that this cloned DNA encodes all or part of an interesting mRNA. In order to determine the frequency with which this sequence is present in moose genomic DNA, you label a preparation of the cloned DNA with ^{32}P by the technique known as nick translation. You then mix the labeled cloned DNA (1 pg) with 10-mg samples of moose genomic DNA) shear the DNAs to 400-bp fragments, denature the mixtures by heating, and renature the mixtures to different C_0t values (C_0 = initial total DNA concentration; t = time). You measure renaturation of the samples by column chromatography on hydroxylapatite and hybridization of the labeled DNA by resistance to S1 nuclease. The data are shown in Figure 6-4.

a. Are the sequences in the cloned DNA fragment repeated in the moose genome?

b. Is the 2-kb fragment likely to encode for an mRNA found only in a differentiated cell?

C H A P T E R 7

RNA Synthesis and Gene Control in Prokaryotes

PART A: Reviewing Basic Concepts

Fill in the blanks in statements 1–24 using the most appropriate terms from the following list:

activator	hydrolysis
antiterminators	induction
autogenous control	inverted repeat
attenuation	leader sequence
cI	lysogeny
catabolite repression	negative control
cis-active	one
consensus sequence	operon
constitutive synthesis	phosphorylation
coordinate control	polycistronic
coupled transcription-translation	positive control
	Pribnow box
Cro	promoter
dimeric	protein kinase
dyad symmetry	regulatory genes
effector	regulon
epigenetic control	repressor

rho	three
sigma (σ)	trans-active
stringent response	trimeric
structural genes	two

1. _____ code for enzymes, and _____ code for proteins that modulate gene expression.

2. The bacterial DNA segment that includes conserved sequences centered at the -35 and -10 positions and is the recognition sequence for RNA polymerase constitutes the _____. The -10 sequence is called the _____.

3. A(n) _____ is one whose nucleotide sequence and/or known function is the same or similar for several genes or operons.

4. A(n) _____ is a low-molecular-weight molecule that alters gene expression by interacting with a regulatory protein.

5. Interaction of a regulatory protein with DNA to prevent transcription is termed _____.

6. _____ sequences affect only genes on the same chromosome.

7. _____ sequences regulate genes on the same chromosome or different chromosomes.

8. Regulation by a protein of the synthesis of mRNA for that protein is called _____.

9. The process by which a virus integrates into the bacterial chromosome and replicates as bacterial genes is called _____.

10. A(n) _____ is a group of genes that are controlled together but are not located together on the bacterial chromosome.

11. A transcription unit containing several genes under the control of one promoter is a(n) _____. The genes are under _____ through synthesis of a _____ mRNA. Rapid and simultaneous synthesis of the proteins coded by the genes is achieved by _____.

12. Correct initiation of transcription requires binding of _____ factors to RNA polymerase.

13. Inhibition of the synthesis of rRNA, tRNA, and some mRNAs when a required amino acid is in short supply is called, the _____.

14. When tryptophan is abundant, synthesis of the complete *trp* operon mRNA is prevented by _____, caused by alternative stem-loop structures at the 5' end of the message.

15. The _____ protein terminates transcription for about half of all *E. coli* mRNAs.

16. When bound to DNA at operator sequences, a(n) _____ prevents synthesis of mRNA.

17. Low expression of genes for sugar metabolism in medium containing glucose is called _____.

18. Interaction of regulatory proteins with DNA to activate, or "turn on," transcription is termed _____ _____.

19. The continuous, unregulated production of regulatory proteins is called _____.

20. Two proteins, called _____ and _____, control transcription from P_{RM} and P_R of lambda DNA.

21. _____ override the effect of rho protein.

22. The inactive form of the NR_1 regulator of the *ntr* operon is converted to the active form by _____.

23. The attenuation site in the *trp* operon is located in the sequence that encodes the _____ of *trp* mRNA.

24. The active form of many regulatory proteins is _____; that is, it contains _____ subunit(s).

PART B: *Linking Concepts and Facts*

Circle the letters corresponding to the most appropriate terms/phrases that complete or answer items 25–33; more than one of the choices provided may be correct in some cases.

25. The *lac* repressor

    a. is a DNA-binding protein.

    b. is induced by exposure of a bacterial cell to lactose.

    c. uses the same promoter as the *lacZ* gene.

    d. changes shape in the presence of inducer.

    e. can form alternative stem-loop structures.

26. Which of the following phrases applies to *all* operons?

    a. encode polycistronic mRNAs

    b. are responsive to CAP

    c. exhibit coordinate control

    d. exhibit negative control

    e. can be induced by sugars

27. Control of transcription termination

    a. always occurs at the 3' end of mRNA.

    b. can be regulated by a corepressor.

    c. must always involve rho protein.

    d. may involve consensus sequences.

    e. can be regulated by complementary sequences in nucleic acid.

28. Catabolite repression

    a. is mediated through cAMP.

    b. is mediated through CAP.

c. results in production of a positive activator protein.

d. affects enzymes involved in catabolic reactions.

e. is caused by several sugars.

29. The stringent response

a. is caused by lack of a required amino acid.

b. prevents all mRNA synthesis.

c. requires UTP for expression.

d. leads to formation of ppGpp.

e. prevents rRNA synthesis.

30. Sigma factors

a. are required for RNA polymerase to bind to DNA.

b. can be produced in response to environmental stress.

c. bind to DNA and promote binding of RNA polymerase.

d. mediate recognition of promoters.

e. are transiently associated with RNA polymerase.

31. Addition of inducer would not greatly affect the synthesis of β-galactosidase in bacteria of genotype

a. $Z^-Y^+A^+$.

b. $I^+O^cZ^+$.

c. $I^-O^cZ^+$.

d. $I^-O^cZ^+/I^+O^+Z^+$.

e. $I^+O^+Z^+$ in the presence of glucose.

32. Positive control of transcription

a. occurs in operons also subject to negative control.

b. can affect several operons through one regulatory protein.

c. can be mediated by an effector molecule.

d. includes the stringent response.

e. is controlled by coupled transcription-translation.

33. mRNA for the *lac* operon would hybridize to

a. the *lacI* gene

b. the operator sequence.

c. the *lacY* gene.

d. the *lac* promoter.

e. a *lacZ* gene with a single amino acid substitution.

34. Listed below are several operons. In the spaces provided, write in the letter(s) indicating which type(s) of control can regulate each operon as follows: N = negative control; P = positive control; and CR = catabolite repression.

a. *lac* ____   c. *ara* ____

b. *trp* ____   d. *gal* ____

35. For each of the phrases below concerning gene regulation, select the gene(s) or operon(s) to which the property applies from the following:

lambda *cI* gene   *hut* operon

gal operon   *arg* regulon

ara operon

a. Involves multiple operons _____

b. Involves multiple promoters _____

c. Subject to autogenous control _____

d. Subject to negative control by cAMP and CAP ___

36. After each of the phrases below concerning termination of transcription, write in the letter(s) indicating to which termination mechanism the property applies as follows: A = attenuation; RD = rho-dependent termination; RI = rho-independent termination; AN = antitermination; and N = none of these.

a. Involves proteins of different molecular weights ____

b. Acts by protein binding to DNA to prevent further progression of RNA polymerase ____

c. Acts through a consensus sequence ____

d. Acts through protein-RNA interaction ____

e. Acts through protein-protein interaction ____

f. Is overcome by protein synthesis ____

g. Overcomes rho-dependent termination ____

37. Listed below are several *E. coli* partial diploids. In the spaces provided, write in the letter(s) indicating whether each diploid synthesizses β-galactosidase inducibly (I), constitutively (C), or not at all (N).

a. $I^+O^cZ^-/I^-O^+Z^+$ ____

b. $I^-O^+Z^+/I^+O^cZ^+$ ____

c. $I^-O^+Z^+/I^-O^-Z^-$ ____

d. $I^-O^cZ^+/I^+O^cZ^-$ ____

PART C: *Putting Concepts to Work*

38. Distinguish between constitutive synthesis of a regulatory protein and constitutive synthesis of a structural protein as the result of an O^c mutation.

39. Summarize the evidence that the *lacI* gene is trans-active and that the *lac* operator (O) is cis-active.

40. Using Figure 7-1 as a guide, propose a mechanism other than an O^c mutation that would result in constitutive

No inducer present

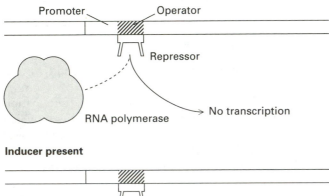

Inducer present

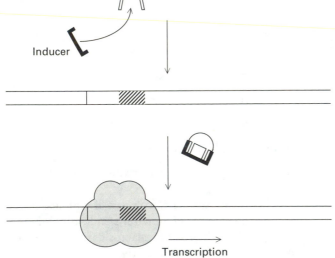

 Figure 7-1

synthesis of the *lac* operon and another mechanism that would result in lack of inducibility by lactose.

41. The ability of nucleic acids to form loops is a feature of the transcriptional control of some operons.

    a. Explain how loop formation in DNA affects expression of the *ara* operon.

    b. Explain how loop formation in RNA affects synthesis of the mRNA encoded by the *trp* operon.

42. For some operons under negative control, the absence of an effector causes inhibition of transcription, whereas for other operons under negative control, the effector is required for this inhibition. Explain the role of the effector in each case and give an example of each type of operon.

43. During feedback inhibition, the presence of an amino acid prevents its own synthesis by inhibiting the activity of the first enzyme uniquely within the synthetic pathway. This inhibition is relieved when the amino acid is in short supply. What type(s) of mutation in the *E. coli arg* regulon would give a phenotype that mimicked feedback inhibition (i.e., low activity for the first enzyme in the arginine pathway and normal activity for the others)? Would this phenotype be reversible in low arginine?

44. Give examples of how protein-protein interactions regulate initiation and termination of transcription.

45. Sigma factors for different promoters contain conserved and divergent amino acid sequences. What do you hypothesize are the functions of both of these types of sequences?

46. Two critical points in the lambda life cycle are (1) the decision to follow the lysogenic rather than the lytic pathway and (2) induction of the lambda prophage. Describe the events controlling the establishment of lysogeny and the induction of prophage.

47. Describe the differences in response of a tryptophan prototroph and a tryptophan auxotroph to depletion of tryptophan, as examples of specific and "global" control of gene expression.

PART D: *Developing Problem-solving Skills*

48. Synthesis of nucleic acids and proteins occurs with defined directionality. Predict the effect on ^{32}P incorporation if RNA chain growth is (1) in the $5' \rightarrow 3'$ direction or (2) in the $3' \rightarrow 5'$ direction under the following experimental conditions:

    a. RNA synthesis initiated in the presence of $^{32}P_\gamma$-labeled ATP followed by addition of a large excess of cold ATP.

    b. RNA synthesis initiated in the presence of cold ATP followed by the addition of a $^{32}P_\alpha$-labeled 3'-deoxyadenosine triphospahte.
    Note: In nature, RNA chain growth is in the $5' \rightarrow 3'$ direction.

49. You wish to visualize transcription units in bacteria and have prepared an electron microscope spread of rapidly growing *E. coli*. How would you distinguish rRNA and tRNA transcription units from mRNA transcription units and from each other?

50. You observe that both before and after addition of glucose to an *E. coli* culture in lactose medium, the turbidity of the culture doubles every 30 min. What effect would you expect glucose addition to have on the *rate* of β-galactosidase synthesis per cell and on the *amount* of β-galactosidase per cell and per total culture?

51. One of the classic studies contributing to our understanding of the *lac* operon was the PaJaMo experiment. In this experiment, merozygotes were formed by conjugation of I^+Z^+ (donor) cells with I^-Z^- (recipient) cells in the absence of inducer. Levels of β-galactosidase activity were monitored as a function of time and of inducer addition. The basic experimental protocol and the results observed are summarized in Figure 7-2. Explain why an increased rate of β-galactosidase synthesis was observed initially in the merozygotes and why at later times inducer was required for rapid β-galactosidase synthesis. What does this explanation imply about the nature of the inducer?

52. Both CAP and *lac* repressor are sequence-specific regulators of *lac* operon expression. Propose two types of biochemical experiments to determine whether the binding sites of these two proteins overlap.

53. For the *lac* operon, mutations in either the repressor gene (*I*) or the operator region (O) can lead to constitutive expression of all three structural genes (Z, β-galactosidase; Y, permease; and A, transacetylase) of the

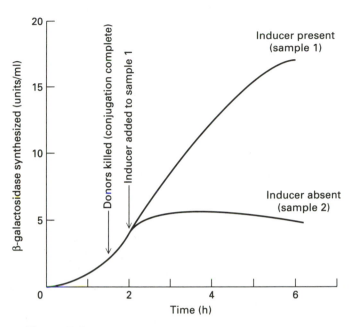

▲ **Figure 7-2**

operon. Compare the effects of repressor and operator mutations on the expression of *arg* regulon structural genes with the effects of *I* and O mutations on expression of the *lac* operon.

54. The expression of the *trp* operon is controlled by a number of genetic elements including an attenuator, an operator, a promoter, and a repressor gene. Describe how the phenotype of mutations that affect only the promoter or the attentuator can be distinguished biochemically.

55. Expression of *araBAD* genes increases by 400-fold when *E. coli* are grown in the presence of arabinose as a carbon source. This expression is dependent on the *araC* gene product. In hopes of understanding how the *araC* gene product acts, you have constructed a series of *E. coli* strains containing mutations in *araC*, O_2, I_1, and I_2. The effect of these mutations on *araBAD* expression is shown in Table 7-1. Genetic mapping data indicate that O_2, I_1, and I_2 are located between *araC* and *araBAD* and that O_2 and I_1 are separated by about 250 base pairs.

    a. Which data in Table 7-1 illustrate that the *araC* gene product can function both as a positive and negative regulator of *araBAD* expression?

    b. What is the likely mechanism through which the I^c and O^c mutations affect *araBAD* gene expression?

Table 7-1

Strain name	Genotype*	araBAD expression	
		− Inducer	+ Inducer
Wild type	$C^+O_2^+I_1^+I_2^+$	Very low	High
Mutant 1	$C^+O_2^cI_1^+I_2^+$	Moderate	High
Mutant 2	$C^-O_2^+I_1^cI_2^c$	Moderate	Moderate
Mutant 3	$C^-O_2^+I_1^+I_2^+$	Moderate	Moderate
Mutant 4	$C^+O_2^+I_1^cI_2^c$	Moderate	High

* X^+ displays a wild-type trait for that sequence. X^c is constitutive for *araBAD* gene expression.

56. *E. coli* lysogenic for the bacteriophage lambda are "immune" to superinfection by the homologous phage. Why?

57. You have isolated a mutant with increased activity of cAMP phosphodiesterase and mapped the mutation to the phosphodiesterase gene. Your research advisor asks you to check the effect of this mutation on the inducibility of the *lac, gal,* and *ara* operons in glucose-cultured *E. coli.* You argue that the experiments are not worth doing because the results are so predictable. Your advisor asks you to explain. What do you say?

58. Early arguments for repressors being proteins centered on the temperature sensitivity of some repressor mutations and the relief of other repressor mutations by nonsense suppressors. How do these mutations support the contention that repressors are protein rather than nucleic acid?

59. Certain chemical treatments can lead to the lytic expression of the prophage lambda in lysogenous cultures of *E. coli.* This has been proposed as a possible screening assay for carcinogens. Explain. (*Hint:* Most carcinogens are mutagens.)

60. Mutation of the lambda P_L promoter produces a "weak" promoter, which supports minimal transcription of the N gene. What are the consequences of this mutation for RNA transcript size early in lambda infection of *E. coli* and for the life cycle of the virus?

61. *RelA* mutants in *E. coli* relieve or relax the stringent response. You have prepared a merozygote containing both *relA⁻* and *relA⁺* genes. How would the level of ppGpp change as these cells are transferred from rich to poor media? Is the *relA⁻* mutation recessive or dominant?

PART E: *Working with Research Data*

62. As a graduate student in Schleif's laboratory at Brandeis, you have constructed a series of plasmids containing additions or deletions in the regulatory region of the *araBAD* operon. In these constructs, the spacing between the O_2 and I_1 segments of the regulatory region is altered by −16, −5, 0, 5, 11, 15, 26, and 32 base pairs. To analyze the effects of these + and − insertions, you introduce each of the plasmids into an *E. coli* strain containing a deletion of the *araBAD* operon. After streaking the transduced *E. coli* on indicator plates for gene expression from the *araBAD* promoter, you observe some dark lines on the plates, indicating high expression, and some light lines, indicating low expression. The data for the eight plasmid constructs tested are summarized in Table 7-2.

    a. Does this assay measure expression of *E. coli* genes, plasmid genes, or both?

    b. Describe how expression of the *araBAD* operon varies with the size of the plasmid insertion.

    c. Explain the results in terms of AraC-mediated repression of the *araBAD* operon. (*Hint:* The pitch of B-form DNA is 10.5 bp per turn.)

    d. What levels of expression would be expected for insertions of −11 and +21 base pairs?

63. Virions in the MS2 family of RNA phages (f2, R17, and MS2) contain a single-stranded, polycistronic RNA molecule, which codes for three major proteins: maturation protein, coat protein, and replicase, in order of coding arrangement. The f2 mRNA has been the subject of extensive investigation and was one of the first purified mRNAs to be translated in a cell-free system. Table 7-3 includes data illustrating the effect of added coat protein and of nonsense mutations on f2 mRNA translation in vitro.

    a. What is the effect of coat protein upon translation of f2 mRNA?

Table 7-2

Size of plasmid insertion (bp)*	Expression of *araBAD* (− inducer)
−16	High
−5	High
0	Low
+5	High
+11	Low
+15	High
+26	High
+32	Low

* Negative numbers indicate deletions.

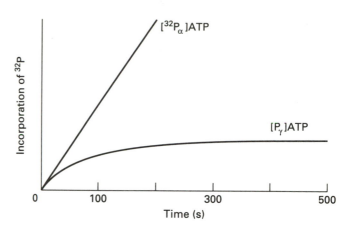

▲ **Figure 7-3**

b. Propose a plausible mechanism through which coat protein exerts its effect on synthesis of replicase and an experiment to test your hypothesis.

c. What is the effect of coat-protein nonsense mutations on the translation of the replicase cistron?

d. How can the nonsense mutation effect and the unequal production of translation products be explained? Propose a model consistent with this explanation.

64. One of the questions that has been intensively investigated in in vitro transcription systems is how long a length of double-stranded DNA is unwound during the elongation phase of RNA synthesis. Gramper and Hearst have approached this problem using *E. coli* RNA polymerase to transcribe closed, circular, double-stranded viral DNA.

   a. The overall kinetics for incorporation of label from $[^{32}P_\alpha]$ATP and $[^{32}P_\gamma]$ATP in this system are shown in Figure 7-3. What can you conclude from these incorporation data about the extent of reinitiation by RNA polymerase in this system?

b. In this system, transcription of a closed circular DNA template results in a change in the number of superhelical twists per DNA molecule. This change occurs in discrete steps. Each unit change in the number of superhelical twists corresponds to the melting of one turn of the DNA double helix. Figure 7-4 shows a schematized gel pattern for the migration of closed circular viral DNA being transcribed by various numbers of RNA polymerase molecules. Figure 7-5 is a replot of the gel migration results.

   What is the effect of greater negative supercoiling of viral DNA on its apparent spatial volume? How many base pairs are melted by one *E. coli* RNA polymerase in the act of transcription? Remember that B-form DNA has a pitch of 10.5 bp per helical turn.

65. The SOS regulon is repressed by the *lexA* gene product, a repressor protein. Unlike the *lac* repressor, no small effector molecule regulates the DNA-binding properties of LexA protein. Also unlike the *lac* operon, repression of the SOS operon can be reestablished only in bacteria actively synthesizing protein. Inactivation of LexA requires the *recA* gene product (RecA). The *recA* gene is subject to repression by LexA.

Table 7-3

mRNA	Relative amount of radioactive amino acid incorporated		
	Coat protein	Replicase	Maturation protein
f2	70	25	1
f2 + coat protein	70	3	1
f2 with nonsense mutation at codon 6 of coat (mutant 1)	0.5	7	1
f2 with nonsense mutation at codon 50 of coat (mutant 2)	26	25	1

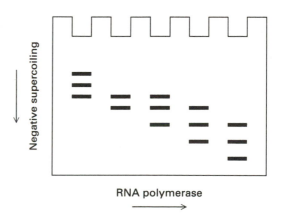

▲ **Figure 7-4**

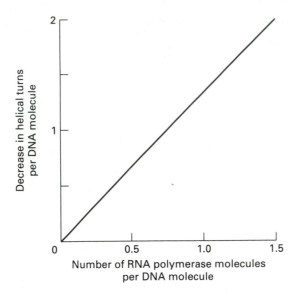

▲ **Figure 7-5**

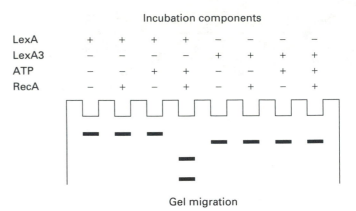

▲ **Figure 7-6**

Little and colleagues investigated the mechanism by which the *recA* gene product inactivates LexA by incubating in vitro LexA or LexA3 (a mutant LexA protein) with RecA and then analyzing the mixtures by gel electrophoresis. Their results are presented in Figure 7-6.

a. What is the probable mechanism through which RecA inactivates LexA? How does this mechanism explain the observation that chloramphenical, an inhibitor of bacterial protein synthesis, blocks rerepression of the SOS genes?

b. Would you expect the *lexA3* mutation to be recessive or dominant? Why?

c. Propose a mechanism by which SOS repression could be reestablished in bacteria actively synthesizing RecA.

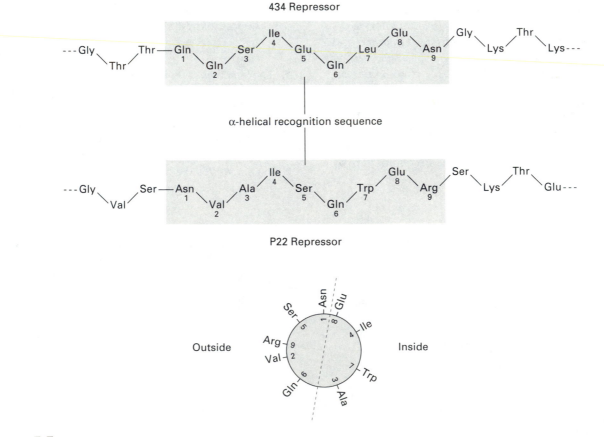

▲ **Figure 7-7**

66. A general feature of bacterial regulatory proteins is a helix-turn-helix motif. The x-ray structure of bacteriophage 434 repressor bound to its operator DNA is known. This structure shows an α-helical segment of the protein that fits in the major groove of the DNA where it presumably makes sequence-specific contacts. These sequence-specific contacts occur on the solvent-facing surface of the protein α helix; exposed amino acid residues on the "inside" surface are important in repressor folding. The amino acid sequence of the *Salmonella* bacteriophage P22 repressor is known. The repressor also contains a helical segment that is a presumed recognition helix. Shown in Figure 7-7 are the known "recognition" helix sequences for 434 repressor and P22 repressor. At the bottom of the Figure is the putative arrangement of P22 repressor amino acids with respect to outside and inside surfaces of the α helix, as proposed by Wharton and Ptashne.

    a. Based on the sequences shown in Figure 7-7, amino acids at which positions are most likely important for binding of each repressor to its operator?

    b. Assuming that plasmids containing the genes for each of these repressors are available, how could you redesign the 434 repressor so that it binds to the P22 repressor? How could you demonstrate that the redesigned 434 repressor (434R) is specifically altered in both its in vitro and in vivo DNA-binding properties?

RNA Synthesis and Processing in Eukaryotes

PART A: *Reviewing Basic Concepts*

Fill in the blanks in statements 1–20 using the most appropriate terms from the following list:

5S RNA	histones
5.8S RNA	introns
18S RNA	lariat
28S RNA	leader
32S RNA	methylated G cap
45S RNA	mRNA
AAUAAA	nuclear pores
CCAAT	nucleolus
GGGCG	poly A tail
GU	splicing
AG	self-splicing
3′ intron-exon junction	snRNA
	spacer RNA
5′ intron-exon junction	TATA box
branch point	thymine-thymine dimers
chromatin	tRNA
enhancers	
exon	

1. Eukaryotic DNA associates with proteins called _____ _____ to form _____.

2. The best known eukaryotic mRNAs without a poly A tail code for _____.

3. Transcription of rRNA genes and assembly of ribosomal subunits occurs in the _____.

4. The four RNAs contained in mammalian ribosomes are _____, _____, _____, and _____. The precursor of three of them is _____.

5. The _____ decreases in length as a mRNA molecule is translated repeatedly.

6. Ultraviolet-damage analysis depends on the production of _____ during UV irradiation.

7. Within a 45S pre-rRNA molecule, the sequences encoding the mature rRNAs are separated by _____.

8. A consensus sequence at −30 that is important in initiation of eukaryotic transcription is called the _____.

9. An RNA from which internal sequences can be excised in the absence of protein is said to be _____.

10. _____ are DNA sequences that can act over long distances and in either orientation to control transcription.

11. The direct precursor of 28S rRNA is _____.

12. During hnRNA processing to produce functional mRNA, _____ reactions excise intervening sequences, called _____, but coding sequences, called _____, are retained.

13. The three eukaryotic RNA polymerases catalyze formation of different products. Polymerase I produces _____; polymerase II produces _____; and polymerase III produces _____.

14. Ribosomes and mRNA may enter the cytoplasm through _____.

15. An untranslated region upstream of an initiation codon in mRNA is called a _____.

16. The 5′ end of an mRNA consists of a _____ structure.

17. The nucleotide sequence in mRNA that signals the location of poly A addition is _____.

18. The generally accepted splicing signals in mRNA are _____ at the upstream border of an intron and _____ at the downstream border.

19. The upstream (5′) end of an hnRNA intron forms a phosphodiester bond with an A at the _____; the excised intron has the form of a _____.

20. During splicing of hnRNA, a U1 snRNP binds to the _____, and a U2 snRNP binds to the _____.

PART B: *Linking Concepts and Facts*

Circle the letters corresponding to the most appropriate terms/phrases that completes items 21–30; more than one of the choices provided may be correct in some cases.

21. Evidence that 45S RNA is a precursor of rRNA includes

    a. similarity of nucleotide sequence.

    b. nucleoplasmic localization of RNA labeled for 5 min with [³H]uridine.

    c. kinetic analysis of the species formed after 15-min labeling with [³H]uridine and treatment with actinomycin D.

    d. presence of methylated sequences in both molecules.

    e. location of 32S RNA in the nucleoplasm.

22. Evidence that hnRNA is a precursor of mRNA includes

    a. similarity of base composition.

    b. electron microscope analysis of introns.

    c. presence of caps in both molecules.

    d. presence of poly A tails in both molecules.

    e. kinetic analyses using nucleoli.

23. Splicing reactions occur in the formation of

    a. all mRNAs.

    b. some 28S rRNAs.

    c. some 5.8S rRNAs.

    d. some 5S rRNAs.

    e. some tRNAs.

24. Self-splicing occurs in the formation of

    a. some 18S rRNAs.

    b. some 28S rRNAs.

    c. some 5.8 rRNAs.

    d. some mRNAs.

    e. some tRNAs.

25. TATA boxes are used for initiation of transcription of

    a. 45S pre-rRNA DNA.

    b. 5S-rRNA DNA.

    c. tRNA DNA.

    d. protein-coding genes that are rapidly transcribed.

    e. genes that are transcribed at low rates.

26. A functional eukaryotic mRNA usually contains

    a. a poly A tail.

    b. a methylated G cap.

c. the GU-AG consensus sequence for splicing.

d. methylated G residues.

e. the AAUAAA poly A signal.

27. Any given piece of eukaryotic

   a. may contain information for multiple proteins.

   b. may contain both an intron and an exon.

   c. may contain both an intron and an enhancer.

   d. may contain a region that is transcribed but not translated.

   e. may contain a region that both acts as an initiator and is transcribed.

28. An eukaryotic transcription unit

   a. contains a stretch of thymine residues to code for the poly A tail of mRNA.

   b. is monocistronic.

   c. produces RNA that is translated while transcription occurs.

   d. directly produces a functional mRNA molecule.

   e. can be transcribed by only one type of RNA polymerase.

29. Splicing

   a. is used to excise the spacers in 45S pre-rRNA.

   b. can occur in vitro without protein.

   c. can occur by more than one biochemical mechanism.

   d. occurs in all mRNAs.

   e. can occur between RNAs produced from different transcription units.

30. Both prokaryotes and eukaryotes exhibit the following characteristics:

   a. modification of RNA polymerase by additional protein factors.

   b. upstream location of TATA sequences.

   c. splicing of transcripts.

   d. interaction of transcripts with nonribosomal proteins to prevent tangling.

   e. presence of leader sequences in RNA.

For items 31–39, indicate the protein(s) having each characteristic by writing in the corresponding letter(s) from the following:

a. RNA polymerase I
b. $TF_{II}B$
c. $TF_{II}D$
d. $TF_{II}E$
e. RNA polymerase II
f. $TF_{III}A$
g. $TF_{III}B$
h. $TF_{III}C$
i. RNA polymerase III
j. SP1

31. Contains "zinc fingers" _____

32. Binds to DNA sequences within a gene _____

33. Binds to DNA _____

34. Required for 5S-RNA synthesis _____

35. Required for tRNA synthesis _____

36. Binds to the promoter-proximal sequence GGGCG _____

37. Contains conserved amino acid sequences in L and L′ subunits _____

38. Is most sensitive to α-amanitin _____

39. Activity is inhibited by actinomycin D _____

40. Indicate the order in which the following steps in the production of a mature mRNA occur (1 = earliest; 4 = latest)

   a. Initiation of transcription _____

   b. Addition of 5′ cap _____

   c. Splicing _____

   d. Addition of poly A tail _____

For items 41–45, indicate the RNA molecule(s) having each property by writing in the corresponding letter(s) from the following:

a. 5S rRNA
b. mRNA
c. hnRNA
d. 45S pre-RNA

41. Can differ from primary transcript at 3′ end _____

42. Can differ from primary transcript at 5′ end _____

43. Contains a 3′ hydroxyl group at the 5′ end _____

44. Contains a 3′ hydroxyl group at the 3′ end _____

45. Encoded by genes that are tandemly arranged _____

PART C: *Putting Concepts to Work*

46. All eukaryotic transcription units produce one primary transcript, but some encode one protein, whereas others encode more than one protein. Explain the difference between the two types of units, called simple and complex.

47. Both pulse-chase and nascent-chain analysis involve labeling of RNA for brief periods of time. In what ways do the two methods differ and what types of information are obtained from each?

48. Describe the available evidence indicating the order in which poly A addition, splicing, and cap addition occurs in the formation of a mature mRNA.

49. What components make up snRNPs, hnRNPs, and spliceosomes?

50. Eukaryotic genomic DNA containing the β-globin gene can be cloned into plasmids that replicate in *E. coli*, but the globin polypeptide is not expressed from this plasmid in *E. coli*. Why? How can expression of eukaryotic genes in bacteria be accomplished?

51. The TATA box, promoter-proximal sequences, and enhancers are all DNA elements that function in controlling transcription.

    a. Describe the location of each element relative to the genes they control.

    b. What is the role of these control elements in transcription?

52. Mammalian ribosomes contain one copy each of 5S, 5.8S, 18S, and 28S RNA. What mechanisms are used by the cell to insure equimolar amounts of each species in ribosomes?

53. What are the similarities and differences between the initiation of transcription of 5S-rRNA genes by RNA polymerase III and of bacterial genes by bacterial RNA polymerase?

54. Describe the mechanisms for processing introns in pre-tRNA and pre-mRNA and the roles of snRNAs in this processing.

55. Which features of mRNA are coded in DNA and which are the result of processing of hnRNA?

56. Give two examples of how methylation plays a controlling role in processing of RNA primary transcripts.

57. Compare the mechanisms of chain termination for RNA molecules synthesized by RNA polymerase I and II.

58. Deletion analysis and footprinting analysis are two methods for investigating the control of transcription. What types of information are obtained from each assay?

PART D: *Developing Problem-solving Skills*

59. Autoradiography of pulse-labeled cells can identify the sites of biosynthetic activity and product accumulation in cells. Would autoradiographic grains be localized over the nucleus or cytoplasm in eukaryotic cells subjected to the following treatments? Why?

    a. 5-min [³H]uridine pulse

    b. 5-min [³H]thymidine pulse

    c. 2-h [³H]uridine pulse

    d. 2-h [³H]thymidine pulse

    e. 5-min [³H]uridine pulse followed by a 2-h chase in precursor-free media

    f. 5-min [³H]thymidine pulse followed by a 2-h chase in precursor-free media

60. In a typical RNA pulse-chase experiment, the sedimentation profile of RNA labeled by [³H]uridine incorpora-

tion changes over time, whereas the RNA distribution assayed by UV absorption stays constant. How do you explain these data?

61. Human 5.8S, 18S and 28S rRNA are, respectively, 160 bases, 1.9 kb, and 5.1 kb in length. For many naturally occurring RNA molecules, the sedimentation coefficient of the molecule, the S value, can be related to molecular weight (MW) by the equation $S = (\text{constant})(\text{MW})^{1/2}$.

   a. Calculate the length in kilobases of 45S pre-rRNA assuming that the ratio between the number of bases and molecular weight is the same for 28S rRNA and 45S per-rRNA.

   b. Calculate the percentage of the 45S pre-rRNA molecule that is present in each of the rRNAs derived from it.

   c. What percentage of a 45S pre-rRNA molecule is metabolically stable and what percentage is rapidly degraded in the nucleus?

62. You have determined that HeLa cell RNA labeled during a 1- to 5-min incorporation period has a DNA-like base composition, whereas that labeled during a 30- to 60-min incorporation period has a rRNA-like base composition. In a pulse-chase experiment, you incubate HeLa cells with a radioactive RNA precursor for 5 min and then add actinomycin D. After 0-, 15-, and 60-min chases, you assay whole-cell extracts for RNA species by sedimentation analysis and for total incorporated radioactivity by liquid scintillation spectroscopy.

   a. What results do you expect to observe?

   b. Explain the difference in base composition of the RNA labeled during a 5-min period and a 60-min period.

63. As shown in Figure 8-1, the production of 18S and 28S rRNA and 45S pre-RNA shows differential sensitivity to UV irradiation, even though all of these RNA species are encoded in a single tandemly repeated transcription unit. Explain the relative sensitivity to UV irradiation of the synthesis of these RNA species.

64. Two species-specific factors, B and S, are required for the initiation of 45S pre-rRNA gene transcription.

   a. Which of these factors binds to DNA?

   b. What are the minimum number of structural domains that these proteins must have and what are their functions?

   c. Which of the domains is species-specific and which is nonspecific?

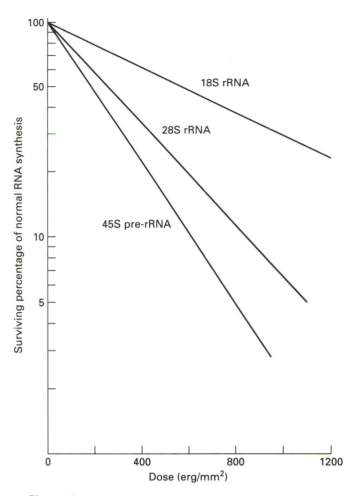

▲ **Figure 8-1**

65. UV-absorbance measurements indicate that in the cytoplasm 18S and 28S rRNA are present in equimolar amounts, whereas in the nucleus the amount of 32S RNA, a precursor to 28S rRNA, is about fivefold greater than the amount of 45S RNA, the precursor to both 18S and 28S rRNA. Since both the 18S and 28S rRNAs are the product of the same transcription unit, how can this be?

66. Analysis of the effects of deletions upon gene transcription is a major approach for identifying DNA control elements. Figure 8-2 summarizes the effects of various deletions upon the transcription of the eukaryotic 5S-rRNA gene. Mutants that are transcribed are marked with a + in part (a) of the figure; those that are not transcribed are marked with a −. Part (b) of the figure depicts a typical autoradiographic electrophoretogram used in the analyses of these deletions.

   a. Based on these data, is production of a 5S-RNA molecule dependent on upstream sequences?

   b. What is the putative location of the 5S gene control region?

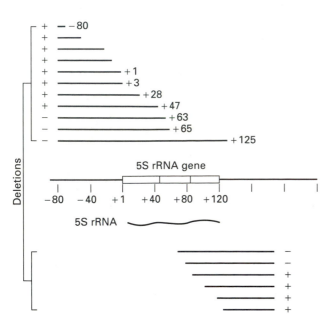

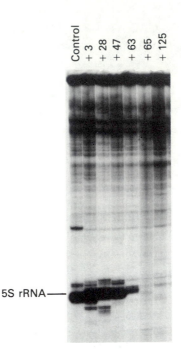

▲ **Figure 8-2**

c. What would be the likely effect on RNA production and the size of the RNA produced of moving the 5S control region 60 bp upstream or downstream?

67. Preinitiation complex formation for both 5S-rRNA and tRNA genes requires multiple transcription factors. Which of these factors protects DNA in footprinting experiments and why?

68. An R-loop analysis of fetal globin mRNA hybridized to genomic DNA indicates the existence of one intron, but a complete sequence comparison of the cDNA and genomic DNA indicates the existence of two introns.

    a. How do you reconcile the conclusions reached based on each method?

    b. Why is the sequence comparison made between cDNA and genomic DNA rather than directly between mRNA and genomic DNA?

69. One of the major objectives of the human genome sequencing project is to identify all protein-coding genes. Analysis of the immense amount of sequence data generated will require excellent computer software. What rules would you give to computer programmers to guide their development of algorithms for the identification of protein-coding genes? Remember these rules must be as complete as possible or many spurious protein genes will be identified.

70. You are asked to determine the effect of a nucleotide change in the CCAAT promoter-proximal sequence on the transcription of the mammalian thymidine kinase gene. Plasmids containing the normal or mutant gene are available. Your laboratory has available in vitro transcription systems and cultured mammalian cells deficient in thymidine kinase activity. Down the hall are practiced investigators who could give advice on the microinjection of plasmids into mammalian cells and *Xenopus* eggs, a third experimental system.

    a. Which system would you choose for transcription assays? Why?

    b. Describe how you would introduce the plasmid into cultured cells and *Xenopus* eggs?

    c. What experimental data would you collect and what could you conclude from these data?

71. For genes transcribed by RNA polymerase II, what effect would sequence alterations near the termination site probably have on the sequence of the corresponding mRNAs?

72. Describe the intermediate(s) in the splicing of pre-mRNA that can be immunoprecipitated by anti-U5 snRNP?

73. Guanosine in the form of free guanosine (G), GMP, GDP, or GTP functions as a cofactor for self-splicing of *Tetrahymena* ribosomal RNA.

    a. Does guanosine serve as an energy source for the splicing reaction?

b. Self-splicing exhibits a K_m of 32 μM for G and is competitively inhibited by inosine, a nucleoside analog. What do these properties suggest regarding the interaction of G and *Tetrahymena* rRNA?

PART E: *Working with Research Data*

74. The 3′ region of one of the two introns in human β-globin pre-mRNA has the following sequence (the * indicates the normal intron boundary):

(5′)CCUAUUGGUCUAUUCUUCCACCCUUAG*GCUGCUG(3′)
    ↑
    A

Within the human population, a point mutation sometimes results in the substitution of an A for the bold-faced G. This substitution results in a clinical condition known as β^+-thalassemia. In homozygous β^+-thalassemic individuals, the production of β-globin chains is depressed to 5–30 percent of normal, but the β-globin chains that are produced are normal. Why does this G → A substitution result in decreased production of normal β-globin?

75. Individuals with Sandhoff disease, a rare genetic defect, are severely deficient in the activity of the lysosomal enzyme β-hexosaminidase and hence are unable to degrade gangliosides, a membrane lipid component. Such patients die early because of impaired neurological function. The defect is due to mutations that affect the B subunit of the enzyme. Each enzyme molecule contains two B subunits.

a. Frozen fibroblasts from patients with Sandhoff disease (GM2144 cells) are available today for research analysis. Using B-subunit cDNA prepared against mRNA from GM2144 cells frozen in 1975, Nakano and Suzuki established a molecular defect in juvenile-onset Sandhoff disease; all 12 B-subunit cDNA clones prepared from GM2144 cells had a 24-base insertion, although only a point mutation was found within the genomic DNA. What is the most likely mechanism through which a point mutation could result in a 24-base insertion in the cDNA?

b. In a diploid human fibroblast, what would be the expected product of a heterozygous β-hexosaminidase allele? Would you expect heterozygotes expressing the Nakano and Suzuki mutation to be severely deficient in β-hexosaminidase activity? Do you expect GM2144 cells to be homozygous or heterozygous?

c. Using the polymerase chain reaction to amplify the B-subunit gene of GM2144 cell DNA followed by sequence analysis, Nakano and Suzuki found the cells to be heterozygous for the B-subunit gene. Propose an explanation for this unexpected observation consistent with the observed genotype and phenotype of GM2144 cells.

76. Attardi and colleagues used a nucleic acid hybridization approach to determine which nucleotide sequences are shared among various rRNAs and how many rRNA genes there are in the human genome.

Competitive RNA-DNA hybridization experiments were done to determine which sequences are common to 18S rRNA, 28S rRNA, 32S pre-rRNA, and 45S pre-rRNA. In these experiments, increasing amounts of unlabeled RNA were added to compete with [³H]-labeled RNA for hybridization to HeLa cell DNA. Representative data from these experiments are presented in Figure 8-3.

Saturation RNA-DNA hybridization experiments were done to determine how many rRNA genes there are per human genome. In these experiments, increasing amounts of rRNA were added to hybridization mixes containing a constant amount of HeLa cell DNA. Figure 8-4 shows data for such an experiment using 45S pre-rRNA.

a. Based on the data in Figure 8-3, what is the extent of sequence homology between 18S and 28S rRNA? Is 32S pre-rRNA a processing intermediate in the formation of both 18S and 28S rRNA? Propose a processing pathway for the generation of 18S and 28S rRNA that is consistent with the competitive hybridization data.

b. The molecular weight of 45S pre-rRNA is 4.6×10^6, and there are 15 pg of DNA per HeLa cell. From these values and the data in Figure 8-4, calculate the number of rRNA genes per HeLa cell genome.

77. You have cloned a 10-kb piece of genomic DNA that includes the full transcription unit for your prize protein. You wish to determine the approximate locations of the initiation and termination sites for the transcription unit by nascent-chain analysis of labeled RNA.

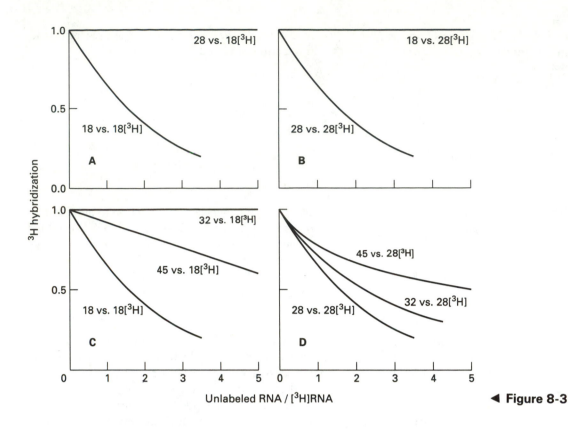

◂ Figure 8-3

a. Your first step is to generate an ordered set of DNA fragments from the genomic clone. After excision of the 10-kb piece with the restriction enzyme *Sal* I, a set of fragments are obtained by digestion with other restriction enzymes. Table 8-1 lists the length of some of these fragments. Diagrammatically summarize the order of restriction sites in the 10-kb DNA clone.

Table 8-1

Restriction fragment	Length (bp)
A. *Sal* (left end) — *Pvu*	8000
B. *Sal* (left end) — *Bam*	2200
C. *Eco* — *Bam*	700
D. *Eco* — *Pvu*	6500
E. *Eco* — *Hind*	400
F. *Pvu* — *Sal* (right end)	2000
G. *Hind* — *Bam*	300
H. *Pst* — *Pvu*	3200

b. Next you test each of these restriction fragments for hybridization to a population of nascently labeled RNA chains. Your results are shown in Table 8-2. Where are the initiation and termination sites for the transcription unit located?

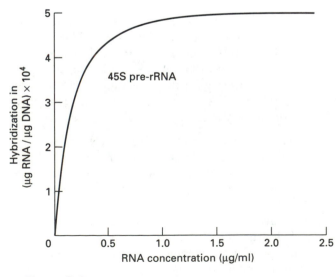

▲ Figure 8-4

Table 8-2

Fragment	Hybridization	Fragment	Hybridization
A	+	E	−
B	−	F	−
C	−	G	−
D	+	H	+

c. What is the maximum length of the mRNA that could be coded by this piece of genomic DNA?

d. How could the precision of the assignment of the initiation and termination sites be improved?

78. You have screened a cDNA library from the F9 cell line, a mouse teratocarcinoma line, and have isolated a 3.0-kb fragment bounded by *Eco*RI digestion sites; this fragment hybridizes to a developmentally regulated mRNA. (F9 cells are a germ-line tumor and have frequently been used as model cells for early steps in mammalian embryogenesis.) You have completed a restriction map of the fragment, as shown below (indicated lengths in kilobases):

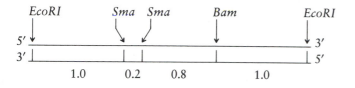

a. You want to detemine whether the protein-coding sequence in the fragment is oriented left-to-right or right-to-left relative to the restriction map. Your approach is to (1) isolate the *Eco*RI-bounded DNA fragment, (2) end-label the 5′ ends with ^{32}P using polynucleotide kinase, (3) cut the labeled DNA with the restriction enzyme *Bam*, (4) isolate the labeled DNA fragments, and (5) hybridize these fragments to a Northern blot of F9 mRNA. You find that the 1.0-kb fragment hybridizes to a 3.5-kb mRNA, whereas the 2.0-kb piece does not hybridize to any RNA. What is the 5′-to-3′ orientation of mRNA on the restriction map? What is the left-to-right orientation of the mRNA on the restriction map?

b. Your cDNA clone is shorter than the corresponding mRNA, and you wish to determine whether the clone includes the 5′ end of the mRNA. You hybridize 5′ end-labeled 3.0-kb cDNA to a mRNA preparation and then add reverse transcriptase to extend the 3′ end of the primer. You find that the end-labeled DNA now migrates as a 3.2-kb molecule. How does this result allow you to conclude whether your cDNA includes the 5′ end of the mRNA? Does your cDNA include the 5′ end?

c. You clone your *Eco*RI fragment into M13 and prepare a uniformly radiolabeled, *single-stranded* DNA corresponding to the cDNA fragment. You hybridize this DNA to total F9 cell RNA and to cytoplasmic mRNA. You then incubate the hybridization mixes with S1 nuclease to digest single-stranded nucleic acid. Upon gel electrophoresis of the hybrization mix with total F9 cell RNA, you see *four* bands —one at 3 kb, a second at 750 bp, a third at

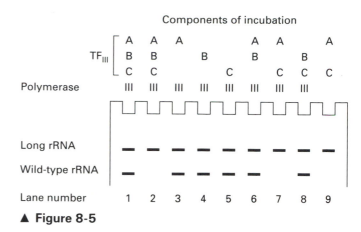

▲ **Figure 8-5**

1050 bp, and a fourth at 1200 bp. The hybridization with cytoplasmic mRNA, produces only a single band at 3 kb. How can these results be explained?

79. A template competition assay for transcription factor specificity includes the following steps: (1) gene A is incubated with limiting amounts of transcription factor, (2) competing gene B is added, and (3) transcription of each gene is assayed by gel electrophoresis. Isolated human transcription factors for RNA polymerase III genes were assayed with this procedure, using a 5S-rRNA gene with an insert (termed long gene) and a wild-type 5S gene. The gel patterns of the reaction products (long rRNA, wild-type rRNA) are shown diagrammatically in Figure 8-5; the components incubated in step 1 are indicated at the top (A, B, and C refer to TF_{III} A, B, and C, and III to RNA polymerase III). In all reaction mixtures except that shown in gel lane 1, the genes were added sequentially with the long gene added first. In the reaction shown in lane 1, the two genes were added simultaneously.

a. What conclusions regarding the role of TF_{III} A, B, and C and RNA polymerase III in the formation of protein-RNA complexes are suggested by these data?

b. If TF_{III} A acted catalytically rather than stoichiometrically, would further TF_{III} A be required for transcription of long gene added to a reaction mix containing excess TF_{III} B and C, RNA polymerase, wild-type 5S gene, and limiting amounts of TF_{III} A?

c. When Roeder and colleagues performed the experiment described in part (b), they found that more TF_{III} A was required for transcription of the long gene. In parallel experiments with TF_{III} C, similar results were obtained. Based on these data, what is the role of TF_{III} A and C in the transcription of 5S-rRNA genes?

C H A P T E R 9

The Structure of Eukaryotic Chromosomes

PART A: Reviewing Basic Concepts

Fill in the blanks in statements 1–14 using the most appropriate terms from the following list:

Barr body	lyonization
C value	metaphase
C-value paradox	minute chromosome
centromere	nuclear scaffold
cistron	nucleosome(s)
chiasma	polytenization
chromatid(s)	recon
chromatosome	solenoid
complementation	synapsis
euchromatin	telomere
heterochromatin	topoisomerase I
histone(s)	topoisomerase II
karyotype	

1. The number, size, and shape of the chromosomes at the _____ stage of cell division define the _____.

2. The _____ is the total amount of chromosomal DNA in a haploid cell. The observation that this amount does not increase with increasing complexity of organisms is called the _____.

3. A heterochromatic, inactivated X chromosome in female mammals is called a(n) _____.

4. The _____ is required for normal segregation of chromosomes.

5. In an interphase nucleus, dark-staining chromosomal material is called _____; light-staining material is called _____.

6. The _____ is the smallest unit of a chromosome composed of DNA in association with protein. When packed, these units form a(n) _____.

7. The basic proteins found associated with DNA in eukaryotic chromosomes are _____.

8. The heterochromatin at the tip of a chromosome is the _____.

9. During meiosis, homologous chromosomes pair in a process called _____. For recombination to occur, nonsister chromatids must cross over. The X-shaped crossing point is called a(n) _____.

10. A _____ is a unit of DNA encoding one polypeptide chain.

11. _____ tests define the outer limits of a gene.

12. Random formation of heterochromatin in one of two X chromosomes, called _____, results in inactivation of the genes on that chromosome.

13. Formation of additional copies of DNA molecules that remain closely associated in parallel is termed _____.

14. The support structure for large loops of DNA is called the _____. One of the nonhistone proteins present in this structure is _____.

PART B: *Linking Concepts and Facts*

Circle the letters corresponding to the most appropriate terms/phrases that complete or answer items 15–24; more than one of the choices provided may be correct in some cases.

15. The components of an artificial yeast chromosome include

    a. all the histones except H1.

    b. at least 50 kb of DNA.

    c. an autonomously replicating sequence (ARS).

    d. a centromere (CEN) sequence.

    e. a telomere (TEL) sequence.

16. A eukaryotic chromosome

    a. is also called a chromatid.

    b. contains regions that remain condensed throughout interphase.

    c. contains sequences that are never transcribed.

    d. is bound to the spindle during mitosis.

    e. most likely contains one molecule of DNA.

17. The haploid amount of DNA in the cells of most eukaryotic species

    a. is about 10^4 base pairs.

    b. is about 10^5 base pairs.

    c. is about 10^6 base pairs.

    d. ranges from 5×10^8 to 10^{10} base pairs.

    e. is greatest among some members of the Amphibia.

18. The average amount of DNA in a chromosome is

    a. the C value divided by the total number of chromatids.

    b. the C value divided by the haploid number of chromosomes.

    c. about half the amount of protein present.

    d. equivalent to about 5×10^9 base pairs for humans.

    e. about ten times greater in humans than in *Drosophila*.

19. Restriction fragment length polymorphisms (RFLPs) in human DNA

    a. can be located on a chromosome by in situ hybridization.

    b. are detected by probing with cloned human DNA.

    c. arise because of a mutation in human DNA that causes a genetic abnormality.

    d. arise because a mutation in DNA changes a recognition site for a restriction endonuclease.

    e. can be present in the heterozygous state in an individual.

20. Bands within chromosomes that are visible with the light microscope

    a. are constant from generation to generation within a species.

    b. can be seen in metaphase after staining.

    c. can be seen easily in *Drosophila* salivary glands.

    d. can contain repetitive DNA.

    e. can be used to confirm recombination.

21. Heterochromatin

    a. consists only of inactive genes.

    b. can contain repetitive DNA.

    c. is visible only during metaphase.

    d. can be activated by methylation.

    e. contains no chromosomal proteins.

22. Assuming that a cistron encodes one polypeptide chain (i.e., one enzyme), which of the following statements are true of "cis-trans" tests carried out with mutants within a eukaryotic transcription unit?

    a. The results can be interpreted assuming the mutations lie only within coding sequences.

    b. Complementation can occur between strains with mutations for two enzymes of a biosynthetic pathway if the mutations are introduced trans (i.e., on different chromosomes).

    c. Complementation can occur between strains with mutations for two enzymes of a biosynthetic pathway if the mutations are introduced cis (i.e., on the same chromosome).

    d. Complementation can occur between strains with mutations in nonoverlapping exons of two different proteins.

    e. Complementation can occur between strains with mutations in an exon common to two different proteins.

23. Telomeres

    a. contain regions with a high G content.

    b. are required for replication of linear artificial chromosomes in yeast.

    c. contain short repetitive sequences, which vary in different organisms.

    d. contain non-Watson–Crick base pairing.

    e. are synthesized by an RNA-enzyme complex.

24. According to the molecular definition of a gene, which of the following elements can be part of a eukaryotic gene?

    a. TATA box　　　　　d. poly A tail

    b. enhancer　　　　　e. poly A signal

    c. sequences not translated into protein

For items 25–31, indicate the protein(s) having each property by writing in the corresponding letter(s) from the following:

    a. H1　　　　　　　g. nucleoplasmin

    b. H2A　　　　　　h. N1

    c. H2B　　　　　　i. HMG protein

    d. H3　　　　　　　j. topoisomerase II

    e. H4　　　　　　　k. microtubules

    f. H5

25. Two copies present in a nucleosome _____

26. May be required for nucleosome assembly _____

27. Present in solenoids in most cells _____

28. Present in solenoids in duck erythrocytes _____

29. Nonhistone protein required for growth _____

30. Attaches chromosome to spindle _____

31. Prevents long loops of DNA from twisting _____

32. Write in the correct number (1–8) specified by each of the phrases below.

    a. Nucleosomes per turn in a solenoid _____

    b. Protein molecules per nucleosome _____

    c. Generations needed to map a RFLP _____

    d. Copies of H1 per chromatosome _____

    e. Linkage groups in *Drosophila* _____

    f. Turns of DNA per nucleosome _____

    g. Major types of histones _____

PART C: *Putting Concepts to Work*

33. Banding can be observed in metaphase chromosomes and in some interphase chromosomes. What are the underlying structural differences between these two types of bands?

34. What is the evidence that a chromosome contains a single DNA duplex?

35. There is a special mechanism for replicating the termini of chromosomes. Why is this mechanism needed and how does it work to insure chromosomal integrity between generations?

36. What differences have been found between regions of chromosomes being transcribed and untranscribed regions?

37. Can complementation tests be applied equally well to proteins derived from operons and proteins derived from transcription units encoding polyproteins?

38. You have two different haploid yeast mutant strains. Strain A has a TATA box mutation that eliminates initiation of transcription for gene 3 in a biosynthetic pathway and a second, coding mutation in gene 4 in the same pathway. Strain B has both types of mutation in gene 3. If strain A and strain B were mated, would complementation be observed in the diploid A/B yeasts?

39. In the 1940s, the conventional wisdom was "one gene, one enzyme." Discuss the validity of this phrase in terms of current knowledge about DNA and protein structure.

40. Describe how the chromosomal location (cis as opposed to trans) can be used to define the limits of a gene.

41. Is a restriction fragment length polymorphism (RFLP) a gene? Justify your answer.

42. Overlapping protein-coding genes may not be detectable by complementation analysis. Describe one strategy for demonstrating their presence.

PART D: *Developing Problem-solving Skills*

43. Animals and higher plants are composed of several different cell types. Would you expect to find karyotypic differences in number, shape, and size of metaphase chromosomes or in the G, Q, and R banding patterns for metaphase chromosomes from *different* tissues of the *same* plant or animal? To what extent are differences in chromosome arrangement likely to be major determinants of cellular differentiation? Describe an exception to the general rule.

44. You have been asked to determine the sex of a fetus using primary human cells derived from amniotic fluid. How could you do this without determining the actual karyotype of the cells? In which cases would this procedure lead to an incorrect sex assignment?

45. The DNA in a human chromosome is about 10 cm in length. A typical human cell has a diameter of 15 μm, and its nucleus occupies 30 percent of the cell volume. What is the minimum number of times the DNA in this human chromosome must be folded to fit within the cell?

46. The establishment of paternity and/or relatedness is a frequent problem in forensic medicine. At present, blood-group relatedness and other such parameters are the standard criteria for determining paternity. How can restriction fragment length polymorphisms provide a way to increase the precision of these determinations?

47. You have introduced a herbicide-resistance gene into corn, a grass plant, by transferring an artificial chromosome containing the resistance gene. The artificial chromosome is inherited as a stable extrachromosomal element.

    a. Which of the elements included in the artificial chromosome would *not* be required if the resistance gene were introduced as part of a closed circular DNA molecule?

    b. You have also introduced the same engineered chromosome into wheat, another grass plant. It is inherited as a stable extrachromosomal element. Analysis of the artificial chromosome as it is replicated in wheat and corn shows that its terminal sequence differs in the two species. How can this be?

48. After completing a gentle nuclease digestion of HeLa cell chromatin, you electrophorese the digest in an agarose gel and then stain the gel with ethidium bromide to visualize the DNA. What gel pattern would you expect? What gel pattern would you expect after a more extensive digest?

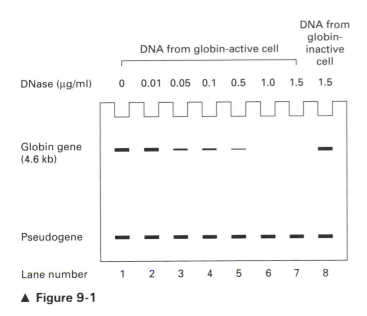

	DNA from globin-active cell						DNA from globin-inactive cell
DNase (µg/ml)	0 0.01 0.05 0.1 0.5 1.0 1.5						1.5

Globin gene (4.6 kb)

Pseudogene

Lane number 1 2 3 4 5 6 7 8

▲ **Figure 9-1**

49. Nuclease digestion has been a classic approach to probing the comparative structure of transcribed and non-transcribed genes in eukaryotes. In one such experiment, nuclei from two different cell types—one expressing globin and the other not—were digested with DNase and then treated with the restriction endonuclease *Bam* I. The resulting digests were subjected to Southern blot analysis with a radioactive globin DNA probe. The Southern blot patterns for the two cell types are shown in Figure 9-1.

   a. How sensitive to DNase digestion is the globin gene in the two cell types.

   b. What does this difference imply about chromatin structure in active and inactive genes?

50. The major molecular components of eukaryotic chromatin include DNA, histones, scaffold proteins, HMGs, and various DNA-binding proteins. Assuming that there are an average of 2×10^8 nucleotide base pairs per human chromosome, what is the molar ratio of histones and scaffold proteins to DNA in the human chromosome?

51. After a month's work, you have isolated eight adenine-requiring Chinese hamster ovary (CHO) cell auxotrophs.

   a. How could you determine if these CHO auxotrophs arose from mutations in separate genes?

   b. You find that two of the auxotrophs have mutations that map to the same chromosome and indeed to the same gene. Would the intragenic recombination approach that Seymour Benzer used to analyze the *rII* region of bacteriophage T4 be practical for determining the distance between these two CHO cell mutations?

52. You have prepared a panel of 83 monoclonal antibodies to a fibrous protein of 95 kDa. You find in immunoprecipitation experiments that 63 of the antibodies precipitate a second protein of 65 kDa, but the remaining 20 do not. Upon automated microsequencing of the N-termini of the 65- and 95-kDa proteins, you find the two proteins to be identical for the first 23 amino acids. Because of the small amount of protein isolated, further amino acid sequence information could not be obtained. Propose a plausible explanation for these results.

PART E: *Working with Research Data*

53. In an electron micrograph of a human chromosome spread, you observe a thick fiber with an apparent diameter of 30 nm, which is expected for the solenoid structure of condensed chromatin, and a length of about 900 nm.

   a. What length (in base pairs) of double-helical DNA is present in this fiber?

   b. What would be the effect on fiber length and diameter of preparing the chromosome spread under low-salt and low-magnesium conditions?

54. Sister chromatid exchange, which occurs in cultured mammalian cells, results in reciprocal linear exchanges between sister chromatids of single chromosomes in cells undergoing mitosis. Exchanges can be visualized by preparing a chromosome spread from a cell culture incubated in a medium containing the thymidine analog bromodeoxyuridine (BrdU) just long enough for the DNA to replicate once followed by growth in normal medium for a second round of replication. When the chromosomes are stained with Giemsa (a DNA stain) and a fluorescent dye, the BrdU-containing chromatids

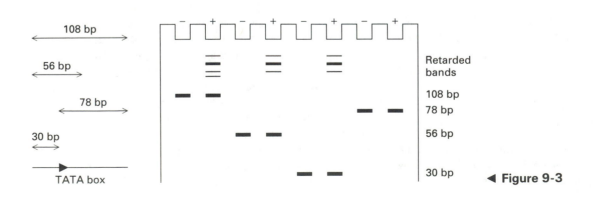

▲ **Figure 9-2**

fluoresce more brightly and hence look lighter than the unsubsitituted chromatids under the microscope. A chromosome spread prepared in this way is shown in Figure 9-2; the arrows indicate points of exchange.

a. Sketch a diagram showing how sister chromatid exchange leads to the chromosome staining pattern in Figure 9-2.

b. Does sister chromatid exchange appear to be a frequent event in mitotic cells? Is this event likely to produce mutations?

55. Because eukaryotic DNA-binding proteins are present in minute amounts, their detection requires very sensi-

tive assays. One suitable technique is a gel-retardation assay, in which crude cell extracts are incubated with radioactive DNA fragments and then the mixture is electrophoresed in a polyacrylamide gel. Binding of protein to DNA results in retarded migration of radioactive fragments in the gel. The position of the radioactive DNA fragments is detected by autoradiography.

The octameric sequence ATGCAAAT is found about 70 bp upstream from the +1 position in all human and mouse immunoglobulin heavy-chain genes. Much effort has been devoted to establishing if cell type–specific DNA-binding proteins bind to this sequence. In one such study, four different radioactive DNA fragments mapping to the octamer region were incubated in the absence (−) or presence (+) of a nuclear extract from the immunoglobulin-producing cell line BCL. The incubation mixtures were then electrophoresed and the migration of the radioactive DNA visualized by autoradiography. Figure 9-3a is a map of the DNA fragments with respect to the TATA box of the immunoglobulin heavy-chain gene. Figure 9-3b is the autoradiogram of the gel pattern.

a. Based on its location, what is a likely function of the octameric sequence?

b. Based on the gel-retardation assay, is binding of BCL nuclear protein to DNA fragment specific?

c. What is the maximal number of DNA-binding proteins in the nuclear extract revealed in this assay and are they present in equal amounts?

d. Do all these proteins bind to the same DNA site? If not, what is the minimal number of protein-binding sites within these fragments?

56. β-Galactosidase in *E. coli* is a multimeric protein containing identical subunits; the subunit is coded by the Z gene of the *lac* operon. Mutations in the Z gene termed Z⁻ result in complete loss of β-galactosidase activity. However, in trans complementation tests, some of these mutations complement each other, with as much as 25

(a) Fragment map   (b) Gel-retardation assay

108 bp

56 bp

78 bp

30 bp

TATA box

Retarded bands

108 bp
78 bp
56 bp

30 bp   ◀ **Figure 9-3**

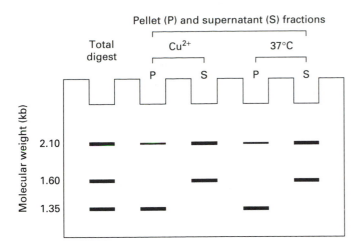

Pellet (P) and supernatant (S) fractions

▲ **Figure 9-4**

percent of the normal enzyme activity being detected. How can mutations that map to the same cistron within an operon complement each other? What are the implications of such mutations for the relationship between genes and proteins? This phenomenon is termed interallelic or intragenic complementation. Should the phenomenon be restricted to bacterial proteins only?

57. Histone genes in *Drosophila* are arranged as a tandem repeat of clustered H1, H2A, H2B, H3, and H4 genes.

About 100 copies of this repeat are found at the cytological 39DE locus of the polytene chromosome. Laemmli and colleagues investigated how this repeat is bound to the nuclear scaffold by determining whether histone gene fragments were retained in a scaffold preparation following digestion of the preparation with a mixture of restriction enzymes. In brief, a scaffold preparation was obtained from nuclei containing polytene chromosomes, digested with restriction enzymes, and pelleted; the DNA from the pellet and supernatant was then analyzed by Southern blotting. Typical results are shown in Figure 9-4 in which P = pellet fraction and S = supernatant fraction. In these experiments, the nuclei were incubated with Cu^{2+} or at 37°C to stabilize the scaffolds. The actual nuclear scaffolds were then prepared by extracting the pellet fraction with lithium diiodosalicylate (LIS).

a. In this experimental protocol, what result indicates that a histone-repeat fragment is bound to the scaffold?

b. What can you conclude from the data in Figure 9-4 about the location of the scaffold-binding site in the histone gene repeat?

c. How could you more precisely locate the scaffold-binding site in the histone gene repeat?

10

Eukaryotic Chromosomes and Genes: Molecular Anatomy

PART A: Reviewing Basic Concepts

Fill in the blanks in statements 1–19 using the most appropriate terms from the following list:

Ac	Q
Alu family	rearrangement
amplification	reciprocal crossing over
copia	7SL RNA
$C_0t_{1/2}$	5.8-S RNA
Ds	retroposon(s)
deletion	selfish DNA
faster	single-copy
gene conversion	simple-sequence
genetic assimilation	SINES
genetic drift	slower
insertion sequences	Tn
intermediate repeat	transposase
LINES	transposon(s)
LTR	Ty
mobile genetic element(s)	unequal meiotic crossing over
P	

1. Sequences that can propagate within the genome but do not seem to have a function in the organism are called _____.

2. Two possible mechanisms for gene duplication are _____ and _____.

3. The general term for sequences that can change location within the genome is _____.

4. A measure of the repetitiveness of a DNA species based on its reassociation rate is the _____ value.

5. _____ move within a genome through an RNA intermediate.

6. Formation of functional immunoglobulin genes involves _____ of DNA.

7. _____ of DNA occurs in the formation of the gene for the variable surface glycoprotein (VSG) antigens of *Trypanosoma brucei*.

8. _____ of DNA sequences accounts for the development of resistance to the anticancer drug methotrexate.

9. _____ DNA is contained within the most rapidly reassociating eukaryotic DNA fraction.

10. Bacterial elements that have inverted terminal repeats and do not contain coding sequences are called _____.

11. Bacterial elements that encode proteins and duplicate the target site for insertion are termed _____. One of the encoded proteins is a _____.

12. Human DNA contains 300-bp repeats, which collectively are called _____. They also are referred to as the _____, based on the presence of a particular endonuclease restriction site within many of these repeats. A small RNA species, _____ RNA, which is contained within the signal recognition particle, has sequence similarity to these repeats.

13. The abbreviation for a yeast transposable element is _____ and for a bacterial transposable element is _____.

14. In eukaryotic DNA, repetitive sequences that are 1–5

kb long and flanked by terminal repeats are collectively called _____.

15. Two well-studied genetic elements in maize, which show similarity to bacterial mobile elements and move by DNA duplication, are _____ and _____.

16. _____ DNA is contained in the most slowly reassociating fraction of eukaryotic DNA.

17. Sequences at the ends of viral retroposons are similar to the _____ of retroviruses.

18. _____ is a gradual process by which functional genes become nonfunctional.

19. The lower the $C_0t_{1/2}$, the _____ the rate of DNA reassociation.

PART B: *Linking Concepts and Facts*

Circle the letter(s) corresponding to the most appropriate terms/phrases that complete or answer items 20–29; more than one of the choices provided may be correct in some cases.

20. Simple-sequence DNA

   a. can be separated from chromosomal DNA by isopycnic centrifugation.

   b. can be found in centromeres.

   c. can be heterochromatic.

   d. all has the same function.

   e. can move in the genome.

21. The DNA within transcription units accounts for what percentage of the total DNA in eukaryotic cells?

   a. 0.01 percent   d. 10 percent

   b. 0.1 percent   e. 90 percent

   c. 1 percent

22. The DNA that encodes polypeptides accounts for what percentage of the total DNA in eukaryotic cells?

   a. 0.001 percent   d. 1.0 percent

   b. 0.01 percent   e. 10.0 percent

   c. 0.1 percent

23. Noncoding DNA in eukaryotic cells may include

   a. introns.   d. mobile genetic elements.

   b. pseudogenes.

   c. simple-sequence DNA.   e. spacer DNA.

24. Related genes in tandem arrays

   a. may all be functional.   d. may be nonidentical.

   b. may have copies that are nonfunctional.   e. may code for rRNA and tRNA.

   c. may be identical.

25. *Alu* sequences

   a. are required for intron processing.

   b. are flanked by short, repeated sequences.

   c. may constitute 5–10 percent of the human genome.

   d. are related to an RNA required for protein transport across the endoplasmic reticulum.

   e. are found in regions that flank protein-coding genes.

26. Duplicated elements that contain protein-coding sequences include

a. bacterial insertion sequences (ISs).

b. bacterial transposons.

c. SINES.

d. *Drosophila* P elements.

e. yeast transposable elements.

27. The presence of mobile genetic elements can be detected by

a. mutation and reversion.

b. in situ hybridization.

c. C_0t analysis.

d. constitutive expression of a normally inducible gene.

e. Southern blotting.

28. Unequal crossing over

a. requires DNA synthesis by DNA polymerase.

b. requires DNA synthesis by reverse transcriptase.

c. is proposed as a mechanism for removing rRNA genes with lethal mutations from genomes.

d. may cause duplication of eukaryotic transcription units.

e. may result in gene amplification in mammalian cells.

29. An example of a duplicated gene is that encoding

a. human δ-hemoglobin.

b. lysozyme.

c. tRNA.

d. 45S rRNA.

e. fibroin.

30. For each type of DNA listed below, indicate the relative $C_0t_{1/2}$ value as follows: 1 = lowest (<0.01); 2 = intermediate (0.01–10); and 3 = highest (100–10,000).

a. Simple-sequence DNA _____

b. Intermediate repeat DNA _____

c. Eukaryotic single-copy DNA _____

d. *E. coli* DNA _____

31. Indicate the mechanism of movement for each type of DNA element listed below by writing in the corresponding abbreviation from the following:
UCO = unequal crossing over
TRP = transposase
RT2 = reverse transcription of polymerase II transcript
RT3 = reverse transcription of polymerase III transcript

a. Bacterial transposon (Tn) _____

b. *Alu* sequence _____

c. Maize activator (Ac) element _____

d. Simple-sequence DNA _____

e. Yeast transposable element (Ty) _____

32. For each type of DNA element listed below, indicate all the sequences present at or near the termini as follows: IR = inverted repeat; DR = direct repeat; LTR = long terminal repeat; and N = none of these.

a. Insertion sequence (IS) _____

b. *Alu* sequence _____

c. Simple-sequence DNA _____

d. Bacterial transposon _____

e. Eukaryotic transposon _____

f. Processed gene _____

PART C: *Putting Concepts to Work*

33. Fingerprinting of human DNA is becoming increasingly important as a forensic tool. Describe the molecular basis of this technique.

34. In the 1960s, differentiation was thought to entail conservation of the genome in all somatic cells and differential gene expression in different tissues. Do you believe this hypothesis is still true?

35. What evidence is lacking that would confirm the importance of mobile genetic elements in evolution?

36. You are given two unlabeled tubes, one containing yeast DNA and the other containing DNA of a bacterial plasmid. How could you distinguish these DNAs by C_0t analysis?

37. Describe the similarities and differences between pseudogenes and processed pseudogenes.

38. Describe the mechanism proposed for retrotransposition of polymerase III transcripts using *Alu* sequences as an example.

39. The mechanism of yeast mating-type switching has been described as similar to a cassette player. Describe the mechanism. Is the analogy accurate?

40. What is the evidence that *copia* and Ty are related to retroviruses?

41. What sequence feature is diagnostic for the presence of transposons and how is this feature produced?

PART D: *Developing Problem-solving Skills*

42. The chicken lysozyme gene is a classic example of a solitary gene. This 15-kb transcription unit contains four exons and three introns and is bounded on either side by flanking regions, each about 20 kb long, upstream and downstream from the transcription unit. When chick DNA is fragmented into 5- to 10-kb pieces and then analyzed for repetitious DNA and single-copy DNA, some of the DNA fragments are found to contain both lysozyme-specific sequences and repetitious DNA. Explain.

43. The best-educated guess today is that not more than about 1 percent of the human genome codes for proteins. Considering that a somatic human cell contains 5 pg of DNA and a rapidly growing *E. coli* cell contains 0.017 pg of DNA, how much more protein might be potentially expressed by a somatic human cell than by an *E. coli* cell. Note that a rapidly growing *E. coli* cell contains, on the average, four DNA genomes.

44. Albumin and α-fetoprotein are present-day members of the same protein family. Shown in Figure 10-1 is the exon-intron map of the genomic DNA coding for each of these proteins.

    a. Draw a diagram depicting the relationships between the present-day albumin and α-fetoprotein genes shown in Figure 10-1 and the first primordial gene that coded for a related primordial protein.

    b. Which DNA sequences must have been altered during evolution to result in the structures of the present-day genes?

45. Explain why in situ hybridization techniques can readily localize satellite DNA sequences on mouse or human chromosomes but are of little value for the localization of *Alu* sequences.

46. Hybrid dysgenesis is a term used to describe the outcome of a mating in which the progeny are sterile heterozygotes. In *Drosophila melanogaster*, hybrid dysgenesis is common when P-strain males are mated with females that lack the P element. Describe how hybrid dysgenesis could be used as a test of the geographic spread of the P element in the *D. melanogaster* population.

47. A rolling circle replication model has been proposed as the mechanism for the production of extrachromosomal, amplified rRNA genes in amphibian oocytes. How much variation in the nontranscribed spacer regions between individual gene repeats would you expect if this model is correct?

48. Any explanation of sequence constancy in transcribed regions of tandemly repeated genes must also explain the comparative lack of sequence constancy in nontranscribed spacer regions. How do the high-frequency unequal crossing over and gene conversion models account for the high variability of spacer DNA?

49. Methotrexate, an inhibitor of dihydrofolate reductase (DHFR), has been used as a chemotherapeutic agent for cancer patients. Unfortunately, patients often develop resistance to methotrexate, but physicians have found that the development of drug resistance is decreased if periods of medication are alternated with drug-free periods.

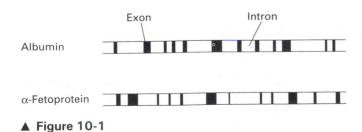

▲ **Figure 10-1**

a. Describe the two mechanisms for methotrexate resistance that have been proposed.

b. Why is the development of methotrexate resistance decreased by use of an intermittent drug schedule?

50. Various simple-sequence DNAs can be readily isolated as satellite DNA. How can the melting temperature of an isolated satellite DNA be used to predict its base composition?

51. Antibiotic-resistance genes are often propagated through bacterial populations as plasmid-resident transposons. Why might transposition be a particularly important mechanism for the development of bacterial plasmids containing multiple drug-resistance genes?

52. Biologists have long assumed that evolution is an efficient process in which nonfunctional structures are not preserved. Knowledge of the molecular anatomy of eukaryotic chromosomes has shaken this belief. Explain.

53. Ds elements in maize have been considered as carriers for the stable integration of genes in corn. Why is a "helper" Ac element required as part of this strategy? Why may Ds elements be a better choice than plasmids for the stable integration of new genes in corn?

54. Many apparent Mendelian mutations in eukaryotes including yeast and *Drosophila* have turned out to be caused by transposons. Describe two experimental approaches — one genetic and one biochemical — for determining whether an apparent Mendelian mutation is actually a transposition event.

55. Compare the effect of mutations in putative promoter regions on the movement of retroposons and DNA transposons within the genome.

PART E: *Working with Research Data*

56. Heating causes dissociation of double-stranded DNA and is accompanied by an increase in ultraviolet absorbance by the DNA bases. Shown in Figure 10-2 are melting curves for three different calf thymus DNA samples: a freshly prepared native DNA, reassociated intermediate repeat DNA, and reassociated single-copy DNA. The total amount of DNA in each sample is the same.

a. Why does the intermediate repeat DNA have a higher absorbance value at 37°C than either the single-copy or native DNA?

b. As the temperature is increased, the DNA dissociates, as evidenced by hyperchromicity, in all the samples. However, denaturation occurs both at lower temperatures and over a much broader temperature range for the intermediate repeat sample than for the other two samples. Why?

c. Based on the DNA melting profiles in Figure 10-2, does the repetitious fraction consist of exact DNA copies?

57. One experimental approach to establishing the function of a repetitive DNA sequence has been to determine whether it is transcribed by use of cloned DNA as an affinity matrix to "fish" transcripts from a cell extract.

a. Using this general experimental approach and assuming you have an *Alu*-family DNA matrix available, how would you proceed to determine if *Alu*-family transcripts are synthesized in vivo and present in the cytoplasm of cultured human cells?

b. Assume that you find in the cytoplasm of cultured cells RNA that hybridizes to the *Alu*-family DNA matrix. How could you determine whether it is an *Alu* transcript or 7SL RNA, a normal cytoplasmic RNA that contains within its primary sequence an *Alu* sequence?

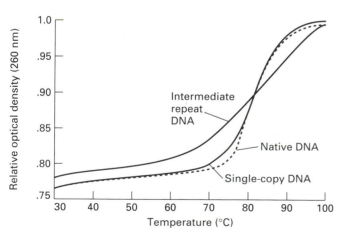

▲ **Figure 10-2**

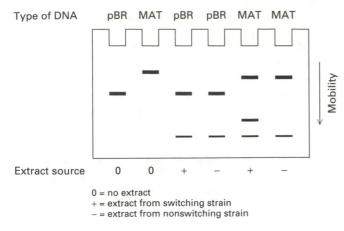

▲ **Figure 10-3**

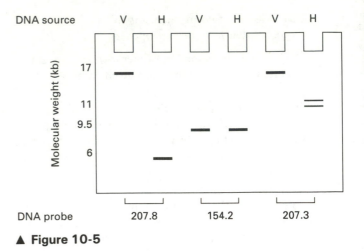

▲ **Figure 10-5**

58. Switching between **a** and α mating types in yeast involves site-specific cleavage of DNA. To understand the biochemical mechanism of this step, Kostriken and colleagues have screened for a specific endonuclease activity. The substrates used for the assay were either radiolabeled pBR322 DNA, a bacterial plasmid, or radiolabeled MATa DNA. The results from one assay are shown in Figure 10-3, which depicts an autoradiogram of the electrophoresed radiolabeled DNA substrates following incubation with extracts from switching (+) and nonswitching (−) yeast strains.

    a. How is endonuclease activity detected in this assay? How many endonuclease activities are present and which of these activities is mating specific?

    b. What types of mutations would result in trans-active or cis-active variations in mating-specific DNA cleavage in yeast?

59. You have just completed β sequencing a genomic DNA segment that you hope is a gene for human β-tubulin. Several features of this segment are illustrated in Figure 10-4. To your surprise, the sequence has few, if any, of the features expected for a tubulin gene. What type of DNA segment have you sequenced and what is its likely origin?

60. Nitrogen fixation by the cyanobacterium, *Anabena*, occurs in specialized, nondividing cells, which differen-

tiate under conditions of nitrogen starvation. These nondividing cells, termed heterocysts, are derived from undifferentiated vegetative cells. The arrangement of nitrogen-fixation (*nif*) genes in vegetative *Anabena* in relation to *Eco*RI restriction sites and three different DNA probes is as follows:

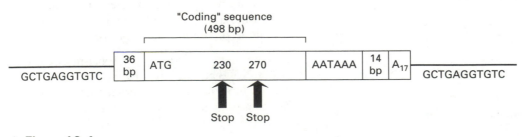

    a. When DNA was extracted from vegetative (V) and heterocyst (H) cells, digested with *Eco*RI restriction endonuclease, and then electrophoresed in an agarose gel, the Southern blot pattern shown in Figure 10-5 was obtained with the 207.8 DNA probe. When the 154.2 DNA probe was used, a single 9.5-kb band was obtained with both vegetative and heterocyst DNA. Based on these results, what can you conclude about the change(s) that occurs in the *nif* gene region during differentiation of vegetative cells into heterocysts?

    b. As also shown in Figure 10-5, when the 207.3 DNA probe was used in the Southern blot analysis, a single

▲ **Figure 10-4**

17-kb fragment was revealed with the vegetative cell DNA and two closely spaced bands (a doublet) at about 11 kb were revealed with the heterocyst DNA. Are these results consistent with those obtained with the other two probes described in part (a)?

c. The 11-kb doublet revealed by probing of *Eco*RI-digested DNA from heterocysts was subjected to further digestion with *Kpn*I restriction endonuclease, which is known to make only one cut in this DNA. Electrophoresis of the digest resulted in a single band of about the same mobility (11 kb) as the doublet. Based on these results, is the doublet DNA a closed circular or linear molecule? (*Note:* A circular DNA with a single-strand cut, called a nick, migrates differently in a gel system than does unnicked circular DNA.)

d. Would you expect sequencing of the doublet DNA to reveal an inverted repeat within this DNA? If so, where would the repeat be located in the doublet DNA sequence?

CHAPTER *11*

Gene Control and the Molecular Genetics of Development in Eukaryotes

PART A: *Reviewing Basic Concepts*

Fill in the blanks in statements 1–16 using the most appropriate terms from the following list:

acidic domain	nuclease
amphiphatic helices	pair rule
AAUAAA	pluripotent
$A(U)_nA$	poly A tail
basic domain	processing
chromatin structure	proteins that bind to cell-surface receptors
cytochrome P-450	pulse-chase
differentiation	ribosome
effector	RNA polymerase II
environmental factors	run-off
fat-soluble small molecules	run-on
helix-turn-helix	second messenger
homeobox	$TF_{II}D$
homeotic	transcription
inhibitor-chase	translation
ligand	unipotent
methylation	xenobiotics

zinc fingers   1000
    100

1. In prokaryotes, gene control occurs in response to _____, whereas in eukaryotes, gene control causes _____.

2. Two chemical classes of molecules are signals for gene control; they are _____ and _____.

3. Three structural motifs found in DNA-binding proteins are _____, _____, and _____.

4. DNA-binding proteins probably interact with _____ and/or _____ to activate transcription.

5. The enzymes induced in the liver in response to oxidation products of toxic compounds called _____ are _____ proteins.

6. The half-life of a mRNA is determined by the length of its _____, which may be affected by a _____ associated with the _____.

7. Gene control in eukaryotes can be affected at three levels: _____, _____, and _____.

8. The rate of transcription of a specific gene can be measured by a _____ assay.

9. Regulatory genes that control developmental pathways in *Drosophila* are called _____ genes.

10. In addition to transcription factors, transcription is also controlled by two chromosomal features— _____ and _____.

11. The nucleotide sequence that controls mRNA half-life is _____.

12. A protein that may control segmentation during development in insects and mammals contains a conserved amino acid sequence called a _____.

13. Eukaryotic cells contain about _____ easily identifiable cell type–specific proteins.

14. A _____ cell has the potential to develop into more than one differentiated cell type.

15. Cell-surface receptors, after interaction with a _____, can initiate transcription through an intermediate called a _____.

16. A(n) _____ assay accurately measures the half lives of mRNAs with rapid turnover times.

PART B: *Linking Concepts and Facts*

Circle the letters corresponding to the most appropriate terms/phrases that complete items 17–26; more than one of the choices provided may be correct in some cases.

17. Environmental signals that affect expression of specific genes in eukaryotic cells include

    a. heat.

    b. light.

    c. cell-cell contact.

    d. starvation for cysteine.

    e. presence of toxic compounds.

18. Structural characteristics that a DNA-binding protein may have include

    a. the presence of a heavy metal ion.

    b. the presence of a "face" of basic amino acids.

    c. conserved sequences shared with other proteins.

    d. conserved secondary structure shared with other proteins.

    e. dyad symmetry.

19. Functional characteristics of DNA-binding proteins may include

    a. acting as a monomer.

    b. acting as a homodimer.

    c. acting as a heterodimer.

    d. participating in pre-initiation complexes.

    e. binding to enhancers.

20. Control of translation in eukaryotes

    a. results in synthesis of specific proteins after fertilization in the surf clam.

    b. occurs by decreasing the size of polysomes during mitosis.

    c. requires secondary structure at the 5′ end of a mRNA for translation to occur.

    d. is mediated by the length of the poly A tail.

    e. usually requires a cap-binding complex for translation initiation.

21. Processing of eukaryotic pre-mRNA to produce alternative mRNAs from a single primary transcript

    a. is one factor in sex determination in *Drosophila*.

    b. can be regulated by sequence differences in snRNA.

    c. can be regulated by the proteins found in snRNPs.

    d. can result in different mRNAs in different cell types.

    e. is directly controlled by transcription factors.

22. DNA-binding proteins attach to upstream sequences

    a. that are present only once in yeast.

    b. that are present several times in higher eukaryotes.

    c. and may cause loop formation in DNA.

    d. and this binding may be useful for distinguishing promoter-proximal sequences from enhancers.

    e. which may show dyad symmetry.

23. Transcription of hepatocyte-specific mRNA

    a. occurs in liver slices in culture.

b. requires transcription factors unique to liver cells.

c. occurs in disaggregated liver cells in culture.

d. requires transcription factors found in other cells.

e. influences transcription of tubulin mRNA.

24. Hormonal control of transcription

    a. requires a receptor normally found in nuclei.

    b. requires a cytoplasmic receptor that can move to the nucleus.

    c. requires two domains in the receptor.

    d. can result in the appearance of chromosomal puffs.

    e. occurs by receptors activating a pre-existing transcription factor.

25. Factors that control the stability of mRNA include

    a. hormones.

    b. translation products of mRNA.

    c. sequences in the poly A tail.

    d. rate of transcription.

    e. ligands.

26. Synthesis of rRNA in eukaryotes

    a. is mediated through ppGpp in lower eukaryotes.

    b. is controlled by the rate of initiation of pre-rRNA transcripts.

    c. ceases almost totally during starvation of yeast.

    d. can be controlled by the rate of ribosomal protein synthesis.

    e. can be controlled by the same transcription factors that activate protein-coding genes.

For each DNA-binding protein listed in items 27–31, write in the letters corresponding to all the properties it has from the following:

    a. contains helix-turn-helix

    b. contains zinc fingers

    c. can form heterodimer

    d. can form homodimer

    e. contains helix-loop-helix

    f. contains leucine zipper

27. AP1 _____

28. C/EBP _____

29. Ultrabithorax protein _____

30. MyoD _____

31. Steroid receptors _____

Indicate all the control mechanisms that regulate each gene or protein listed in items 32–37 by writing in the corresponding letter(s) from the following:

    a. alternative splicing

    b. alternative poly A site selection

    c. transcription termination

    d. autoregulation of translation

    e. posttranscription processing

    f. silencing

32. *c-myc* _____

33. Tubulin _____

34. Troponin _____

35. *HMLα* and *HMLa* _____

36. Calcitonin-CGRP _____

37. rRNA _____

38. Figure 11-1 is a schematic diagram of proteins bound to DNA and mRNA during the initiation of transcription and translation. Identify the proteins on the indicated labels.

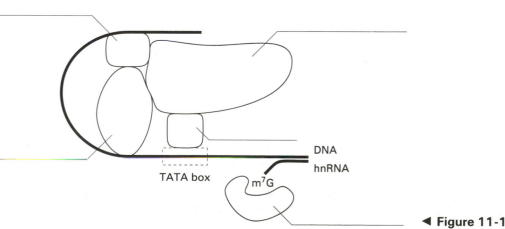

TATA box

m⁷G

DNA

hnRNA

◀ **Figure 11-1**

PART C: *Putting Concepts to Work*

39. By footprinting analysis, you identify the nucleotide sequence to which a particular transcription factor binds. You then synthesize an oligonucleotide with this sequence and link it to an inert support. Affinity chromatography of a cell extract on this support and analysis of the eluate by gel electrophoresis reveals several proteins. Explain.

40. In 1961, just after they had elucidated the regulatory circuits of the *lac* operon, Francois Jacob and Jacques Monod applied the principles of induction and repression to the problem of differentiation. They developed "circuit diagrams" to illustrate how combinations of sequential activation and/or repression could cause expression or lack of expression of specific genes. Is their hypothesis supported by the gene-control patterns observed in *Drosophila*?

41. What features of the regulatory elements of heat-shock genes insure that they can be expressed rapidly under adverse conditions?

42. Genes related to cellular genes have been found in viruses that cause cancer. What do you think is the normal function of the products of these cellular genes? What does the existence of these homologous viral and cellular genes suggest about oncogenesis?

43. SP1 is a transcription factor that was first identified by its binding to the genome of the virus SV40. It also binds to cellular DNA. What can you infer about the cellular DNA sequences with which SP1 interacts and the number of genes SP1 can affect?

44. Why is the ability of transcription factors to forms dimers advantageous to the cell?

45. How did the discovery of dimeric transcription factors help explain the functioning of enhancers, which are located some distance away from the genes they influence?

46. What is the functional difference between the basic and acidic amino acid domains found in DNA-binding proteins?

47. Describe three mechanisms for regulating the activity of transcription factors.

48. What principles about cellular differentiation are demonstrated by the sequence of gene expression leading to the synthesis of muscle proteins?

49. Certain DNA viruses can enter cells but do not replicate in these cells. Explain this observation.

50. Two functional domains have been identified in several hormone receptors. What are the functions of these domains? How have recombinant DNA techniques been used to define them.

51. Does the identification of cell type–specific transcription factors that activate genes provide an explanation of differentiation? Justify your answer.

52. Although gene control can operate by many mechanisms, both positive and negative, some genes can be silenced and not transcribed at all.

    a. If you wanted to study this phenomenon, which gene(s) would you choose as a model system and why?

    b. Assuming that upstream DNA sequences are involved in the silencing mechanism, describe an experimental strategy to identify them. How might you confirm the silencing function of these sequences?

53. Many transcription factors have been identified by their effects on the expression of the genes of viruses that infect eukaryotic cells. Why have viruses been used so extensively in these studies?

PART D: *Developing Problem-solving Skills*

54. You propose to correct a genetic defect in mice by introducing the DNA sequences encoding the missing protein into mouse embryos. To be useful, the protein must be expressed only in the liver. What DNA sequences must be introduced along with the protein-coding segment to insure liver-specific expression of this protein?

55. Shown in Figure 11-2 are the partial amino acid sequences of 10 different proteins. The *Drosophila* pro-

Gene product		1	2	3	4	5	6	7	8	9	10	11	12	13	14	15	16	17	18	19	20
Drosophila	ftz	Arg	Ile	Asp	Ile	Ala	Asn	Ala	Leu	*Ser*	Leu	Ser	Glu	Arg	Gln	Ile	Lys	Ile	Trp	Phe	Gln
	Antp	Arg	Ile	Glu	Ile	Ala	His	Ala	Leu	Cys	Leu	Thr	Glu	Arg	Gln	Ile	Lys	Ile	Trp	Phe	Gln
	Ubx	Arg	Ile	Glu	Met	Ala	His	Ala	Leu	Cys	Leu	Thr	Glu	Arg	Gln	Ile	Lys	Ile	Trp	Phe	Gln
Frog		Arg	Ile	Glu	Ile	Ala	Asn	Ala	Leu	Cys	Leu	Thr	Glu	Arg	Gln	Ile	Lys	Ile	Trp	Phe	Gln
Yeast	MATa1	*Lys*	Glu	*Glu*	*Val*	*Ala*	Lys	Lys	Cys	Gly	*Ile*	*Thr*	Pro	Leu	*Gln*	*Val*	*Arg*	*Val*	*Trp*	*Phe*	Ile
	MATα2	Leu	Glu	Asn	*Leu*	Met	Lys	Asn	Thr	*Ser*	Leu	Ser	Arg	Ile	*Gln*	Ile	Lys	Gln	*Trp*	Val	Ser
Bacteria	λ Cro	Gln	Thr	Lys	Thr	*Ala*	Lys	Asp	*Leu*	Gly	*Val*	Tyr	Gln	Ser	Ala	*Ile*	Asn	Lys	Ala	Ile	His
	P22 Cro	Gln	Arg	Ala	*Val*	*Ala*	Lys	*Ala*	Leu	Gly	*Ile*	Ser	Asp	Ala	Ala	*Val*	Ser	Gln	*Trp*	Lys	*Gln*
	CAP	*Arg*	Gln	*Glu*	*Ile*	Gly	Gln	Ile	*Val*	Gly	Cys	*Ser*	Arg	Glu	Thr	*Val*	Gly	Arg	Ile	Leu	Lys
	AraC	Ile	Ala	Ser	*Val*	*Ala*	Gln	His	*Val*	Cys	*Leu*	Ser	Pro	Ser	Arg	*Leu*	Ser	His	Leu	*Phe*	Arg

▲ **Figure 11-2**

teins are Fushi tarazu (Ftz), Antennapedia (Antp), and Ultrabithorax (Ubx); the yeast proteins are mating-type proteins (MATa1 and MATα2); and the bacterial proteins are the Cro protein from λ and P22, the cyclic AMP-binding protein (CAP), and the arabinose-operon positive activator (AraC). One frog protein is also shown.

a. What, if any, evidence of amino acid sequence conservation among species is present in these 20-aa segments?

b. How much more related are the three *Drosophila* proteins to each other than to *E. coli* CAP?

c. These amino acid sequences are in the DNA-binding region of the proteins. The *Drosophila* sequences are from homeobox proteins, which determine body segmentation. How important may apparently small differences in amino acid sequence, as exemplified by the *Drosophila* examples, be in determining tissue organization?

56. Encoded in the DNA of eukaryotic cells are enhancer and promoter sequences, which exert important effects on the rate of gene transcription. These sites differ considerably in their proximity to the transcription start site. Is the nature of the proteins that bind to these sites different? Is the mechanism through which these sites affect transcription different?

57. In a Northern blotting experiment, cDNAs to the C-terminal portion of calcitonin and CGRP hybridize to different poly A–containing cytoplasmic RNA species. The calcitonin-specific hybridization is observed only with cytoplasmic RNA prepared from thyroid tissue, and the CGRP-specific hybridization is observed only with cytoplasmic RNA from neurons. When the same cDNAs are hybridized to a nuclear poly A–containing RNA preparation from neurons, the CGRP cDNA detects high-molecular-weight RNA that codes for both CGRP and calcitonin. When the nuclear poly A–containing RNA preparation is from thyroid tissue, the only major nuclear RNA species found is an intermediate-molecular-weight species that is detected only by the calcitonin-specific cDNA probe. Explain.

58. Biologists have long hoped that differentiation in plants and animals could be explained by one or at most a small number of mechanisms. Is such a hope tenable today?

59. You have isolated a new hormone that is soluble in chloroform and acetone and has a molecular weight of about 300 daltons. It is only very sparingly soluble in water. Elemental analysis indicates the presence of C, H, and O. No S, I, or N is detected. Your company has assigned a registry number of RO12347 to the hormone. To what hormone family do you think RO12347 belongs and what might be the mechanism by which it affects gene expression?

60. Measurements of metallothionein levels in fish have been proposed as a sensitive biological assay for environmental contamination of lakes and streams. What is the molecular basis of the assay and what type of contamination could be detected?

61. You have completed a run-on assay to determine the relative expression of ovalbumin in an oviduct nuclei preparation. When you used [^{14}C]ATP in the assay, considerable incorporation was observed, whereas when you used β-labeled [^{32}P]ATP, little, if any, incorporation was observed. Explain these results. You plan to use this system for determining the effect of a steroid-receptor complex on transcription by adding the isolated complex to the nuclear preparation. Will the planned experiment work?

62. The application of molecular approaches to the analysis of *Drosophila* embryonic lethal mutations has revealed

that many of these mutations are defects in cognate DNA-binding factors. Why might DNA-binding factors be the expected molecular basis of an embryonic lethal phenotype?

63. Leucine-zipper proteins are a class of transcription activator proteins in eukaryotes. They share the common design of a dimeric structure with each of the two subunits having two functional domains: a region rich in basic amino acids, which comprises the cognate DNA-binding site, and a region rich in hydrophobic amino acid leucine, which comprises the monomer interaction domain. C/EBP was the first leucine-zipper protein to be extensively studied. Although there are likely to be several structurally similar leucine-zipper proteins in hepatocytes, the cell from which C/EBP was first isolated, C/EBP was found to be a homodimer. What does this suggest regarding the interaction of leucine-zipper protein subunits?

64. Diploid yeast produce a1, a negative regulator of haploid-specific genes, and α2, a negative regulator of a-specific genes, as well as PRTF, the pheromone and receptor transcription factor. These proteins together regulate the transcription of α-specific genes, a-specific genes, haploid-specific genes, and the α1 gene. What would be the effect of a mutation in the a1 gene that gave rise to a defective a1 protein on the mating-type phenotype of a diploid yeast? Would you expect the mutant to be capable of mating?

65. Cycloheximide, an inhibitor of eukaryotic protein synthesis, is frequently added to tissue culture media to test whether new protein synthesis is required for a metabolic response. In *Drosophila* cell cultures repeatedly cycled between normal and elevated temperatures, would cyclic activation of heat-shock genes and synthesis of heat-shock proteins be expected after cycloheximide treatment?

66. Cultural fibroblasts can be stained with a fluorescent antibody against interferon-activated transcription factor. What would be the expected staining pattern of cells fixed and stained before and after interferon treatment?

67. MyoD-1 and myogenin can be microinjected into growing cultured mouse 10T1/2 fibroblasts. What changes, either positive or negative, in gene expression would be expected following microinjection of either substance.

68. During the characterization of a ribosomal protein mutant, you make the fortuitous observation that the $t_{1/2}$ of histone mRNA is 30 h in this mutant. What putative ribosomal activity does this mutation appear to affect? What effect on the $t_{1/2}$ of other mRNAs would this mutation have? How would such a mutation affect the phenotype of the cells?

PART E: *Working with Research Data*

69. Under defined cell culture conditions, mouse F9 stem cells will give rise to various embryonic tissues. For example, when these cells are treated with retinoic acid, endodermal tissue is formed. About 10 percent of the protein in the stem cells and the differentiated cells consists of actin, a 45-kDa protein. About 0.01 percent of the protein in the differentiated cells is ERgp76, a 76-kDa glycoprotein localized to the endoplasmic reticulum. No ERgp76 is detectable in the stem cells.

   a. Assuming that both the stem cells and the differentiated cells contain 200 μg of protein per 1×10^6 cells, what is the average number of molecules of actin and ERgp76 per cell in the stem cells and the retinoic acid–treated cells?

   b. What probe(s) is needed to determine if the amount of ERgp76 mRNA is increased with retinoic acid treatment and how would the probe(s) be used?

70. The NFκB transcription factor binds to the kappa light-chain enhancer, the human immunodeficiency virus (HIV) enhancer, and an upstream sequence of the interleukin-2 receptor α-chain gene. Binding of NFκB confers transcriptional activity and phorbol ester inducibility to genes controlled by these cis-acting elements. This factor can exist in active and inactive forms. Baeuerle and Baltimore have used electrophoretic mobility shift analysis (EMSA) to investigate the nature of these two forms. EMSA can detect the ability of a transcription factor to retard the mobility of enhancer DNA in a gel.

    In one such experiment, isolated NFκB factor was incubated with ^{32}P-labeled enhancer DNA in the presence or absence of varying amounts of complete cytosol or NFκB-depleted cytosol from pre-B cells. The migration of the labeled DNA in the gel following incubation is shown in Figure 11-3. Free DNA migrates rapidly and

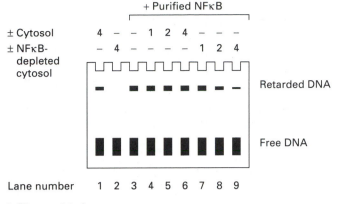

+ Purified NFκB

| ± Cytosol | 4 | – | – | 1 | 2 | 4 | – | – | – |
| ± NFκB-depleted cytosol | – | 4 | – | – | – | – | 1 | 2 | 4 |

Retarded DNA

Free DNA

Lane number 1 2 3 4 5 6 7 8 9

▲ **Figure 11-3**

is found at the bottom of the gel. The addition of increasing amounts of cytosol or depleted cytosol to the incubation mix are indicated by the numbers 1 → 4 at the top of the figure. As controls, incubation is with cytosol or depleted cytosol alone (lanes 1 and 2).

a. Based on lanes 1 and 2 of the EMSA pattern, what is the effect of NFκB on enhancer DNA migration?

b. Does cytosol have any apparent effect on the migration of enhancer DNA incubated with purified NFκB?

c. Does NFκB-depleted cytosol have any apparent effect on the migration of enhancer DNA incubated with purified NFκB? If so, does depleted cytosol appear to contain an activator or inhibitor of NFκB activity?

71. You have constructed a plasmid set containing a series of nucleotide insertions spaced along the length of the glucocorticoid-receptor gene. Each insertion encodes three of four amino acids. The map positions of the various insertions in the coding sequence of the receptor gene is as follows:

```
0  Glucocorticoid-receptor coding sequence  783
```

Insertion: A C E G I K M O Q   S
            B D F H J L N P R

The plasmids containing the receptor gene can be functionally expressed in CV-1 and COS cells, which contain a steroid-responsive gene. Using these cells, you determine the effect of each of these insertions in the receptor on the induction of the steroid-responsive gene and on binding of the synthetic steroid dexamethasone. The results of these analyses are summarized in Table 11-1.

a. From this analysis, how many different functional

Table 11-1

Insertion	Induction	Dexamethasone binding
A	++++	++++
B	++++	++++
C	++++	++++
D	0	++++
E	0	++++
F	0	++++
G	++++	++++
H	++++	++++
I	+	++++
J	++++	++++
K	0	++++
L	0	++++
M	0	++++
N	+	++++
O	++++	++++
P	++++	++++
Q	0	0
R	0	0
S	0	0

domains does the glucocorticoid receptor have? Indicate the position of these domains relative to the insertion map.

b. Which domain is the steroid receptor—binding domain?

c. How could you determine which of the domains is the DNA-binding domain?

72. Eukaryotic elongation factor 2 (EF2) is phosphorylated by an EF2 protein kinase. You would like to determine if phosphorylation has any direct effect on the ability of EF2 to support protein synthesis. You find by analytical biochemistry that the extent of phosphorylation of an EF2 preparation can be readily established by the separation of pEF2 from EF2 by isoelectric focusing. In your laboratory, you have available an EF2-depleted translational system prepared from rabbit reticulocytes; polyU, a synthetic RNA that codes for polyphenylalanine; radioactive phenylalanine; and alkaline phosphatase.

a. Using the indicated reagents, how could you establish whether EF2 phosphorylation affects translation?

b. Assuming that EF2 phosphorylation does affect translation, would its effect be mRNA specific?

73. Gene expression can be controlled at a number of levels in eukaryotic cells. Shapiro and colleagues investigated the balance between the rate of synthesis and degradation of vitellogenin mRNA in a *Xenopus laevis* liver cell

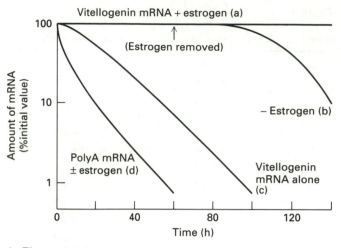

▲ **Figure 11-4**

culture system. They found that the transcription of the vitellogenin gene is greatly enhanced upon addition of estrogen, a steroid hormone, to the culture. They also studied the effect of estrogen treatment on the stability of mRNA in *Xenopus* liver cells. The results of such an experiment are shown in Figure 11-4.

a. What is the effect of estrogen on the stability of vitellogenin mRNA in *Xenopus* liver cells?

b. Is the estrogen effect mRNA specific?

c. Is the estrogen effect reversible? What does this suggest about the nature of the associations that mediate this effect?

DNA Replication, Repair, and Recombination

PART A: *Reviewing Basic Concepts*

Fill in the blanks in statements 1–25 using the most appropriate terms from the following list:

3′	gamma (γ)
5′	growing fork
3′ → 5′	helicase
5′ → 3′	Holliday
alpha (α)	lagging
aphidicolin	leading
beta (β)	ligase
catenanes	linking number
conservative	Meselson-Stahl
delta (δ)	Okazaki fragments
deoxyribonucleoside diphosphates	PCNA
	primase
deoxyribonucleoside monophosphates	primer
	processivity
deoxyribonucleoside triphosphates	proofreading
	replication origin
G_1	replicon
G_2	

replisome	topoisomerase I
ribonucleoside triphosphates	topoisomerase II
	twist
S	writhe
SOS response	
semiconservative	

1. The direction of DNA synthesis is _____.

2. DNA polymerases require a _____ with a free _____ hydroxyl group.

3. The substrates needed for DNA synthesis are _____ and _____.

4. The strand of DNA that is synthesized as a continuous piece is the _____ strand; the strand that is synthesized in fragments is the _____ strand.

5. The DNA polymerase found in mitochrondria is called _____.

6. During _____ replication, a preexisting strand is paired with a newly made strand.

7. A(n) _____ is a stretch of DNA necessary and sufficient to insure replication of a circular DNA in the appropriate host cell.

8. DNA synthesis occurs during the _____ phase of the cell cycle in eukaryotes.

9. Nucleic acid segments containing RNA and DNA that are intermediates in DNA replication are called _____.

10. An enzyme called _____ joins newly synthesized DNA fragments.

11. The RNA fragments that are necessary for discontinuous synthesis of the lagging strand are produced by an enzyme called _____.

12. Replication of both parental DNA strands occurs at the _____, which progresses in the 5′ → 3′ direction.

13. A drug called _____ specifically inhibits DNA polymerases α and β, thus blocking DNA synthesis.

14. The complex of all the proteins required for DNA synthesis is referred to as a(n) _____.

15. _____ is required for full activity of DNA polymerase δ.

16. The DNA region between two adjacent replication origins is called a(n) _____.

17. The enzyme _____ is required to separate newly replicated chromosomes.

18. The linked forms obtained after replication of circular genomes are called _____.

19. The ability of the replication apparatus to correct errors in replication is called _____. This is carried out by an exonuclease that works in the _____ direction.

20. The enzyme _____ unwinds the DNA double helix during replication.

21. The ability of DNA polymerase to remain on the template and continue synthesis is its _____.

22. The number of times one strand of a DNA double helix crosses the other in a molecule with fixed ends is the _____.

23. Gene activation resulting from exposure of *E. coli* to ultraviolet light is the _____.

24. The parameter that describes the distance from a position where one strand of the DNA helix crosses the other to the next similar position is _____.

25. A cross-stranded intermediate in recombination is called a _____ structure.

PART B: *Linking Concepts and Facts*

Circle the letters corresponding to the most appropriate terms/phrases that complete items 26–35; more than one of the choices provided may be correct in some cases.

26. DNA synthesis begins

   a. at a single location in *E. coli*.

   b. at a single location in the SV40 genome.

   c. at a single location in the adenovirus genome.

   d. at a single location in yeast.

   e. at a site(s) that is G-C rich in *E. coli*.

27. A protein that interacts with the origin of replication

   a. in SV40 is T antigen.

   b. in *E. coli* is dnaA.

   c. in adenovirus is terminal protein.

   d. unwinds DNA prior to replication.

   e. nicks DNA prior to replication.

28. Form I DNA

   a. is supercoiled.

   b. sediments faster than form II DNA.

   c. sediments faster than form III DNA.

   d. can be converted to form II by a nick in a single strand.

   e. can have more than one linking number.

29. Topoisomerase I activity

   a. cuts one strand of a DNA double helix.

   b. cuts both strands of a DNA double helix.

   c. changes the linking number by 1.

   d. changes the linking number by 2.

   e. requires energy supplied by ATP.

30. Topoisomerase II activity

    a. cuts one strand of DNA double helix.

    b. cuts both strands of a DNA double helix.

    c. changes the linking number by 1.

    d. changes the linking number by 2.

    e. requires energy supplied by ATP.

31. The nucleotide sequences of replication origins

    a. are conserved across bacteria.

    b. contain repeated elements.

    c. can contain a conserved sequence present in *Alu* sequences and SV40 DNA.

    d. can be palindromes.

    e. can be functional when cloned into plasmids.

32. Negative supercoiling in DNA

    a. can be relieved by topoisomerase I nicking followed by religation.

    b. is necessary for binding of *E. coli* RecA protein.

    c. results from underwinding in a circular molecule.

    d. occurs in front of a replication fork.

    e. occurs in some but not all naturally occurring plasmids.

33. Damage to DNA

    a. can prevent replication and transcription.

    b. can occur spontaneously because of the nature of the chemical bonds in DNA.

    c. can be repaired during normal replication of DNA.

    d. caused by the same environmental agents can be repaired by different mechansims.

    e. caused by different environmental agents can be repaired by the same mechanisms.

34. DNA replication in eukaryotic cells

    a. fills the same proportion of the generation time as it does in prokaryotic cells.

    b. is normally restricted to the S phase even in rapidly dividing cells.

    c. can be induced in G_1 by factors from S-phase cells.

    d. can be induced in G_2 by factors from S-phase cells.

    e. is not amenable to genetic analysis.

35. Features in various models of recombination include

    a. branch migration.

    b. single-strand nicks.

    c. double-strand breaks.

    d. synaptanemal complexes.

    e. explanation of reciprocal exchanges close to the site of recombination.

For each protein listed in items 36–43, indicate its function(s) by writing in the corresponding letter from the following:

    a. repairs DNA

    b. unwinds DNA

    c. cleaves nucleic acids

    d. acts in recombination

    e. prevents tangling of DNA strands

36. *E. coli* dnaA ____

37. *E. coli* DNA polymerase I ____

38. RecA ____

39. Mammalian DNA polymerase β ____

40. Single-stranded DNA-binding protein ____

41. UvrC ____

42. RAD3 ____

43. RecBCD ____

For each protein listed in items 44–50, indicate to which protein(s) it is similar in sequence, function, or mechanism of action by writing in the corresponding letter from the following:

    a. primase

    b. eukaryotic DNA polymerase α

    c. photolyase

    d. RecA

    e. eukaryotic DNA polymerase δ

    f. UvrABC

    g. eukaryotic DNA polymerase β

Also briefly describe the similarity between the matched proteins.

44. *E. coli* DNA polymerase III ____

45. UvrABCD ____

46. RAD10 ____

47. PCNA ____

48. Topoisomerase II ____

49. Yeast DNA polymerase II ____

50. Single-stranded DNA-binding protein ____

PART C: *Putting Concepts to Work*

51. Adenoviruses, which contain double-stranded linear DNA genomes, require a protein primer to begin DNA replication. In contrast, parvoviruses, which contain single-stranded linear DNA genomes, can begin DNA replication in the absence of a protein primer. Propose a mechanism for initiation of DNA replication in parvoviruses consistent with this observation. Sketch a diagram illustrating this mechanism.

52. How does one replisome copy both strands simultaneously and in the same direction at the growing fork?

53. Why are temperature-sensitive mutants often used in studies of the enzymes involved in DNA replication?

54. As a general rule, RNA viruses accumulate mutations at a more rapid rate than DNA viruses. Propose a hypothesis to explain this observation.

55. What parallels exist in the structure and function of the prokaryotic and eukaryotic DNA polymerases involved in replication?

56. Viruses of mammalian cells have been widely used to study the biochemistry of DNA replication. Give one reason why these systems may be imperfect models for chromosome replication.

57. A hypothetical origin of replication is diagrammed in Figure 12-1. Based on the mechanism involved in replication of bacterial circular DNA and eukaryotic DNA, indicate the location and direction of synthesis for each new strand. Use solid lines for leading strands and dotted lines for lagging strands.

58. The homology between *Alu* sequences and the SV40 origin of replication has prompted the suggestion that *Alu* sequences may function as chromosomal origins. Indeed, some *Alu* sequences have been shown to function as the origin for a plasmid introduced into cells that contain T antigen.

    a. Does this observation support the hypothesis that *Alu* sequences are chromosomal origins? Explain.

    b. What experiments could be performed to confirm or refute this hypothesis about *Alu* function?

59. Explain why electron-microsope "bubble analysis" of replicating viral genomes does not distinguish unidirectional from bidirectional replication. Describe the experimental approach using restriction endonucleases that has been used to discriminate between these two mechanisms.

60. Describe two findings that link DNA repair with recombination.

61. According to the original model of Robin Holliday, which two steps in the formation and resolution of a Holliday structure are critical in the formation of progeny that are both heteroduplexes and recombinants?

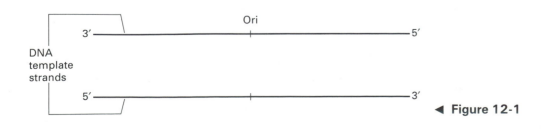

▲ **Figure 12-1**

PART D: *Developing Problem-solving Skills*

62. The Meselson-Stahl experiment demonstrated that semiconservative replication occurs over long stretches of *E. coli* DNA. The protocol of this experiment and the results are summarized in Figure 12–2. In cultured mammalian cells, sister chromatid exchange is a frequent phenomenon, which involves strand breakage, exchange, and rejoining of sister chromatids. What would be the effect of repeated sister chromatid exchange on the results obtained with this experimental approach using cultured mammalian cells?

63. You carry out a series of incubations to compare the properties of *E. coli* DNA polymerase I and III. After incubating a DNA template prepared from bacteriophage T7 with one or the other polymerase for 20 min, you then add a large amount of a second template, bacteriophage T3 DNA, and permit the reaction to continue for another 40 min. You then determine how much of the DNA synthesized is T3 DNA and how much is T7 DNA. You find that most of the DNA in the polymerase I incubations is T3 DNA, but almost all of

the DNA in the polymerase III incubations is T7 DNA. Explain these results.

64. You have briefly labeled a set of coverglass cultures of human HeLa cells with [³H]thymidine. After a pulse-labeling period, you add nocodazole, a drug that blocks cell-cycle progression at mitosis by disrupting the mitotic spindle. Explain how the minimum duration of the G_2 period of the HeLa cell cycle can be determined from such a protocol.

65. The duration of the S phase in plant cells is 12 h, not very different from the 10 h of human cells. Plant cells also replicate DNA at about the same rate as human cells, roughly 100 bp/s/fork. Broad bean, *Vicia faba*, has about 3.5×10^{10} bp of DNA per genome. What is the minimum number of bidirectional forks required to replicate the *Vicia faba* genome during the S phase?

66. The effects of various inhibitors of DNA synthesis have been investigated in vitro. In an in vitro replication sys-

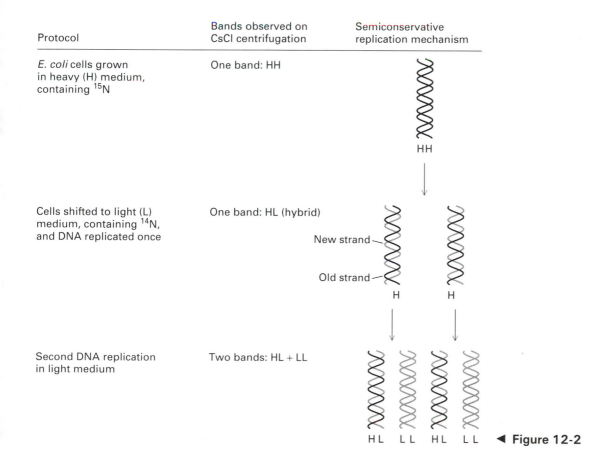

Protocol	Bands observed on CsCl centrifugation	Semiconservative replication mechanism
E. coli cells grown in heavy (H) medium, containing ¹⁵N	One band: HH	HH
Cells shifted to light (L) medium, containing ¹⁴N, and DNA replicated once	One band: HL (hybrid)	New strand / Old strand / H H
Second DNA replication in light medium	Two bands: HL + LL	HL LL HL LL

◀ Figure 12-2

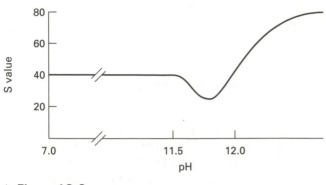

▲ **Figure 12-3**

▲ **Figure 12-4**

tem, synthesis of bacteriophage M13 DNA, a 6.4-kb molecule, is sensitive to rifampicin, an inhibitor of host-cell RNA polymerase, whereas the ongoing synthesis of *E. coli* cellular DNA is not inhibited by rifampicin.

a. What do these results imply about the priming enzymes involved in replication of M13 DNA and *E. coli* DNA?

b. What would be the effect of a mutation in the $5' \rightarrow 3'$ exonuclease activity of DNA polymerase I on the replication of *E. coli* DNA? Would you expect such a mutation to be lethal?

67. Mitochondrial DNA isolated from some mammalian cell lines has a high percentage of catenated DNA molecules. Despite this oddity in their mitochrondrial DNA, these cell lines have a normal generation time. In what enzyme are these cells likely to be deficient? Is this enzyme activity likely to be coded by more than one gene?

68. The results of a sedimentation velocity experiment with oak tree virus DNA are shown in Figure 12-3.

a. Based on these data, is this viral DNA a linear or closed circular molecule?

b. Propose two different experimental approaches to confirm your conclusion.

69. DNA contains thymine rather than uracil as a base. Why is the use of thymine rather than uracil in DNA essential for the correct repair of deaminated cytosine in DNA?

70. DNA molecules during recombination can be joined at either homologous or nonhomologous sites. Shown in Figure 12-4 is a chi form, or Holliday structure, a recombination intermediate. What visible feature of this chi form indicates that it is the result of a homologous recombination event? Prepare diagrams illustrating the allelic structure of chi forms resulting from homologous and nonhomologous recombination events.

71. Explain why many mutations in DNA replication functions are lethal, whereas most mutations in DNA-repair functions are not.

PART E: *Working with Research Data*

72. Shown in Figure 12-5 are chromosome homologues from an experiment designed to determine if different portions of eukaryotic chromosomes replicate in different portions of the S phase of the cell cycle. In this experiment, white blood cells, which will divide a few times outside of the body, were cultured with bromodeoxyuridine (BrdU) for 48 h and then placed in a medium containing [³H]thymidine for a brief period of time. Cells were then incubated in isotope-free medium and chromosome spreads were prepared. The regions of [³H]thymidine incorporation were detected by autoradiography (black dots in Figure 12-5a). The chromosomes were also stained with a fluorescent dye that binds well to DNA contaning thymidine and poorly to DNA substituted with BrdU (Figure 12-5b). The chromosome homologues shown are from the *earliest* post-labeling time point to show [³H]thymidine incorporation.

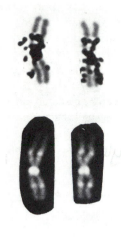

▲ Figure 12-5

a. To which portions of the chromosome is [³H]thymidine incorporation restricted?

b. The chromosome staining pattern also reflects the chromosome replication pattern. Based on the staining pattern in Figure 12-5b, which portions of the chromosome are replicated in early, mid, and late S phase?

c. Which of the two methods — fluorescent dye staining or autoradiography following tritium incorporation — has the higher resolution?

d. Based on the labeling and staining patterns, do the telomeric regions of the chromosome replicate early or late in the S phase?

73. The mechanisms that repair DNA can themselves be *error-prone*. For example, during the replication of single-stranded ΦX174 DNA treated briefly with acid to cause removal of one or two purine residues, repair does produce, albeit with low efficiency, gap filling of the infecting apurinic DNA molecules. However, because there is no information by which the choice of a correct base can be made, the probability of mutation is high, about three out of four. In *E. coli,* the two major genes involved in error-prone repair are *umuC* and *umuD* (for UV non*mu*table). These are SOS genes and are inducible in response to severe DNA damage. UmuC and UmuD proteins cause a base to be inserted into a nascent DNA chain, even when no base exists in the template strand.

a. Propose a rationale for why error-prone repair might be of benefit to a damaged organism or virus.

b. Propose a mechanism by which error-prone repair might occur in *E. coli.*

c. The test bacteria for the Ames assay, a procedure in which the mutagenicity of chemicals is assessed, carry a plasmid that efficiently expresses *umuC* and *umuD*. Why engineer the test bacteria to carry this plasmid?

74. When mammalian DNA is replicated in the presence of a protein synthesis inhibitor (e.g., emetine) and the nucleoside analog 5-bromodeoxyuridine (BrdU), the segregation of nucleosomes on newly synthesized DNA is conservative. The parental nucleosomes are associated with the leading portion of the growing fork, and no new nucleosomes are formed under these conditions. The lagging-strand DNA with no associated nucleosomes is accessible to micrococcal nuclease digestion, whereas the leading-strand DNA is protected and released as nucleosome-associated DNA; the BrdU-containing leading strand DNA can then be isolated by isopycnic centrifugation. Potentially, if strand-specific probes are available, the results of blot hybridization to the BrdU single-stranded DNA can be used to map the origin of replication within a replicon.

The results of applying this approach to the dihydrofolate reductase region of hamster chromosomal DNA is shown in Figure 12-6. The upper portion of this figure shows the position of various (+) and (−) strand probes relative to a kilobase pair ruler. The lower portion shows the dot hybridization patterns.

a. Indicate the direction of replication at each probe site by placing appropriate arrows on the map of the probes in Figure 12-6a. Assuming that replicons are bidirectional, how many different replicons are involved in replication of this 70- to 80-kb DNA segment? Also indicate on the probe map the position of the origin(s) of replication.

b. Based on these results, are origins always centrally located within a chromosomal replicon? If not, propose an explanation for their noncentral location.

(a) Position of probes

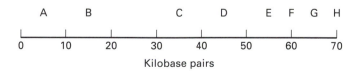

(b) Blot hybridization patterns

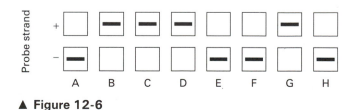

▲ Figure 12-6

a) ^{32}P-labeled (*) Holliday structures

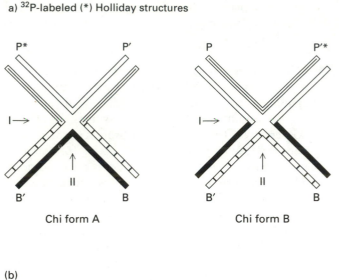

(b)

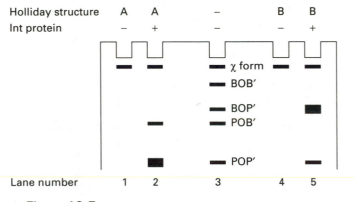

▲ **Figure 12-7**

75. Both integration and excision of the bacteriophage λ in *E. coli* requires the phage-coded protein Int. Int-dependent recombination does not require any high-energy cofactors or involve any degradation or synthesis of DNA. Genetic experiments suggest Int-dependent recombination proceeds via a Holliday structure. Holliday structures can be resolved into recombination products by a cycle of nicking and ligation.

Experimentally, synthetic Holliday-structure analogs for λ integration and excision containing a single radioactive ^{32}P atom can be constructed. Shown in Figure 12-7a are two examples (A and B) of such structures. At the center of each of these forms is the O or "overlap region." During recombination, this region is cut at position $-3/-2$ (from the center) in one strand and at position $+4/+5$ in the other strand to generate a 7-bp overlap. Homology within the overlap region is necessary for efficient recombination.

Figure 12-7b shows the gel electrophoretic patterns, revealed by autoradiography, of the products obtained by incubating Holliday structures A and B with Int protein. The components of the incubation mixtures are shown at the top; lane 3 is the migration pattern of a set of standards corresponding to the segments labeled in Figure 12-7a and the unresolved chi form.

a. Are Holliday structures A and B preferentially cleaved at site I or II?

b. What are the nonradioactive resolution products from Holliday structures A and B?

c. Draw a diagram showing the derivation of the radioactive and nonradioactive resolution products resulting from cleavage at site I and site II in chi form A and B.

C H A P T E R *13*

The Plasma Membrane

PART A: *Reviewing Basic Concepts*

Fill in the blanks in statements 1–17 using the most appropriate terms from the following list:

α-helical	lateral mobility
ankyrin	leaflet(s)
apical	lipid anchors
basolateral	membrane protein(s)
cardiolipin	Meselson-Stahl
chloroplast(s)	myelin sheath
cholesterol	N-linked
connexin	patching
cytoplasm	peripheral
cytoplasmic face	phase transition
cytoskeletal	phospholipid
electron spin resonance (ESR)	plasma membrane
	plasmadesmata
erythrocyte(s)	polarized cell(s)
exoplasmic face	spectrin
fluidity	sphingolipid(s)
gap junctions	sucrase-isomaltase
glycophorin	thylakoid membrane
Gorter and Grendel	tight junctions
lactoperoxidase	

1. Biological membranes consist of a _phospholipid_ bilayer having two _leaflets_ in which _membrane proteins_ are embedded.

2. Most integral membrane proteins have at least one _α-helical_ transmembrane domain.

3. Integral membrane proteins may also be held to the membrane by _____.

4. _____ is a planar lipid, chemically similar to steroid hormones, that affects the _____ of the plasma membrane in a complex manner.

5. Glycolipids and glycoproteins are located preferentially in the _____ of the _____, or cell surface.

6. The _Chloroplast_, a plant organelle, is rich in galactolipids, which form spontaneous inverted micelles in the absence of proteins.

7. Mitochondria are rich in the lipid called _____.

8. _____, a major integral membrane protein in mammalian red blood cells, contains both _____ and O-linked oligosaccharides.

9. Because membranes are impermeable to proteins, _____-mediated radioiodination is an effective approach to establishing the orientation of membrane proteins.

10. Loosely attached, _____ membrane proteins can often be released by alterations in salt conditions.

11. The _____ experiment, although marred by compensating errors, was important in establishing the bilayer nature of biological membranes.

12. The plasticity of red blood cell shape is an outcome of the interaction of Cytoskeletal elements with the plasma membrane.

13. _____ is a physical approach to monitoring the effect of lipid composition on the _____, or melting, of biological membranes.

14. The interconnections between adjacent plant cells, called _____, allow exchange of small molecules and may contain a tube of endoplasmic reticulum.

15. Pancreatic acinar cells are an example of _____ in which _____ delimit the plasma membrane into _____ and _____ domains.

16. In mammalian tissues, _____ allow metabolic coupling between adjacent cells.

17. erythrocytes are a common source of pure plasma membrane because these cells contain no internal membranes and may be readily obtained.

PART B: *Linking Concepts and Facts*

Circle the letters corresponding to the most appropriate terms/phrases that complete or answer items 18–29; more than one of the choices provided may be correct in some cases.

18. Common chemical features of all membrane lipids include

    (a.) a polar head group.

    b. a glycerol backbone.

    c. sugar constituents.

    (d.) a hydrophobic domain.

    e. a phosphate group.

19. Experimental support for the universality of the phospholipid bilayer in biological membranes comes from

    a. FRAP analysis of lipid mobility in membranes.

    b. low-angle x-ray analysis.

    c. the Gorter and Grendel molecular-film experiment.

    d. osmium tetroxide–staining patterns of membranes.

    e. the asymmetry of phospholipid distributions in biological membranes.

20. Lipid movement in phospholipid bilayers is

    (a.) dependent on temperature.

    (b.) affected by the presence of cholesterol and the length of the fatty acyl chain.

    (c.) rapid with lateral diffusion rates of about 1 μm/s in liposomes.

    d. frequent for flip-flop as well as lateral diffusion.

    e. essentially uniform for the entire population of phospholipids in a biological membrane.

21. Integral membrane proteins are

    a. abundant and readily purified.

    b. removed from the membrane by solutions of high ionic strength or chemicals that bind divalent cations.

    c. soluble in the absence of detergents.

    d. extremely rich in hydrophobic amino acids in comparison with other classes of proteins.

    e. amphipathic molecules.

22. Lipid-anchored membrane proteins are

    a. an example of one class of peripheral membrane proteins.

    b. restricted to the cell surface.

    c. sometimes covalently attached to complex glycosylated phospholipids.

    d. sometimes transforming proteins such as v-src, which is anchored by an amide linkage to myristic acid.

    e. sometimes transforming proteins such as p21-ras, which is anchored by a thioester bond to palmitic acid.

23. Typical traits of membrane-embedded segments of integral membrane proteins include

a. an α-helical segment long enough, about 20–25 amino acids, to span the phospholipid bilayer.

b. exposure of polar C=O and NH groups on the outside.

c. placement of charged residues on the interior of the molecule.

d. a highly conserved sequence of hydrophobic amino acids.

e. associations with membrane lipids that restrict their mobility.

24. The orientation of proteins or lipids within membranes

a. can be probed with degradative enzymes such as proteases or phospholipases.

b. is symmetric.

c. with respect to the arrangement of hydrophobic and hydrophilic domains is altered by nondenaturing detergents.

d. is statistical with glycosylated molecular species distributed between both faces of the plasma membrane.

e. varies continuously for each molecule during the cell cycle.

25. Cytoskeletal interactions with the plasma membrane

a. are very limited in extent and have little effect on the mobility of integral membrane proteins.

b. limit the loss of glycophorin and band 3 when the erythroblast loses its nucleus.

c. are mediated by weak, noncovalent bonds, which often involve "linking" or "connecting" proteins.

d. consist principally of direct interactions between spectrin and band 3 protein.

e. are sensitive to salt conditions.

26. Polarized cells in animals

a. are common among free-floating cells of the bloodstream.

b. are a general feature of epithelial tissue, which lines body cavities.

c. have a plasma membrane divided into specialized regions of different function.

d. have different molecules exposed on their apical and basolateral surfaces.

e. may be sealed together by tight junctions to form a barrier to the passage of most small molecules.

27. Which of the following traits is (are) exhibited by higher plant cells?

a. presence of major glycolipids only in plasma membranes and biosynthetic precursors

b. little polarity of cell structure and function in multicellular arrays

c. absence of junctional proteins because cell walls separate cells

d. absence of cytoskeletal proteins such as actin

e. absence of metabolic coupling between cells because no gap junctions are present

28. Typical traits of tight junctions and gap junctions in animal cells include

a. a sensitivity to Ca^{2+} levels.

b. the presence of junction-specific proteins.

c. a distinct morphological appearance in freeze-fracture electron micrographs.

d. a role in linking epithelial cells together into a functional unit.

e. the presence of highly abundant proteins.

29. The unique properties of different biological membranes are related to

a. variations in their phospholipid composition.

b. qualitative differences in their glycolipids.

c. the presence or absence of cardiolipin.

d. the relative abundance of cholesterol.

e. the distinctive properties of their protein species.

30. Figure 13-1 is a diagram showing the structural components of a typical biological membrane.

a. Fill in each label in Figure 13-1 with the correct term(s) from the following list:

peripheral protein	phospholipid
integral protein	exoplasmic
glycoprotein	cytoplasmic
glycolipids	

b. Indicate which of the molecular species labeled in Figure 13-1 are susceptible to the following treatments by writing in the corresponding capital letter next to the label:

C = solubilization by divalent cation-binding agents

D = solubilization by detergents

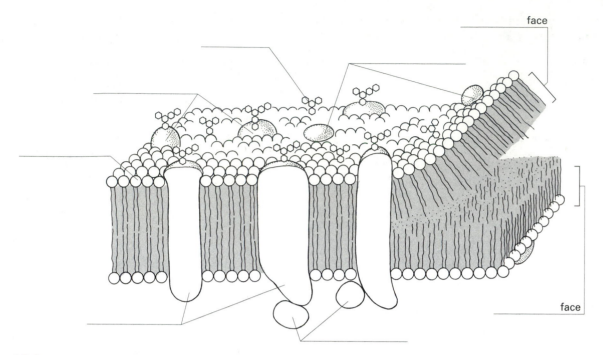

face

face

▲ **Figure 13-1**

R = radioiodination of intact cells

P = digestion by externally added phospholipase

Items 31–36 describe various cellular roles or functions. After each item, indicate which of the following substances has that role by writing in the corresponding letter(s) from the following list:

a. actin
b. ankyrin
c. band 3 protein
d. connexin
e. galactolipid
f. glycophorin
g. spectrin
h. sucrase-isomaltase

31. Cytoskeletal protein ____

32. Bridging protein between cytoskeleton and plasma membrane ____

33. Gap-junction protein ____

34. Glycolipid in chloroplast thylakoid membranes ____

35. Integral membrane protein of erythrocytes ____

36. Integral membrane protein of apical surface of intestinal epithelial cells ____

PART C: *Putting Concepts to Work*

37. Compare the physical state of membrane proteins when a biological membrane is exposed to detergent at a concentration below and above the critical micellar concentration (CMC) of the detergent.

38. The photosynthetic reaction center of the bacterium *Rhodopseudomonas viridis* was the first integral membrane protein to be characterized at atomic resolution. Based on the structure of this and other less well studied cases, membrane proteins have been described as being "inside-out" compared to water-soluble proteins. How appropriate is this description?

39. Which parameters were measured in the Gorter and Grendel experiment? Explain how the observed values of these parameters led to the conclusion that the erythrocyte membrane is a phospholipid bilayer.

40. Why do fragmented biological membranes tend to reseal to form sealed vesicles? Describe an example of this phenomenon involving erythrocytes.

41. Why do membranes rich in short-chain, unsaturated fatty acids generally have a comparatively low phase-transition temperature?

42. Why is the mobility of lipids in liposomes greater than that of lipids in biological membranes?

43. Explain how membrane anchoring of transforming proteins, which frequently have protein kinase activity, might promote transformation of mammalian cells.

44. Both ionic detergents (e.g., sodium dodecyl sulfate) and nonionic ones (e.g., octyl glucoside) can solubilize membrane proteins. If you want to maintain the enzymatic activity of a membrane protein after solubilization, which type of detergent would be preferable and why?

45. The glycocalyx is a particularly prominent feature of the microvilli of the intestinal epithelial cell. What function does this structure serve?

46. How do membrane proteins such as band 3 and connexin create a local, aqueous, hydrophilic environment within a phospholipid bilayer?

47. What are the physiological advantages of sealed epithelial layers to animals?

48. In a metabolically coupled epithelium, how is cell injury resulting from damage to one cell limited to the damaged cell?

PART D: *Developing Problem-solving Skills*

49. The Frye and Edidin experiment is a classic demonstration of the fluid-mosaic nature of biological membranes. In this experiment, spherical mouse and human cells were fused and the cell-surface distribution of species-specific proteins determined at various times after fusion by staining the binucleate cells with red and green fluorescent antibodies. Complete intermixing of the proteins was observed.

    a. Assuming that the radius of the individual cells contributing to the binucleate fused cell is 10 μm, calculate the minimum net linear protein diffusion rate required to achieve complete intermixing in a 40-min time period?

    b. How does this diffusion rate compare with that of lipids in a liposome?

50. Erythrocyte ghosts can be readily prepared free of soluble proteins such as hemoglobin. These ghosts contain several polypeptides (see Figure 13-2a). After treatment of ghosts with Triton X-100 in hypotonic buffer, most of these proteins are solubilized to leave so-called shells (Figure 13-2b). Why is band 3, an immobile protein, not retained within the shells following Triton X-100 treatment?

51. The amino acid sequence of a short membrane protein is as follows:

1 → Glu-Arg-Arg-Gln-Leu-Lys-His-His-Lys-Ser-Glu-Pro-Glu-Ile-Leu-Leu-Ile-Ile-Phe-Gly- (10, 20)

21 → Val-Met-Ala-Gly-Val-Ile-Gly-Gly-Ile-Leu-Ile-Leu-Ile-Ser-His-Gly-Ile-Arg-Arg-Leu- (30, 40)

41 → Ile-Lys-Lys-Ser-Pro-Ser-Asp-Val-Lys-Pro-Leu-Pro (50)

    a. Which sequence features suggest that this is an integral membrane protein?

    b. Draw a diagram showing the arrangement of this 52-aa protein with respect to the membrane and indicate the exoplasmic and cytoplasmic face if possible.

52. You have been asked to develop a computer algorithm based on the hydrophilicity of amino acids to predict which proteins are soluble/peripheral proteins and which are integral membrane proteins. You argue that such an algorithm will incorrectly assign a major class of integral membrane proteins. What are these proteins, to what class would they be assigned, and why does the misassignment occur?

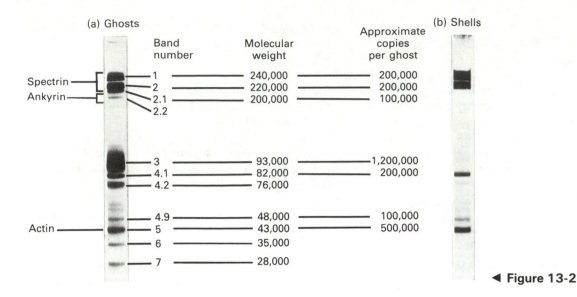

(a) Ghosts

Band number	Molecular weight	Approximate copies per ghost
1	240,000	200,000
2	220,000	200,000
2.1	200,000	100,000
2.2		
3	93,000	1,200,000
4.1	82,000	200,000
4.2	76,000	
4.9	48,000	100,000
5	43,000	500,000
6	35,000	
7	28,000	

Spectrin — bands 1, 2
Ankyrin — band 2.1
Actin — band 5

(b) Shells

◄ **Figure 13-2**

53. For a liposome made of a single phospholipid containing two 18-carbon, monounsaturated fatty acyl chains, varying the position of the single double-bond relative to the ester linkage causes a distinct change in the phase-transition temperature, as shown in Figure 13-3. Explain the effect of double-bond position on the transition temperature.

54. Hereditary pyropoikilocytosis (HPP) is a form of hemolytic anemia. Under well-defined preparative conditions, about 40 percent of the spectrin extracted from red blood cells of patients with HPP is in a dimeric form; the rest exists as higher oligomers. Under the same preparative conditions, about 95 percent of the spectrin extracted from normal red blood cells is in the form of tetramers and higher oligomers.

a. What is the apparent molecular defect associated with HPP?

b. How might this defect lead to anemia in HPP patients?

55. The placement of electrodes on either side of a ESCK epithelial cell layer can be used to determine a conductance value. Based on this value, ESCK cell lines fall into several classes ranging in conductance from high to low.

a. How would high- and low-conductance ESCK epithelial cell layers differ in their lipid and protein distributions between the apical and basolateral surfaces?

b. Differences in the conductance of ESCK epithelial layers should correlate with differences in what cellular structure? How could your speculation be investigated?

56. In freeze-fracture electron micrographs, gap junctions appear as large clusters of connexons, which are hexagonal units of connexin. What does this observation suggest about how gap-junction complexes are held together and how gap junctions might be isolated?

57. The isolation of tight junctions and characterization of the substances forming them have been much more difficult than the isolation and characterization of gap junctions. Why is this so and what does it suggest about the structure of tight junctions?

58. Considerable evidence suggests that gap junctions and connexin are conserved across fairly large evolutionary distances. For example, metabolic coupling between

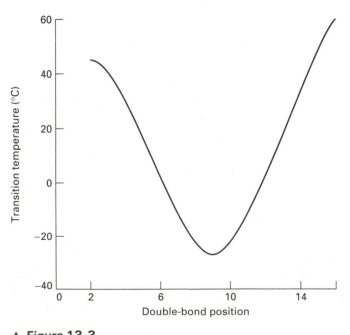

▲ **Figure 13-3**

mouse cells can be demonstrated in cocultures of mutant and wild-type cells and also in heterologous cultures of human and hamster or rat cells. In addition, earthworms have gap junctions that appear morphologically similar to those in mammals.

a. What three structural and functional features of

connexin, the gap-junction protein, would you expect to be most conserved among these species?

b. What do these conserved features suggest about the sequence homology among connexins from different species?

PART E: *Working with Research Data*

59. Richard Pagano and his coworkers have synthesized a series of fluorescent lipid analogs, which can be used to trace lipid distribution in fixed and living cells by fluorescence microscopy. Three types of analogs have been prepared. In one, the polar head group of a lipid is

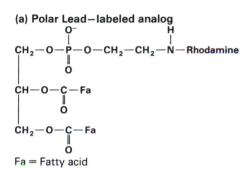

(a) Polar Lead—labeled analog

Fa = Fatty acid

(b) C₆— NBD-glycerolipids

Fa = Fatty acid
Y = choline or ethanolamine

(c) C₆— NBD—sphingolipids

If Y = H, then compound is C₆—NBD—ceramide.
If Y = phosphocholine, then compound is C₆—NBD—sphingomyelin.

▲ **Figure 13-4**

labeled with a fluorescent group (e.g., the free amino group of phosphatidylethanolamine has been labeled with rhodamine). In the second type of analog, one of the fatty acids of a phospholipid is replaced by a six-carbon chain attached to a fluorescent NBD group. In the third type, a similar fatty acid substitution is made in a sphingolipid. Examples of the three analog types are shown in Figure 13-4.

The rates of transfer of these analogs and two naturally occurring lipids (phosphatidylcholine and sphingomyelin) from carrier liposomes to another bilayer (i.e., a cell membrane) were determined. Table 13-1 shows the half-time for movement of these lipids and indicates whether flip-flop was observed.

a. How many days are required for 50 percent of the sphingomyelin in a carrier liposome to transfer to another bilayer? How does this value compare with the rate of transfer of membrane lipids in the presence of carrier proteins?

b. Why does transfer of the C₆ analogs from carrier liposomes to another bilayer occur so much more rapidly than transfer of naturally occurring lipids and N-rhodamine-phosphatidylethanolamine?

Table 13-1

Lipid	Half-time for transfer between bilayers (min)	Flip-flop
C₆-NBD-sphingomyelin	0.04	No
C₆-NBD-ceramide	0.42	Yes
C₆-NBD-phosphatidylcholine	0.73	No
C₆-NBD-phosphatidylethanolamine	1.54	No
Phosphatidylcholine	2.9×10^3	No
N-rhodamine-phosphatidyl-ethanolamine	$>1 \times 10^4$	No
Sphingomyelin	1.2×10^5	No

c. Explain the observation that only one of the tested lipids (C₆-NBD-ceramide) flip-flopped during transfer.

60. You have immunized mice with purified glycophorin and have succeeded in preparing a monoclonal antibody against glycophorin. On testing the specificity of this antibody, you are surprised to find that it reacts with numerous membrane proteins from fibroblasts and with thylakoid membranes from plants.

    a. Based on the observed reactivity pattern of this antibody, with which moiety in glycophorin does it react? Explain.

    b. Would such widespread crossreactivity be expected for other monoclonal antibodies prepared against purified membrane proteins?

61. A commonly used assay and preparative approach for peripheral and integral membrane proteins is based on the partitioning of a protein between an aqueous and detergent phase. Membranes are solubilized at 4°C with the detergent Triton X-114, and then the temperature is shifted to 30°C; at this temperature, the solution separates into a detergent and an aqueous phase. Electrophoresis of the phases reveals if there is a preferential partitioning of any membrane proteins. The results of treating erythrocyte membranes in this way is illustrated by the electrophoretograms depicted in Figure 13-5a, which shows the protein-staining pattern, and in Figure 13-5b, which shows the radiolabeled galactose-staining pattern for the same preparation.

    a. Which of the bands in Figure 13-5a correspond to integral membrane proteins and which to peripheral membrane proteins?

    b. Which proteins revealed in Figure 13-5b are glyco-

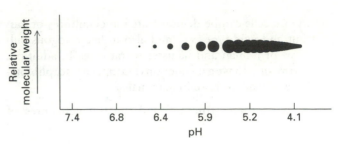

▲ Figure 13-6

sylated and what is the relationship of these proteins to those revealed in Figure 13-5a?

    c. Based on the phase-partition approach, would lipid-anchored proteins be classified as integral or peripheral membrane proteins?

62. Another common approach for identifying and separating integral and peripheral membrane proteins is to expose membranes to pH 11 and then centrifuge the mixture. This treatment releases peripheral proteins, which thus remain in the supernatant and can be identified by electrophoresis. The integral membrane proteins remaining in the pellet can then be solubilized and identified. What are some advantages of the phase-partition approach described in problem 61 compared with the pH 11 treatment for the identification and preparation of integral and perpipheral membrane proteins?

63. mLAMP-1 is a mouse membrane glycoprotein that was identified by its reactivity with a monoclonal antibody. In a one-dimensional polyacrylamide gel, the protein migrates as a single, somewhat diffuse, band. In a two-dimensional polyacrylamide gel, the protein exhibits marked heterogeneity, as shown in Figure 13-6.

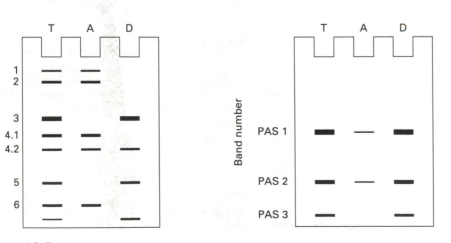

(a) Protein-staining pattern   (b) Galactose-staining pattern

T = total membrane preparation
A = aqueous phase
D = detergent phase

▲ Figure 13-5

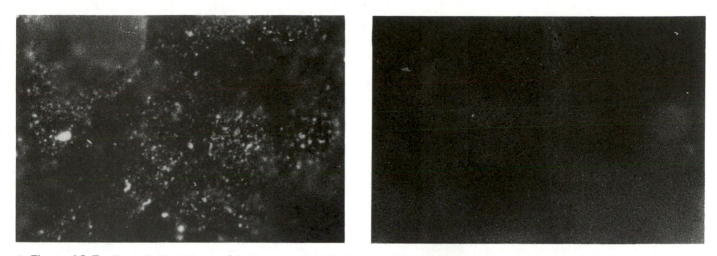

▲ **Figure 13-7**   From G. Van Meer and K. Simons, 1986. The function of tight junctions in maintaining differences in lipid composition between the apical and the basolateral cell surface domains of MDCK cells. *EMBO J.* 5(7):1458.

a. Describe two features of mLAMP-1 that could account for this observed heterogeneity.

b. Propose an experimental approach for establishing
the source of this heterogeneity.

64. Van Meer and Simons have studied the function of tight
junctions in maintaining differences in the lipid composition of the apical and basolateral surfaces of MDCK
cells. They have used liposome fusion with the apical
surface as a means to introduce fluorescent lipid analogs
into either (a) the exoplasmic face or (b) the exoplasmic
and cytoplasmic face of the plasma membrane. In these
experiments, the lipid analog was polar head–labeled
with rhodamine, a red fluorescent dye (see Figure
13-4a).

   Shown in Figure 13-7 are MDCK fluorescent patterns
for *N*-rhodamine-ethanolamine (*N*-rh-PE) fused into
the exoplasmic face of the apical surface. In Figure
13-7a, the focal plane is the apical surface; in Figure
13-7b, the focal plane is the lateral surface of the MDCK
cell. Fluorescent labeling is apparent only on the apical
surface. Similar focal plane views of MDCK cells labeled
by fusion of *N*-rh-PE into both the exoplasmic and
cytoplasmic leaflets of the apical surface are shown in
Figure 13-8. In this case, labeling is apparent on both the
apical and lateral surfaces.

a. Based on these results, does *N*-rh-PE have equal access to the lateral surface from both leaflets?

b. If tight junctions had a role in lipid diffusion in
MDCK cells, what would be the effect of removing
Ca^{2+} on *N*-rh-PE access from either leaflet to the
apical surface?

c. Sketch a diagram showing the positioning of the
tight junctions relative to exoplasmic and cytoplasmic leaflets that would account for the observed
results.

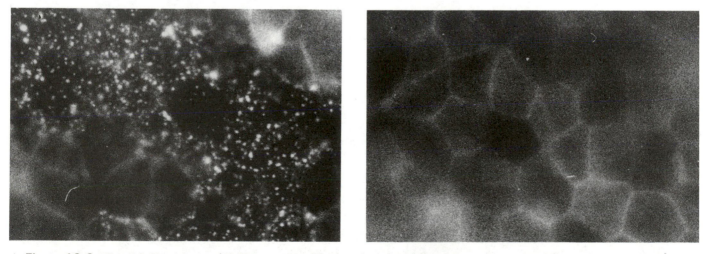

▲ **Figure 13-8**   From G. Van Meer and K. Simons, 1986. The function of tight junctions in maintaining differences in lipid composition
between the apical and the basolateral cell surface domains of MDCK cells. *EMBO J.* 5(7):1460.

CHAPTER 14

Transport across Cell Membranes

PART A: Reviewing Basic Concepts

Fill in the blanks in statements 1–21 using the most appropriate terms from the following list:

active ion transport	group translocation
antiport	H$^+$ ATPases
apotransferrin	human immunodeficiency virus (HIV)
asialoglycoprotein	
ATP	ion concentration gradient
Ca^{2+} ATPase	ion pumps
chloroplast	
clathrin	K_m
coated pit(s)	lysosome(s)
CURL	Michaelis equation
endocytosis	mitochondria
endosomes	Na$^+$-K$^+$ ATPase
enveloped viruses	Nernst equation
facilitated diffusion	osmotic
ferrotransferrin	partition coefficient
Fick's law	passive transport
Δ G (change in Gibb's free energy)	phagocytosis

pH-dependent	transcytosis
pinocytosis	transferrin
prelysosomal	turgor
receptor-mediated endocytosis	van't Hoff equation
	V_{max}
symport	

1. In simple diffusion, the transport of a small molecule across a lipid bilayer is proportional to its hydrophobicity, which is measured by its _____ K.

2. _____ provides a quantitative description of the diffusion rate of uncharged molecules through a biological membrane.

3. The transport of molecules is against a concentration gradient in both _____ and _____. Because of transport proteins, the rate of equilibration is much higher for _____ than for _____.

4. At ligand concentrations substantially less than the _____ of a permease, transport via facilitated diffusion is not much faster than transport via simple diffusion.

5. The _____ describes the dependence of electric potential across a membrane on ion distribution.

6. The total _____ for the movement of Na^+ ions across a selectively permeable membrane is the sum of both a membrane electric potential contribution and a(n) _____ contribution.

7. _____ results in movement of ions against a concentration gradient and is directly coupled to the consumption of _____.

8. In animal cells, the action of _____ maintains the intracellular concentration of Na^+ and Ca^{2+} low and of K^+ high. Specific examples of such proteins include the _____, which couples movment of one ion into the cell to the outward transport of a second ion, and the _____, which in muscle cells is concentrated in the sarcoplasmic reticulum.

9. _____, which maintain the low pH of plant vacuoles and of lysosomes, belong to the V class of proton pumps.

10. In _____ and _____, ATP consumption is indirectly coupled to the movement of ions or small molecules into the cell. In a(n) _____, the transported molecule and the cotransported ion move in the same direction. In a(n) _____, the transported molecule and the cotransported ion move in opposite directions.

11. Some nutrients are chemically modified by a(n) _____ process during their transport into prokaryotic cells. This process does not occur in eukaryotic cells.

12. The shape of guard cells and the opening and closing of leaf stomata depend on variations in _____, or _____, pressure.

13. Macromolecules (e.g., LDL or horseradish peroxidase) and particles (e.g., bacteria) are internalized into animal cells by fundamentally different processes termed _____ and _____, respectively.

14. _____ is the nonspecific uptake of small droplets of extracellular fluid by vesicles, whereas _____ is a selective, uptake process.

15. The selective transfer of circulating proteins such as maternal antibodies to the human fetus occurs across a cell layer and is termed _____.

16. Most endocytosed ligands are delivered to _____, highly acidic organellar compartments.

17. Low-density lipoprotein (LDL), a major cholesterol carrier in the mammalian bloodstream, binds to a cell-surface receptor that is localized in _____, which are membrane indentations lined by the fibrous protein _____. As the LDL is internalized, it is found progressively in _____ and the subsequent site of receptor recycling termed _____.

18. The _____ receptor mediates the removal of abnormal serum glycoproteins from the mammalian bloodstream.

19. Most recycling of endocytosed receptors is a _____ event, which requires a _____ dissociation of ligand and receptor.

20. _____, an iron-carrying protein, is internalized by receptor-mediated endocytosis as _____ and recycled to the bloodstream as _____.

21. The _____, which binds to CD4 as a receptor, is transferred into the cell by fusion with the plasma membrane. However, the entry of many other _____ into mammalian cells requires transfer into a low-pH endocytic compartment.

PART B: *Linking Concepts and Facts*

Circle the letters corresponding to the most appropriate terms/phrases that complete items 22–36; more than one of the choices provided may be correct in some cases.

22. Rapid passive diffusion of ions and small molecules across a membrane requires

   a. low temperatures to create an ordered lipid array.

   b. a high partition coefficient.

   c. the uphill concentration gradient needed for a positive ΔG value.

   d. a favorable electric potential gradient in the case of ions.

   e. high water solubility.

23. Facilitated diffusion can be distinguished from passive diffusion by

   a. the equilibrium concentration of solute reached.

b. comparison of the transport rate of the molecule to that predicted by Fick's law.

c. determination of K_m, the Michaelis constant of the process.

d. determination of V_{max}, the maximum velocity of the process.

e. determination of P, the permeability coefficient of the molecule.

24. Common properties of permeases include

a. the ability to distinguish structurally similar molecules such as D-glucose and D-mannose.

b. a permanent open pore through the membrane.

c. multiple transmembrane α-helical segments.

d. K_m values much higher than the typical physiological concentration of ligand.

e. an ATPase activity for the direct coupling of ATP hydrolysis to transport.

25. The generation and maintenance of a membrane electric potential requires

a. a selectively permeable membrane.

b. ion-specific membrane channel proteins.

c. metabolic energy.

d. cyclic conformational changes in ion channels.

e. active pumping of ions.

26. The magnitude of the electric potential for a single ion across a biological membrane can be derived from

a. the Nerst equation as applied to a single ion.

b. the universal gas equation.

c. the concentration inside and out of the ion and its ΔG value for movement across the membrane.

d. the Michaelis equation.

e. measurements from electrodes placed on either side of the membrane.

27. ATP hydrolysis is directly coupled to the movement of

a. glucose across the animal-cell plasma membrane.

b. protons into the plant vacuole.

c. Cl^- and HCO_3^- across the erythrocyte membrane.

d. Na^+ and K^+ across the plasma membrane.

e. Ca^{2+} to the extracellular medium or endoplasmic reticulum lumen.

28. Properties of P-class ion-motive ATPases include

a. localization in vacuoles.

b. localization in chloroplasts and mitochondria.

c. the ability to transport protons and/or other ions depending upon the enzyme.

d. an $\alpha_2\beta_2$ tetrameric structure.

e. a conserved amino acid sequence about the covalent phosphorylation site of the enzyme.

29. Differences between cotransport and active transport include

a. the mechanism of ATP coupling to the transport process.

b. the directed nature of molecule movement.

c. the restriction of active transport to ions such as Na^+, K^+, and Ca^{2+}.

d. the inability of cotransport to move a molecule against a concentration gradient.

e. the restriction of cotransport to small organic molecules such as glucose and amino acids.

30. The uptake of nutrients by bacteria

a. occurs only by group translocation.

b. may be driven by a H^+ gradient.

c. occurs against relatively small concentration gradients in freeliving bacteria.

d. may involve enzymes that are specific for individual sugars and others that are comparatively nonspecific.

e. is an ATP-independent process.

31. The osmotic properties of plant and animal cells

a. are very similar.

b. may be approximated by the van't Hoff equation.

c. are subject to little regulation.

d. are chiefly the result of transport events into and out of the vacuole.

e. may involve large-scale movement of water molecules.

32. The internalization of macromolecules and particles occurs

a. via transport enzymes.

b. via similar mechanisms.

c. via membrane-limited vesicles.

d. as a multistep process that frequently involves the binding of ligand to receptor.

e. with equal specificity in all mammalian cell types.

33. Receptor-mediated endocytosis functions to transport

    a. nutrients of limited water solubility into cells.

    b. egg proteins into chicken eggs.

    c. peptide hormone receptors to lysosomes where they can be degraded.

    d. small molecules such as sucrose into cells.

    e. influenza virus to a subcellular site where HA is activated to a fusogenic protein.

34. The localization of receptor in a coated pit requires

    a. binding of ligand to receptor.

    b. a cytoplasmic domain on the receptor protein.

    c. assembly particles.

    d. clathrin.

    e. spectrin and other cytoskeletal proteins such as actin.

35. Coated vesicles

    a. are large relative to CURL.

    b. can be uncoated in an enzyme-catalyzed, ATP-requiring reaction.

    c. are long lived.

    d. can carry solutes such as horseradish peroxidase as well as ligands.

    e. play a role in the internalization of all ligands.

36. Typical ion/ligand concentrations are

    a. 1×10^{-23} M in the plant vacuole.

    b. $< 1 \times 10^{-6}$ M Ca^{2+} in the cytosol of an animal cell.

    c. 150 mM Na^+ in the mammalian bloodstream.

    d. $\leq 1 \times 10^{-8}$ M (0.4 μg/ml) asialoglycoprotein in the mammalian bloodstream.

    e. 10 mM K^+ in the cytosol of a mammalian cell.

37. Listed below are several transport ATPases. Indicate to which class each belongs by writing in the class letter— P, F, or V.

    a. H^+ ATPase located in the thylakoid *Chlamydomonas* chloroplast ___

    b. H^+ ATPase located in the lysosome of a cultured HeLa cell ___

    c. H^+ ATPase located in the plasma membrane of a bacterium ___

    d. H^+/K^+ ATPase located in the apical membrane of parietal cells (oxyntic cells) ___

    e. Ca^{2+} ATPase located in the plasma membrane of erythrocytes ___

38. Listed below are several cotransport proteins. Write an A after those that are antiports and a S after those that are symports.

    a. Glucose-Na^+ cotransport protein of intestinal microvilli ___

    b. Na^+-H^+ cotransport protein of fibroblast plasma membranes ___

    c. Anion cotransport protein of erythrocytes (band 3) ___

    d. H^+-sucrose cotransport protein of plant vacuoles ___

39. Listed below are several ligands internalized by receptor-mediated endocytosis. Place a D after those that are degraded; R after those that are recycled; and a T after those that are transcytosed.

    a. Transferrin ___

    b. Immunoglobins in epithelial cells of neonatal animals ___

    c. Insulin ___

    d. Galactose-terminal glycoproteins in mammalian hepatocytes ___

    e. Phosphovitellogenins in chick ___

PART C: *Putting Concepts to Work*

40. Carrier models in which membrane transport proteins shuttle from one membrane face to the other to move molecules such as glucose from the cell exterior to interior currently have little acceptance. Why?

41. How is glucose transported across the apical and basolateral surfaces of intestinal epithelium?

42. What clues does the reversibility of Ca^{+2} ATPases and

Na$^+$-K$^+$ ATPases provide about the mechanism by which ATP may be generated in chloroplasts and mitochondria?

43. Tumor cells frequently become simultaneously resistant to several chemotherapeutic agent (e.g., colchicine, vinblastine, and adriamycin). How does the multidrug-resistance protein appear to function?

44. What is the relationship between cotransport proteins (i.e., symports and antiports) and ATP consumption?

45. The lumen of the mammalian stomach is 0.1 M in hydrochloric acid. What is the pH of the stomach lumen and how many times higher is the H$^+$ concentration in the stomach lumen than in the cytosol of the adjacent cells?

46. The concentrations of solutes in the cytosol and vacuole of plant cells are generally much higher than those typical in the cytosol and lysosome of mammalian cells. Why does the plant-cell plasma membrane not undergo osmotic lysis?

47. Why are phagocytosis and endocytosis thought to be fundamentally different processes?

48. The K_m for asialo-orosomucoid binding to the asialoglycoprotein receptor is 8×10^{-9} M. Assuming that this value is typical for formation of receptor-asialaoglycoprotein complexes, what is the expected molar concentration of asialoglycoproteins (i.e., galactose-terminal glycoproteins) in the mammalian bloodstream?

49. Why does the transfer of the nucleocapsid of many enveloped viruses into the mammalian cytosol require an acidic pH?

50. Describe two types of evidence indicating that ligand binding to an endocytic receptor and internalization are two separate steps in the process of receptor-mediated endocytosis.

51. What is the molecular basis of familial hypercholesterolemia, one of the most common human genetic diseases? What evidence does this disease provide about the effect of high cholesterol levels in the circulation on human health?

52. What is the fate of most ligands internalized by receptor-mediated endocytosis?

53. How might receptor-mediated endocytosis be used to target pharmacological agents to specific cell types?

PART D: *Developing Problem-solving Skills*

54. The glucose permease in mammalian erythrocytes has a K_m of 1.5 mM for glucose and a V_{max} of 500 μmol/glucose/ml packed cells/h. Blood glucose concentration is normally 5 mM (0.9 g/L).

    a. Calculate the velocity of glucose transport when the blood glucose concentration is low (3 mM, a typical level after several days of starvation); normal (5 mM); and high (7 mM, typical level after a feast).

    b. How does the variation in glucose concentration compare with the variation in instantaneous transport velocity over this concentration range?

    c. If the K_m of a mutant permease was 5 mM, would it rapidly transport glucose over the entire range of physiological extremes in blood glucose?

55. Chlorea toxin, produced by *Vibrio cholerae*, a pathogenic intestinal bacterium, causes an indirect reduction in the activity of the Na$^+$-K$^+$ ATPase in intestinal epithelial cells. This results in reduced uptake of small sugars and amino acids from the intestine. How is this reduction in uptake coupled to impaired Na$^+$-K$^+$ ATPase function?

56. For mammalian cells, the concentration of K$^+$ in the cell interior is 140 mM and in the extracellular medium is 5 mM.

    a. Assuming that the plasma membrane is permeable only to K$^+$, compute the membrane electric potential E_K.

    b. Assuming again that the membrane is permeable only to K$^+$ and also that the membrane electric potential is -70 mV, compute the free-energy change ΔG for movement of 1 mol K$^+$ outward (from the cell interior to the extracellular medium) and for movement inward. Which of these transport processes is energetically favored?

57. Ouabain is an inhibitor of the Na^+-K^+ ATPase when placed in the medium but not when microinjected into the cell. Ouabain treatment causes the α subunits of the enzyme to become more negatively charged but does not alter the charge properties of the β subunit.

    a. On which side of the plasma membrane are the ouabain- and ATP-binding sites of the enzyme located?

    b. During the transport process, the Na^+-K^+ ATPase becomes transiently phosphorylated. During normal transport, which of the two subunits, α or β, is most likely to be phosphorylated? What is the likely mechanism by which ouabain inhibits the enzyme?

    c. For intact cells, what would be the effect of removing extracellular ouabain on the charge of the Na^+-K^+ ATPase?

58. The Ca^{2+} ATPase of sarcoplasmic reticulum transports two Ca^{2+} cations for each ATP consumed. A functional Ca^{2+} pump can be reconstituted by the insertion of the Ca^{2+} ATPase into a liposome. In such a liposome, the Ca^{2+} transport rate is 30 Ca^{2+} cations/s/enzyme molecule.

    a. What is the rate of ATP consumption/enzyme molecule/s?

    b. If the direction of Ca^{2+} movement were reversed by loading the liposome with a high concentration of Ca^{2+}, how many ATP molecules would be generated for each Ca^{2+} cation transported from the lumen of the liposome to the extravesicular medium?

59. Antibody-coated bacteria or cells are efficiently phagocytosed via the F_c receptor, which can interact with the exposed F_c segment on the antibody. You have prepared a lymphocyte population coated with an IgG antibody that has induced a redistribution of antibody-antigen complexes into large cell-surface patches, or caps, over one hemisphere of the cells, as diagrammed in Figure 14-1.

    a. Would such a coated lymphocyte bind to a macrophage? If so, how would the lymphocyte and antibody "cap" be positioned with respect to the macrophage surface at 4°C?

    b. Would the "capped" lymphocyte be internalized by the macrophage at 37°C?

60. You have genetically reengineered the LDL receptor (LDLR) and asialoglycoprotein receptor (AGR) by swapping their exoplasmic domains to create two hybrid receptors. One hybrid contains the LDLR exoplas-

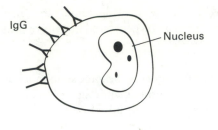

IgG — Nucleus

Antibody-coated lymphocyte

▲ **Figure 14-1**

mic domain + the AGR transmembrane and cytoplasmic domains; the other hybrid contains the AGR exoplasmic domain + the LDLR transmembrane and cytoplasmic domains. The altered receptors are introduced into cultured cells by transfecting receptor-negative cells with the engineered hybrid receptor gene.

    a. Describe the expected cell-surface distribution of the two hybrid receptors in the absence of bound ligand.

    b. Assuming that the two hybrid receptors function normally, under what conditions should the transfected cells internalize each receptor?

61. Assuming that endosomes are spherical and have an average diameter of 0.2 μm, how many protons would need to be pumped into an endosome to change its pH from 7.0 to 6.0 and subsequently to 5.0? A pH of 5.0 approximates that of a lysosome.

62. The surface glycoprotein CD4, whose primary amino acid sequence is known, acts a receptor for binding of HIV to T lymphocytes. You have available a series of synthetic peptides corresponding to specific portions of CD4.

    a. How might these peptides be used to determine which portion of the CD4 molecule is the actual binding site for HIV?

    b. Might these peptides have therapeutic value for HIV-positive individuals?

63. Strategies for screening cultured mammalian cells mutant in endosome acidification often are based on selection for simultaneous resistance to different endocytosed toxins (e.g., *Pseudomonas* exotoxin A and diphtheria toxin). Why is such a strategy more likely to select for cells mutant in endosome acidification than is selection for resistance to a single toxin?

PART E: *Working with Research Data*

64. You have conducted a series of experiments measuring the rate of glucose uptake by red blood cells from two different patients (KD and DZ) and a control group. The data are shown in Figure 14-2 in which the uptake rate (in μmol glucose/ml packed cells/h) is plotted against the glucose concentration.

    a. Estimate the V_{max} and K_m for glucose uptake by red blood cells from the control population and each of the patients.

    b. Is the defect in each of the patients the same?

    c. Propose two explanations for the defect in patient KD. What is the nature of the defect(s) in patient DZ?

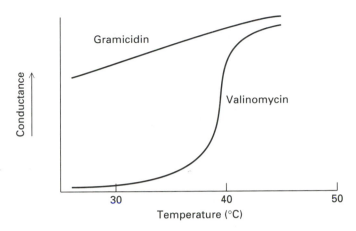

▲ **Figure 14-3**

65. Transport antibiotics are a class of drugs that increase the ionic permeability of membranes. Based on their mode of action, these drugs fall into two classes — channel formers and carriers. Channel formers generate pores that traverse the membrane, and carriers shuttle ions across the membrane. In order to determine whether valinomycin and gramicidin are channel formers or carriers, the temperature dependence of the conductance change in an artificial liposome membrane caused by gramicidin and valinomycin was determined. The data are shown in Figure 14-3.

    a. Based on these data, classify each drug as a channel former or carrier. Explain.

    b. Would you expect the temperature dependence of the valinomycin-"doped" membrane to be as sharp if the experiment were performed with a more natural membrane such as an erythrocyte ghost?

66. The color of the anthocyanin pigments in the vacuoles of petunia flowers and other plants suggests that the plant vacuole is an acidic compartment. In more recent years, detailed evidence has been accumulated on the mechanism of acidification of the plant vacuole. Data from a study designed to determine the ion requirements of the acidification process are shown in Figure 14-4. In this study, isolated maize tonoplasts were used; these are typical plant vacuoles. The $MgCl_2$ and $MgSO_4$ concentration was 5 mM; all other salts were 50 mM.

    a. Is tonoplast acidification ATP dependent? If so, on which side of the tonoplast membrane should the ATP-binding site be?

    b. What are the ion requirements for tonoplast acidification?

▲ **Figure 14-2**

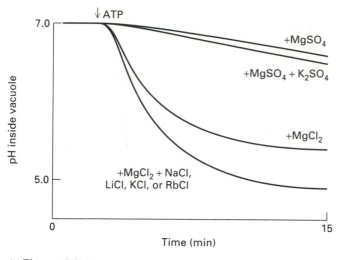

▲ **Figure 14-4**

c. In vivo, tonoplasts would be expected to accumulate lipid-soluble bases, such as ammonia or chloroquine. Propose a mechanism for accumulation of lipid-soluble pigments in plant vacuoles?

67. Cultured mammalian fibroblasts endocytose both sucrose and LDL. Experimentally, sucrose uptake generally is monitored as the intracellular accumulation of [^{14}C] sucrose and LDL uptake as the intracellular accumulation of [^{125}I] LDL. The concentration and time dependence of [^{14}C] sucrose and [^{125}I] LDL uptake by mouse 3T3 cells are shown in Figure 14-5. In mice, sucrose can only be degraded by intestinal cells, which are positive for sucrase-isomaltase.

    a. Are sucrose and LDL internalized by receptor-mediated endocytosis or by pinocytosis in mouse 3T3 cells?

    b. What is the K_m for uptake of sucrose and LDL by 3T3 cells?

    c. Why does uptake of both sucrose and LDL exhibit apparent saturation with time?

68. The gaseous air pollutant sulfur dioxide (SO_2) can affect plant stomates. This effect has been correlated, in part, with the effects of sulfur dioxide on phosphoenolpyruvate carboxylase, an enzyme of the malate biosynthetic pathway. Malate acts as a counterion to K^+ in guard cells. It is produced in photosynthetic plants as the result of CO_2 addition to phosphoenol pyruvate catalyzed by phosphoenolpyruvate carboxylase to form oxalacetate. Oxalacetate is then converted to malate. The effect

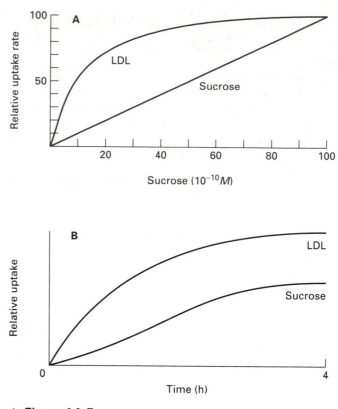

▲ **Figure 14-5**

of SO_2 on guard-cell phosphoenolpyruvate (PEP) carboxylase activity is illustrated by the data in Figure 14-6.

    a. Based on these data, what should be the effect of SO_2 on K^+ levels in guard cells?

    b. How would the change in guard-cell K^+ levels induced by SO_2 affect the status of stomates in leaf tissue? How might this affect plant growth?

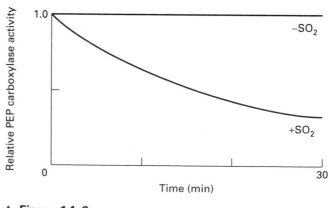

▲ **Figure 14-6**

CHAPTER *15*

Energy Conversion: The Formation of ATP in Mitochrondria and Bacteria

PART A: *Reviewing Basic Concepts*

Fill in the blanks in statements 1–30 using the most appropriate term(s) from the following list:

acetyl CoA	F_0
anions	F_0F_1
antiport	four (4)
ATP	glucose
carbon dioxide	glucose 6-phosphate
cations	glycerol
cristae	hexokinase
cytoplasmic	kinase
electric gradient	Krebs cycle
electron	lipids
Embden-Myerhof	membrane
endergonic	mitochondria
exergonic	Nernst
ethanol	oxygen
facultative aerobes	phosphofructokinase
facultative anaerobes	porin

proton	standard free energy ($G°'$)
proton gradient	sun
pyruvate	symport
pyruvate dehydrogenase	thermogenin
respiratory control	two (2)
spectroscopic	

1. _____, a six-carbon sugar, is the primary source of energy for animal cells.

2. Glycolysis results in the production of the energy-rich molecule _____, which contains a phosphoanhydride bond.

3. The _____ generates the energy necessary to drive ATP formation in plants.

4. Glycolysis is more formally known as the _____ pathway.

5. The number of ATP molecules produced per molecule of glucose in glycolysis is _____. This number does not take into account the two ATP molecules that are used in the reactions involving hexokinase and phosphofructokinase.

6. The hydrolysis of ATP has a standard free energy $\Delta G^{\circ\prime}$ of -7.3 kcal, which indicates that this reacton is _____.

7. Any enzyme that can transfer a phosphate group to a target enzyme or substrate is classified as a(n) _____.

8. Yeast cells can grow in either the presence or absence of O_2 and therefore are considered to be _____.

9. Yeast cells can ferment various substrates, forming the two-carbon compound _____, which is found in both wine and beer.

10. Lactic acid in mammalian muscle can be transported to the liver where it is reoxidized to _____, which can be transported back to muscle mitochondria for further breakdown to release carbon dioxide.

11. _____ is the outer mitochondrial membrane protein that forms a transmembrane channel.

12. The _____ of the mitochondria are infoldings of the inner mitochondrial membrane, which serve to increase the surface area and also house the electron transport system.

13. The enzyme _____, which converts pyruvate to acetyl CoA, has an unusually high molecular weight of 7 million.

14. _____, which are stored in adipose tissue, can generate much more ATP than glycogen on a gram for gram basis. The oxidation of these compounds to generate ATP occurs in _____.

15. The hydrolysis of triacylglycerols gives fatty acids and _____.

16. Both pyruvate and fatty acids are converted to _____ in the matrix of the mitochondrion.

17. NADH and $FADH_2$ are oxidized by _____ in mitochondria.

18. The membrane particle that acts as the ATP synthase in chloroplasts is called the _____ complex.

19. Mitochondria, chloroplasts, and bacteria have their F_1 complex attached to a _____.

20. Of the two components of the proton-motive force, the _____ makes a more important contribution than the _____ to ATP production in chloroplasts.

21. Because the inner mitochondrial membrane is relatively more impermeable to _____ than to protons, the transmembrane electric potential is the most important contributor to the proton-motive force in mitochondria.

22. The transmembrane electric potential generated by chloroplasts and mitochondria can be calculated using the _____ equation.

23. The _____ component of the F_0F_1 particle is an integral membrane protein and contains the binding site for oligomycin.

24. The ATP-ADP transport system that pumps ATP out of the mitochondrion and ADP into the matrix can be classified as a(n) _____ system.

25. The export of ATP from mitochondria is powered by the _____ gradient.

26. Fewer ATP molecules are generated from $FADH_2$ than from NADH because $FADH_2$ has a lower _____ than does NADH.

27. The order of the electron carriers can be determined with _____ techniques because the oxidized and reduced forms of each cytochrome absorb light at different wavelengths.

28. The regulation of mitochondrial respiration by the available pool of ADP is called _____.

29. _____, an inner mitochondrial membrane protein, can uncouple oxidative phosphorylation and thus increase the generation of heat; it is present in the mitochondria of brown fat.

30. The amount of ATP produced in mitochondria is indirectly monitored by the cytosolic enzyme _____, which can, in turn, regulate glycolysis in response to the respiratory activity of mitochondria.

PART B: *Linking Concepts and Facts*

Circle the letters corresponding to the most appropriate terms/phrases that complete or answer items 31–46; more than one of the choices provided may be correct in some cases.

31. The products that result from the complete aerobic respiration of glucose are

    a. glycerol.   d. oxygen.

    b. water.   e. carbon dioxide.

    c. urea.

32. ATP is synthesized in the

    a. mitochondria.   d. lysosomes.

    b. cytoplasm.   e. trans-Golgi network.

    c. chloroplasts.

33. Based on the observation that ATP is present in archae-bacteria, eubacteria, plants, and animals, one can conclude that

    a. all these life forms have mitochondria where ATP is produced.

    b. all these life forms use fucose for their primary energy source.

    c. ATP evolved in an early life form.

    d. the complete hydrolysis of ATP in all organisms will drive gluconeogenesis.

    e. ATP was originally an extracellular, secreted product, which was endocytosed and then subsequently used as an energy source.

34. The proton-motive force

    a. is generated by the hydrolysis of ATP.

    b. is, in part, a proton gradient.

    c. is generated by ATP synthase.

    d. is, in part, an electric potential.

    e. drives the ADP-ATP antiport system in mitochondria.

35. The first step in glycolysis involves

    a. the activation of mitochondrial enzymes.

    b. the enzyme hexokinase.

    c. hydrolysis of ATP to ADP.

    d. an input of energy.

    e. the reduction of molecular oxygen and production of carbon dioxide, which is evolved.

36. The generation of ATP by substrate-level phosphorylations

    a. requires the F_0F_1 complex.

    b. is blocked when hexokinase is inhibited.

    c. depends on a transmembrane electric gradient.

    d. depends on a transmembrane pH gradient.

    e. occurs in the cytoplasm.

37. During glycolysis, the hydrolysis of 1,3-bisphosphoglycerate is coupled to the substrate-level phosphorylation of ADP to ATP. What is the $\Delta G°'$ for hydrolysis of the phosphoanhydride bond in 1,3-bisphosphoglycerate when this hydrolysis is *not* coupled to ADP phosphoylation?

    a. $+7.3$ kcal/mol   d. -7.3 kcal/mol

    b. 0 kcal/mol   e. -11.8 kcal/mol

    c. -1.5 kcal/mol

38. Triosephosphate isomerase interconverts which of the following two compounds?

    a. dihydroxyacetone phosphate

    b. fuctose 6-phosphate

    c. phosphoenolpyruvate

    d. glyceraldehyde 3-phosphate

    e. glucose 6-phosphate

39. Both mitochondria and chloroplasts contain

    a. DNA.

    b. an outer and inner membrane.

    c. proton gradients.

    d. thylakoid disks.

    e. ATP synthase.

40. The matrix of mitochondria contains

    a. a high concentration of cardiolipin.

    b. dark reaction intermediates.

    c. Krebs cycle intermediates.

    d. $FADH_2$.

    e. electron transport chain intermediates.

41. Which of the following parameters has a *higher* value in metabolically active cells than in inactive cells?

    a. surface area of the plasma membrane

    b. number of mitochondria

    c. affinity of phosphoenolpyruvate for pyruvate kinase

    d. glucagon synthase activity

    e. surface area of mitochondrial cristae

42. Which of the following energy-related compounds is *not* present in both plants and animals?

    a. starch   d. pyruvate

    b. glucose   e. ATP

    c. glycogen

43. Since bacteria can generate ATP in a similar manner as do eukaryotic cells, which of the following features must they share?

    a. Embden-Meyerhof pathway

    b. mitochondria

    c. ATP synthesis enzymes

    d. proton gradients

    e. cristae

44. Protons are thought to affect the F_0F_1 particle by

    a. changing the conformation of the F_1 component.

    b. initiating the phosphorylation of ADP.

    c. causing the dissociation of ATP from the F_0F_1 particle.

    d. moving through a pore in the F_0 particle.

    e. causing the addition of two δ subunits to the F_1 component immediately after phosphorylation.

45. Bacteriorhodopsin, a protein found in photosynthetic bacteria, is most functionally analogous to which *one* of the following molecules or pathways?

    a. F_0 subunit   d. electron transport chain

    b. F_1 subunit

    c. Krebs cycle   e. porin

46. Phosphofructokinase can be allosterically regulated by

    a. glucose.   d. citrate.

    b. ADP.   e. NADH.

    c. ATP.

PART C: *Putting Concepts to Work*

47. Explain the statement that ATP is the energy "currency" of the cell.

48. Why are the protons and electrons generated during glycolysis important to the energy considerations of the cell and what relationship do they have with mitochondria?

49. The standard free energy of many individual reactions in glycolysis is positive, indicating that these steps are not thermodynamically spontaneous. Why is it that glycolysis proceeds, nonetheless?

50. Why is it critical to seal the container during the fermentations that produce wine and beer?

51. What is the similarity between lactic acid accumulation in mammalian muscle cells and ethanol accumulation in fermenting yeast cells?

52. It is possible to freeze-fracture an entire mitochondrion, thus splitting both the outer and inner mitochondrial membranes. How would you be able to distinguish the outer from the inner membrane assuming that both are present in a given freeze-fracture electron micrograph?

53. Many of the metabolites of glycolysis and energy intermediates such as pyruvate, ADP, and fatty acids are present in mitochondria. However, only CO_2 and O_2 are able to diffuse across the mitochondrial membrane. How do these metabolites and intermediates get into mitochondria?

54. One of the intriguing mysteries about the Krebs cycle is that it is a cascade of reactions all of which are closely linked. Soluble enzymes, as they are known to exist in the cytosol, would not be physically close enough to allow these reactions to proceed in such a coordinated fashion. What feature of the Krebs cycle enzymes per-

mits these coordinated reactions to occur in the mitochondrial matrix and even in dilute buffer?

55. What attributes of the cytochromes allow them to transport electrons in the electron transport system?

56. Why is it important that the reduction potential of the electron carriers in mitochondria increase steadily from NADH to O_2?

PART D: *Developing Problem-solving Skills*

57. In early experiments testing the chemiosmotic hypothesis, it was critical to demonstrate that synthetic phospholipid bilayers (i.e., liposomes) could maintain two distinct pH domains, one inside the liposome and the other outside. Why were these experiments critical to the theory of chemiosmosis and how could they have been developed to show that this gradient *alone* can generate ATP from ADP?

58. Facultative anaerobes can survive in environments with or without molecular oxygen. This ability would seem to be advantageous and is exhibited by yeast cells and muscle cells as well as some annelids and mollusks. In addition, flooded roots have this metabolic flexibility. Why is this ability to switch metabolic routes *not* common in all eukaryotes?

59. As a chemical engineer working with polycyclic compounds, you find that a particular compound (X) is especially useful in polymerizing plastic. However, some X is left as a residue on containers to be used in the food industry. Testing at FDA laboratories showed that X was toxic to cells in culture. When this compound was tested in a metabolism laboratory, it mildly inhibited the formation of ATP in eukaryotic cells, but it did not compromise the proton or electric gradient in mitochondria and did not act as a poison to the F_0F_1 particle. Surprisingly, it had no direct affect on glycolysis. Suggest a mechanism of action of compound X.

60. It has been conclusively demonstrated that related cells can differ in their metabolic activities under various conditions. For instance, dividing cells have much more active mitochondria than do interphase (nondividing) cells. Transformed cells also have more active mitochondria than do their normal (nontransformed) counterparts. To select clones that differ in the metabolic states of their mitochondria, you can first treat cells in culture with a chemical carcinogen to mutate the DNA in both the nucleus and the mitochondria and then select for mutants that are highly active and for those that are less active.

   a. Suggest an assay that could distinguish these two cell types in vitro.

   b. How could you select for less active mutants.

61. Critical to the support of Peter Mitchell's chemiosmotic hypothesis was the ability to produce inside-out submitochondrial vesicles, as illustrated in Figure 15-1. When you prepare similar vesicles and assay them for ATP production, you find to your dismay that they produce little ATP in the presence of ADP, O_2, and a physiological buffer at pH 7. Assuming that the electron transport system and the F_0F_1 particle are functioning normally in your vesicular preparation, what relatively minor change in your assay system would be likely to increase oxidative phosphorylation?

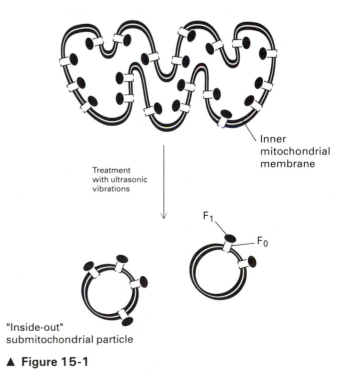

Inner mitochondrial membrane

Treatment with ultrasonic vibrations

F_1

F_0

"Inside-out" submitochondrial particle

▲ **Figure 15-1**

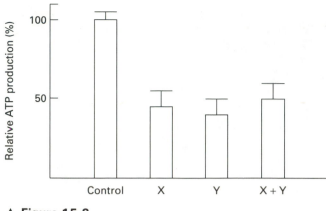

▲ **Figure 15-2**

62. In a preparation of isolated mitochondria, the phosphorus to oxygen (P-O) ratio was determined to be 2.5. In a prepartion of vesicles harvested from the inner mitochondrial membrane, the P-O ratio was found to be 1.0. The latter vesicular preparation originated from the same cells as did the whole mitochondrial preparation, and the moles of O_2 consumed was the same for both preparations adjusted to equal ATP synthase concentrations. Suggest possible reasons why the P-O ratios for the two preparations differed.

63. In one study, antibodies against electron transport chain complexes such as NADH-CoQ reductase and cytochrome *c* oxidase were found to bind to both inside-out and right-side out vesicles derived from the inner membrane of mitochondria. However, each antibody preferentially bound to one or the other type of vesicle. Although the antibodies were initially dispersed on each vesicle, after a 10-min incubation period, they were clustered at one pole. This clustering was associated with decreased ATP production. What is the significance of both the preferential binding and asymmetric clustering of antibodies in relation to the electron transport chain and the chemiosmotic hypothesis?

64. In a series of experiments, the ability of two mitochondrial poisons (X and Y) to inhibit ATP production by isolated, whole mitochondria was studied. In all cases, the inhibitor compounds were added in excess to the reaction mixtures.

    a. From the data shown in Figure 15-2, what is the *most* likely conclusion about the mechanism(s) of action of compound X and Y?

    b. When excess oligomycin was added to the three reaction mixtures, ATP production decreased further (to 5 percent of the control level) in the mixtures containing compound X and compound X + Y but not in the mixture with compound Y alone. Based on these additional data, what can you conclude about the inhibitory mechanisms of compound X and Y?

    c. Are the conclusions based on the data in parts (a) and (b) consistent? If not, which conclusion is most persuasive and why?

65. Dinitrophenol has a greater inhibitory effect on oxidative phosphorylation in chloroplasts than in mitochondria, whereas the relative effects of valinomycin are just the opposite. Based on the mechanisms of action of these two inhibitors, what can you conclude from their relative effects on mitochondria and choloroplasts about the underlying mechanism of oxidative phosphorylation in each of these organelles.

PART E: *Working with Research Data*

66. Lan bo Chen at the Dana Farber Cancer Institute in Boston has pioneered the use of the laser dye rhodamine 123, which partitions to mitochondria in living cells. The dye is specific for mitochondria and does not stain other organelles, such as lysosomes and endosomes, that like mitochondria have acidic, cisternal domains. To determine which attribute of mitochondria is responsible for this unpredicted behavior of the dye, Lan bo Chen's group measured the effect of several drugs on the ability of mitochondria to retain the dye. The drugs tested, their action on mitochondria, and the effect on rhodamine 123 retention are summarized in Table 15-1.

    a. Based on the data in Table 15-1, which parameters govern the ability of mitochondria to retain rhodamine 123?

    b. Based on the action of the drugs that precede oligomycin in Table 15-1, why would oligomycin be expected to cause a "slight increase" in rhodamine 123 retention?

    c. In the course of these experiments, it was noticed that neighboring cells often varied extensively in their uptake of rhodamine 123. For instance, mitotic cells often had one daughter cell that stained well

Table 15-1

Drug	Action	Rhodamine 123 retention
Valinomycin	Decreases pH and electric gradient (K^+ ionophore)	Diminished
FCCP	Decreases pH and electric gradient (H^+ ionophore)	Diminished
DNP	Decreases pH and electric gradient (H^+ ionophore)	Diminished
Nigericin	Decreases pH gradient and increases electric gradient (Na^+—H^+ exchange ionophore)	Increased
Oligomycin	Inhibits ATP synthase	Slightly increased
Sodium azide	Inhibits electron transport	Diminished
Chloramphenicol	Inhibits mitochondrial protein synthesis	No effect
Anaerobic conditions	Prevents access to O_2	Diminished

with rhodamine 123 and one that did not stain at all. In the presence of nigericin, however, all cells took up rhodamine 123 equally well. What can you conclude about the cause of the variation in rhodamine 123 uptake observed in some cells?

d. Transformed and nontransformed cells have very different abilities to retain rhodamine 123. In general, nontransformed cells retained the dye for a few hours, whereas genetically similar transformed cell lines retained the dye up to several days. Design an experiment to determine if this difference is caused by the higher metabolic activities of mitochondria in transformed cells compared with those in nontransformed cells.

67. Because rhodamine 123 can distinguish some transformed cells from their nontransformed counterparts (see question 66d), Lan bo Chen decided to determine if rhodamine 123 could selectively kill transformed cells in culture. Temperature-sensitive (ts) mutants were used for these studies. At 39°C the mutants demonstrate the normal, nontransformed phenotype, but at 37°C they demonstrate the transformed phenotype. Cells were initially labeled with rhodamine 123, the excess dye was washed out, and the number of living cells was periodically determined for 1 week. Representative data from this type of study are shown in Figure 15-3.

a. Why were temperature-sensitive mutants used in this study?

b. Do the data in Figure 15-3 indicate that mitochondria from transformed cells are more sensitive than those from normal cells to the lethal effects of rhodamine 123?

c. What change in the experimental protocol would

help to determine whether the lethal effect of rhodamine 123 on the transformed phenotype of the ts mutants resulted from their increased retention of or increased sensitivity to the dye?

68. A proton gradient can be analyzed with fluorescent dyes whose emission intensity profiles depend on pH. One of the most useful dyes for measuring the pH gradient across mitochondrial membranes is 2′,7′-bis-(2-carboxyethyl)-5(and 6)-carboxyfluorescein (BCECF). When the lipid-soluble acetoxymethyl ester form of this dye (BCECF-AM) is incubated with isolated mitochondria, it passes easily into the mitochondrial matrix. Once there, nonspecific mitochondrial esterases cleave the acetoxymethyl group, yielding BCECF, which is membrane impermeable and consequently is trapped in the

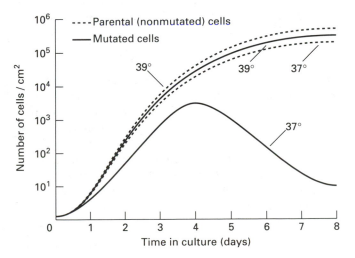

▲ **Figure 15-3**

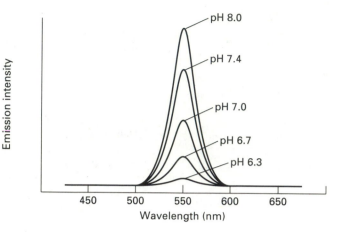

▲ Figure 15-4

matrix. The effect of pH on the emission intensity of BCECF, excited at 505 nm, is shown in Figure 15-4.

a. Vesicles containing the electron transport system were made from isolated inner mitochondrial membranes and treated with BCECF-AM. After excess dye was removed, the vesicles were incubated in a physiological buffer containing ADP and O_2. The fluorescence of BCECF trapped inside the vesicles gradually decreased in intensity, suggesting a decrease in cisternal pH. Explain why an increase in emission intensity would be expected and what the observed decrease might indicate about this vesicular preparation.

b. When the same vesicles described in part (a) are incubated with ATP rather than ADP and dilute hydrochloric acid is added to the buffer (on the outside of the vesicles), ATP hydrolysis is observed. Explain why this observation is predictable.

c. When the uncouplers dinitrophenol and valinomycin are added to the vesicular preparation described in part (a), dinitrophenol causes a decrease in BCECF fluorescence when BCECF is on the outside of the vesicles, whereas there is only a transient effect with valinomycin. Why is this the case?

69. Mitochondria have a very high concentration of Ca^{2+} ions. Indeed, Ca^{2+} granules can be easily distinguished as electron-dense particles in the matrix of many mitochondria. The emission intensity of the fluorescent dye Fluo-3 increases with the concentration of Ca^{2+}, as shown in Figure 15-5, when the excitation wavelength is 505 nm. Like BCECF discussed in the previous problem, the acetoxymethyl ester of Fluo-3 is lipid soluble and passes easily into the mitochondrial matrix, whereas the free form of the dye is membrane impermeable.

a. When drug X, whose function is unknown, is added to right-side out vesicles originating from inner mi-

tochondrial membranes, the fluorescence of Fluo-3 trapped inside the vesicles decreases in intensity. In similar experiments, the emission intensity of BCECF trapped inside the vesicles also decreases when this drug is added. What is the predicted effect of drug X on ATP production?

b. When the isolated vesicles described in part (a) are pretreated with proteinase K, a protease, the effect of drug X on the fluorescence of both BCECF and Fluo-3 trapped inside the vesicles decreases significantly. What information does the protease data add about the possible mechanisms underlying the movement of Ca^{2+} and ions and protons across the vesicular membrane?

70. Wolfgang Junge and his coworkers in West Germany have accomplished extensive work delineating the roles of the CF_0 and the CF_1 portions of the ATP synthases found in thylakoids. Of particular interest have been their attempts to elucidate the activities of the five different polypeptides that compose CF_1. These polypeptides are named α, β, γ, δ, and ϵ in order of their decreasing molecular weights. All can be distinguished from each other using SDS gel electrophoresis. The subunit stoichiometry is $3:3:1:1:1$. The composite molecular weight of all five subunits (i.e., the CF_1 portion) is approximately 400,000. It is known that the β subunits probably bind ADP and ATP and contain the catalytic sites. The γ polypeptide acts as an inhibitor of ATP hydrolysis; the function of the δ subunit is unknown.

In order to explore the functions of these subunits, Junge's group prepared thylakoid membranes and tried four different extraction methods (A, B, C and D) to remove only the CF_1 portion from the CF_0CF_1 particle. They monitored both the ability of the membrane preparations to dissipate voltage and to diminish the proton gradient. They used a single flash of light to induce voltage changes in the thylakoid membrane prepara-

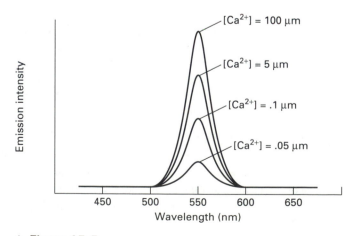

▲ Figure 15-5

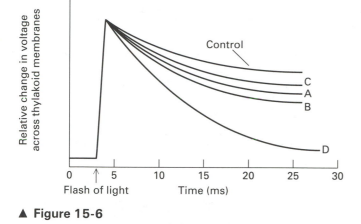

▲ Figure 15-6

tions. Figure 15-6 depicts the decrease in induced voltage across thylakoid membranes prepared with the four extraction techniques. These data are similar to those obtained by Junge's group. The data show that extraction methods A, B, and C produced membranes that dissipated voltage only slightly more than the no-extraction control. In contrast, method D resulted in leaky membranes, which rapidly dissipated voltage.

a. Why was flashing light used in these experiments?

b. Since extraction method D made the membrane leakier, it is possible that either CF_1 or CF_0CF_1 was removed from the thylakoid membranes by this procedure. Suggest a method to determine if CF_1 was removed by this extraction method.

c. The material extracted by each method was subjected to Western blot analysis with an antibody to the δ subunit used as a probe. The resulting blots are depicted in Figure 15-7. What do these data suggest about the function of the δ subunit in the CF_0CF_1 particle?

d. Phenol red is a colorimetric dye that is purple under alkaline conditions and yellow under acidic conditions. How could this dye be useful in further delineating the possible role of the δ subunit?

e. When the CF_0 inhibitor N,N'-dicyclohexylcarbodiimide (DCCD) is added to membrane D it exhibits the same profile for dissipation of voltage and changes in the pH gradient as do the control and membranes A–C. However, DCCD does not affect the Western blot in Figure 15-7. Why is this experiment critical to understanding the possible role of the δ subunit in regulating voltage and proton flux through the CF_0 particle?

f. Given the data suggesting that removal of the δ subunit results in membranes that are leaky for both protons and voltage, propose more definitive experiments to test the hypothesis that the δ polypeptide acts as a "stopcock" of the CF_0 portion?

71. A current objective of oxidative phosphorylation research is to understand how F_0 and F_1 are associated with each other in the inner mitochondrial membrane. A number of experiments have suggested that sulfhydryl (thiol) groups may play an important role in the function of the mitochondrial inner membrane. Treatment of vesicles from this organelle with agents that disrupt thiol groups results in the malfunction of several integral membrane proteins. The permeases, including the ADP-ATP antiport system, are particularly sensitive to drugs, such as diamide, that disrupt thiol groups. Using diamide and dithiothreitol (DTT), which can reverse the effects of diamide, Giovanna Lippe of the Universita di Padova in Italy tried to determine if thiol groups are critical for the link between the CF_0 and the CF_1 in beef heart mitochondria.

Lippe and colleagues added isolated F_0 and F_1 to synthetic phospholipid vesicles and determined the effect of diamide and DTT on the oligomycin sensitivity of the reconstituted vesicles. Oligomycin sensitivity was monitored using a pH electrode. ATP hydrolysis was also measured. The data in Table 15-2 are representative of Lippe's work.

a. Why were oligomycin sensitivity and ATP hydrolysis monitored in these experiments?

b. What do the data in Table 15-2 indicate about the effect of diamide?

c. What do the results with samples 2 and 4 suggest about the association of F_0 and F_1?

d. What is the importance of sample 3 and what does it suggest about the effect of diamide in this system?

e. The samples described in Table 15-2 were centrifuged and the ATP hydrolysis activity in the supernatants and pellets were determined. For samples 1,

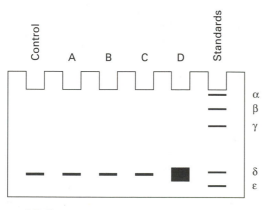

▲ Figure 15-7

Table 15-2

Sample	Addition order				ATP hydrolysis rate	Relative oligomycin sensitivity (%)*
	First	Second	Third	Fourth		
1	F_0	F_1	—	—	14.5	95
2	F_0	diamide	F_1	—	15	30
3	F_0	diamide	F_1	DTT	15	90
4	F_0	F_1	—	diamide	14	95

* As percentage of oligomycin sensitivity of isolated mitochondria.

3, and 4, most of the original ATP hydrolysis activity (82–93 percent) remained in the pellet; very little was released into the supernatant. In contrast, 43 percent of the original ATP hydrolysis activity of sample 2 was released into the supernatant, and 55 percent remained in the pellet. What do these results indicate about the effects of diamide and DTT on the association of F_0 and F_1?

16

Photosynthesis

PART A: *Reviewing Basic Concepts*

Fill in the blanks in statements 1–20 using the most appropriate terms from the following list:

680

700

bundle sheath

calcium (Ca^{2+})

Calvin

cellulose

chlorophyll *a*

chlorophyll *b*

chloroplast(s)

cytochrome

Emerson

epidermis

glyceraldehyde 3-phosphate

heme

high

Krebs

light-harvesting complex (LHC)

low

magnesium (Mg^{2+})

manganese (Mn)

mitochondrial

NADPH

$NADP^+$

photorespiration

photosynthesis

plasma

quinone

ribulose 1,5-bisphosphate (RBPase)

sucrose

stroma

triazine(s)

water (H_2O)

1. Photosynthesis occurs in plant organelles called _____ _____ .

2. _____ is one of the chief products of photosynthesis. Its final synthesis occurs in the cytosol, and it is shipped to other cells where it is used as an energy source.

3. Chlorophyll is similar in structure to the oxygen-carrying molecule _____ .

4. Evolved O_2 originates from _____ in chloroplasts.

5. During photosynthesis the proton gradient across the thylakoid membranes is generated, in part, by the splitting of _____ .

6. The final electron acceptor in photosynthesis is _____ _____ .

7. The primary chlorophyll molecule of the reaction center is _____ .

8. In purple bacteria the proton-motive force, which drives photosynthesis, is generated across the _____ membrane.

9. In purple bacteria a positively charged electron donor, _____, on the exoplasmic surface of the plasma membrane and a reduced electron acceptor, _____, on the cytoplasmic surface represent the charge separation that is critical to photosynthesis.

10. Light of two different wavelengths has a greater effect on photosynthesis than the same-intensity light of an intermediate wavelength. This phenomenon is called the _____ effect.

11. The _____ resides in thylakoids and consists of several transmembrane proteins that capture light energy for PSI and PSII.

12. The oxygen-evolving complex has four _____ ions, which are thought to have various oxidation states that facilitate the splitting of H_2O to release O_2.

13. Light of wavelength _____ nm causes cytochromes b_6 and f to become more oxidized, whereas light of wavelength _____ nm causes these cytochromes to become more reduced.

14. _____ is a herbicide that blocks electron transfer in PSII.

15. The fixation of CO_2 into hexose sugars is often called the _____ cycle after the researcher who discovered this pathway.

16. Fixation of CO_2 occurs in the _____ of chloroplasts.

17. The three-carbon sugar that is synthesized as a consequence of the Calvin cycle is _____, which is shipped to the cytoplasm for synthesis into sucrose.

18. The light-dependent process called _____ consumes O_2 and releases CO_2.

19. The dark reactions are favored by _____ concentrations of O_2 and _____ concentrations of CO_2.

20. The concentration of CO_2 in C_4 plants is higher in _____ cells than it is in the atmosphere.

PART B: *Linking Concepts and Facts*

Circle the letters corresponding to the most appropriate terms/phrases that complete or answer items 21–31; more than one of the choices provided may be correct in some cases.

21. Two sugars produced directly during photosynthesis are

    a. glycogen.
    b. starch.
    c. glycerol.
    d. sucrose.
    e. cellulose.

22. Porin is located in

    a. the plasma membrane.
    b. the membranes of microbodies.
    c. the outer membrane of chloroplasts.
    d. the inner membrane of chloroplasts.
    e. the thylakoids.

23. Photosynthesis occurs in

    a. the plasma membrane of green plants.
    b. the membranes of lysosomes.
    c. the outer membrane of chloroplasts.
    d. the grana.
    e. the thylakoids.

24. Which of the following terms is (are) often used to refer to light energy?

    a. quanta
    b. particles
    c. photons
    d. electromagnetic radiation
    e. waves

25. Chlorophyll fluoresces if it is separated from

    a. β-carotene.
    b. $NADP^+$.
    c. an acceptor molecule.
    d. ATP.
    e. ADP.

26. When chlorophyll is in its oxidized state,

    a. it fluoresces in the thylakoids.
    b. H_2O is reduced.
    c. O_2 is generated from sucrose.
    d. it directly reduces NADP.
    e. it results in protons being generated from H_2O.

27. Photosynthetic bacteria and higher plants differ in that

    a. the former do not release evolved O_2.

    b. the former do not use chlorophyll.

    c. the former utilize only one photosystem.

    d. the former do not use light as an energy source.

    e. the former do not use pigments associated with membranes.

28. Picosecond absorption spectroscopy

    a. uses a laser light.

    b. can determine electron movement.

    c. can track photosynthetic processes that occur in less than 1 ms.

    d. is also called optical absorption spectroscopy.

    e. can give insight into the molecular mechanisms of photosynthesis.

29. P_{680} and P_{700} are

    a. carotenoids.

    b. specialized chlorophyll *a* molecules.

    c. both in PSII.

    d. in the "reaction centers."

    e. the only two types of chlorophyll that can absorb light.

30. The oxidation and reduction of copper (Cu) is critical to the function of

    a. ferredoxin.

    b. plastocyanin.

    c. the cytochrome *b/f* complex.

    d. NADPH.

    e. pheophytin.

31. Cyclic electron flow

    a. involves only PSII.

    b. involves only PSI.

    c. results in O_2 being evolved.

    d. results in ATP photophosphorylation.

    e. is inhibited by a high concentration ratio of NADPH to $NADP^+$.

PART C: *Putting Concepts to Work*

32. What distinctive feature of the chemical structures of chlorophyll and other porphyrins accounts for their ability to release electrons?

33. Why does O_2 accumulate in the luminal surface rather than the stromal surface of thylakoids?

34. During photosynthesis, electrons are responsible for the reduction of $NADP^+$ to NADPH and are moved through photosystem I (PSI) and photosystem II (PSII) of the thylakoid membranes. What is the sequence in which electrons flow through these components during noncyclic photophosphorylation?

35. What complex contains the ATP-synthesizing enzyme of chloroplasts?

36. Sometimes the phrase "dark reactions" is used to refer to carbon fixation. Why is this phrase somewhat inappropriate?

37. Why would photons of blue light be expected to have a greater impact on photosynthesis than photons of red light? Assume that both types of photons can be absorbed by the photosynthetic system.

38. What does the comparison of the absorption spectrum of chlorophyll and the action spectrum of photosynthesis suggest about chlorophyll?

39. Under certain conditions, electron flow in photosynthetic purple bacteria is noncyclic. In this case, $NADP^+$ is reduced, but O_2 is not evolved. How do purple bacteria accomplish this?

40. Mild detergents can extract PSI but not PSII from thylakoid membranes. What does this differential solubility suggest about the nature of the two photosystems?

41. Why are mobile electron carriers critical to the function of noncyclic photophosphorylation?

42. The ratio of PSI to PSII is not 1 : 1 in all plants. Some reports indicate that the ratio may be as high as 1 : 3. Yet in noncyclic photophosphorylation there is a functional 1 : 1 relationship between these two photosystems. How could protein kinase activity resolve this apparent paradox? (Do not consider cyclic photophosphorylation in your answer).

43. In certain plants, the level of CO_2 in bundle sheaf cells falls below the K_M of ribulose 1,5-bisphosphate carboxylase for CO_2. What is the implication of this phenomenon? What special mechanisms do such plants have?

PART D: *Developing Problem-solving Skills*

44. Numerous compounds are known to inhibit photosynthesis by affecting specific reactions or components of the photosynthetic apparatus. When one such agent (X) was added to isolated purple bacteria cells, formation of ATP ceased. The treated cells did not evolve O_2 or reduce $NADP^+$ to NADPH. Based on these observations, what can you conclude about the mechanism of action of this agent?

45. Although the current concept that PSI and PSII are coupled in noncyclic photophosphorylation is generally accepted, certain observations appear to be inconsistent with this theory. For example, some researchers have shown that the ratio of PSI to PSII is not always 1 : 1 (see problem 42). Similarly, the distance between the two photosystems may be too great to allow for electron transfer between them. What single technique could you use to accurately measure both of these parameters?

46. Antibodies directed against many of the photosynthetic membrane components are available. Using antibodies against plastocyanin and the CF_o complex, you find preferential binding of these antibodies on the lumenal side of thylakoids. On the other hand, antibodies directed against the cytochrome b/f complex, ferredoxin, and NADP reductase preferentially bind on the stromal side of thylakoids. Why is this asymmetrical binding critical to the current model of photophosphorylation?

47. Daniel Arnon has been a principal contributor to current concepts about the relationship of PSI and PSII in photosynthetic plants. In experiments performed a few years ago, he showed that an inhibitor of plastiquinone did not block the reduction of ferredoxin. In this case, both O_2 and NADPH were produced. What particular significance do these experiments have for our current understanding of noncyclic photophosphorylation?

48. When a suspension of intact, isolated thylakoids is incubated in the presence of NH_4Cl, there is a significant reduction in the amount of ATP produced but only a slight difference in the amount of O_2 evolved in treated versus untreated preparations. The difference in O_2 evolution is not enough to account for the reduced levels of ATP production. What is the probable effect of NH_4Cl on these preparations?

49. Thylakoids can be sonicated into membrane fragments, which then reseal into vesicles. In one experiment, these vesicles were incubated for 30 min with an antibody to the oxygen-evolving complex (OC). Next, protein A–linked Sephadex beads, which recognize and bind to the antibody-OC complex, were added to the preparation and then spun out of solution at $500 \, g$ for 15 min. This procedure left other vesicles in the supernatant. Next, the isolated OC vesicles were separated from the antibody and protein A with high salt treatment. When these OC vesicles were incubated in the presence of light, the medium became slightly acidic relative to the lumen of the vesicles. However, when the vesicles that remained in the supernatant were similarly analyzed, the external medium became slightly basic relative to the lumen. What could account for this difference? What would be the effect of adding 0.1% Triton X to both vesicle preparations?

50. If thylakoids are sonicated to a greater extent than in the experiments described in problem 49, both larger and smaller vesicles are produced. When such vesicles were separated by size using centrifugation techniques, the smaller vesicles were found to produce ATP but did not form O_2 or NADPH when stimulated with light; in contrast, the larger vesicles were able to form O_2, NADPH, and ATP. What is the most plausible reason for these observations?

51. Investigators have used 3-(3,4-dichlorophenyl)-1,1-dimethylurea (DCMU), an inhibitor of photosynthesis, in studies of noncyclic photophosphorylation. Comparisons of DCMU-treated and untreated chloroplasts have demonstrated the following:

Parameter	DCMU-treated chloroplasts versus untreated controls
Electron flow from H_2O to ferricyanide	Reduced
NADPH production under conditions that promote cyclic photophosphorylation	Similar
PSI cyclic photophosphorylation	Similar
Reduction of plastiquinone	Similar

Based on these data, what is the probable site of action of DCMU?

52. Many blockers of photosynthesis can act either as uncouplers of photophosphorylation or as inhibitors of PSI, PSII, or the CF_oCF_1 complex. When compound Y is added to chloroplasts, ATP production is reduced and electron flow is similarly reduced in noncyclic photophosphorylation. One possible mechanism of Y is that it binds to the chlorophyll *a* of the reaction centers of either or both PSI and PSII. How could this hypothesis be tested?

53. Mutants of the duckweed *Lemna perpusilla* have been used to investigate how different components of PSI and PSII interact to accomplish photophosphorylation. When light-dependent ATP production is measured in chloroplasts from wild-type and mutant plants, data similar to those in Figure 16-1a are obtained. When the pH gradient is reversed and hydrolysis of ATP is measured, both wild-type and mutant chloroplasts exhibit similar phosphatase abilities, as shown in Figure 16-1b. What do these data indicate about the locus of the defect in this duckweed mutant?

54. A chloroplast suspension was prepared and analyzed for its ability to fix CO_2 in the presence of compound Z. The suspension was exposed to light while being continually incubated in the presence of excess ^{14}C-labeled CO_2. Monitoring of the amount of labeled triosephos-

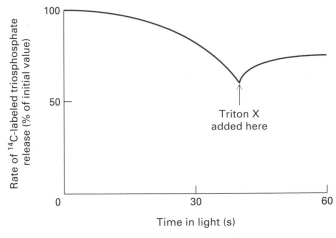

▲ **Figure 16-2**

phate released into the surrounding medium after addition of compound Z showed that there was a time-dependent decrease in the rate of triose release until it eventually fell to undetectable levels. Increasing the light intensity or changing the wavelength had no effect on the rate of triose release.

a. When 0.1 percent Triton X was added to the suspension, the rate of triosephosphate release from the chloroplasts was slightly enhanced, as shown in Figure 16-2. Assume that both ATP and NADPH levels in the chloroplasts remained constant during the study period. What is the most probable cause of the decrease in the rate of triosephosphate release?

b. Assume that ATP and NADPH accumulated in the chloroplasts, rather than remaining constant as in part (a). In this case, what is the most likely cause for the decrease in the rate of triosephosphate release and would Triton X partially reverse this effect as described in part (a)?

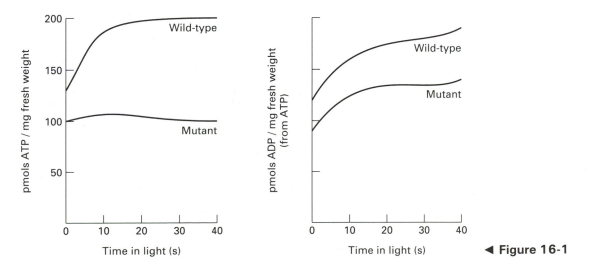

◀ **Figure 16-1**

PART E: *Working with Research Data*

55. Samuel Robinson and Archie Portis of the U.S. Department of Agriculture Research Service and the Department of Agronomy, University of Illinois, Urbana, have been working on ribulose 1,5-bisphosphate carboxylase (RBPase), which has several unusual properties. First, it is a rather enzymatically sluggish enzyme, which occurs in great abundance in the biosphere. Second, during the dark reactions, it exhibits carboxylase activity, but it also has oxygenase activity, which is the first reaction in photorespiration. Paradoxically, RBPase activity declines precipitously during in vitro CO_2 fixation (a phenomenon termed fallover) but not in vivo during in vivo CO_2 fixation. Robinson and Portis have recently discovered a RBPase activase that prevents the in vitro decline in RBPase activity. The effect of this putative activase on in vitro CO_2 fixation measured in a one-step assay system is illustrated in Figure 16-3. All substrates were present in excess.

    a. Is the oxygenase or the carboxylase activity of RBPase measured in this assay?

    b. What is the effect of the activase on the initial rate of CO_2 fixation by RBPase?

    c. What do the data in Figure 16-3 indicate about the ability of the activase to reverse (but not prevent) the fallover effect?

    d. When the assay mixture without activase was subjected to gel filtration after 30–40 min, ribulose 1,5-bisphosphate (RuBP) was removed, and the RBPase activity was restored to its initial value when the preparation was reincubated, with RuBP. What do

these observations suggest about the relationship between RBPase and RuBP.

    e. Other data collected by this group indicate that an inhibitor may bind to the activated form of RBPase, since gel filtration of the sample with activase as described in part (d) did not restore RBPase to its fully activated state. Alkaline phosphatase was found to reverse some of this inhibition in RBPase activity. What do these data indicate about the nature of the inhibitor?

    f. Although this inhibitor has not yet been purified, how could the future development of a photoaffinity analog of this inhibitor be useful in characterizing this system?

56. Green plants and algae contain several types of pigments. Most of these auxiliary pigments funnel their energy to chlorophyll *a/b* in the photosynthetic reaction center. A prerequisite for understanding this energy-shuttling system is to characterize these auxiliary pigments. In recent experiments by Marvin Fawley of the Department of Botany, North Dakota State University, a new form of chlorophyll *c* was discovered in the alga *Pavlova gyrans*. In this work, cells were extracted with three different agents: acetone, methanol, and liquid-nitrogen freezing of pellets followed by methanol extraction. A reverse-phase, high-pressure liquid chromatographic (HPLC) system was used to separate the pigments extracted from the cells, as shown in Figure 16-4.

    a. Why were acetone and methanol used to extract the pigments?

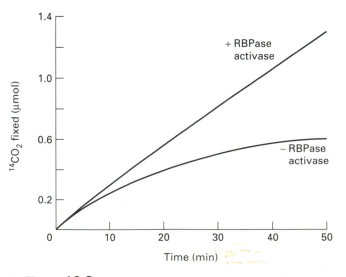

▲ **Figure 16-3**

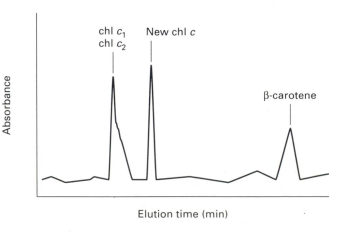

▲ **Figure 16-4**

b. Why might freezing and thawing be useful here, whereas it might be counterproductive if enzymatic activity were being monitored?

c. Using only the reverse-phase HPLC fractions illustrated in Figure 16-4, how could you determine that the peak labeled "new chl c" is not merely a chromatographic separation artifact of chl c_1 or c_2?

d. The absorption spectra of new chl c, chl c_1, and chl c_2 were determined to be different. What does this imply about their possible structural and functional similarities?

e. A comparison was made of the retention characteristics of these three pigments in cellulose thin layer chromatography and polyethylene thin layer chromatography. The R_f values for new chl c differed from those for chl c_1 or chl c_2 in both systems. In light of the question posed in part (c), why are these additional techniques useful in describing a newly discovered photosynthetic pigment, such as new chl c.

57. Chlorophyll a/b light-harvesting complex proteins are the most plentiful proteins of the thylakoid membranes. They help capture light and transfer it to the reaction centers. Chlorophyll b is present in both the light-harvesting complex of PSI (LHC-I) and the light-harvesting complex of PSII (LHC-II). Tomio Terao and Sakae Katoh of Japan have produced chemical and x-irradiated mutants of rice and have found that if these mutants are defective in chlorophyll b, they also have diminished levels of LHC-I and LHC-II proteins. In fact, the greater the degree of deficiency in chlorophyll b, the greater the deficiency in the amounts of LHC-I and LHC-II in the thylakoid membranes. These researchers suggest that chlorophyll b may serve to stabilize the LHC-I and LHC-II apoproteins.

Chloroplasts from wild-type rice, chlorina 2 (mutant with no chlorophyll b), and chlorina 11 (mutant with only a little chlorophyll b) were incubated in [^{35}S]methionine for 30 min, and then the thylakoid membranes were extracted and the proteins analyzed by gel electrophoresis/autoradiography. The resulting autoradiograms are shown in Figure 16-5. In the lanes labeled Cap and Chl, the cells were treated with chloramphenicol and cycloheximide, respectively, for 30 min before labeling; in the lanes labeled Unt, the cells were not treated. The protein labeled LS is the large subunit of RPBase; Qb is a protein known to be synthesized very rapidly in the light; LHC-II and LHC-I are the proteins of the light-harvesting complexes.

a. How could you determine whether a particular mutant was deficient in chlorophyll b?

b. Why is the appearance of LS in the gels unexpected?

c. What do the diminished levels of LS in the Cap lanes but not in the Chl lanes indicate about this subunit of RBPase?

d. Why was chloramphenicol especially useful in examining LHC-II and LHC-I?

e. What do the autoradiograms in Figure 16-5 indicate about the synthesis rates of LHC-I and LHC-II in the wild-type strain and mutant strains?

f. How could the rates of synthesis of LHC-I and LHC-II be determined in a quantitative manner from the gels?

58. In experiments subsequent to those described in problem 57, leaves from wild-type and mutant rice strains were pulse-labeled with [^{35}S]methionine for 30 min followed by a chase of nonradioactive methionine for various time periods. After the chase period, the thylakoid

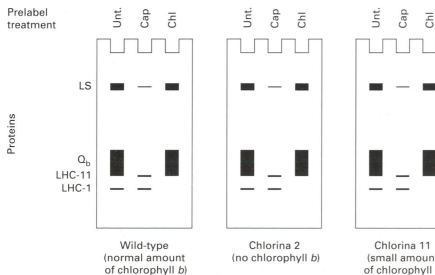

Wild-type (normal amount of chlorophyll b)

Chlorina 2 (no chlorophyll b)

Chlorina 11 (small amount of chlorophyll b)

◀ **Figure 16-5**

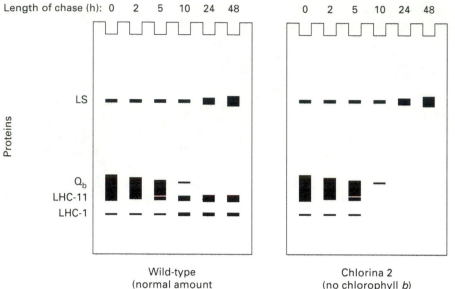

Length of chase (h): 0 2 5 10 24 48   0 2 5 10 24 48

Proteins

LS

Q_b
LHC-11
LHC-1

Wild-type
(normal amount
of chlorophyll *b*)

Chlorina 2
(no chlorophyll *b*)

◄ **Figure 16-6**

membranes again were extracted and the proteins analyzed by gel electrophoresis/autoradiography. The resulting autoradiograms for the wild-type strain and chlorina 2, which contains no chlorophyll *b*, are shown in Figure 16-6. The same number of counts were loaded on each lane.

a. The intensity of the LHC-II band in Figure 16-6 increases with time in the wild-type strain. Does this indicate that the amount of labeled LHC-II in the cells increased during the chase period?

b. What do the data in Figure 16-6 indicate about the nature of the LHC-I and LHC-II proteins?

c. Chlorina 11, which contains a small amount of chlorophyll *b*, exhibited a pulse-chase profile intermediate to those shown in Figure 16-6 for the wild-type strain and chlorina 2. Does this additional information support the hypothesis of Terao and Katoh that chlorophyll *b* stabilizes the LHC-I and LHC-II apoproteins?

Plasma-membrane, Secretory, and Lysosome Proteins: Biosynthesis and Sorting

PART A: Reviewing Basic Concepts

Fill in the blanks in statements 1–23 using the most appropriate terms from the following list:

amphipathic

ATP

bound ribosomes

capacitance

cis

conformation

cytoplasm

cytoplasmic

degradation

endoplasmic reticulum (ER)

free

Golgi

hormones

hydrophilic

lumenal

N-linked

NSF

O-linked

pH

phosphorylation

propeptide

regulated

resistance

smooth endoplasmic reticulum

tertiary

tonicity

trans

tunicamycin

unfolded

1. The primary source of energy for the transport of proteins through biological membranes is _____.

2. Phospholipids are _____ a term indicating that they have both hydrophilic and hydrophobic ends.

3. Most phospholipid biosynthesis occurs in the _____.

4. All newly synthesized phospholipids are located on the _____ side of the endoplasmic reticulum.

5. Addition of terminal sugars to lipids occurs in the _____ apparatus.

6. Ribosomes in the cytoplasm are functionally equivalent; however, because of their association with the endoplasmic reticulum, they can be divided by cell fractionation procedures into bound and _____ ribosomes.

7. The rough endoplasmic reticulum fraction isolated by centrifugation techniques contains _____.

8. Protein synthesis is initiated in the _____.

9. The role of the 70,000-Da heat-shock protein as studied in yeast protein synthesis is to keep the protein in a(n) _____ conformation.

10. Formation of disulfide bonds occurs in the ———, a membrane-bound compartment.

11. If sugar residues are linked to hydroxyl groups of proteins they are said to be ——————.

12. The —————— of the protein determines whether or not a particular N-linked oligosaccharide becomes "complex" or remains "high-mannose."

13. Fucose and galactose are incorporated into complex N-linked oligosaccharides in the —————— Golgi region.

14. An antibiotic useful for blocking N-linked oligosaccharide addition to proteins is ——————.

15. Glycosylation of fibronectin is important because it inhibits —————— of this extracellular matrix component.

16. One enzyme mediating the fusion of the membrane of a transport vesicle to that of an "acceptor" Golgi vesicle is called ——————.

17. Targeting of hydrolytic enzymes to lysosomes results from the —————— of mannose residues.

18. The mannose 6-phosphate receptor and its ligand can be dissociated by lowering the —————— of the environment.

19. Sorting of precursors of vacuolar enzymes in yeast cells is dependent on the ——————.

20. Insulin secretion is not constitutive, but rather is ——————.

21. Both calcium and various —————— are known to trigger the release of regulated, secretory vesicles.

22. A vesicle whose release is under —————— control generally is electron dense (dark staining).

23. The key technique that has been used to determine if osmotic swelling of vesicles occurs before or after vesicular release is to measure the —————— of the plasma membrane.

PART B: *Linking Concepts and Facts*

Circle the letters corresponding to the most appropriate terms/phrases that complete or answer items 24–37; more than one of the choices provided may be correct in some cases.

24. Membranes expand by

   a. de novo synthesis.

   b. expansion of existing membranes.

   c. exocytosis of membranes actively transported from outside the cell.

   d. breakdown of endocytotic vesicles.

   e. still unknown mechanisms.

25. Newly synthesized phospholipids can be tracked by incorporation of

   a. [^{3}H]thymidine.

   b. [^{35}S]methionine.

   c. [^{14}C]methane.

   d. [^{32}P]phosphate.

   e. [^{3}H]leucine.

26. Which of the following statements are true?

   a. All organelles split via fission.

   b. Membranes grow by expansion of pre-existing membranes.

   c. Endoplasmic reticulum grows by expansion of existing organelles.

   d. Plasma membranes are permeable to most proteins.

   e. The transport of proteins across membranes requires expenditure of energy.

27. Once a protein leaves the endoplasmic reticulum and reaches the Golgi complex, which part of the complex does it enter first?

   a. trans

   b. CURL

   c. cis

   d. medial

   e. trans Golgi reticulum (TGR)

28. Regulated secretion is often controlled by the intracellular concentration of

   a. sodium.

   b. calcium.

   c. manganese.

d. chloride.

e. zinc.

29. During protein maturation in most cells, the sequence of organelles through which proteins move is

    a. endoplasmic reticulum (ER) → Golgi vesicles → trans Golgi reticulum.

    b. trans Golgi reticulum → ER → Golgi vesicles.

    c. Golgi vesicles → trans Golgi reticulum → ER.

    d. Golgi vesicles → ER → trans Golgi reticulum.

    e. not known yet.

30. Which of the following statements is true about translocation of *most* nascent proteins across the endoplasmic reticulum membrane?

    a. Transported proteins are complete with disulfide bonds.

    b. Transported proteins are glycosylated.

    c. Transported proteins are in an unfolded form.

    d. Transport depends on the presence of ATP.

    e. The N-terminal end of the protein enters and leaves the membrane first.

31. The ability of 0.5 *M* NaCl to strip many proteins from the microsomes of a preparation of rough ER was critical in the discovery of which of the following?

    a. mannose 6-phosphate receptor (MPR)

    b. signal peptide

    c. signal peptidase

    d. mRNA

    e. signal recognition particle

32. Common modifications of proteins in the lumen of the endoplasmic reticulum include

    a. glycosylation.

    b. formation of disulfide bonds.

    c. conformational folding and formation of quaternary structure.

    d. proteolytic cleavage.

    e. condensation.

33. The defect in α_1-antiprotease results from a mutation in

    a. mannose glycosylation.

    b. its primary sequence of amino acids.

    c. phosphorylation.

    d. fucose glycosylation.

    e. improper signal sequences.

34. Misfolded proteins can

    a. be secreted but not retained.

    b. cause protein synthesis to stop.

    c. be retained by the endoplasmic reticulum.

    d. be digested by the endoplasmic reticulum.

    e. be digested by the trans Golgi complex.

35. Oligosaccharides *N*-linked to proteins are

    a. characterized by the presence of *N*-acetylglucosamine.

    b. generally shorter than *O*-linked oligosaccharides.

    c. covalently linked to threonine.

    d. covalently linked to serine.

    e. characterized by the presence of *N*-acetylgalactosamine.

36. Lysosomal enzymes are activated by which of the following events either in the sorting vesicle of the trans Golgi reticulum (TGR) or in the lysosomes?

    a. addition of mannose

    b. addition of fucose

    c. decrease in pH

    d. proteolytic cleavage of the proprotein to the mature enzyme

    e. addition of phosphate to mannose

37. Prosequences are known to be located in which part of proteins?

    a. C-terminus

    b. N-terminus

    c. internally

    d. where a mannose is attached

    e. not known yet

PART C: *Putting Concepts to Work*

38. What technique is used to separate rough microsomes (i.e., bound ribosomes) from free ribosomes (polysomes)? Why was this separation important to our understanding of protein synthesis?

39. What technique has been the most useful in analyzing the transit of a particular protein through the Golgi complex from the cis to the trans sides?

40. Ultrastructural autoradiographic studies with thin slices of pancreas in culture have demonstrated the movement of secretory proteins from their site of synthesis in the rough ER to mature secretory vesicles. In this type of experiment, why is [^{3}H]leucine commonly used rather than other labeled amino acids?

41. Why are the zymogen granules present in pancreatic cells more electron dense than the Golgi complex, even though the protein within these granules originated from the Golgi before its association with the zymogen granules?

42. How have chimeric proteins been useful in elucidating the role of signal peptides in the translocation of secretory proteins across the ER membrane, and what have such chimeric proteins indicated about translocation?

43. How can proteases be used to determine the topology of proteins in cell membranes?

44. Justify the argument that a signal sequence can also be a topogenic sequence.

45. Why are mammalian cells, rather than bacteria, the preferred hosts for insertion of mammalian cDNA in the production of secreted proteins of commercial value?

46. Why are proteins with Lys-Asp-Glu-Leu at the C-terminus rarely secreted from cells?

47. The Golgi complex has been described as a "processing plant" or an "assembly line." Why are these terms appropriate?

48. How is glycosylation important to the stability of the low-density lipoprotein (LDL) receptor?

49. What internal safeguard prevents insulin stored in pancreatic cells from interacting with insulin receptors, which may be associated with the vesicular membrane compartmentalizing this hormone?

PART D: *Developing Problem-solving Skills*

50. A new approach for studying phospholipid turnover in cells is to add fluorescently tagged phospholipids to the plasma membrane and then to determine their fate in living cells. What instrument(s) is required in this approach and what is the most likely major problem with it? Where would you expect such tagged phospholipids to be localized in cells?

51. Phospholipid synthesis in the smooth endoplasmic reticulum can be easily tracked with radioactive tracers. Design an experiment to determine the time it takes a newly synthesized phospholipid to be incorporated into the plasma membrane of cells after its synthesis in the endoplasmic reticulum.

52. There are several methods for determining if a particular protein is synthesized as a cytoplasmic protein or within a vesicle. These techniques include ultrastructural autoradiography and immunocytochemistry as well as differential centrifugation. For example, the incorporation and movement of [^{3}H]leucine in pancreatic cells can be tracked by ultrastructural autoradiography. In such studies, the cells are exposed either to a continuous "pulse" of labeled leucine or to a short "pulse" of labeled leucine followed by a "chase" of excess unlabeled leucine. What information can be obtained with the pulse-chase approach that cannot be obtained with the continuous pulse method?

53. Although secretory proteins are synthesized in the lumen of the endoplasmic reticulum, you cannot detect a particular secretory protein in a non-protease-treated, detergent-lysed preparation of rough microsomes (i.e., bound ribosomes) using an enzymatic assay. However, you can detect the protein with immunodetection techniques such as Western blotting and immunoprecipitation. Assuming no methodological problems, discuss the discrepant results with the two types of assay.

54. In some temperature-sensitive (ts) yeast mutants, all

protein secretion is blocked at high temperatures but not at low temperatures. How could such ts mutants be useful in elucidating the protein maturation pathway?

55. You have obtained a protein-synthesizing preparation by high-speed centrifugation of lysed cells. After incubating this cytoplasmic supernatant with a single mRNA and radioactive amino acids, you isolate the labeled protein fraction and analyze it by gel filtration, which separates proteins based on their size. As shown in Figure 17-1, two peaks are found.

    a. Explain the presence of two labeled protein peaks following incubation in this reaction mixture, which contained only one mRNA.

    b. How could monoclonal antibodies be used to determine whether the two proteins revealed in the gel filtration are related to each other?

56. The signal recognition particle (SRP) is involved in the three-way function of initiation, terminating, and reinitiating synthesis of secretory proteins. Describe an experiment in which these functions of SRP have been demonstrated.

57. A cDNA for a membrane protein of industrial interest is cloned into a neuronal cell line grown in culture on microcarrier beads. After immunoprecipitation, the

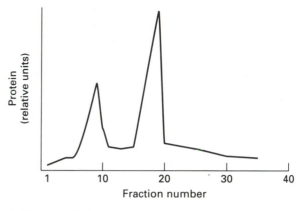

▲ **Figure 17-1**

protein is detected in significant quantities in the particulate fraction of the cytoplasm from cells into which the cDNA has been introduced. This approach would be of more potential value if the protein were released into the culture medium where it could be collected. How might this objective be achieved by manipulation of the cDNA?

58. How can ultrastructural autoradiography be useful in determining the cellular sites where glycosylation occurs during protein synthesis and processing?

PART E: *Working with Research Data*

59. The signal recognition particle (SRP) consists of six polypeptide subunits (9, 14, 19, 54, 68, and 72 kDa), which are organized into three different functional entities surrounding a 7S-RNA molecule. After protein synthesis is initiated in the cytoplasm, the 54-kDa subunit of the SRP binds to the signal sequence shortly after it is synthesized on the ribosome. This SRP-ribosome unit then associates with the SRP receptor (docking protein) on the endoplasmic reticulum where synthesis continues and the newly synthesized protein is translocated across the endoplasmic reticulum.

    In experiments designed to sort out the factors necessary to promote translocation, a complete cell-free translational system was incubated with a mRNA encoding a 50-kDa secretory protein, [^{35}S]methionine to monitor protein synthesis, and various preparations containing different factors. The seven preparations assayed contained the following components:

Preparation number	Components
1	SRP-ribosome complex + microsomes + ATP + GTP
2	SRP-ribosome complex + microsomes + ATP + GTP + 5mM EDTA
3	SRP-ribosome complex + microsomes + AMP-PNP + GTP
4	SRP-ribosome complex + microsomes + ATP + GMP-PNP
5	SRP-ribosome complex + microsomes + ATP + GTP + more time for incubation
6	SRP-ribosome complex + microsomes
7	SRP-ribosome complex + microsomes + ATP

GMP-PNP and AMP-PNP are nonhydrolyzable analogs of GTP and ATP, respectively. After each of the above preparations was incubated with the translational system in appropriate buffers, the sample was incubated with a protease (proteinase K). Then all proteins were precipitated, denatured, and separated on an SDS gel. Autoradiography of the gel revealed the pattern shown in Figure 17-2. Each lane of the autoradiogram is labeled with the corresponding preparation number.

a. In Figure 17-2, what is the significance of the discrete bands in lanes 1, 4, and 5 and of the diffuse bands in the other lanes.

b. What can you conclude from this experiment about the factors required for protein translocation and the mechanisms involved in this process?

60. As indicated in problem 59, GTP is an important co-requisite factor for translocation of proteins across the ER membrane. Connolly and Gilmore have tried to identify the subunit of the SRP receptor to which GTP binds using the technique of photoaffinity labeling. A photoaffinity label contains a group that binds to a receptor in the normal way via weak bonds; however, when the ligand-receptor complex is irradiated with light of the proper wavelength (e.g., 254 nm), a covalent linkage forms between the receptor and the photoaffinity ligand.

In order to determine whether either of the two known subunits of the SRP receptor (the 68-kDa subunit and the 30-kDa β subunit) binds ATP or GTP, you incubate the isolated SRP receptor with photoaffinity analogs of $[^{32}P]$ATP or $[^{32}P]$GTP in the presence and absence of unlabeled GTP and ATP. After incubation, the samples are irradiated, denatured, and then subjected to gel electrophoresis and autoradiography. Figure 17-3 shows which photoaffinity label and which competitor are present in each sample and the corresponding autoradiograms.

a. What can you conclude from this experiment about the binding of GTP and ATP to the SRP receptor?

	Lane number							
---	1	2	3	4	5	6	7	8
Photoaffinity analog added (5 µM):	GTP	GTP	GTP	GTP	GTP	ATP	ATP	ATP
Competitor added (µM):	None	GTP (20)	GTP (50)	ATP (20)	ATP (50)	None	GTP (50)	ATP (50)
α subunit (68 kDa)	▬			▬		▬		
β subunit (30 kDa)	▬	▬		▬				

▲ **Figure 17-3**

b. In response to a manuscript reporting the data in Figure 17-3, a reviewer questions whether GTP binds to each of the subunits separately or whether the photoaffinity analog appears to label both subunits because of their molecular proximity to each other. Another possibility posed by the reviewer is that the two subunits share a single GTP-binding site. Devise an experiment to distinguish these possibilities, so as to allay the reviewer's concerns.

61. A specific secretory protein (SP), which can be quantified by immunoprecipitation, has been isolated. This protein, which has enzymatic activity, is a tetramer with four different subunits. A sensitive enzyme assay has been developed that can detect nanomolar concentrations of SP when it is constitutively exocytosed from cells. The formation of this oligomeric protein during synthesis and modification has been tracked using immunoprecipitation, SDS and native gel electrophoresis, and ultrastructural immunocytochemistry.

When microsomes of endoplasmic reticulum are isolated, lysed, and the protein contents separated on a SDS gel, the four subunits of SP (180, 150, 50, and 20 kDa) are revealed by immunoprecipitation with a monoclonal antibody specific for SP, as depicted in Figure 17-4 (lane 1). When the microsomal protein is first treated with glutaraldehyde and then run on a gel, only one band corresponding to a molecular weight of 400 kDa is present (lane 2). Surprisingly, when this aldehyde-treated protein is analyzed in the enzymatic assay system, it has 80 percent of the full enzymatic activity of SP. When the non-glutaraldehyde-treated microsomal protein is treated with papain, a mild protease, and then run on a gel, six different bands appear (lane 3). When native secreted tetrameric SP is isolated and treated with papain, the same pattern is obtained (lane 4).

a. Based on these experiments, what can be inferred about the conformational organization of SP as it relates to the protein maturation pathway?

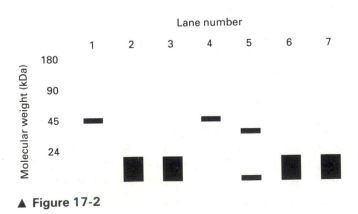

▲ **Figure 17-2**

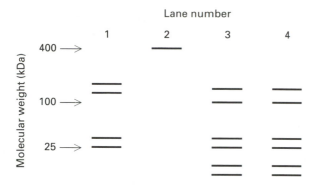

Lane number

Lane 1 = microsomal protein
Lane 2 = microsomal protein treated with glutaraldehyde
Lane 3 = microsomal protein treated with papain
Lane 4 = isolated SP treated with papain

▲ **Figure 17-4**

b. When a monoclonal antibody against SP is used in conjunction with a secondary antibody covalently linked to a 5-nm gold particle, ultrastructural immunocytochemistry demonstrates specific binding of SP in the endoplasmic reticulum, but *not* in Golgi vesicles. Assuming no procedural problems, explain this apparent discrepancy.

62. Several mutant cell lines that accumulate glycolipids and proteins in their lysosomes have been identified. Such lysosome-deficient cells, which lack various hydrolytic enzymes in the lysosomes, can be distinguished from their wild-type (normal) counterparts by phase microscopy because they are larger and have swollen lysosomes.

    When two different types of lysosome-deficient cells (B and C) are incubated in a common culture with normal cells (A), the morphology of the B cells changes from the diseased phenotype to one resembling that of normal cells, whereas the morphology of C cells does not change. Most surprisingly, a chromatographic/gel filtration analysis of selected lysosomal enzymes in A, B, and C cells shows that those from A and C cells comigrate and are immunologically indistinguishable, whereas those from B cells are distinctly different in their separation pattern and antigenicity behavior. Suggest possible reasons why B cells are "corrected" in the presence of A cells, whereas C cells still demonstrate a diseased phenotype, even though they contain wild-type hydrolytic enzymes.

63. Kai Simons and Gareth Griffiths have examined the relationship between the Golgi complex and the trans Golgi network (TGN), also known as the trans Golgi reticulum (TGR). One particularly interesting avenue of investigation originating with Simons and Karl Matlin has been to examine newly synthesized vesicular stoma-

titis virus (VSV) glycoproteins, which are synthesized in the endoplasmic reticulum, are processed in the Golgi complex, and accumulate in the TGN. After infecting baby hamster cells with VSV, they monitored the accumulation of the viral membrane G protein (VSV-G protein) using ultrastructural immunocytochemistry. At 39°C, sorting and release of the protein in the TGN occurs normally. When the temperature is lowered to 20°C, the cells produce ATP and endocytosis and exocytosis occur normally, but the VSV-G protein accumulates in the TGN. When released to a permissive temperature of 32°C, the VSV-G protein is again transported normally to the plasma membrane, and the accumulated VSV-G protein in the TGN disappears within 10 min of reversing the cold-induced block.

Simons and Griffiths observed that the TGN seemed to increase in size during the accumulation of the VSV-G protein at 20°C and to decrease rapidly when cells were returned to the permissive temperature of 32°C. Using mathematical approaches, they calculated the volume of both the Golgi complex and the TGN in VSV-infected hamster cells during different phases of this incubation procedure. Results similar to their values are shown in Table 17-1. These workers also calculated the surface area of the Golgi stack and TGN; values resembling theirs are shown in Table 17-2.

Table 17-1

Incubation conditions	Golgi volume/cell volume	TGN volume/cell volume
2 h at 20°C	0.58 ± 0.09	0.44 ± 0.87
2 h at 20°C, then 5 min at 32°C	0.76 ± 0.13	0.33 ± 0.12
2 h at 20°C, then 15 min at 32°C	1.30 ± 0.16	0.34 ± 0.12
2 h at 20°C, then 30 min at 32°C	1.79 ± 0.41	0.27 ± 0.32

Table 17-2

Incubation conditions	Surface area (μm^2)	
	Golgi stack	TGN
2 h at 20°C	580 ± 133	250 ± 82
2 h at 20°C, then 5 min at 32°C	780 ± 182	190 ± 62
2 h at 20°C, then 15 min at 32°C	1320 ± 361	200 ± 65
2 h at 20°C, then 30 min at 32°C	1820 ± 495	150 ± 49

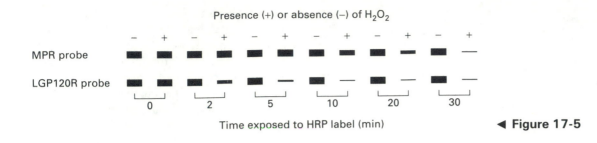

Presence (+) or absence (−) of H_2O_2

◄ Figure 17-5

a. Why is the volume of both the Golgi complex and TGN expressed as a ratio to the total cell volume in Table 17-1.

b. What do the values in Table 17-1 indicate about the relationship between the Golgi and TGN volumes and how can that relationship be expressed in a quantitative manner?

c. Based on the surface area values in Table 17-2, what is the relationship between the Golgi complex and TGN?

d. Based on your answer to part (c) as well as your knowledge of protein processing in animal cells, what would be a reasonable experiment to do next?

64. The core glycosylation of lysosomal proteins occurs in the endoplasmic reticulum, but the Golgi-specific addition of mannose 6-phosphate directs the mannose-containing proteins to lysosomes. The enzymes then sort into coated buds at the trans Golgi reticulum and are transported to the acidic lysosome vesicles, where they dissociate from the mannose 6-phosphate receptor (MPR). The receptor is recycled to the Golgi complex. Deficiencies in this "address tag" result in the secretion of lysosomal enzymes to the outside of the cell.

Hans Geuze and Ira Mellman have demonstrated that another class of lysosomal glycoproteins are synthesized and processed via the ER → Golgi → TGN route; these proteins are major constituents of the lysosome. Antibodies against the receptor for one of these proteins (LGP120R) and against the MPR are available. The LGR receptor is thought to go directly from the Golgi to lysosomes without incorporation into the plasma membrane, whereas the MPR recycles to the Golgi complex and is present in the plasma membrane as well.

In an unusual and elegant experiment to determine if both the MPR and LGP120R are present in the same endocytic vesicles, H_4S cells were incubated for various time periods with horseradish peroxidase (HRP), which coats the exterior of cells and can gain access to the cell interior during endocytosis of other proteins. After this treatment, the cells were incubated a second time in the presence (+) and absence (−) of hydrogen peroxide (H_2O_2). In the presence of hydrogen peroxide and HRP, the antigenicity of proteins is inhibited; in the absence of hydrogen peroxide, antigenicity is not compromised. After these incubations, the cells were lysed and the proteins subjected to Western blot analysis using antibodies against MPR and LGP120R as probes. The data are shown in Figure 17-5.

a. Do the Western blots using probes for MPR and MGP120R indicate that these receptors are located in or on endocytic vesicles in these cells?

b. Geuze and Mellman have suggested that four different types of vesicles exist: MPR+/LGP120R−, MPR+/LGP120R+, MPR−/LGP120R+, and MPR−/LGP120R− (the + and − superscripts denote the presence and absence, respectively, of the indicated receptor). Is the Western analysis shown in Figure 17-5 sufficient to demonstrate the four types of vesicles? If not, what other experiment(s) would be necessary to do so?

CHAPTER 18

Organelle Biosynthesis: The Nucleus, Chloroplast, and Mitochondrion

PART A: Reviewing Basic Concepts

Fill in the blanks in statements 1–27 using the most appropriate terms from the following list:

A	fission
aerobically	five
anaerobically	folding
B	fusion
C	gaps
channels	hybridization
charge	intermembrane space
chimeric	introns
condensed chromatin	kinase
conformation	maternal
core	matrix
cytoplasm	mitosis
decondensed chromatin	molecular weight(s)
diameter(s)	mosaic
electron microscopy	nucleus
endosymbiosis	outer mitochondrial membrane
exons	paternal

peroxisomes	signal sequence
phosphatase	tail
phytochrome	ten
pores	translocation of proteins
porin	
ribulose 1,5-bisphosphate carboxylase	unfolding
	uptake-targeting sequence(s)

1. Most proteins located in mitochondria, chloroplasts, and nuclei are synthesized on ribosomes located in the _____.

2. Most proteins are translocated to the nucleus via _____ in the nuclear envelope.

3. A protein that is created as the result of a rearrangement of exons and introns is called a _____ protein.

4. Cytochrome b_2 is targeted first to the matrix and then is directed into the _____.

5. Both histones and sucrose, which have similar _____ _____, can pass easily into the nucleus.

6. Nucleoplasmin is a pentameric protein containing _____ different polypeptides.

7. Most cytoplasmically synthesized proteins destined for the mitochondria have targeting sequences that are cleaved once translocation has occurred. However, the outer mitochondrial membrane protein _____, which acts as a channel through the bilayer of the outer membrane, has a targeting sequence that is preserved.

8. The enzyme _____ contains eight copies of the holoenzyme. Each copy has one large subunit, which is synthesized in the chlorplast, and one small subunit, which is coded for in the nucleus.

9. The _____ portion of nucleoplasmin specifies entry into nuclei.

10. If the hydrophobic domain of the mitochondrial protein porin is experimentally deleted, this protein will accumulate in the _____.

11. The fact that apocytochrome *c*, but not cytochrome *c*, can be taken up by the mitochondria illustrates the importance of the _____ of a protein to its translocation.

12. The nuclear envelope disappears during _____ in eukaryotes.

13. Type _____ lamin binds to lamin-depleted nuclei.

14. The "short-gene" mutant of yeast is defective and has fewer _____ than the "long-gene" mutant.

15. The 20–60 amino acids that are critical for targeting cytoplasmically synthesized proteins to the mitochondria are called _____.

16. Lamins _____ and _____ bind to chromatin.

17. Nucleoplasmin migration through nuclear pores can be followed experimentally by _____ of gold-conjugated nucleoplasmin.

18. Denaturation of mitochondrial matrix proteins facilitates their translocation because this treatment inhibits _____ of proteins.

19. Chloroplasts are inherited from the _____ parent.

20. Genetic exchanges of mitochondrial DNA with nuclear DNA can be examined using _____ techniques.

21. Phosphorylation of lamins is probably accomplished by a lamin _____.

22. The assembly of the nuclear envelope is initiated by _____, followed by the dephosphorylation of lamins.

23. When yeast cells are grown _____, the levels of cytoplasmic mRNAs coding for subunits of cytochrome *c* decrease.

24. Chloroplast DNA and bacterial DNA code for numerous homologous proteins including RNA polymerase and ribosomal proteins. This fact supports the theory of _____.

25. Recombinant-generated mutant SV40 T viruses have been useful for examining migration of proteins into the _____.

26. Daughter chloroplasts are created through the process of _____.

27. The points of contact of the inner and outer mitochondrial membrane are important for _____ to the matrix.

PART B: *Linking Concepts and Facts*

Circle the letters corresponding to the most appropriate terms/phrases that complete or answer items 28–39; more than one of the choices provided may be correct in some cases.

28. Proteins that are synthesized in the cytoplasm and are incorporated into the mitochondrial matrix and the chloroplast stroma need to cross at least one membrane. The translocation of such proteins is characterized by

   a. prefolding of the proteins into enzymatically active shapes before their incorporation.

   b. an energy requirement.

   c. interaction with a specific receptor.

   d. the involvement of Ca²⁺ ions at millimolar concentrations.

   e. fusion with "carrier vesicles," which in turn fuse to the organelle membrane.

29. The outer membrane of the nuclear envelope is directly continuous with

   a. the Golgi complex.

b. lysosomes.

c. the endoplasmic reticulum.

d. receptosomes.

e. peroxisomes.

30. The lamins are

    a. proteins.

    b. filaments.

    c. located in the nucleus.

    d. hydrolytic enzymes.

    e. regulators of nuclear shape.

31. Which of the following cellular processes do *not* occur when a cell is undergoing mitosis?

    a. proteolytic processing

    b. transport of proteins from the endoplasmic reticulum to the Golgi complex

    c. fusion of secretory vesicles with the plasma membrane

    d. ATP production

    e. endocytosis

32. Lamins repolymerize during

    a. prophase.   d. telophase.

    b. metaphase.   e. the G_2 phase.

    c. anaphase.

33. Which of the following nuclear proteins/complexes are synthesized in the nucleus?

    a. lamins   d. RNA polymerase

    b. histones   e. none of the above

    c. DNA polymerase

34. Which of the following forms of nucleoplasmin can gain access to the nucleus?

    a. the "core" of nucleoplasmin

    b. the "tail" of nucleoplasmin

    c. the entire pentamer (i.e., native nucleoplasmin)

    d. nucleoplasmin plus a cytoplasmic carrier

35. Most mtDNAs code for which of the following proteins or subunits of proteins?

    a. nucleoplasmin   d. ATP synthase (F_0F_1)

    b. cytochrome *c*   e. cytochrome *b*

    c. NADH-CoQ

36. Cytoplasmically synthesized mitochondrial proteins are imported into which of the following mitochondrial domains?

    a. outer membrane   c. inner membrane

    b. intermembrane   d. matrix space

    e. none (i.e., all mitochondrial proteins are made in the mitochondria)

37. The signal that directs most cytoplasmically synthesized mitochondrial proteins is usually found in the

    a. carboxyl end of the protein.

    b. middle of the protein.

    c. amino end of the protein.

    d. associated mannose 6-phosphate.

    e. associated mannose.

38. Matrix-targeting sequences of cytoplasmically synthesized mitochondrial proteins

    a. are not hydrolyzed in the matrix.

    b. bind to receptors in the outer mitochondrial membrane.

    c. are rich in arginine and lysine.

    d. are rich in hydroxylated amino acids such as serine and threonine.

    e. are clipped just prior to binding with receptors on the outer mitochondrial membrane.

39. Which of the following organelles are derived from proplastids?

    a. chloroplasts   d. amyloplasts

    b. chromoplasts   e. peroxoplasts

    c. elaioplasts

PART C: *Putting Concepts to Work*

40. What is the temporal relationship between the phosphorylation of lamins and nuclear organization?

41. Are lamins critical for the condensation of chromatin during cell division?

42. What is the embryological significance of the observation that an unfertilized frog egg contains enough histones and lamins for 20,000 cells?

43. Explain why nuclear reassembly is not blocked when lamin antibodies are introduced into the cytoplasm.

44. Describe the "symbiotic relationship" between the core and tail portions of nucleoplasmin in the nuclear targeting and retention of this protein.

45. Mild protease treatment of isolated nuclei results in reduced uptake of nucleoplasmin. What does this observation suggest about the relationship of nucleoplasmin with the nuclear envelope?

46. Even though some yeast petite mutants are missing most of their mitochondrial DNA (mtDNA), they still contain mitochondria. What does this observation imply about mitochondrial biogenesis?

47. Why was the discovery of genes encoding the large subunit of ribulose 1,5-bisphosphate carboxylase (RDPase) in plant mitochondria unexpected and what is the significance of this discovery?

48. The soluble unfolding proteins that are bound to mitochondrial proteins have also been called "chaperone" proteins. Explain why this is an appropriate name.

49. Why does proteinase K treatment of isolated mitochondrial and chloroplast vesicles inhibit the uptake of many cytoplasmically synthesized mitochondrial matrix and chloroplast stromal proteins?

50. In some experiments, an antibody specific for the C-terminal end of a cytoplasmically synthesized mitochondrial protein has been shown to prevent completion of translocation of the precursor protein into the mitochondrial matrix, although a portion of the precursor is translocated. Explain these observations.

51. What is the salient difference between the energy requirement for translocation of proteins to mitochondria and to chloroplasts?

PART D: *Developing Problem-solving Skills*

52. In an experiment investigating the roles of lamin in nuclear envelope structure, you add ^{125}I-labeled lamins A and B to lamin-stripped nuclear envelopes. You find that lamin A + lamin B bind well to the nucleus, but lamin A alone does not, as illustrated in Figure 18-1. Propose two explanations consistent with these results. What additional experiment(s) could distinguish between these explanations?

53. You wish to study lamins using 3T3 cells synchronized so that more than 90 percent are going through mitosis at the same time. With antibodies to lamins A, B, and C, you immunoprecipitate all lamins and characterize them using gel electrophoresis. You find that only lamins A and C are detected. Assuming that you in fact are using an active anti-B antibody in this experiment, why is

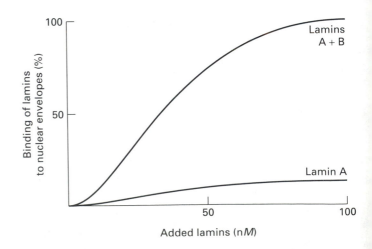

▲ **Figure 18-1**

lamin B not detected? How could you demonstrate that the inability to detect lamin B does not result from the lack of specificity of the presumed anti-B antibody you are using.

54. In an experiment to examine the role of phosphorylation and dephosphorylation in nuclear membrane assembly, you label cultured cells with inorganic phosphate tagged with ^{32}P but observe no phosphorylation of the appropriate lamins. However, in vitro polymerization experiments indicate that lamins can be phosphorylated. How could you change the cell-culture experiment to rectify the lack of labeling?

55. Microinjection of mammalian DNA into an unfertilized frog egg elicits the formation of an "artificial" nuclear envelope complete with lamins around the injected DNA. This observation suggests that decondensed DNA initiaes nuclear reorganization and also implies that there are excess lamins in the cytoplasm. What control experiments could you perform to demonstrate that it is not (a) the microinjection technique or (b) the presence of foreign biological material that elicits this response?

56. Design an experiment to determine whether small proteins (<9 nm in diameter) can pass into the nucleus in vitro. Next design an additional experiment(s) to determine if a soluble cytoplasmic factor(s) is necessary for the translocation of such small proteins to the nucleus.

57. In eukaryotes, cycloheximide inhibits protein synthesis in the cytoplasm, and chloramphenicol inhibits protein synthesis in mitochondria. In contrast, protein synthesis in bacteria is inhibited by chloramphenicol but not by cycloheximide. What do these observations suggest about the origin of mitochondria?

58. In an in vitro translation system, mRNAs encoded by both nuclear DNA and mtDNA can support protein synthesis. What does this finding imply about the relationship of both types of mRNA and the ribosomes in the cytoplasm?

59. After isolating a chloroplast protein that is synthesized in the cytoplasm and transported to the chloroplast, you radiolabel this protein with ^{125}I and add it to a buffered in vitro preparation of chloroplasts. After several minutes, you determine the fraction of protein that is translocated to the stroma of the chloroplast under various conditions. The results are shown in Figure 18-2. What can you tell from these data about the factor(s) required for this translocation event?

60. The translocation of most precursor proteins to mitochondria depends on the transmembrane electric po-

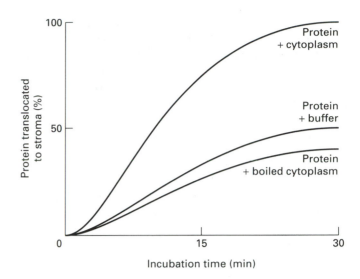

▲ **Figure 18-2**

tential. The ability of "unfolding proteins" to maintain precursor proteins in a partially unfolded state depends on ATP. Design an experiment that can distinguish the different energy inputs required for these two processes. How could you determine whether ATP is acting as an energy source or merely facilitates binding of precursor proteins to the outer mitochondrial membranes?

61. After isolating and characterizing three different matrix enzymes from mitochondria, you conclude that each must use a receptor on the outer membrane for translocation, as proteinase K pretreatment of mitochondria

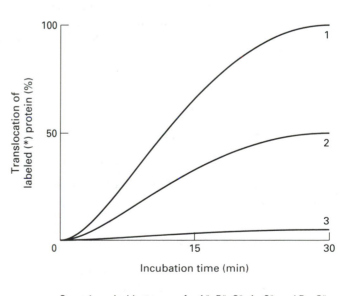

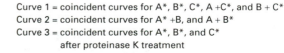

Curve 1 = coincident curves for A*, B*, C*, A +C*, and B + C*
Curve 2 = coincident curves for A* +B, and A + B*
Curve 3 = coincident curves for A*, B*, and C*
　　　　　after proteinase K treatment

▲ **Figure 18-3**

decreases translocation of all three proteins. The three proteins—A, B, and C—have similar targeting sequences. To determine if translocation of these proteins involves the same or different receptors, you tag all three with a radioactive label, incubate them separately and together with a second unlabeled protein in an in vitro translocation system, and measure the extent of translocation of the labeled protein. The resulting data are shown in Figure 18-3. In all cases, the proteins are present in excess amounts; labeled proteins are indicated by an asterisk (A*, B*, and C*). Based on these data, how many receptors probably are involved in translocating proteins A, B, and C?

62. Drugs like nigericin can artificially increase the electric potential across inner mitochondrial membranes relative to the matrix. Although nigericin treatment increases the rate and extent of translocation of matrix proteins, the rate of translocation reaches a plateau at a lower drug concentration than does the drug-induced change in membrane potential, as shown in Figure 18-4.

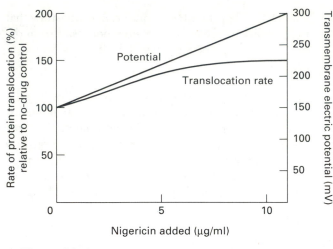

▲ **Figure 18-4**

Explain these data based on your knowledge of the mechanisms involved in protein translocation to the mitochondrial matrix.

PART E: *Working with Research Data*

63. Nucleoplasmin, a 165-kDa pentameric protein, is synthesized in the cytoplasm of frog oocytes and translocated to the nucleus through nuclear pores. The 33-kDa monomers can be carefully digested to 23-kDa cores and 16-kDa tails. When the native pentamer is digested with pepsin, it is possible to obtain five different types of nucleoplasmin differing in the number of tails that are attached to the cores. For instance, if C5T5 represents the native pentamer, C5T4 represents a pentamer that is missing one tail. By using this pepsin degradation procedure, the following species, in addition to native, undigested nucleoplasmin (C5T5), have been generated: C5T4, C5T3, C5T2, C5T1, and C5T0 (i.e., cores only with no tails). This notation is used in the following questions.

    a. When ^{125}I-labeled C5T5 and C5T0 were microinjected into the cytoplasm of an egg cell, only C5T5 was found in the nucleus. Do these data suggest that the full complement of five tails is necessary for translocation of nucleoplasmin?

    b. To further explore the question of whether the number of tails is proportional to the ability of nucleoplasmin to translocate, the six species listed above were radiolabeled, adjusted to equal molar concen-

trations, and injected simultaneously into the cytoplasm of oocytes using liposome microcarriers. After 30 min, the cells were fractionated into cytoplasmic and nuclear fractions, which were analyzed using SDS-PAGE and autoradiography. The results are shown in Figure 18-5. Do these gel profiles support the hypothesis that the efficiency of nucleoplasmin translocation is proportional to the number of tails?

    c. Design an experiment to determine if "protection" of nucleoplasmin cores from degradation in the nucleus is increased with a corresponding increase in the number of tails.

    d. Relatively recent ultrastructural immunocytochemical experiments have revealed the sites at which mitochondrial translocation occurs and have demonstrated that this process depends on ATP hydrolysis. In earlier experiments, the question of the energy dependency of translocation was unresolved. For example, when the effects of ATPase inhibitors on translocation were tested in vivo, the results were equivocal. When you perform similar whole-cell experiments you find that translocation of nucleoplasmin to oocyte nuclei is not inhibited by the addition of an ATPase inhibitor, but translocation of other

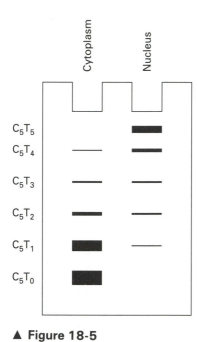

▲ **Figure 18-5**

The figure columns are labeled **Cytoplasm** and **Nucleus**, with rows from top to bottom: C_5T_5, C_5T_4, C_5T_3, C_5T_2, C_5T_1, C_5T_0.

Table 18-1

Carboxyl-terminal amino acid sequence	Localization in peroxisomes
—R-E-I-L-I-K-A-K-K-G-G-K-S-K-L	+ (native luciferase)
—R-E-I	—
—R-E-I-L-I-K-A-K-K-G-G-K	—
—R-E-I L	—
—R-E-I G-G-K-S-K-L	+
—R-E-I S-K-L	+
—R-E-I-L-I-K-A-K-K-G-G-K-S-K	—
—R-E-I K-L	—
—R-E-I-L-I-K-A-K-K-G-G-K-S	—
—R-E-I K-S-K-L	+
—R-E-I-L-I-K-A-K-K-G-G-K-S-K-L-S	—
—R-E-I-L-I-K-A-K-K-G-G-K-S-K-L-S-L	—

$$\uparrow$$
$$550$$

nuclear proteins in other types of cells grown in culture is completely stopped by an ATPase inhibitor. Explain the discrepancy in these data.

e. When oocytes are cooled to 4°C, translocation of nucleoplasmin is diminished. What does this observation suggest about the ATP dependency of translocation in vivo?

64. It is clear that pre-sequences that act as targeting domains are the rule in mitochondrial, chloroplast, and lysosomal biogenesis. Suresh Subramani and his co-workers in the Department of Biology and Center for Molecular Genetics in LaJolla, California, have been examining enzymes that sort to peroxisomes. This group introduced the gene for firefly luciferase into recipient cells. This protein of 550 amino acids normally sorts into peroxisomes, and the proper targeting of the enzyme can be examined using immunofluorescent microscopy. Cells in which proper synthesis and sorting of this enzyme occur have punctate spots that are highly fluorescent and correspond to peroxisomes.

To clarify which part of luciferase aids its translocation to peroxisomes, these researchers constructed a series of mutant luciferase genes encoding, carboxyl-end-defective enzymes. The sequence at the carboxyl-end of these proteins are listed in Table 18-1 in which each letter corresponds to a different amino acid (the code is unimportant for this problem). Also indicated in the table is whether punctate spots indicative of peroxisome localization are (+) or are not (−) observed in recipient cells containing the various genes.

a. Based on the data in Table 18-1, what is the probable targeting, or sorting, domain in luciferase?

b. What is the importance of the last two entries in Table 18-1?

c. What can you conclude from this experiment about how peroxisomal targeting of luciferase differs from targeting of proteins to mitochondria and chloroplasts?

d. Given the technique used in these experiments, how could the researchers determine that the site-directed mutants that did not localize luciferase in the peroxisomes failed to do so because of sorting problems rather than synthesis defects?

e. Design an experiment to demonstrate that only the sequence identified in part (a), and not any other domain(s), is necessary to target proteins to peroxisomes.

f. Other peroxisomal enzymes contain the same sorting domain as that identified in luciferase. In all cases, the targeting sequences are not cleaved. Why is this an unexpected finding?

65. Lamins play a major role in mitosis in most eukaryotic cells. In contrast to higher eukaryotic cells, the yeast *Saccharomyces cerevisiae* undergoes "closed" mitosis. That is, during mitosis, the parental yeast nucleus does not disassemble but rather stays intact and buds off; a separate daughter nucleus appears after elongation of the preexisting nucleus. Studies in Gunter Blobel's laboratory in Rockefeller University have indicated that turkey erythrocyte lamin B is associated with a 58-kDa integral membrane protein residing in nuclear envelopes. With antibodies available to this lamin B receptor as well as to lamins A and B, Blobel and colleagues have

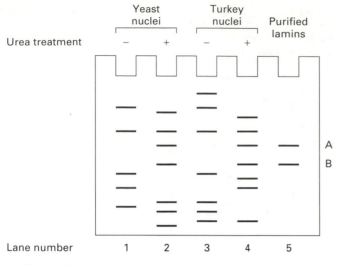

▲ **Figure 18-6**

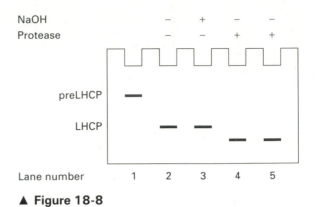

▲ **Figure 18-8**

tried to determine if lamins are present in yeast and if their biological roles in higher eukaryotes and yeast are similar.

In one series of experiments, yeast and turkey nuclear envelopes were incubated with antibodies to lamins A and B in the presence or absence of urea. After incubation, the mixtures were separated by SDS-PAGE electrophoresis and stained with Coomassie blue. The gel profiles shown in Figure 18-6 are similar to those obtained by Blobel's group.

In subsequent experiments, ^{125}I-labeled lamin B was incubated with urea-extracted turkey nuclear envelopes, urea-extracted yeast nuclear envelopes, or yeast plasma membranes, and the amount of lamin B binding to each membrane sample was measured. The results are shown in Figure 18-7.

a. What do the gel profiles in Figure 18-6 suggest about the presence of lamins A and B in yeast cells?

b. In the antibody experiments illustrated in Figure 18-6, why was it critical to include turkey nuclear envelopes (lanes 3 and 4) even though the presence of lamins A and B in turkey cells had been demonstrated previously?

c. What additional experiment would be necessary to show that lamins A and B are present in yeast cells?

d. What do the curves in Figure 18-7 suggest about the presence of a lamin receptor in yeast nuclei? What additional data is necessary to determine whether a specific lamin B receptor is present in yeast nuclei?

e. In Figure 18-7, why is there a slight increase in the binding of labeled lamin B to yeast plasma membranes at high lamin B concentrations? How could you determine if this binding is due to a low-affinity lamin B receptor?

66. Kenneth Cline at the University of Florida in Gainesville has done extensive work on the light-harvesting complex protein (LHCP) and its insertion into the thylakoid membranes of the chloroplast. Like many other chloroplast stromal and thylakoid proteins, LHCP is synthe-

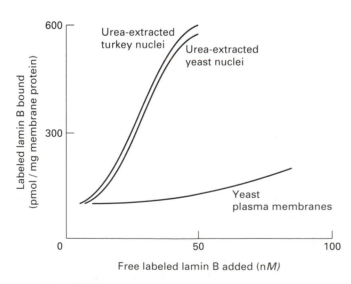

▲ **Figure 18-7**

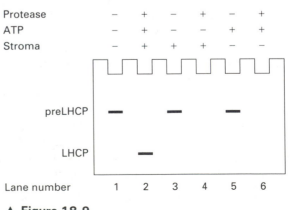

▲ **Figure 18-9**

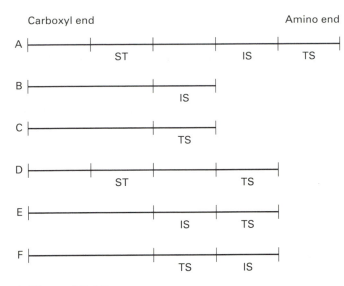

▲ **Figure 18-10**

sized as preLHCP in the cytosol and imported into the chloroplasts where it is subsequently processed to a mature form.

In an attempt to understand the forces governing translocation of preLHCP, Cline measured the in vitro translocation of [^{3}H]preLHCP to isolated chloroplasts. After an appropriate time of incubation, the chloroplasts were lysed and a fraction containing both thylakoid and envelope proteins was prepared and analyzed by SDS-PAGE autoradiography, as depicted in Figure 18-8, lane 2. Lane 1 is a control incubation of labeled preLHCP in the absence of chloroplasts. Next, the thylakoid membranes were treated with NaOH (lane 3) or a protease; preparations in the sample in lane 4 was treated with thermolysin, and the sample in lane 5 with trypsin.

a. What does the difference in the migration patterns in lanes 1 and 2 in Figure 18-8 indicate about preLHCP? Why is this a critical part of the experiment?

b. Why were NaOH and proteases used in this experiment?

c. In Figure 18-8, what does the slight decrease in molecular weights in lanes 4 and 5 compared with lanes 2 and 3 suggest about the location of LHCP?

d. In other experiments, purified thylakoid membranes were prepared on a sucrose gradient, and the effects of both ATP and stroma on binding of labeled preLHCP were analyzed by SDS-PAGE autoradiography. The resulting gel profiles are shown in Figure 18-9. What conclusions can be drawn from these data?

67. Proteins destined for the mitochondria are sorted into four different subcompartments: the outer mitochondrial membrane (OMM), the inner mitochondrial membrane (IMM), the intermembrane space (IMS), and the matrix (M). Precursor proteins destined for the matrix contain a matrix-targeting sequence (TS), which directs their translocation to the matrix. Research with Fe/S cytochrome bc_1 complex indicates that another sorting domain next to the TS signal is responsible for subsequent sorting from the matrix to the intermembrane space (IS). Additional sorting domains that act as stop-transfer (ST) sequences also have been discovered; these anchor mitochondrial proteins in place in a specified membrane.

In one study testing the ability of these three types of signals (TS, IS, and ST) to direct targeting and sorting of mitochondrial proteins, various hybrid proteins were constructed and labeled with [^{35}S] methionine. The 19-aa sequence from the vesicular stomatitis virus G (VSV-G) protein, which stops its transfer from the cytoplasm to the cisternal space during protein synthesis in the endoplasmic reticulum, was used as the ST sequence. The Fe/S cytochrome bc_1 IS and TS sequences were used as well. The general structures of the resulting hybrid proteins (A – E) are shown in Figure 18-10. Each hybrid was incubated with isolated beef heart mito-

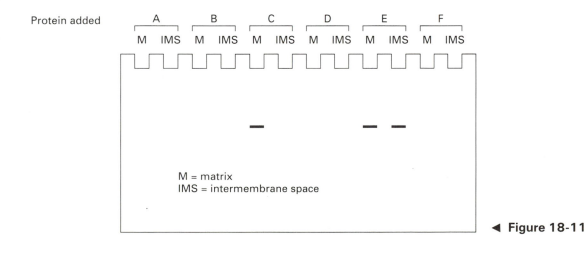

M = matrix
IMS = intermembrane space

◀ **Figure 18-11**

chondria, and the presence of the [^{35}S] methionine-labeled hybrid protein in the intermembrane space (IMS) or the matrix (M) was determined by SDS-PAGE autoradiography. The results are depicted in Figure 18-11.

a. Is the TS alone sufficient to direct matrix entry?

b. Can the IS alone direct sorting to the intermembrane space?

c. Is the VSV-G protein stop-transfer sequence recognized in mitochondria as it is in the endoplasmic reticulum?

d. Is the position of the TS critical for matrix targeting?

CHAPTER 19

Cell-to-Cell Signaling: Hormones and Receptors

PART A: Reviewing Basic Concepts

Fill in the blanks in statements 1–24 using the most appropriate terms from the following list:

affinity
chromatography

amino

autocrine

calcium

calmodulin

cAMP

carboxy

catechol

cholesterol

cloning

cytokinin(s)

dopamine

down regulation

endocrine

estrogen(s)

fluorescent

follicle-stimulating
hormone (FSH)

gibberellins

glucagon

glycolipids

higher

hydrophilic

oncoproteins

peroxisomes

insulin

integral

ion exchange
chromatography

leuteinizing hormone

lipophilic

lower

lysosome(s)

paracrine

peripheral

phosphate

practolol

protease(s)

receptor-mediated
endocytosis

terbutaline

tumor promoters

up regulation

1. The interaction of hormones with receptors on distant target cells is referred to as _____ signaling.

2. Neurotransmitter stimulation of adjacent neurons is referred to as _____ signaling.

3. Hormones can be classified as _____ or _____ based on their abilities to traverse a phospholipid bilayer.

4. All steroids are synthesized from _____.

5. Catecholamines and some peptide hormones are inactivated by extracellular _____.

6. The hormone that influences the growth of the ovarian follicle is _____.

7. The release of FSH from the anterior pituitary is regulated by _____.

8. The higher the K_M of a receptor, the _____ is affinity for ligand.

9. Most plasma membrane receptors are _____ membrane proteins and thus must be solubilized with detergents for full characterization.

10. The most useful protein separation technique for purifying the insulin and β-adrenergic receptors is _____.

11. The binding of insulin to a specific subunit of the insulin receptor was demonstrated by affinity chromatography using a chemical that cross-links _____ groups.

12. Receptors that are present in very low amounts (≤1000 molecules/per cell) are difficult to purify by conventional approaches. Such receptors (e.g., the erythropoetin receptor) can be obtained in purified form by _____ the corresponding cDNAs.

13. Studies with agonists and antagonists of epinephrine have indicated that the _____ group is responsible for the hormone-specific increase in cAMP.

14. _____ is a selective agonist of the β_2-adrenergic receptors on cells of the bronchioles.

15. Compounds that bind specifically to Ca^{2+} ions and are _____ have been used to quantify intracellular Ca^{2+} levels. Examples include fura-2 and quin-2.

16. Cell division can be elicited through the activation of protein kinase C. The compounds that can initiate this cycle are called _____.

17. Diabetes can result from a deficiency in the release of the hormone _____.

18. Excessive exposure to hormones, especially those that are released in a regulated fashion, can cause a change in receptor number called _____.

19. The process known as _____ involves receptor-specific membrane invagination and is responsible for the cellular uptake of ligands at concentrations many times greater than those in the extracellular fluid.

20. Many hormone receptors are engulfed by the cell and destroyed in _____.

21. The addition of a _____ group to receptors can decrease their affinity for their ligand.

22. The second messenger _____ is responsible for aggregation of the unicellar free-living ameba form of *Dictyostelium discodeum*.

23. The plant hormones that exert a primary effect on cell division are called _____.

24. _____ are plant hormones that induce germination of seeds.

PART B: *Linking Concepts and Facts*

Circle the letters corresponding to the most appropriate terms/phrases that complete or answer items 25–39; more than one of the choices provided may be correct in some cases.

25. Many transformed cells release growth factors that subsequently stimulate the same cells from which they were released. This type of signal pathway is termed

a. autocrine. d. paracrine.

b. endocrine. e. mesocrine.

c. exocrine.

26. The process of communication by extracellular signals does *not* include

a. proteolytic cleavage of a hormone immediately prior to interaction with its receptor.

b. detection of a hormone by its receptor.

c. glycosylation of a hormone initiated by binding to its receptor.

d. a change in the metabolism of the target cell.

e. removal of a hormone by its target cell or extracellular degradative enzyme.

27. Hormones whose receptors are located in the nucleus of the cell include

a. progesterone. d. insulin.

b. nerve growth factor. e. follicle-stimulating hormone.

c. thyroxine.

28. Which of the following effects are mediated by acetylcholine?

a. contraction of striated muscle

b. decrease in the rate of contraction of heart muscle

c. release in secretory granules

d. initiation of cell division by releasing from DNA synthesis

e. none of the above

29. Which one of the following plasma membrane–active hormones is the most structurally *dissimilar* from the others?

   a. insulin
   b. growth hormone
   c. epinephrine
   d. epidermal growth factor
   e. interleukin 2

30. Which of the following are *not* considered to be a second messenger?

   a. Ca^{2+}
   b. Na^+
   c. GMPPNP
   d. diacylglycerol
   e. inositol 1,4,5-trisphosphate

31. Hormones that demonstrate a positive feedback mechanism include

   a. estrogen.
   b. leutenizing hormone.
   c. progesterone.
   d. estrogen.
   e. follicle-stimulating hormone.

32. Cell-fusion experiments have demonstrated that

   a. adenylate cyclase can be linked to more than one receptor.
   b. adenylate cyclase can diffuse laterally through the plasma membrane.
   c. all G_s proteins are specific for certain hormone receptors.
   d. the G_i and G_s proteins are functionally different but biochemically indistinguishable.
   e. epinephrine cannot stimulate the production of cAMP.

33. Cholera toxin increases cAMP levels by

   a. inhibiting cAMP phosphodiesterase.
   b. binding to G_s protein.
   c. binding to G_i protein.
   d. binding to adenylate cyclase.
   e. binding to hormone.

34. The enzyme that inactivates cAMP is

   a. adenylate cyclase.
   b. cAMP-dependent protein kinase.
   c. cAMP phosphodiesterase.
   d. a phosphatase.
   e. cAMP-dependent phospholipase.

35. The responses to Ca^{2+} ions acting as a second messenger include

   a. contraction of skeletal muscle.
   b. activation of cAMP-dependent phosphodiesterase.
   c. breakdown of glycogen to glucose 1-phosphate.
   d. exocytosis of secretory vesicles.
   e. down regulation of LDL receptors.

36. Inositol 1,4,5-trisphosphate initially causes Ca^{2+} to be released into the cytoplasm from

   a. mitochondria.
   b. lysosomes.
   c. the endoplasmic reticulum.
   d. the plasma membrane (from extracellular to intracellular).
   e. Ca^{2+}-calmodulin complexes.

37. Protein kinase C is activated by

   a. magnesium ions.
   b. calcium ions.
   c. diacylglycerol.
   d. calmodulin.
   e. zinc and iron ions in combination.

38. Insulin's ability to enhance glucose transport is primarily due to

   a. phosphorylation of glucose.
   b. dephosphorylation of glucose.
   c. an increase in the number of transporters in the plasma membrane.
   d. a change in the affinity of the transporters for glucose.
   e. a decrease in the activity of the Na^+-K^+ pump.

39. Apical dominance in plants is governed by

   a. cytokinins.
   b. auxin.
   c. gibberellins.
   d. ethylene.
   e. abscisic acid.

PART C: *Putting Concepts to Work*

40. Explain why some ligands (hormones) can induce modifications in cells within seconds, whereas others may take several days to effect changes.

41. Why do few, if any, extracellular polypeptide hormones bind to nuclear sites?

42. Why are compounds like thyroxine and cholesterol usually bound to a protein as they travel through the vascular system?

43. Aspirin should not be taken by those who have blood-clotting problems. What is the basis for this advice?

44. Some hormones (e.g., thyroxine and those produced in the adrenal cortex) are stored in a precursor form and then processed to the mature form immediately before release. What is the physiological importance of this phenomenon?

45. In cells transfected with cDNA coding for the insulin receptor, the number of insulin receptors per cell can be much higher than in nontransfected cells. Nonetheless, the physiological response of such transfected cells to insulin often differs very little from that of normal non-transfected cells. Why is this the case?

46. Why is norepinephrine secreted by two different tissues, the adrenal gland and differentiated neurons?

47. What characteristic of the G_s protein has allowed researchers to isolate it by affinity chromatography?

48. Many different receptor types can activate the same adenylate cyclase. What is the most plausible physiological reason for this overlap?

49. What is the relationship between autophosphorylation of tyrosine and the action of insulin?

50. What would be the effect of isobutylmethylxanthine, a cAMP phosphodiesterase inhibitor, on the aggregation of the slime mold *D. discoideum*?

51. A mutant of *Arabidopsis* that is refractory to the effects of ethylene compared with its wild-type parents has been shown to have a defective gene. What does this observation suggest about the mode of action of ethylene?

PART D: *Developing Problem-Solving Skills*

52. Many transformed (i.e., cancerous) cell lines are known to be "detransformed" (i.e., they no longer exhibit their cancerous phenotype or behavior) if placed in culture with parental cells that do not demonstrate the transformed phenotype. For instance, when nontrasformed C3H10 cells, a T^1/2 cell line, are placed together with a transformed mutant offspring cell line, MC4AB10, the latter become normal in phenotype. Design an experiment to determine if this effect results from cell-to-cell contact between the two types of cells or from the release of a hormone-like factor by the nontransformed cells.

53. Binding of nerve growth factor (NGF) to its target cell, the sympathetic neuron, exhibits a bimodal curve, as shown in Figure 19-1, suggesting that the receptor has two K_M values. How can these data be interpreted?

54. In situ hybridization can be used to detect small levels of mRNA in cells. In this technique, a radioactively labeled cDNA is inserted into cells where it hybridizes to its corresponding mRNA. The labeled cDNA-mRNA hybrid can then be visualized autoradiographically in the light microscope.

    a. How could this technique be used to determine which cells in a very heterogenous tissue contain target cells for a particular hormone (X)?

    b. Describe the positive controls and negative controls that should be performed when using this technique? Why are both important?

55. In several cell types, acidic fibroblast growth factor (aFGF) has biochemical and morphological effects similar to those of nerve growth factor (NGF). In an attempt

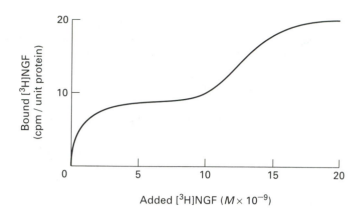

▲ **Figure 19-1**

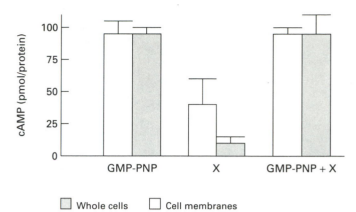

▲ **Figure 19-3**

to determine if NGF and aFGF act through the same receptor, you measure intracellular cAMP accumulation during stimulation of responsive cells with these factors. The amount of NGF and aFGF present in all cases is sufficient to saturate their receptor systems. Representative data are shown in Figure 19-2. What can you conclude from these data concerning the receptors utilized by NGF and aFGF?

56. GMPPNP (a nonhydrolyzable analog of GTP) has been extremely useful in investigating the role of the G_s protein in the adenylate cyclase system. An investigator is studying a new chemical toxin (X), which is thought to act similarly to GMPPNP and cholera toxin by binding to the G protein and stimulating adenylate cyclase. When toxin X is assayed with a preparation of cell membrane vesicles, it is half as effective as GMPPNP; when living cells are used in the assay, toxin X has almost no activity, whereas GMPNP has the same activity as in the membrane assay. Typical data are shown in Figure 19-3. Explain these results.

57. In a newly engineered cell line, the K_M for epinephrine binding is much lower than the K_M for epinephrine-stimulated cAMP accumulation, as illustrated in Figure 19-4. What do these data suggest about the mechanism of action of epinephrine in this new cell line?

58. Purified adenylated cyclase, G_s protein, and hormonal receptor can be reconstituted in liposomes (synthetic phospholipid membrane–bounded vesicles). After treatment of such liposomes with chemical cross-linkers, six different protein species can be identified on denaturing gels. What are the nature of these six species?

59. The dependence of cAMP accumulation on added GMPPNP, a nonhydrolyzable analog of GTP, in cell membrane preparations from differentiated and undifferentiated cells is shown in Figure 19-5. For undifferentiated cells, the kinetics are similar to a typical hormone-ligand binding curve. For differentiated cells, however, the kinetics are an unusual parabolic curve. Based on the nature of the regulating forces that can influence adenylate cyclase activity, how could these results be explained?

60. Cyclic AMP–dependent protein kinases can be separated into two types based on their order of elution from an ion exchange column. The two types — termed cAMP-PKI and cAMP-PKII — can be quantified by measuring their ability to phosphorylate an exogenous protein such as histone. When you chromatograph a

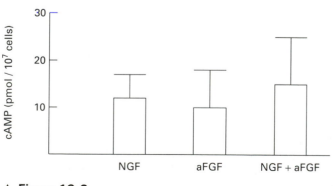

▲ **Figure 19-2**

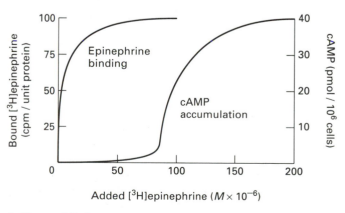

▲ **Figure 19-4**

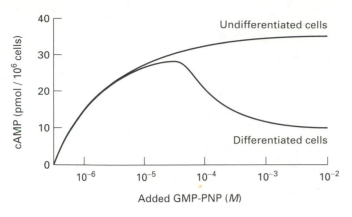

▲ **Figure 19-5**

kinase preparation from Purkinje cells, you find two peaks of catalytic activity, which you presume are these two enzymes. Using a ^{32}P-labeled photoaffinity analog of cAMP, 8-azido cAMP, which can covalently attach to the regulatory units of these two enzymes, design an experimental protocol to ensure that these two peaks of activity are indeed cAMP-PKI and cAMP-PKII.

61. What features of the Ca^{2+} concentration gradient across the cell membrane and of Ca^{2+} binding to calmodulin are particularly favorable for regulation of various cellular activities by Ca^{2+} and Ca^{2+}-calmodulin?

PART E: *Working with Research Data*

62. K. Chou's laboratory has obtained good evidence that the autophosphorylation of tyrosine amino acids in the cytoplasmic domain of the insulin receptor (IR) is essential for insulin action. In their research, Chou and his colleagues have used a variety of techniques including site-directed mutagenesis of the insulin receptor and introduction of antibodies to the IR kinase domain into whole cells.

In the case of site-directed mutagenesis, insulin genes were created in which a lysine codon in the active site of the IR kinase domain was converted to an alanine. Because no ATP could bind to the receptors coded for by the mutant gene, autophosphorylation of the receptor was prevented. When these mutant receptors were expressed in Chinese Hamster Ovary (CHO) cells, they were found to bind insulin normally. The transfected CHO cells had ~20,000 mutant receptors per cell. When these transfected cells were treated with insulin, no enhancement in insulin-stimulated parameters (e.g., glycogen synthesis, thymidine incorporation into DNA, and S6 kinase activation) was detected compared with wild-type controls.

Other investigators studying the necessity of autophosphorylation of tyrosine residues in insulin transduction produced antibodies against the IR cytoplasmic tyrosine kinase domain. These antibodies can gain access to the cytoplasmic surface of the plasma membrane when they are introduced to cells via liposomes. When transfected CHO cells expressing ~20,000 wild-type receptors per cell were treated with antibodies in this manner, insulin-stimulated thymidine incorporation was depressed below that of controls (nontransfected

cells expressing ~2000 wild-type receptors per cell). In contrast, thymidine incorporation by transfected CHO cells expressing the mutant receptors was approximately the same as control levels. These data are summarized in Figure 19-6. Assume that all cells in each sample expressed the receptor types indicated.

a. Assuming that the cell-surface receptors in all four samples have a similar affinity for insulin (i.e., similar K_M values), why is insulin-stimulated thymidine incorporation lower in the TW + AB sample than in

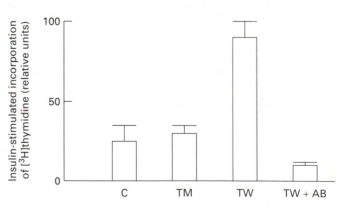

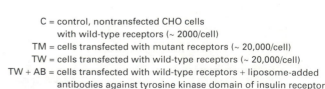

C = control, nontransfected CHO cells
with wild-type receptors (~ 2000/cell)
TM = cells transfected with mutant receptors (~ 20,000/cell)
TW = cells transfected with wild-type receptors (~ 20,000/cell)
TW + AB = cells transfected with wild-type receptors + liposome-added
antibodies against tyrosine kinase domain of insulin receptor

▲ **Figure 19-6**

the C sample but the activity of the TM sample the same as the control?

b. When the phosphorylated cytosolic proteins produced during incubation of the TW cells with insulin were analyzed by two-dimensional gel electrophoresis, only one phosphorylated protein was detected. The expectation was that several phosphorylated proteins would be present because insulin receptors mediate several insulin-specific responses. Since most polypeptide hormones cause an increase in the phosphorylation of several cytosolic and/or particulate proteins, does this observation indicate that the data in Figure 19-6 do not represent an enhanced insulin-specific stimulation of insulin receptors in sample TW?

63. Considerable effort has been expended on determining how different hormones can elicit the same physiological response in cells. It is now known that many receptors share the same pool of adenylate cyclase, thus integrating responses at the level of the plasma membrane. For example, in a celebrated study of this type, it was noted that both nerve growth factor (NGF) and stimulators of adenylate cyclase (dbcAMP) can induce differentiation of the transformed cell line PC12 into a cell resembling a sympathetic neuron. Other reports have indicated that NGF can stimulate the accumulation of cAMP, implying that NGF acts through the cAMP pathway. However, based on studies with mutant cell lines John Wagner of the Dana Farber Cancer Institute in Boston has concluded that the NGF and cAMP pathways are distinct from each other.

Christine Richter-Landsberg and Bernd Jastorff have used a new group of cAMP analogs, which can act as either an agonist or antagonist of the cAMP-dependent protein kinases. The agonist is (Sp)-cAMPS and the antagonist is (Rp)-cAMPS. To further examine the possible overlap of the cAMP and NGF pathways, these workers incubated PC12 cells with combinations of NGF and these analogs and then determined the percentage of cells with neurites, which are a hallmark of differentiated PC12 cells. Thus an increase in the percentage of cells with neurites is a direct, morphological indication of differentiation. The results of this experiment are presented in Figure 19-7.

a. What do the data in Figure 19-7 indicate about the relationship between the NGF and cAMP pathways in these cells?

b. In subsequent experiments, this group monitored the differentiation of PC12 cells in the presence of forskolin alone, a stimulator of adenylate cyclase, and in the presence of forskolin + saturating levels of NGF, as illustrated in Figure 19-8. How do these data affect your answer to part (a)?

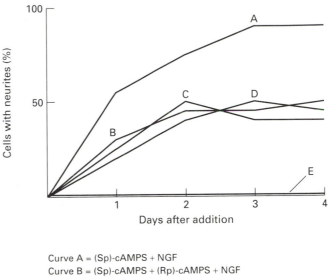

Curve A = (Sp)-cAMPS + NGF
Curve B = (Sp)-cAMPS + (Rp)-cAMPS + NGF
Curve C = (Rp)-cAMPS + NGF
Curve D = NGF
Curve E = no additions

▲ **Figure 19-7**

✱ 64. Many experimental approaches can be used to determine if two or more different second-messenger systems can regulate the same biosynthetic process. For example, specific agonists and antagonists can be used to facilitate dissection of transduction pathways in cells, as described in problem 63. However, highly specific antagonists to all hormones and second-messenger systems are not yet available and alternative approaches must be used in some cases. One such situation involves the ability of NGF and protein kinase C to activate ornithine decarboxylase (ODC) in PC12 cells. As noted in problem 63, NGF can induce differentiation of these cells. Ornithine decarboxylase (ODC) is a key regulatory enzyme in the production of polyamines. Although the latter are associated with cell differentiation, they may not be obligatory to the NGF-induced differentiation of PC12 cells. Nonetheless, ODC is a marker of differ-

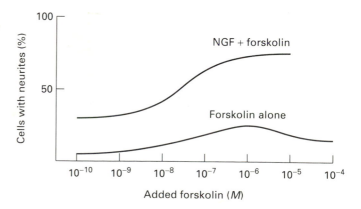

▲ **Figure 19-8**

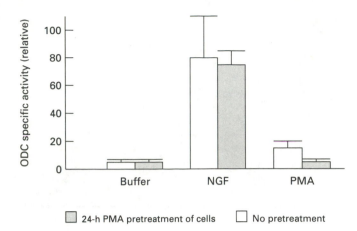

▲ **Figure 19-9**

entiation in this cell line and is induced by both NGF and protein kinase C.

a. In an attempt to sort out the pathways by which NGF and protein kinase C activate ODC, PC12 cells were pretreated with the tumor promoter PMA for 24 h to down-regulate protein kinase C. Next, aliquots of pretreated and untreated PC12 cells were incubated with buffer alone, NGF, or PMA and then the activity of ODC was measured. The results are shown in Figure 19-9. Do these data indicate that the activation of ODC by protein kinase C and NGF occur by the same or different pathways?

b. Lloyd Greene of Columbia University has created a mutant PC12 cell line that has no high-affinity NGF receptors. When presented with NGF, these cells do not increase their ODC activity compared with uninduced control levels. Suppose that samples of these mutant cells are incubated in the presence (+) and absence (−) of PMA and NGF, as indicated in Figure 19-10, and that the ODC activity is then determined. Based on your conclusion in part (a) and the partial results shown in Figure 19-10, predict the levels of

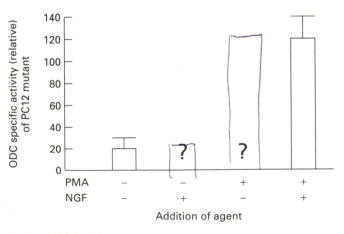

▲ **Figure 19-10**

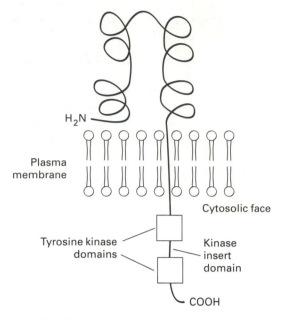

▲ **Figure 19-11**

ODC activity for the PMA⁻/NGF⁺ and PMA⁺/ NGF⁻ samples, which are not shown in the figure.

65. Platelet-derived growth factor (PDGF) is a growth-promoting substance, which is released from platelets when they adhere to the surface of an injured blood vessel. The effects of PDGF binding on fibroblasts and other cell types is extremely diverse and include an increase in DNA synthesis, changes in ion fluxes, alterations in cell shape, and changes in phospholipid metabolism.

As reviewed by Lewis Williams, the structure of the PDGF receptor is intriguing because it shares similarities with both the insulin receptor and immunoglobulin G. The PDGF receptor has intrinsic tyrosine kinase activity and can be down-regulated by endocytosis. Immunoglobulin G and the PDGF receptor have similar external domains, indicating that the PDGF receptor and IgG may be related to each other. A schematic diagram of the PDGF receptor and its relationship to the plasma membrane is presented in Figure 19-11.

The experimental approaches used to elucidate the transduction mechanisms of the PDGF receptor have been similar to those used to unravel the mysteries of the insulin receptor. For example, PDGF receptor mutants have been generated and assayed for various receptor-mediated functons. The mutant subtypes assayed and their functional abilities in the presence of PDGF are summarized in Table 19-1. Note that all the defects are in the cytoplasmic domain of the receptor.

a. Which of the mutants shown in Table 19-1 are deficient in tyrosine kinase activity?

b. Are all the activities listed in Table 19-1 regulated by autophosphorylation?

Table 19-1

Receptor type and defect	PDGF-induced functions*								
	Binding of PDGF	Autophosphorylation	Hydrolysis of phosphoinositol	Increased Ca²⁺ flux	pH change	Down regulation	PI kinase activity	Conformational change	Increased DNA synthesis
WT (wild type, no defect)	+	+	+	+	+	+	+	+	+
ATP (defective ATP-binding site in cytoplasmic domain)	+	−	−	−	−	+	−	−	−
CT (mutation in C-terminal region)	+	−	−			+	−	−	−
TM (substitution in transmembrane domain)	+	−	−	−	−	+	−	−	−
NOKI (deletion of kinase insert domain)	+	+	+	+	+	+	−	−	−

* A + indicates the presence and a − indicates the absence of the indicated functions in the presence of PDGF.

c. Is conformational change of the receptors protein necessary for autophosphorylation?

d. Which activities of the wild-type receptor suggest that PDGF might stimulate protein kinase C? How could this hypothesis be tested?

66. Peter van Haastert and colleagues have investigated the signal transduction mechanisms of *Dictyosteilium discoideum* using mutants defective in the aggregation response. A number of mutants that fail to aggregate have been constructed; one particular group is named "frigid" (fgd) mutants because they are nearly completely unresponsive to exogenous cAMP. Some frigid mutants do not make cAMP receptors, whereas others have the normal complement of receptors with wild-type binding characteristics for cAMP. It is this latter group of mutants that these workers have studied in detail to understand the mechanisms of the transduction systems. For example, they have measured the accumulation of both cAMP and cGMP in two frigid strains, HD85 and HC112, after exposure of cells to a cAMP agonist. The time-dependent accumulation profiles are shown in Figure 19-12.

a. Do the data in Figure 19-12 demonstrate that there are two receptors being stimulated by cAMP, one resulting in the production of cAMP and the other in

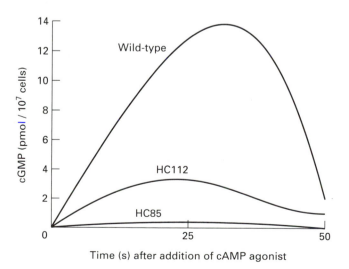

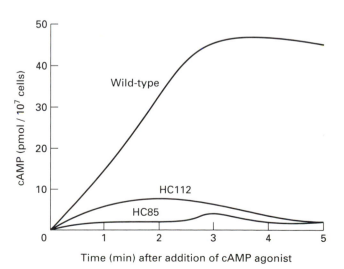

▲ **Figure 19-12**

the production of cGMP? If not, what additional experiment would have to be done to demonstrate this?

b. How could you determine through the use of mutants if the production of cGMP is a critical part of the aggregation process? Do the data in Figure 19-12 indicate that this is the case?

c. Two subclasses of cAMP receptors have been suggested: type A receptors linked to adenylate cyclase and type B receptors linked to guanylate cyclase. The A sites are rapidly dissociating ($t1/2 = 2$ s), whereas the B sites are slowly dissociating ($t1/2 = 15-150$ s). Could this explain the differences in the time courses of cGMP and cAMP accumulation shown in Figure 19-12.

d. When GTP was added to membranes isolated from strains HC85 and HC112, both GTP-dependent stimulation and inhibition of adenylate cyclase were similar for wild-type mutant strains. What do these observations indicate about the mutant locus in these frigid mutants?

CHAPTER 20

Nerve Cells and the Electric Properties of Cell Membranes

PART A: Reviewing Basic Concepts

Fill in the blanks in statements 1–26 using the most appropriate terms from the following list:

acetylcholinesterase	morphine
α-bungarotoxin	myelinated
axon	Nernst
Ca²⁺	nodes of Ranvier
conductivity	permeability
dendrite(s)	K⁺
depolarized	seconds (s)
desensitization	selective permeability
excitatory	slower
facilitator neuron(s)	slow postsynaptic
faster	Na⁺
ganglia	synaptophysin
habituation	tetrodotoxin
hyperpolarized	threshold
inhibitory	voltage-gated
interneuron(s)	voltage-gated Ca²⁺
ionic concentration gradients	channel(s)
	voltage-insensitive
milliseconds (ms)	

1. The portion of a neuron that conducts an electric signal away from the cell body is called the _____.

2. A _____ neuron, which is enveloped by membranes from a glial cell, can conduct an impulse faster than one that is not enveloped by such membranes.

3. A presynaptic neuron can elicit hyperpolarization in a postsynaptic neuron. This type of response is called _____.

4. An action potential is generated in a neuron when the electric potential decreases to a level called the _____ potential.

5. _____ connect one neuron with another neuron in neuronal circuits.

6. The cell bodies of motor neurons in vertebrates are clustered in _____ located immediately outside the spinal cord.

7. When the electric potential of a neuron shifts to a *less* negative state it is said to be _____.

8. When the electric potential of a neuron shifts to a *more* negative state it is said to be _____.

9. The action potential is approximately 1–2 _____ in duration.

10. The properties responsible for the resting potential in a neuron are _____ and _____.

11. The value of P_{Na} can be calculated from the _____ equation if the membrane potential is known.

12. The action potential is initiated by the opening of _____ Na$^+$ channels.

13. The refractory period of a neuron is caused by the inactive state of the _____ channel.

14. The passive spread of current along a nerve results from _____ of the plasma membrane to ions and the _____ of the cytosol.

15. A thicker axon will propagate current _____ than a thinner axon.

16. The largest concentration of voltage-gated Na$^+$ channels in a myelinated nerve is found at the _____.

17. Of the Na$^+$, K$^+$, and Ca^{2+} voltage-gated channels, the _____ channel probably arose first in evolution.

18. _____, an integral membrane protein in synaptic vesicles, may be involved in the uptake of neurotransmitters from the cytosol.

19. The snake venom _____ has been useful in locating and blocking acetylcholine receptors at the nerve-muscle synapse.

20. Prolonged exposure of the acetylcholine receptor to acetylcholine results in _____ of the receptor.

21. Many nerve gases are lethal because of their specific interaction with _____, the enzyme that degrades acetylcholine.

22. Responses that begin after a lag of several milliseconds are called _____ potentials.

23. Met-enkephalin and leu-enkaphalin have effects similar to the drug _____.

24. The behavioral opposite of sensitization is called _____.

25. Habituation in *Aplysia* can result from the decreased opening rate of _____.

26. Sensitization in *Aplysia* can be mediated through _____.

PART B: *Linking Concepts and Facts*

Circle the letters corresponding to the most appropriate terms/phrases that complete or answer items 27–42; more than one of the choices provided may be correct in some cases.

27. Much of the research on neurons has been conducted on invertebrates because

    a. their neurons have very long microvilli that make them easily identifiable in the dissecting microscope.

    b. their neurons generally are larger than mammalian neurons.

    c. they have simpler nervous systems than mammals.

    d. their neurons contain electric junctions, whereas mammalian neurons do not.

    e. genetic studies are easier to accomplish in invertebrates than in mammals.

28. The synthesis of most neuronal proteins occurs in the

    a. dendrites.   d. axon.

    b. cell body.   e. synapses.

    c. axon terminals.

29. The changes in the electric potential of a neuron that constitute the action potential occur in the following order:

    a. resting potential → depolarization → hyperpolarization → resting potential.

    b. depolarization → resting potential → hyperpolarization → resting potential.

    c. resting potential → hyperpolarization → depolarization → resting potential.

    d. resting potential → hyperpolarization → resting potential.

    e. resting potential → depolarization → resting potential.

30. The Na$^+$-K$^+$ ATPase

    a. maintains a higher concentration of K$^+$ ions outside the cell than inside the cell.

    b. maintains a higher concentration of Na$^+$ ions inside the cell than outside the cell.

    c. phosphorylates ADP to ATP.

d. maintains the resting potential of neurons.

e. is the voltage-sensitive channel that initiates an action potential.

31. The value of the resting potential E calculated from the Nernst equation is more negative than observed values of E because

a. there is a slow leak of K^+ ions.

b. there is a slow leak of Na^+ ions.

c. there is a slow leak of Cl^- ions.

d. there is a slow leak of large anions.

e. there are fewer K^+ channels in the membrane than the Nernst equation predicts.

32. The hyperpolarizaton of the resting potential immediately after the depolarization component of the action potential results from

a. activation of voltage-gated K^+ channels.

b. activation of the Na^+ leak current.

c. activation of the K^+ leak current.

d. activation of voltage-gated Na^+ channels.

e. all of the above.

33. Examples of demyelinating diseases are

a. the staggerer mutant.

b. the shiverer mutant.

c. multiple sclerosis.

d. cerebral palsy.

e. polio.

34. Plasma-membrane single-channel recordings can be accomplished using

a. extracellular electrodes.

b. intracellular microelectrodes.

c. liposome fusion technology.

d. lucifer yellow fluorescent dye.

e. voltage-clamp techniques.

35. Which of the following are criteria used to identify a substance as a neurotransmitter?

a. Introducton of the substance to the synaptic cleft must induce a response similar to that of the presynaptic neuron.

b. The substance must be removed or degraded rapidly.

c. Synaptic vesicles must contain the substance.

d. Synaptic vesicles must release the substance when the presynaptic neuron is stimulated.

e. Synaptic vesicles must release a sufficient quantity of the substance to elicit a response.

36. Miniature end-plate potentials induced by acetylcholine at the neuromuscular junction

a. are 1-mV hyperpolarizations of the postsynaptic membrane.

b. result from the spontaneous release of synaptic vesicles.

c. result from the evoked release of neurotransmitter.

d. are caused by the influx of Cl^- ions across the postsynaptic membrane.

e. are blocked by α-bungarotoxin.

37. Acetylcholine binds to which subunits of the acetylcholine receptor?

a. α subunits

b. β subunits

c. γ subunits

d. δ subunits

e. ϵ subunits

38. Mutation of which of the following amino acids in the transmembrane helices of the acetylcholine receptor reduces the rate of the Na^{2+} and K^+ movement through the receptor?

a. lysine

b. glutamate

c. aspartate

d. tyrosine

e. serine

39. The postsynaptic membrane can be returned to the depolarized (inactivated) state by

a. an increase in Ca^{2+} concentraton in the synaptic cleft, which directly inactivates the neurotransmitter.

b. binding of the protein inactivon, which irreversibly binds to the neurotransmitter.

c. uptake of neurotransmitter into the presynaptic processes.

d. degradation of neurotransmitter in the synaptic cleft.

e. diffusion of the neurotransmitter away from the synaptic cleft.

40. Which of the following neurotransmitters are synthesized from tyrosine?

a. acetylcholine

b. epinephrine

c. substance P

d. dopamine

e. norepinephrine

41. The inhibitory effects of GABA and glycine in vertebrate synapses result from the influx of

    a. Na^+ ions.
    b. K^+ ions.
    c. Cl^- ions.
    d. Ca^{2+} ions.
    e. Mg^{2+} ions.

42. The molecular mechanisms underlying the learning deficiencies found in certain *Drosophila* mutants (e.g., dunce, turnip, and rutabaga mutants) include

    a. defects in cAMP phosphodiesterase.

    b. increases in cAMP levels.

    c. decreases in protein kinase C.

    d. alterations in regulation of adenylate cyclase.

    e. changes in subunit composition of receptor molecules.

PART C: *Putting Concepts to Work*

43. Describe the major differences between chemical and electric synapses. Why are chemical synapses more common than electric synapses in most vertebrate and invertebrate brains even though impulse transmission across the latter is faster than across the former?

44. What is the major similarity and difference in the parasympathetic and sympathetic nervous systems?

45. Fibroblasts and most other non-neuronal cells exhibit an inside-negative electric potential. However, when they are depolarized, fibroblasts do not produce an action potential even though the concentrations of Na^+ and K^+ inside and outside fibroblasts are identical to those associated with neurons? Why do fibroblasts not generate an action potential?

46. P_K is much greater than P_{Na} at the resting potential. What does this imply about the transmembrane potential of cells in this state?

47. Describe three different changes in membrane properties that could result in the hyperpolarizaton of the resting potential.

48. Cyanide and carbon monoxide (CO) inhibit the electron transport chain of the mitochondria and can inhibit neuronal firing. The effect of these drugs on impulse transmission takes hours to occur, whereas the effect of Na^+ and K^+ channel blockers occurs within seconds. Why does it take so long for cyanide and CO to inhibit impulse transmission?

49. Why does the action potential travel in only one direction along an axon?

50. What technique has been used to harvest and purify the voltage-gated Na^+ channel from neurons, thus permitting subsequent cloning of this channel protein?

51. According to one current model, the nicotinic acetylcholine receptor molecule has five subunits ($\alpha_2\beta\gamma\delta$) arranged around a central channel, or pore, whose diameter is slightly less than 1 nm, which is much larger than the dimensions of Na^+ or K^+ ions. Why is the channel pore so large compared with the size of the ions that pass through it?

52. Discuss the differential effect of protein synthesis inhibitors on short-term and longer-term memory as demonstrated in *Aplysia*.

53. Explain why stimulation of the rod photoreceptors of the eye can be considered the "reverse" of a typical neuron.

54. What is the dual role of cGMP in rod photoreceptors and how is it involved in the adaptation of rod cells to varying levels of ambient light?

PART D: *Developing Problem-solving Skills*

55. The simple nervous system of the leech, an invertebrate, has been extensively studied, and many of its component circuits are well characterized. As illustrated in Figure 20-1a, one such circuit consists of three presynaptic neurons (A, B, and C) that synapse with a postsynaptic neuron. In situ analyses indicate that each presynaptic neuron contributes an excitatory component to the postsynaptic cell. Tracings of the changes in membrane potential following stimulation of the presynaptic cells are shown in Figure 20-1b; the changes were measured with intracellular microelectrodes located at the arrows in Figure 20-1a. When neuron A *or* B is inhibited using voltage-clamping techniques, the postsynaptic cell does not fire. However, if neuron C is inhibited, stimulation of neurons A and B is sufficient to produce an action potential in the postsynaptic cell. Based on the data in Figure 20-1b, suggest the most plausible reason for this unexpected finding.

56. The theoretical, expected change in the electric potential across the axonal membrane at the time a microelectrode penetrates an axon is depicted in Figure 20-2a. The changes in potential actually observed, however,

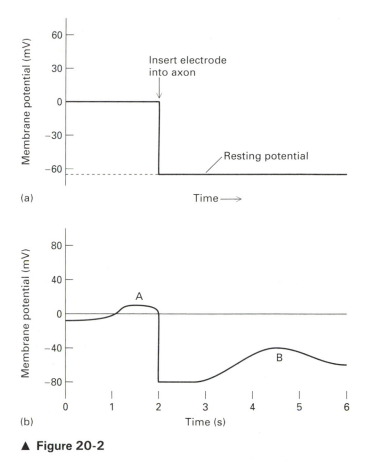

(a)

(b)

▲ **Figure 20-2**

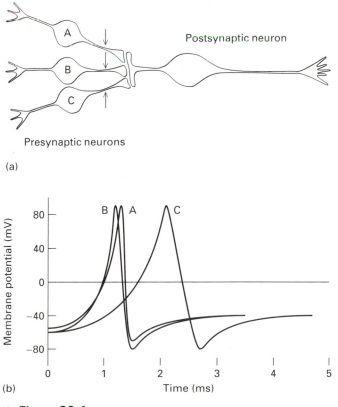

(a)

(b)

▲ **Figure 20-1**

often differ from the theoretical. For example, when the microelectrode touches the outside of the neuron there is a transient increase in the electric potential (point A in Figure 20-2b); this indicates that the tip of the microelectrode is on, but not in, the neuron in question. Once the microelectrode has entered the cell, there is often a slight time-dependent depolarization of the neuron (point B in Figure 20-2b). Assuming there are no problems with the electronics of the equipment and that no voltage changes have been applied to the cell in question, suggest at least two explanations for this slow depolarization.

57. During a summer internship at Woods Hole, you explore some of the aspects of the action potential using giant squid axons. When you impale an isolated neuron with a microelectrode, you observe two curious anomalies: (1) the resting potential is 20 mV more negative than what has been reported in the literature, and (2) no action potential is elicited when the cell is depolarized. You then examine the voltage-sensitive Na^+ channels using a patch-clamping technique and find that they are functional. What might be the cause of *both* the hyperpolarized resting potential and lack of an action potential in your first experiment?

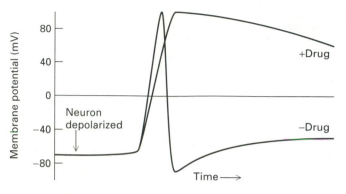

▲ **Figure 20-3**

58. When a neurotoxin isolated from a mud-dwelling fish from the Amazon River is placed in the solution bathing an isolated neuron, it affects the action potential of the neuron as shown in Figure 20-3. What is the probable mechanism of action of this drug on this neuron?

59. A certain neurotoxin depolarizes the resting potential and diminishes the acton potential of an identified neuron from *Heliosoma* as shown in Figure 20-4. How could you re-create the effect of this toxin simply by manipulating the ion concentrations surrounding the neuron in question?

60. *Drosophila* shaker mutants have a defective K^+ channel that causes delayed repolarization of the plasma membrane of axons. However, this type of delay also can result from other membrane-specific defects. Suggest one other possible defect that would contribute to delayed repolarization of neurons. Which techniques are most suitable for demonstrating such a defect?

61. Gap junctions mediate impulse transmission between neurons with electric synapses. When microelectrodes are filled with the fluorescent dye lucifer yellow and iontophoresed (injected using current) into gap junction–coupled neurons, the dye diffuses through the entire cell making it visible in the fluorescent micro-

scope. Usually the fluorescent dye will pass through a gap junction into the adjacent cell. How could this dye be useful in determining both the size and charge characteristics of gap junctions between neurons?

62. In an investigation of the neuronal circuitry of the leech, a commonly studied invertebrate due to its simple nervous system, intracellular microelectrodes were implanted in two nerve cells. An inhibitory neurotransmitter was added to the medium surrounding the cells and the membrane potential of each cell was recorded. The resulting tracings are depicted in Figure 20-5.

   a. Which of the two cells demonstrated an inhibitory response to this neurotransmitter?

   b. How could you determine if the two cells are synaptically connected?

   c. Explain why a single neurotransmitter can have an inhibitory effect on one cell and an excitatory effect on another.

63. The ability to distinguish electric and chemical synapses is critical in determining the function of a particular neural circuit. Describe two experimental approaches for differentiating the two types of synapses.

64. You notice an unusual defect in a mutant cholinergic neuronal cell line, which can be induced to differentiate in culture. When two adjacent mutant neurons form a synapse, they are deficient in their ability to transmit an action potential from the presynaptic neuron to the postsynaptic neuron, whereas wild-type cells do not exhibit a similar defect. How could you determine whether the mutant cells are defective in (a) the amount of neurotransmitter in presynaptic vesicles, (b) the ability of the vesicles to be released into the synaptic cleft, and/or (c) the responsiveness of the postsynaptic receptor to acetylcholine.

65. Cells in the adrenal medulla are embryologically related to neurons. When chromaffin cels from the adrenal

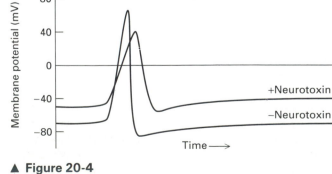

▲ **Figure 20-4**

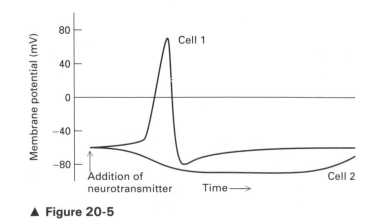

▲ **Figure 20-5**

medulla are removed and placed in cell culture, they have a typical rounded morphology. However, when presented with nerve growth factor, they differentiate into neuronal-like cells. What characteristics would have to be examined to determine if this cell is physiologically, as well as morphologically, a neuron?

66. The dopamine receptor has been implicated in the etiology of both schizophrenia and Parkinson's disease. Research in this area has indicated that subtypes of the dopamine receptor are present in the central nervous system. Binding studies demonstrate that one receptor subtype (A) binds nanomolar concentrations of dopamine, whereas another subtype (B) binds micromolar concentrations of dopamine. When dopamine is added to a vesicular preparation from the retina (a part of the brain), a dose-dependent accumulation of cAMP occurs as indicated in Figure 20-6. Based on these data, which dopamine receptor subtype most likely is present in the retinal preparation?

67. An unusual toxin causes toxin-infected individuals to become less sensitive to light; indeed, these victims are functionally blind. You isolate a pure population of rod photoreceptors from rabbits, a species with similar ret-

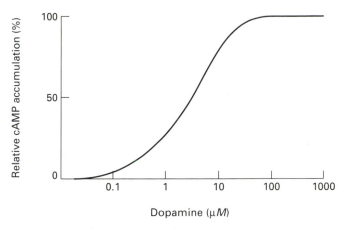

▲ **Figure 20-6**

inas, through sedimentation techniques for the purpose of determining the biochemical basis underlying the action of this toxin. When the toxin is added to a preparation of outer segments from rods, you notice the accumulation of 3′,5′-cGMP in the toxin-treated preparations relative to the untreated controls. What could be the basis for the toxin's ability to cause blindness?

PART E: *Working with Research Data*

68. One of the three glutamate receptor subtypes, the *N*-methyl-D-aspartate (NMDA) receptor, may be involved in the death of hippocampal cells in the central nervous system during episodes of cerebral ischemia (lack of O_2 to the brain). According to this hypothesis, the large increase in extracellular glutamate in the hippocampus that occurs during cerebral ischemia overstimulates the NMDA receptors, leading to a large influx of Ca^{2+} ions into hippocampal neurons and their subsequent death. This mechanism is thought to be the primary reason why hippocampal neurons are among the first to die during cerebral ischemia.

   In order to investigate NMDA-elicited cell death, Carl Cotman of the University of California has studied the effect of NMDA on the survival of embryonic hippocampal cells in vitro. In these studies, hippocampal neurons were isolated from 18-day-old rat embryos and plated in culture; the number of neurons surviving was determined periodically over a 2-week period using a trypan blue exclusion test. Among other things Cotman

analyzed the ability of MK801, a noncompetitive NMDA blocker, to inhibit NMDA-elicited cell death. Data representative of Cotman's results are shown in Figure 20-7.

   a. Which data presented in Figure 20-7 supports the hypothesis that activation of NMDA receptors by NMDA causes cell death?

   b. Did the embryonic hippocampal cells used in these experiments have NMDA receptors present throughout the 2-week study period?

   c. Do the data suggest that MK801 might affect the survival of hippocampal neurons by more than one mechanism?

   d. As shown in Figure 20-7, NMDA did not elicit death of hippocampal cells during the first 7 days of the study period. How could you distinguish between the two following explanations for this observation: (1) no NMDA receptors are present during this pe-

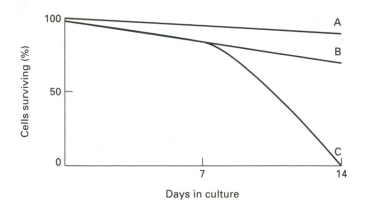

Curve A = MK801 added to culture
Curve B = MK801 and NMDA added to culture,
         coincides with control curve (no additions)
Curve C = NMDA added to culture

▲ **Figure 20-7**

riod and (2) receptors are present but they do not permit entry of Ca^{2+} ions into the cells, which subsequently leads to cell death.

e. From day 7 to day 14 of the study period, NMDA caused a substantial increase in cell death. How could you determine if this effect was caused, directly or indirectly, by the influx of Ca^{2+} ions from the extracellular medium into the hippocampal neurons?

f. Assume that Ca^{2+} is the major contributor to cell death in this system and that NMDA receptors are present at all times during the study period. When the Ca^{2+} ionophore A23187 is added at day 2 and day 10, the cells exhibit quite different sensitivities to this drug, which increases the membrane permeability to Ca^{2+} ions, as shown in Figure 20-8. Based on these data and the stated assumptions, how might the two-phase +NMDA curve in Figure 20-7 be explained?

g. Isolated hippocampal cells are incubated with [³H] glutamate for two different time periods, after culturing—day 7–14 and day 14–21; the cells are then prepared for quantitative autoradiography. When the fixed cells are examined, the same number of autoradiographic grains are found in the preparations from each culture time period. Are these results consistent with the data presented in Figures 20-7 and 20-8?

69. Ira Farber of Boston and his colleagues in Isreal have explored the importance of the Ca^{2+} channel in nerve impulse propagation. As discussed in problem 67, excessive intracellular Ca^{2+} is thought to be a harbinger of cell death. Nonetheless, most nerve cells have voltage-sensitive Ca^{2+} channels at their synapses, and many also have Ca^{2+} channels that are responsible for the major inward current; these channels are replaced by Na^+ channels later in development.

Using intracellular electrodes, Farber has measured the changes in membrane potential in the cell bodies of N1E-115 neuroblastoma cells maintained in monolayer cultures. The tracing shown in Figure 20-9 is representative of his data.

a. What would happen if TTX, a blocker of voltage-sensitive Na^+ channels, was added to this system before recordings were obtained?

b. What would happen if TEA, a blocker of voltage-sensitive K^+ channels, was added to this system?

c. The slow, latent depolarization that occurs immediately after the action potential distinguishes these cells from other types of neurons. When Farber incubated neuroblastoma cells in TEA + TTX, he found the response depicted in Figure 20-10. What does this experiment indicate about the channels responsible for this slow depolarization and its dependence on the passage of Na^+ and K^+ ions during the fast depolarization?

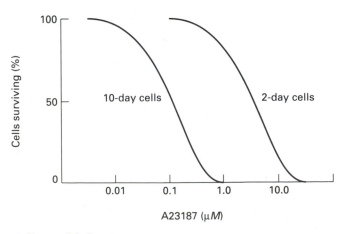

▲ **Figure 20-8**

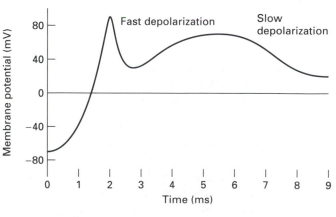

▲ **Figure 20-9**

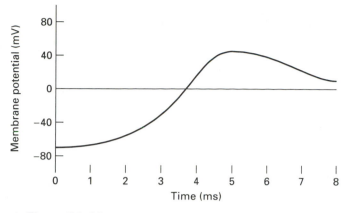

▲ **Figure 20-10**

d. When cobalt or cadmium was added to neuroblastoma cells before recordings were obtained, the slow depolarization shown in Figure 20-9 was abolished. What does this observation indicate about the possible nature of the slow depolarization in these cells?

e. It is possible to produce an action potential in these cells without the slow depolarizing current and in the absence of cadmium or cobalt. How can this be accomplished?

f. When neuroblastoma cells are cultured under depolarizing conditions (high external K^+ concentration), neurite outgrowth at the growth cone occurs. What technique could be used to determine whether this growth results from local, increased levels of free Ca^{2+} in the growth cone?

70. Del Castillo and Katz proposed in 1957 that neurotransmitters, released from presynaptic vesicles into the cleft, are responsible for the electric potential changes in the postsynaptic neuron. A large number of experiments have been performed to explore this hypothesis. One approach involves correlating postsynaptic potentials with the number of vesicles in the presynaptic neuron, the latter being determined by quick-fix transmission electron microscopy. Recently, the shi mutants of *Drosophila* have proved useful in investigating this hypothesis. These temperature-sensitive mutants carry a single-base change in the gene coding for an as yet unidentified protein, making presynaptic neurons defective in endocytosis. At 29°C, this protein is nonfunctional, whereas at 19°C, it is fully functional. J. Koenig and colleagues at the Beckman Research Institute of the City of Hope in Duarte, California, have recently explored these mutants and have tested the hypothesis put forth by Del Castillo and Katz.

Koenig's group examined the dorsal longitudinal flight muscle (DLM) of both wild-type and mutant flies by recording the intracellular excitatory junctional potentials (EJP) in the muscle after stimulation of the DLM fiber. They stimulated DLM fibers in both wild-type flies and shi mutants and recorded the EJP amplitude and the corresponding distribution of vesicles per synapse after stimulation. The data in Figure 20-11 are representative of their results.

a. Why do you think the wild-type *Drosophila* would be less suitable than the shi mutant for these studies?

b. What do the data in Figure 20-11 indicate about the relationship between EJPs and synaptic vesicles?

c. How might you decrease the EJP in the wild-type fibers experimentally? Assume that you do not know the chemical nature of the neurotransmitter.

d. If you did not know that the defect in the shi mutant was in endocytosis, how could you determine if the defect resulted from a habituation (desensitization) response at the level of the postsynaptic receptor or from an inability to release vesicles. Assume you know the identity of the neurotransmitter.

e. If you did not know that the defect in the shi mutant was in endocytosis, how could you determine whether the defect resulted from the inability of the DLM fiber to incorporate recycling neurotransmitter into vesicles?

f. How could you determine whether the protein that is altered in shi mutants affects endocytosis in cells other than neurons?

g. Do the data in Figure 20-11 definitively confirm the Del Castillo and Katz vesicle hypothesis?

71. Paul Greengard's laboratory at The Rockefeller University in New York has done extensive research on synapsin I, a major neuron-specific phosphoprotein located on the cytoplasmic surface of synaptic vesicles. His group has recently been able to separate synapsin I from isolated vesicles and then to reconstitute synaptic vesicles by incubating free synapsin I with the stripped experimental vesicles under various conditions. Figure 20-12 shows data similar to their results from such binding experiments.

a. What can you conclude from the data in Figure 20-12 about the binding of synapsin I to synaptic vesicles?

b. In a similar experiment, protease-treated vesicles were unable to incorporate synapsin I with the same affinity as non-protease-treated vesicles. What do these data and the high-salt curve in Figure 20-12 suggest about the formation of these reconstituted vesicles?

c. Synapsin I has a collagenase-insensitive head domain, which can be phosphorylated by cAMP-dependent protein kinase or Ca^{2+}-calmodulin–dependent pro-

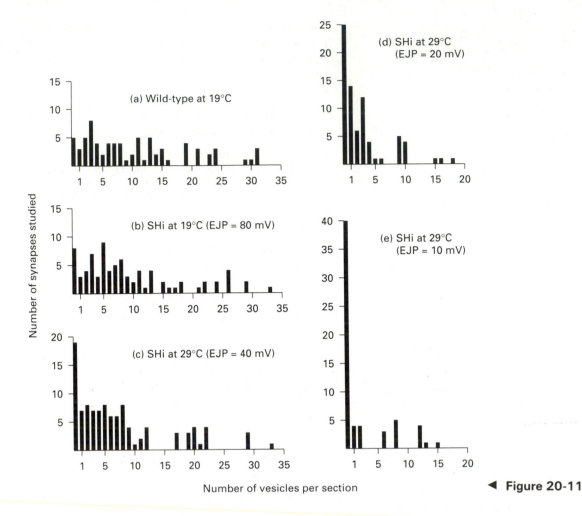

Number of vesicles per section

◀ **Figure 20-11**

tein kinase I. It also has an elongated collagenase-sensitive tail domain, which can be phosphorylated by Ca^{2+}-calmodulin–dependent protein kinase II.

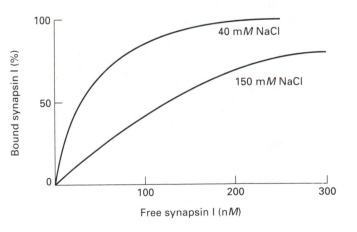

▲ **Figure 20-12**

Experiments with the isolated tails and heads have shown that tails bind less well to protease-treated vesicles than to non-protease-treated vesicles, whereas heads bind equally well to both types of vesicles. What is a possible interpretation of these results?

d. In other experiments, a hydrophobic photoaffinity label was used to label proteins in synaptic vesicles. This probe can gain entry into the hydrophobic domain of proteins. The same labeling pattern was found (using SDS gel electrophoresis and autoradiography) with both endogenous vesicles and reconstituted vesicles (i.e., those that had been stripped of synapsin I and reconstituted by exposure to free synapsin I). What is the significance of this experiment?

e. What type of experiment could be performed to determine whether one or more of the protein bands revealed in the photoaffinity-labeling experiment described in part (d) is indeed synapsin I?

Microtubules and Cellular Movements

PART A: Reviewing Basic Concepts

Fill in the blanks in statements 1–27 using the most appropriate terms from the following list:

acetylation	flagella
alpha (α)	gamma (γ)
ATP	genetic
basal bodies	glycosylation
beta (β)	GTP
central pair	isotypes
centrioles	microtubule-organizing center (MTOC)
chromosome(s)	
cilia	microtubule-associated proteins (MAPs)
cofactors	
colchicine	microvilli
cytocholasin D	minus (−)
degradation	multigene
depolymerize	mutants
discontinuous	omega (ω)
elongation	phase
fast	

polarizing	synthesis
polymerize	tau
plus (+)	triplet
radial spokes	two
serine	tyrosine
shortening	

1. Actin and tubulin share in common with each other the property that these individual monomeric units can _____, forming filaments and tubules, respectively.

2. Many of the cytoskeletal components, such as tubulin, are encoded by _____ families; in such cases, many different protein isotypes differing slightly in structure exist.

3. Tubulin is a dimer containing one ___α___ and one ___β___ subunit and having a total length of 8 nm.

4. Microtubules in nondividing cells extend out from the ___MTOC___.

5. The spindle apparatus of dividing cells is critical for ensuring that ___chromosomes___ are equally spaced and positioned in the cell.

6. Serial sectioning of the long axonic and dendritic processes of neurons has revealed that microtubules present in these structures are _____.

7. The plant alkaloid _____ inhibits cell division at metaphase due to its ability to bind to tubulin and thus prevent polymerization.

8. Proteins that are associated with microtubules in a constant ratio to α- and β-tubulin and are thought to be responsible for regulating the function of microtubules are collectively called _____.

9. The microtubules in axons and the microtubules emanating from the MTOC of most cells radiate from structures called _____.

10. The ___*plus*___ ends of spindle microtubules make contact with chromosomes.

11. Newly formed flagella can be synthesized by *Chlamydomonas* experimentally deprived of their flagella in the absence of tubulin _____, indicating that a large pool of free tubulin monomers exists in the cytoplasm.

12. The dephosphorylation of _____, two molecules of which are bound to αβ-tubulin, occurs during the incorporation of tubulin dimers into a microtubule.

13. Several forms of the microtubule-associated protein called _____ are found in axons; these are formed by alternative splicing of the primary transcript from a single gene.

14. Centrioles and ___*basal bodies*___ are functionally and structurally similar.

15. Structurally different types of tubulins that are encoded by different genes are called tubulin _____.

16. A GDP cap at the end of a microtubule will cause it to _____.

17. The _____ ends of flagellar microtubules point toward the basal body.

18. The covalent posttranslational modification of tubulin that is evident in flagellar α-tubulin but not in cytoplasmic α-tubulin is the _____ of lysine.

19. Microtubules regulate _____ axonal transport of proteins.

20. The number of proteins necessary for the beating of flagella has been determined through _____ analysis of nonmotile *Chlamydomonas* _____.

21. The "9 + 2" array of microtubules is a characteristic structural feature of both _____ and _____.

22. The posttranslational, ATP-dependent addition of the amino acid _____ to α-tubulin may determine whether α-tubulin is incorporated into all classes of microtubules or only into cytoplasmic microtubules.

23. Kinesin moves proteins from the _____ to the _____ end of microtubules.

24. Centrioles and basal bodies consist of nine sets of ___*triplet*___ microtubules.

25. It has been demonstrated through genetic studies that two components of the axoneme, the _____ and _____, are not necessary for beating.

26. Anaphase A involves the _____ of the polar microtubules.

27. The organization and changes in microtubules during mitosis can be visualized best using the _____ light microscope, which has the capacity to detect birefringence of parallel, highly ordered structures.

PART B: *Linking Concepts and Facts*

Circle the letters corresponding to the most appropriate terms/phrases that complete or answer items 28–42; more than one of the choices provided may be correct in some cases.

28. Microtubules are involved in which of the following processes?

   a. covalent association with actin in muscle cells

   b. movement of the mitotic spindle

   c. cell separation during cytokinesis

   d. transport of small vesicles within the cytoplasm

   e. movement of flagella and cilia

29. Microtubules in axons function

   a. in the synthesis of neurotransmitters.

b. as structural components maintaining the shape of the axon.

c. to conduct the electric impulse (action potential) along the course of the axon.

d. to direct axoplasmic transport.

d. to cross-link with myosin thus forming a stable cyto-skeletal network.

30. The initial growth of a microtubule depends on

    a. the ATP activation of the tail portion of the micro-tubule.

    b. the quaternary form of tubulin.

    c. a primer

    d. the ATP activation of the head portion of the micro-tubule.

    e. high concentrations of Cl^- ions.

31. Polymerization of $\alpha\beta$-tubulin occurs

    a. at the (+) end of microtubules.

    b. at the (−) end of microtubules.

    c. at kinetochores.

    d. in the presence of GTP.

    e. in the presence of colchicine.

32. Colchicine, vinblastine, and vincristine

    a. inhibit cell division.

    b. bind to actin filaments.

    c. bind to intermediate filaments

    d. bind to microtubules.

    e. can act as anticancer drugs.

33. MAP1 and MAP2

    a. are microtubule-associated proteins present in many nerve cells.

    b. are nucleoproteins.

    c. inhibit the polymerization of microtubules at the (+) end, but not at the (−) end.

    d. initiate the degradation of microtubules in the presence of increased levels of Na^+ ions.

    e. can cross-link adjacent microtubules.

34. The polymerization of tubulin at the (−) end of micro-tubules is inhibited by

    a. MAP1.   d. tau proteins.

    b. MAP2.   e. basal bodies.

    c. centrioles.

35. Studies on the formation of new flagella in *Chlamydo-monas reinhardtii* following amputation of the original flagella indicate that

    a. new tubulin is added at the (−) end.

    b. new tubulin is added at the (+) end.

    c. new flagella fail to assemble.

    d. new flagella originate from the centriole.

    e. new flagella form but intermediate filaments are used as a temporary scaffold until new tubulin units are synthesized.

36. Which of the following factors regulate the polymeriza-tion and depolymerization of microtubules?

    a. MAPs

    b. concentration of free tubulin

    c. bound GDP

    d. rate of hydrolysis of GTP to GDP

    e. temperature

37. The existence of almost identical α- and β-tubulin iso-types in plants and mice implies

    a. that α- and β-tubulin coexist in a 1 : 1 ratio.

    b. that all MAPs have similar functions.

    c. that the isotypes probably arose early in evolution.

    d. that the various isotypes probably have the same function in cells.

    e. that they must have arisen through gene splicing.

38. The ATPase activity of cilia and flagella is associated with

    a. subfiber A.   d. dynein.

    b. subfiber B.   e. the outer sheath.

    c. nexin.

39. Centrioles in mammalian cells

    a. replicate during the G_1 phase of the cell cycle.

    b. orient themselves perpendicular to each other.

    c. can be synthesized in the absence of a pre-existing centriole.

    d. play important roles in mitosis.

    e. can carry out their own protein synthesis.

40. When purified centrioles are microinjected into eggs,

    a. cell division stops.

    b. cell division is initiated.

    c. microtubules are polymerized.

d. centrioles degenerate.

e. they form flagella within the cytoplasm of the eggs.

41. Which of the following structures is responsible for the delivery of chromosomes to the two daughter cells?

   a. kinetochore fiber

   b. astral fiber

   c. polar fiber

   d. subfiber A

   e. radial spoke

42. Sclerodoma patients produce antibodies against

   a. chromosomes.

   b. kinetochores.

   c. astral fibers.

   d. nexin.

   e. zone of interdigitation.

PART C: *Putting Concepts to Work*

43. Although myosin and actin filaments are present in most eukaryotic cells, most research on these structures has been with filaments from muscle. Why is this the case?

44. What is the structural basis for the generalization that all microtubules are polar.

45. What is the functional reason why some microtubules undergo periodic disassembly and reassembly, whereas others are quite stable and exhibit little cycling between the assembled and disassembled states?

46. Microtubules and other filaments often are visualized by quick-freeze, deep-etching technique. What are the advantages of each part of this technique for visualizing such structures?

47. What is the functional importance of the long lifetime (100 days or more) of microtubules in the axons of neurons?

48. At the critical concentration of $\alpha\beta$-tubulin, the average length of microtubules remains constant. However, if previously labeled microtubules are incubated with the critical concentration of $\alpha\beta$-tubulin, label is often released. Explain this observation.

49. Under identical conditions and with excess amounts of $\alpha\beta$-tubulin, the in vitro polymerization of tubulin dimers occurs more rapidly in the presence of free microtubule primers than in the presence of primers of the same length that are attached to centrioles. Explain these observations.

50. What experimental technique is most suitable for demonstrating the in situ assembly of microtubules from newly synthesized tubulin dimers as occurs during regeneration of flagella in *Chlamydomonas?*

51. What is the functional significance of the observation that when antibodies directed against tyrosine-modified α-tubulin are microinjected into cells, the formation of some microtubular classes but not others is inhibited?

52. Why are *both* video-enhancement Nomarski optics and electron microscopy useful in demonstrating microtubule-dependent axonal transport of structures?

53. AMPPNP, a nonhydrolyzable ATP analog, facilitates binding of kinesin to microtubules, but ATP is necessary for translocation of vesicles along microtubules. What do these observations suggest about the mechanism of the motor protein kinesin?

54. What are the similarities and differences between kinesin and MAP1C, also called cytoplasmic dynein.

55. Individuals who are genetically defective in dynein demonstrate a characteristic cough and are sterile. What is the molecular basis for these clinical symptoms?

56. How do the mitotic spindles in plant and animal cells differ?

57. When chromosomes from premetaphase cells are rotated 180° just prior to metaphase, the attachment of chromatids to their specific kinetochores is broken and the chromosomes are pulled immediately into the opposite daughter cell. What does this experiment demonstrate about the forces governing chromosome movement?

PART D: *Developing Problem-solving Skills*

58. Over the past decade it has become clear that microtubules are very dynamic. As a cell changes shape, microtubules depolymerize and can then reassemble in a different part of the cell. For instance, mitotic cells use microtubules as part of the spindle apparatus during mitosis, although this structure does not exist in this form only a few hours prior to a cell's entry into the mitosis phase of cell division. Design an experiment to demonstrate that mitotic cells use $\alpha\beta$-tubulin synthesized during interphase to construct the spindle.

59. Both α- and β-tubulin have been isolated from brain in polymerized form and purified using centrifugation techniques. Two preparations, A and B, of primer microtubules were prepared; in both, Sephadex beads were covalently linked to one end of the microtubules. The extent of polymerization in the presence of primer A and primer B was then determined with a fluorescent-binding technique; the results are shown in Figure 21-1. Low-resolution ultrastructural observations (whole mounts fixed and sprayed with a thin layer of metal) revealed that primer A and primer B were indistinguishable from each other. Assuming that the polymerization reaction mixtures were identical except for the primer preparation, suggest the most plausible reason why primer A induced polymerization, whereas primer B did not.

60. When free $\alpha\beta$-tubulin dimers are incubated under polymerizing conditions (37°C in the presence of GTP), the polymerization reaction exhibits three phases as illustrated in Figure 21-2. Polymerization plateaus even in the presence of excess free tubulin dimers. Explain all three phases of the polymerization curve.

61. Colchicine binds to tubulin but does not directly cause disassembly of microtubules. In many cells in which

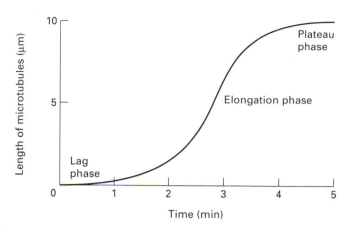

▲ **Figure 21-2**

microtubules seem to be stable elements as revealed by fluorescent immunocytochemistry, addition of colchicine and related drugs results in the shortening of microtubules and the consequential increase in free tubulin dimers. Explain why these seemingly contradictory observations are consistent with current understanding of the mechanisms underlying the assembly and disassembly of microtubules.

62. An in vitro system was developed for the purpose of determining the time it takes for one tubulin dimer to proceed from the (+) end of a microtubule to the (−) end. Nonlabeled microtubules with an average length of 5 μm were incubated in the presence of [^{35}S]methionine-labeled tubulin at the critical concentration. At various times after the addition of the radioactive label at time 0, the relative concentration of radioactive free tubulin was determined using standard scintillation techniques. Based on the data in Figure 21-3, determine the speed of tubulin turnover, that is, the time it takes for one tubulin dimer to add to the (+) end, pass along

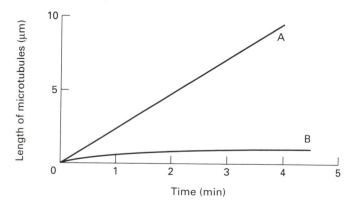

▲ **Figure 21-1**

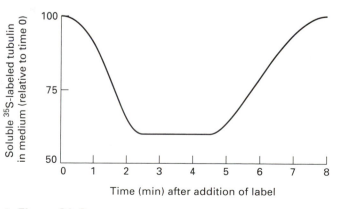

▲ **Figure 21-3**

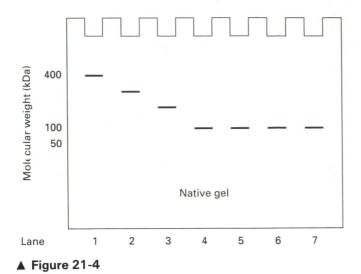
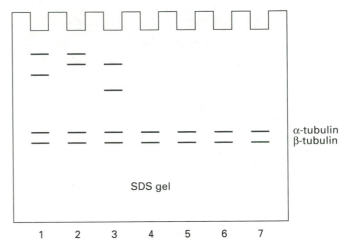

▲ **Figure 21-4**

the length of a microtubule, and be lost at the (−) end. The answer should be expressed in length/time units. Assume that the labeled free tubulin is in large excess of the unlabeled tubulin in the starting microtubules.

63. Tubulin has been purified from cow brain by repeated polymerizations and depolymerizations. After each polymerization-depolymerizations step, the resulting preparation was immunoprecipitated with antitubulin antibodies and analyzed by native gel electrophoresis, a technique that does not involve use of denaturing agents, and by SDS gel electrophoresis in the presence of β-mercaptoethanol and urea, which are denaturing agents. The resulting gel patterns, shown in Figure 21-4, indicate that the molecular weight of tubulin determined by native gel electrophoresis decreased during purification, whereas the molecular weight determined by SDS gel electrophoresis did not. Each numbered lane in the figure represents successive preparations in the purification. Explain the difference in these gel patterns and what it indicates about the molecular characteristics of tubulin.

64. Many advances in research result from combining preexisting techniques into novel experimental proto-

cols. For example, tubulin polymerization in living cells has recently been examined using a combination of light microscopy autoradiography and fluorescent immunocytochemistry. In these experiments, A6 cells were bulk-injected via a liposome carrier with [³H]leucine-labeled tubulin. One minute and 30 min post-injection, the cells were fixed, exposed to fluorescent anti-tubulin antibodies, and examined in a fluorescence microscope. The cells were then washed free of the glycerol-based mounting medium, dehydrated for light microscopy autoradiography, filmed, and exposed for 2 weeks. Based on your knowledge of microtubules, what kind of information could be obtained by this approach and why is this combination of immunocytochemical and autoradiographic approaches important to our understanding of microtubule polymerization in living cells?

65. The ability of an unknown agent to affect microtubule-directed movement of pigment granules has been studied in an experimental system using frog melanophores. When these cells are maintained under dark conditions, the pigment granules cluster near the center of the cells, as illustrated in Figure 21-5a. When cells are exposed to light, the pigment granules move away from the center in a distinctive radiating pattern (Figure 21-5b). When

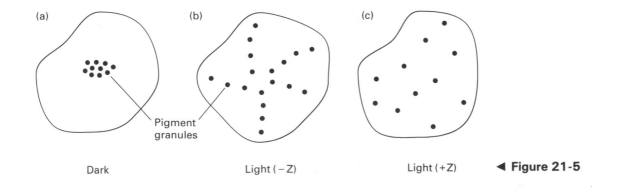

Dark   Light (−Z)   Light (+Z)   ◄ **Figure 21-5**

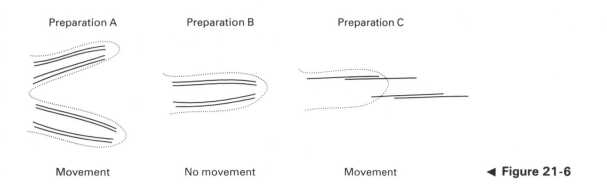

Preparation A   Preparation B   Preparation C

Movement   No movement   Movement   ◀ **Figure 21-6**

dark-adapted melanophores are exposed to agent Z and then moved into the light, a different pattern of pigment movement is observed (Figure 21-5c). Suggest the most probable mechanism of action of Z and design an in vitro experiment to test your hypothesis.

66. An investigator isolated individual flagella in order to study the mechanisms underlying the bending of these structures. He prepared the flagella using one of three different physiologic buffers, A, B, or C. In all three cases, the outer plasma membrane was stripped from the flagella, which were then placed on a glass slide. After adding ATP to activate the system, the researcher observed each preparation with Nomarski optics; the results are illustrated in Figure 21-6 (the dashed line represents the original position of the plasma membrane). Although preparation A appeared to move normally, preparations B and C were clearly defective. What are the probable defects in the B and C flagellar preparations?

67. Many anticancer drugs affect dividing cells by interfering directly with the mitotic apparatus during mitosis. In the search for more specific anticancer drugs, compound D, a derivative of a plant alkaloid was found to inhibit cell division. When compound D is added to cells, the mitotic spindle forms normally in prometaphase, but just prior to metaphase the kinetochore microtubules rapidly depolymerize. Curiously, observations with the polarizing microscope demonstrate that both the astral and polar fibers look normal in the presence of compound D. Give a possible reason for the differential effects of compound D on spindle fibers and compare the action of this drug with that of colchicine.

68. Microtubules are stable yet dynamic structures. Because of the differences in the polymerization and depolymerization rates at their (+) and (−) ends, microtubules can, under certain conditions, constantly add and lose tubulin and yet maintain a constant length. In one study, one area of a highly differentiated, polarized cell was found to contain a population of microtubules that were turning over (see problem 62) and another population of microtubules that were not turning over but nonetheless were maintaining the same length. What is the possible explanation for the latter case? (Assume that the microtubules are not capped at either end.)

PART E: *Working with Research Data*

69. J. Avila in Spain and Barbara Crute at the State University of New York have been examining the enzyme(s) that may be responsible for modulating the polymerization and depolymerization of tubulin. A protein kinase, casein kinase II (CKII), has been implicated as a possible regulatory protein in this regard. CKII is so named because it can phosphorylate exogenously added casein, but its endogenous substrates are still unknown. The activity of CKIII increases during differentiation of a variety of cells, especially cells that differentiate into neurons.

a. When CKII is partially purified from neuronal cells in culture by a two-step procedure involving ion-exchange chromatography followed by gel filtration, two chromatographic profiles are obtained depending on the amount of salt present in the column buffer. These profiles are illustrated in Figure 21-7.

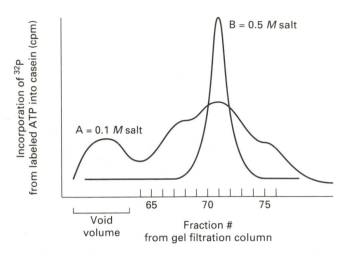

▲ **Figure 21-7**

What is the explanation for the difference in the A and B profiles? What does this suggest about the nature of CKII?

b. When fractions from column A were incubated in the presence of γ-labeled [^{32}P]ATP but without casein, the phosphorylation of a 50-kDa protein was observed. This protein was present in most fractions containing CKII activity and was thought to be tubulin. How could the identity of this 50-kDa protein as tubulin be confirmed?

c. When fraction 70 from column B was incubated with purified tubulin from rat brain and ^{32}P-labeled ATP or GTP, the SDS gel autoradiograms depicted in Figure 21-8 were obtained. Because CKII is unique among the protein kinases in that it is inhibited by heparin and can use ATP or GTP equally well as a phosphate source, lanes C and D were used to test if the observed tubulin kinase activity behaved like CKII. Since the data in Figure 21-7 shows that fraction 70 can phosphorylate casein, why was it critical to run lane A in this experiment?

d. When the experiment described in part (c) was modified by substituting casein for fraction 70 in the samples in lanes B and C, phosphorylated casein appeared in both lanes but with less activity in lane C than in lane B. What is the interpretation of this experiment and what does it suggest about the relationship of CKII and tubulin?

e. Why is CKII's unique utilization of GTP consistent with the hypothesis that this enzyme plays a role in tubulin polymerization?

f. Purified cold-stabilized microtubules were prepared and a series of assembly-disassembly incubations were performed followed by Western blot analysis for tubulin and CKII using anti-tubulin and anti-CKII antibodies. The resulting blots are depicted in Figure 21-9. The number of each lane corresponds to each successive assembly-disassembly cycle of tubulin. What do these data indicate about the relationship of CKII and tubulin?

g. Using the antibodies against tubulin and CKII, how could you demonstrate that CKII and tubulin associate with each other in intact cells and that this association is not merely an artifact of the biochemical separations used to isolate both tubulin and CKII?

70. Both α- and β-tubulin are encoded by six or seven functional genes, which produce tubulin isotypes differing primarily in a 15-aa carboxyl-terminal variable domain. These isotypes have been identified by Don Cleveland at the Johns Hopkins University School of Medicine and are referred to as classes I, II, III, IV, and V. Most investigations have found that these isotypes are functionally equivalent, although some are preferentially used in highly specialized cells such as sperm cells. Recently Cleveland examined the PC12 neuronal cell line, a transformed cell that undergoes differentiation into a neuronal-like cell in the presence of nerve growth factor (NGF). He asked whether tubulin isotypes preferentially accumulated in response to NGF and/or in spe-

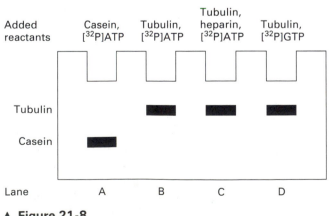

▲ **Figure 21-8**

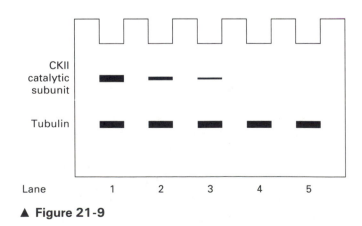

▲ **Figure 21-9**

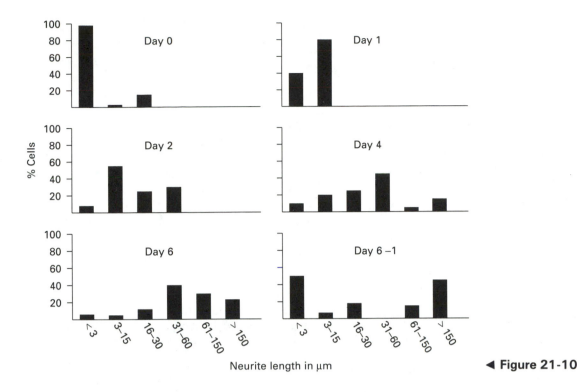

◀ **Figure 21-10**

cialized regions of the cell. Some of the data presented in this problem are representative of his results.

a. PC12 cells were incubated with NGF for up to 6 days and the lengths of the neurites (axonlike processes) were determined at various times. The neurite length serve as a morphological indicator of neuronal differentiation in these cells. At the end of day 6, the NGF was withdrawn from one sample (day 6-1) before determination of neurite length. The percentage distributions of neurite length in the samples are shown in Figure 21-10. Based on these data, what is the apparent effect of NGF on PC12 cells? What does the day 6-1 sample indicate about the function of NGF in the neuronal differentiation of these cells?

b. The samples described in part (a) were subjected to quantitative immunoblot analysis for each of the five

tubulin isotypes. The resulting profiles are depicted in Figure 21-11. Based on these data, what is the effect of NGF on each of the five isotypes?

c. The samples described in part (a) also were analyzed for the microtubule-associated proteins tau and MAP2. The levels of both these proteins was found to increase in parallel with tubulin isotypes II and III and to fall off to non-NGF-stimulated levels in the day 6-1 sample. This finding could suggest that the increased synthesis of isotypes II and III induces the synthesis of the MAPs. How could this hypothesis be tested?

d. The total amount of tubulin (all isotypes) in the cytoskeleton and in the soluble fraction of the samples described in part (a) was determined. An antibody that recognizes all β-tubulin isotypes equally well was used for these analyses. The data are shown in Figure 21-12. What can you conclude from these data about the effect of NGF on tubulin synthesis and polymerization?

e. Indirect immunofluorescent microscopy was used to visualize the relative distribution and amounts of the five tubulin isotypes in specific parts of PC12 cells. With the help of D. Lansing Taylor at the Carnegie Melon Institute, individual neurites were stained with FITC-labeled antibodies specific for one of the five isotypes and with a rhodamine-labeled anti-tubulin antibody that recognizes all isotypes of tubulin. Why was use of the latter antibody critical for this type of analysis.

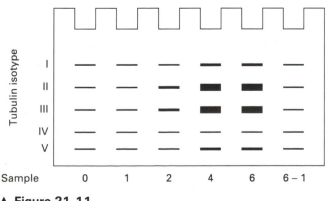

▲ **Figure 21-11**

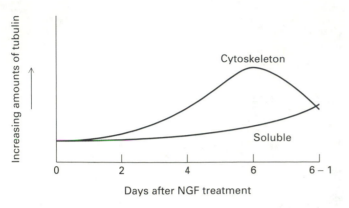

▲ Figure 21-12

71. Evelyn Houliston and Bernard Maro in France have examined the posttranslational modifications of tubulin and the different functional role(s) that such modified tubulin subtypes may play in cells. Antibodies that can distinguish both tyrosinated and acetylated subtypes of tubulin have been produced and have been instrumental in examining tubulin subtypes in cells.

a. Houliston and Maro have examined preimplantation embryos, which are known to undergo an extensive degree of polarization in the inner mass and the trophectoderm (outer layer of cells). The cells flatten on each other and the microtubules undergo a rearrangement. Inhibition of microtubules assembly with colchicine or nocodazole blocks this differentiation. Using specific antibodies to tyrosinated α-tubulin (YL1/2) and acetylated α-tubulin (6-11BG-1), these investigators stained blastomeres for each of these tubulin subtypes. Data representative of their results are presented in Table 21-1. What major changes occur in the localization of the tyro-

Table 21-1

Probe type	Time post-division (h)	Percentage of cells with the microtubule network		
		Enriched in apical cytoplasm	Depleted in cytoplasm near contact areas	Augmented in cortex near contact areas
YL1/2 (against tyrosinated α-tubulin)	2	40	77	0
	5	72	92	0
	9	78	80	0
6-11BG-1 (against acetylated α-tubulin)	2	6.5	0	14.5
	5	6.4	2.6	52.6
	9	6.6	0	59.9

sinated and acetylated α-tubulin subtypes in blastomeres between 2 and 9 h postdivision?

b. In subsequent experiments, these researchers compared the effects of nocodazole, a drug that can inhibit the assembly of microtubules at micromolar concentrations, on the localization of tyrosinated tubulin-containing microtubules and acetylated tubulin-containing microtubules. Blastomeres were treated for 15 s with 1 μM nocodazole, then stained with antibodies against each subtype, and examined in the fluorescent microscope. Diagrams of the images showing the cellular localization of microtubules labeled with each probe are presented in Figure 21-13. What conclusion can be drawn from these results?

c. Taxol, a microtubule-stabilizing agent, was added to blastomeres in the presence of colchicine, which inhibits tubulin polymerization. When these cells were analyzed as described in part (b), the staining pattern with each antibody was similar. What does this suggest about the two tubulin subtypes?

d. When cells were stained for the enzyme acetyltransferase, which acetylates tubulin, the enzyme was found to be uniformly distributed throughout the cells. Why is this observation consistent with the results of the taxol experiment described in part (c).

72. The number of cellular processes thought to involve microtubules has been increasing over the years. It is clear that both kinesin and MAP1C can drive movement of structures and vesicles along microtubular networks. Furthermore, the cytoskeleton is now thought to be responsible for the maintenance of specific proteins on the apical surface of polarized epithelial cells. Recently, Ronald Vale and Hirokazu Hotani have indicated that kinesin and microtubules may be critical for a different type of cytoskeletal network containing membrane components, not tubulin.

It is known that the endoplasmic reticulum and lysosomes move toward the center of a polarized cell when the cell is incubated with colchicine. When colchicine is washed away and taxol added, these organelles move back to their polarized orientation within the cell at the same speed as do small vesicles being transported on microtubules. Electron microscope analysis of these organelles demonstrate cross-bridges linking them to microtubules. This internal network, which may consist of extensions of organelle membranes, is referred to as a membrane network and may be governed by kinesin. The data presented below are similar to those obtained by Vale and Hotani.

a. Kinesin was first isolated from squid axons using microtubules as an affinity system in the presence of the nonhydrolyzable analog AMPPNP. Why is

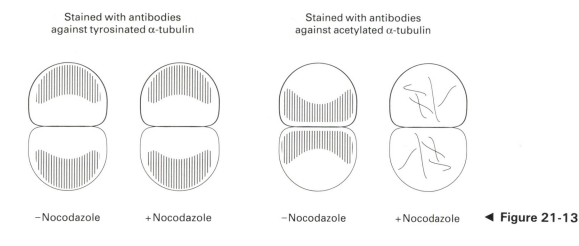

Stained with antibodies against tyrosinated α-tubulin

Stained with antibodies against acetylated α-tubulin

−Nocodazole +Nocodazole −Nocodazole +Nocodazole ◀ **Figure 21-13**

AMPPNP, not ATP, necessary for this purification step?

b. When kinesin was isolated in this manner, a protein was revealed by native gel electrophoresis that had the molecular weight of native kinesin. Since several other proteins were present on the gel as well, how could it be verified that this protein was indeed kinesin? Assume that you do not have an antibody to kinesin at your disposal and thus can not do a Western blot analysis.

c. When microtubule affinity-purified kinesin, taxol-stabilized microtubules, and ATP were applied to a glass slide and covered by a coverslip, a microtubule-like network formed in the solution, attached to the glass, and moved along the surface. The network as revealed by dark-filled microscopy is sketched in Figure 21-14. What feature of this image indicates that something other than tubulin microtubules are present in the preparation?

d. How could the presence of membrane networks not containing tubulin be demonstrated and distinguished from tubulin microtubules (known to be

present since they were in the reaction mixture) in the preparation depicted in Figure 21-14?

e. When Triton-X was added to the preparation described in part (c), the lamellar-like portions of the network disappeared. What does this observation indicate about the nature of this network?

f. When the preparation depicted in Figure 21-14 was infused with a hypotonic solution, more than half of the entire network disappeared. Is this effect consistent with the effect of Triton-X?

g. When ATP was replaced with AMPPNP, the membrane networks described in part (c) failed to form. What does this observation indicate about the mechanisms driving the formation of these membrane networks?

73. Microtubule assembly and disassembly during cell division is a very active area of research. The major questions involve the polymerization and depolymerization sites of kinetochore microtubules and how these activities may relate to chromosome movement. Recently, T. Mitchison at the University of California, San Francisco, synthesized a photoactive tubulin derivative. When this compound is microinjected into cells, it becomes incorporated into microtubules and is converted

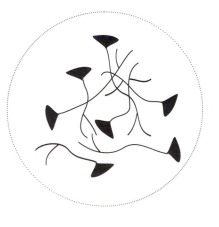

Microscopic field of view

▲ **Figure 21-14**

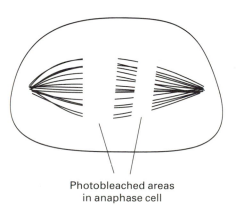

Photobleached areas in anaphase cell

▲ **Figure 21-15**

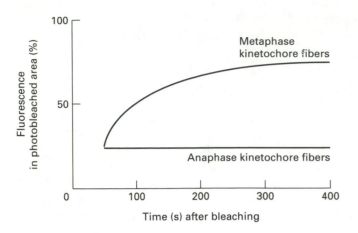

▲ **Figure 21-16**

to a fluorescent form when irradiated with 365-nm light. Thus movement of this probe in the mitotic spindle can be monitored using fluorescence microscopy.

a. A similar approach has been taken by Gorbsky and Borisy at the University of Wisconsin. They have microinjected X-rhodamine tubulin into LLC-PK cells. This probe labels all microtubules. With a defined beam, they then photobleach areas of the mitotic spindle to determine (1) the rate of movement of the photobleached area and (2) the degree of fluorescence recovery of the photobleached area. One of the first experiments this group performed was to develop a buffer (PHEM) that lysed cells and preserved only kinetochore fiber microtubules. Why was this a critical experiment?

b. What does the effect of this buffer suggest about the differences in microtubules in vivo?

c. What other methods might be useful for preserving kinetochore tubules at the expense of other microtubules?

d. When anaphase cells were labeled as described above and then photobleached at the spot indicated in Figure 21-15, the expected absence of fluorescence was noted. How could you demonstrate that the irradiation does not destroy the kinetochore microtubules at this point?

e. When the positions of the photobleached spot was monitored in anaphase cells, it moved ~1 μm/100 s toward the pole, but no fluorescence recovery of the photobleached area was noted. What is the interpretation of these findings?

f. Figure 21-16 shows the fluorescence recovery after photobleaching in anaphase and metaphase kinetochore tubules. What is the significance of these data?

CHAPTER 22

Actin, Myosin, and Intermediate Filaments: Cell Movements and Cell Shape

PART A: Reviewing Basic Concepts

Fill in the blanks in statements 1–29 using the most appropriate terms from the following list:

actin	minus (−)
α-actinin	myosin heads
ATP	myosin light-chain kinase
Ca²⁺ ions	
Ca²⁺ ATPase	pH
creatine phosphate	plus (+)
depolymerization	polymerization
filamin	relaxation
filipodia	rigor
gel	S1
gelsolin	Na⁺
GTP	sol
heavy	stress fibers
involuntary	thick filament
intermediate filaments	thin filament
isoforms	titin
light	tropomyosin
lower	troponin C

troponin I	vimentin
uvomorulin	viscosity
villin	voluntary

★ 1. __ACTIN__ is the single most abundant protein in most mammalian cells and can exist in a polymeric filamentous form or a monomeric globular form.

2. Polymerization of isolated actin causes a large increase in the _____ of the solution.

3. Actin can be purified and separated from other cytosolic proteins that bind to it by a series of _____ and _____ reactions. This strategy is also used during purification of tubulin.

4. _____ bound to an actin monomer enhances addition of actin to microfilaments.

★ 5. The growth of an actin filament occurs 5–10 times faster at the _plus_ end than at the _minus_ end.

6. Myosin varies among species but usually consists of two _heavy_ chains and two _light_ chains.

7. Several hundred myosin molecules can assemble into a _thick filament_, which is the major unit of myosin in muscle cells.

8. The _____ fragment of the myosin molecule contains all of the ATPase activity of myosin.

9. In photomicrographs, S1 myosin fragments bound to an actin filament look like "arrowheads," which all point toward the _minus_ end of the filament.

10. Smooth muscle, unlike skeletal muscle, is under _involuntary_ control.

★ 11. If muscle is depleted of its stores of ATP, it goes into a state known as _rigor_.

12. The two major sources of energy for muscle contraction are ATP and _____.

★ 13. The release of _Ca²⁺ ions_ from the sarcoplasmic reticulum triggers the contraction of muscle.

14. A potent _____ in the membrane of the sarcoplasmic reticulum (SR) pumps Ca^{2+} from the cytosol to the lumen of the SR.

15. _____ is a Ca^{2+}-binding protein associated with the actin filaments of skeletal muscle.

16. Evidence based on x-ray diffraction and synchrotron analysis of striated muscle indicates that during contraction the movement of _____ occurs before the movement of _____ toward the actin filaments.

17. Contraction of smooth muscle is governed by the phosphorylation and dephosphorylation of _____.

18. Different mysoin _____ are found in striated and smooth muscle and can result from alternative splicing of RNA transcripts.

19. _α-actinin_ bundles the ends of actin filaments together and is critical for the attachment of these filaments to Z disks in skeletal muscle.

20. A large (1000-kDa) protein called _____ may serve to center myosin filaments during contraction.

21. Myosin is present in most nonmuscle cells but usually at a much _____ ratio to actin than in muscle cells.

22. At high Ca^{2+} concentrations, _villin_ can act as an actin-severing protein, whereas at low Ca^{2+} concentrations it cross-links actin filaments; the severing activity of this 95-kDa protein is thought to be involved in the disassembly of microvillar actin filaments during starvation, leading to loss of microvilli as small vesicles.

23. _____ is a transmembrane protein that links cells and desmosomes and needs Ca^{2+} to effect cell-to-cell adhesion.

24. The movement of amebas involves transformation of the cytoplasm from a fluidlike _____ to a more solidlike _____; this transition can be induced in vitro by the actin-severing protein _____.

25. The actin-associated protein _____ cross-links actin filaments in the ectoplasm of amebas and is thought to be partly responsible for the gel-like consistency of this cytoplasmic domain.

26. The thin, actin-containing, spikelike projections that emanate from nerve growth cones are called _____.

27. _____ are located adjacent to the plasma membrane and contain actin filaments of different polarities. They are commonly found in endothelial cells and are thought to be important in the strength and resiliency of these cells.

28. _____, which are 8 to 12 nm in diameter, are the main structural proteins of skin and hair.

29. Intermediate filaments composed of the 57-kDa protein _____ are prevalent in mesenchymal cells. These filaments may help keep the nucleus in a particular position in the cell and form a cage around lipid droplets in adipocytes.

PART B: *Linking Concepts and Facts*

Circle the letters corresponding to the most appropriate terms/phrases that complete or answer items 30–43; more than one of the choices provided may be correct in some cases.

30. Polymerization of actin monomers in vitro to form actin filaments is induced by

   a. Na^+ ions.   b. Cl^- ions.

c. K⁺ ions.

e. Ca²⁺ ions.

d. Mg²⁺ ions.

31. Which of the following comparisons of microtubules and actin filaments are true?

a. Actin filaments have a smaller diameter than microtubules.

b. Microtubules have associated proteins, whereas actin filaments do not.

c. Microtubules are less stable than actin filaments.

d. The critical concentration for addition of monomers to the polymers is higher in the case of actin than tubulin.

e. Actin filaments are more sensitive than microtubules to destabilization by high K⁺ concentrations.

32. Actin filaments in motile animal cells can depolymerize and repolymerize in

a. 1 week.

d. minutes.

b. days.

e. 1 to 2 milliseconds.

c. hours.

33. Myosin is

a. a protein with a molecular weight of approximately 10,000.

b. a protein present in both animal and plant cells.

c. a structural protein.

d. a fibrous protein.

e. an asymmetric molecule with a globular head at the N-terminal end.

34. When myosin is treated with low concentrations of chymotrypsin,

a. heavy meromyosin (HMM) is produced.

b. the myosin covalently links to actin.

c. light meromyosin (LMM) is produced.

d. the "hinge" is proteolytically cleaved.

e. the molecule is not cleaved.

35. The ATPase activity of myosin is enhanced in the presence of

a. fodrin.

d. tubulin.

b. vimentin.

e. cytokeratins.

c. actin.

36. Vertebrate muscle can be classified into the following types:

a. striated.

d. cardiac.

b. multilayered.

e. interdigitated.

c. smooth.

37. Which of the following sequences indicates the correct order from largest to smallest structure?

a. myofibers > myofibril > muscle > myosin filament > actin filament

b. myofibers > myofibril > muscle > actin filament > myosin filament

c. myofibril > muscle > myofibers > myosin filament > actin filament

d. muscle > myofibers > myofibril > actin filament > myosin filament

e. muscle > myofibers > myofibril > myosin filament > actin filament

38. Which of the following is (are) the major constituent(s) of the A band in striated muscle?

a. actin

d. tubulin

b. myosn

e. fodrin

c. Z disk

39. Which of the following remain(s) constant in length during muscle contraction?

a. I band

b. A band

c. sarcomere length

d. myosin thick filaments

e. actin thin filaments

40. Actin filaments can be anchored to

a. tubulin.

b. Z disks.

c. intercalated disks.

d. the plasma membrane.

e. cytoplasmic plaques.

41. The filament network within microvilli and their terminal webs include

a. actin filaments.

d. fodrin.

b. calmodulin.

e. villin.

c. fimbrin.

42. In activated platelets, actin networks form following the dissociation of

a. calmodulin from actin.

b. calmodulin from myosin.

c. spectrin from actin.

d. villin from actin.

e. profilin from actin.

43. In contrast to microtubules and actin microfilaments, intermediate filaments

    a. are formed from numerous types of subunit proteins.

b. do not require energy for assembly.

c. remain insoluble when extracted from cells.

d. exhibit no polarity.

e. exist primarily in the polymerized form and undergo little polymerization-depolymerization cycling within cells.

PART C: *Putting Concepts to Work*

44. Several in vitro experiments have indicated that actin from two phylogenetically different sources can copolymerize, forming filaments that resemble those presented in the species used as sources of the actin. What is the significance of this finding?

45. What two features of the structure and assembly of actin filaments can be demonstrated by use of filaments "decorated" with myosin S1 fragments?

46. What change in the average length of actin filaments would be expected in a system in which the concentration of globular actin monomers was maintained at 1 μM at the (+) ends and at 8 μM at the (−) ends? Why?

47. What molecular feature makes the rodlike tail in myosin molecules structurally rigid?

48. If isolated striated muscle cells from which myosin had been selectively extracted were incubated with S1 myosin fragments, would the resulting bound arrowheads point toward or away from the Z disks? Why?

49. The micrograph shown in Figure 22-1 is a longitudinal section of a striated muscle that was in rigor at preparation. Label the indicated structures in this figure. How would this image differ if ATP had been added to the muscle sample just before preparation?

50. What are the structural characteristics of the rigor complex and when does it form?

51. Microinjection of an antibody to myosin light-chain kinase inhibits the contraction of smooth muscle but not of striated muscle. Contraction of both types of

muscle, however, is associated with a rise in cytosolic Ca^{2+}. Explain these findings.

52. The change in standard free energy $\Delta G°'$ for hydrolysis of creatine phosphate is −10.3 kcal/mol, whereas the $\Delta G°'$ for hydrolysis of the terminal phosphate of ATP is −7.3 kcal/mol. In addition, the concentration of creatine phosphate in muscle is about fivefold higher than that of ATP. Assuming that creatine kinase, which can phosphorylate ADP, is present in muscle, what is the implication of these data?

53. Describe the experimental approach used by investigators to harvest relatively pure preparations of actin filaments from just the microvilli of intestinal epithelial cells.

54. Why do actin inhibitors like cytochalasin D affect cell division?

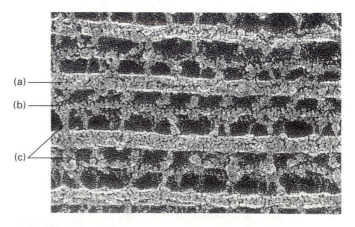

(a)

(b)

(c)

▲ **Figure 22-1**

55. The endoplasmic reticulum (ER) moves in one direction along bundles of actin filaments in the green alga *Nitella*. What is the most probable reason why movement does not occur in both directions?

56. The ratio of monomeric actin to polymerized actin is much higher in some types of nonmuscle cells than in others. What protein might account for this difference? Why?

PART D: *Developing Problem-solving Skills*

57. When myosin isolated from skeletal muscle was treated with a protease recently discovered in a coelenterate from the Gulf of Mexico, it exhibited no actin-stimulated ATPase activity. When isolated actin was pretreated with this protease, washed free of the proteolytic enzyme, and then added to a myosin preparation that was not similarly treated, the myosin demonstrated full ATPase activity. What is the probable site of action of this protease? How would the electron microscope be useful in evaluating the mechanism of action of this protease?

58. The ATP-driven movement of myosin along actin filaments can be assayed in the *Nitella* motility system. In this assay, myosin or myosin fragments are coated on Sephadex beads, which have covalently bound protein A on their surface, by antimyosin antibodies that also bind to protein A. When these preparations are added to polarized actin filaments adhering to a glass slide, attachment of the beads to actin and their movement can be followed microscopically. Table 22-1 presents the results of this assay for three different myosin preparations in the presence of ATP or AMPPNP, a nonhydrolyzable ATP analog. Based on these data, what is the nature of the three myosin preparations?

59. Much of ultrastructure of skeletal muscle has been revealed using standard thin sections and the quick-freeze, deep-etching technique. To accurately demonstrate the entire end-to-end ultrastructure of a single sarcomere 2 μm in length, it is necessary to obtain and view serial sections 90 nm thick.

   a. How many serial sections would be necessary to reconstruct an entire sarcomere end to end?

   b. Figure 22-2 depicts one section from this type of three-dimensional analysis. Label the indicated structures in this figure. From which region of the sarcomere was this section most likely taken?

60. Jeff Cook has noted that a parallel array of filaments, believed to be actin, are found in the basal portion of MDCK cells differentiated in culture. The appearance of these putative stress fibers coincides with the deposition of a basement membrane by these cells. A sketch of an electron micrograph of these cells is shown in Figure 22-3.

   a. In order to determine if these fibers contain actin filaments, Cook tried to produce a polyclonal antibody against actin by injecting the fibers into rabbits, but antibodies failed to develop. Assuming that there were no procedural problems, what is the most plausible reason for the inability of these fibers to elicit antibody production? How could this inability be overcome?

Table 22-1

Myosin preparation	Attachment to actin	Movement along actin	
	+AMPPNP	+ATP	+AMPPNP
A	++++	++++	0
B	++++	0	0
C	0	0	0

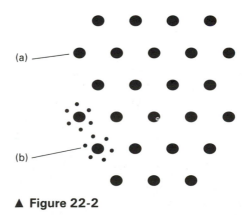

▲ **Figure 22-2**

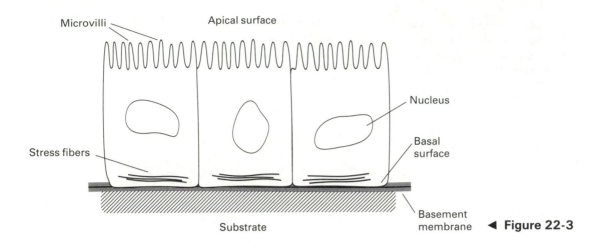

Microvilli   Apical surface

Nucleus

Stress fibers

Basal surface

Substrate

Basement membrane   ◀ **Figure 22-3**

b. Describe two other approaches that could be used to determine whether these basal fibers contain actin filaments.

61. In the experimental system described in problem 60, the functional relationship between the appearance of the basal fibers and the basement membrane is of interest. Design an experiment to determine whether deposition of the basement membrane is dependent on the formation of these fibers within MDCK cells. (Assume that the fibers have been shown to contain actin filaments.)

62. Many experiments have been conducted over the years to determine which factors influence muscle contraction. In order to reproduce some of these experiments, you measure the amount of isometric tension produced when contraction is elicited in three different skeletal muscle preparations. The results are shown in Figure 22-4. All three preparations were obtained by homogenizing skeletal muscle in the same buffer. The amounts of actin, myosin, and ATPase activity were identical in all three preparations. Furthermore, no molecular differences in the actin and myosin in the preparations were evident. Suggest a reason for the observed differences in the abilities of these three preparations to contract.

63. When you added compound X, isolated from a rare fish found in the Amazon River, to a synchronously beating preparation of isolated chick cardiac muscle cells growing in culture, it produced a state of rigor. Removal of compound X reversed rigor, and the cells began beating again. In order to investigate the mechanism of action of this poison, you carried out two in vitro experiments. First, the ability of compound X to inhibit the ATP-driven movement of myosin S1 fragments along actin filaments was assayed with the *Nitella* motility system. Remarkably, compound X did not inhibit myosin movement in this assay. Second, the effect of compound X on the chymotrypsin digestion of myosin was determined. The poison appeared to inhibit the digestion of myosin by chymotrypsin.

a. In the proteolytic digestion study, what control experiment(s) is necessary?

b. Based on the results of both experiments, what is the most probable mechanism of action of compound X?

64. The functions of myosin in nonmuscle cells have been studied in detail only in recent years. In such studies,

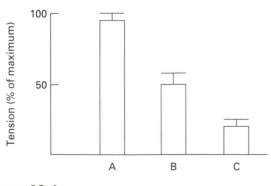

▲ **Figure 22-4**

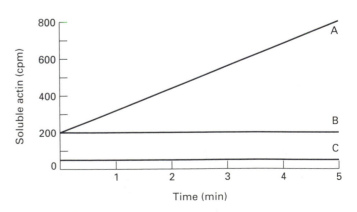

▲ **Figure 22-5**

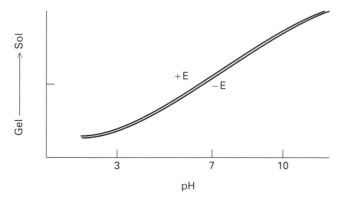

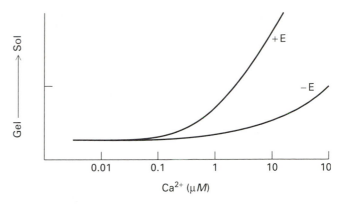

▲ **Figure 22-6**

researchers have used genetic mutants, generally of lower vertebrates, that either do not express myosin or express defective myosin. Such mutant cells seem to be defective in cytokinesis but otherwise are apparently normal.

a. Since these genetic mutations might affect the expression of other proteins, suggest an alternative approach for demonstrating that myosin plays a key role in cytokinesis.

b. Would in situ decoration of actin filaments with S1 fragments be a useful technique for analyzing these mutants?

65. An in vitro system was developed to study actin assembly and disassembly in nonmuscle cells. In this study, C3H10 T$^{1}/_{2}$ cells were labeled for several hours with [^{35}S]methionine, so that all actin filaments were completely labeled; that is, all the actin monomers in each filament were labeled. Actin filaments were then collected by differential centrifugation and put into a buffer containing one of three different cytosolic extracts (A, B, or C). The amounts of soluble actin in each sample was monitored over time; the results are shown in Figure 22-5. What do these data indicate about the effects of A, B, and C on the assembly and disassembly of actin filaments?

66. The regulation of the sol to gel transition in amebas has been investigated by several investigators including Robert Allen and Lans Taylor. The viscosity of cytosol isolated from amebas exhibits a pH and Ca^{2+} dependency as shown in Figure 22-6. When a highly purified extract (E) is added to these preparations, there is a change in the Ca^{2+} dependency but not in the pH dependency. Based on your knowledge of the factors that affect gel-sol transitions, which protein could have this type of effect?

67. In electron micrographs of thin sections of spot desmosomes connecting epithelial cells, bundles of filaments can be seen to radiate out from the darkly staining cytoplasmic plaques that line the inner surface of adjacent cell membranes. These connecting fibers contribute to the structural rigidity of the cells and cell layer.

a. How could you demonstrate that these fibers contain intermediate filaments and not actin filaments?

b. How could you demonstrate that these filaments connect cells via desmosomes in the interconnecting network — desmosomes → keratin   filaments → desmosomes → keratin   filaments → demosomes, etc.?

PART E: *Working with Research Data*

68. Actin filaments are involved in a variety of cellular functions including cell motility, structure, and adhesion. Actin also may anchor some, but not all, plasma membrane proteins in specific areas in polarized epithelial

cells. George Ojakian and Randi Schwimmer at the State University of New York have used the kidney cell line MDCK, which demonstrates a polarized phenotype in culture, to explore the possible relationship of actin to the distribution of the plasma membrane glycoprotein gp135. MDCK cells have an apical surface with numerous microvilli and a basal region, which is planar, contains stress fibers, and has a basal lamina between the plasma membrane and the microporous substrate. A schematic diagram of these cells is shown in Figure 22-3 (see problem 60).

a. The Ojakian and Schwimmer research team produced a monoclonal antibody to gp135 that was labeled with ^{125}I. The labeled antibody was applied to the apical surface of a monolayer of MDCK cells and the amount of antibody bound to the cells was determined at various times. The data shown in Figure 22-7 are representative of the results obtained by Ojakian and Schwimmer. The cells were subconfluent at 24 h and confluent by 48 h. Do the data in Figure 22-7 indicate that the number of gp135 molecules per cell increased during the study period?

b. In order to determine quantitatively the relative distribution of gp135 molecules on the surfaces of MDCK cells, ultrastructural immunocytochemistry was performed with anti-gp135. The resulting distribution of 5-nm gold particles over a typical MDCK cell is depicted in Figure 22-8. Based on this figure, which is the ratio of gp135 molecules bound at apical and basal surfaces? Why is the technically difficult ultrastructural approach more suitable for determining this ratio than fluorescent immunocytochemistry, which is technically easier to perform?

c. What assumption is critical in the quantitative approach described in part (b)?

d. In other experiments, cytochalasin was added to MDCK cells, and the subsequent distribution of gp135 was determined by ultrastructural and fluorescent immunocytochemistry. In these cytochalsin-treated cells, gp135 did *not* exhibit the uniform apical distribution shown in Figure 22-8. What control experiment(s) is necessary to show that this drug-

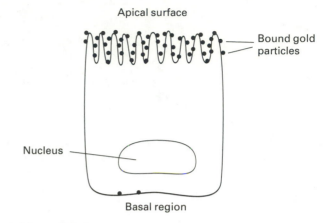

▲ **Figure 22-8**

induced change in gp135 distribution is mediated through an actin-dependent event and is not the result of a direct effect of the drug on gp135?

e. Assuming that cytochalasin does not act directly on gp135, how could immunocytochemistry be used to test the hypothesis that actin filaments are responsible for the distribution of gp135 on the apical surface of MDCK cells?

f. When placed in a low-calcium medium, the monolayer of MDCK cells becomes leaky to membrane-impermeable probes and the transepithelial resistance declines precipitously. What question could be answered by determining the apical-basal distribution of gp135 in cells in a low-calcium medium? (Assume that the low-calcium medium does not significantly affect the intracellular concentration of free Ca^{2+} ions.)

69. The study of human epidermis in culture is now yielding a wealth of data concerning the mechanisms underlying such diverse diseases as squamous-cell carcinoma and psoriasis. Eugene Bell and others have shown that when cultured normal human epidermal keratinocytes are present on floating microporous substrates, the cells can be manipulated with external Ca^{2+} to differentiate into a 20-layered synthetic epidermis. In this system, basal cells are the only ones that divide; apical cells cornify and terminally differentiate. These "squames" slough off the top surface and are replaced by dividing cells in the basal layer of the epidermis, a process similar to what occurs in vivo.

A growing number of keratins have been isolated from epidermis and characterized using Western blot analysis with antibodies that can distinguish the different subtypes of keratins from each other. The expression of the large-molecular-weight keratins K1 and K10 is indicative of terminal differentiation. The expression of K6 and K16 does not normally occur in epidermis at

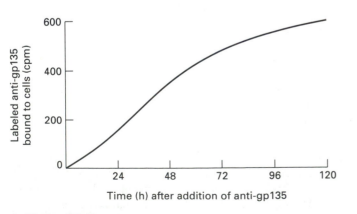

▲ **Figure 22-7**

all, as these proteins are synthesized suprabasally. Raphael Kopan of the Howard Hughes Institute and Elaine Fuchs of the University of Chicago have studied the ability of retinoic acid to regulate the expression of keratins in epidermis in the normal human epidermal keratinocyte (NHEK) system described above and in a squamous-cell carcinoma cell line (SCC-13). Some of the data presented below are representative of their results.

a. The effects of retinoic acid on the incorporation of [^{3}H]thymidine by NHEK cells and SCC-13 cells are shown in Figure 22-9. These data appear to indicate that SCC-13 cells proliferate more rapidly than NHEK cells and that retinoic acid stimulates cell division in both cell types. Are these valid conclusions based on these data?

b. How could the location of proliferative cells be demonstrated in these cell cultures?

c. When the expression of various keratins in NHEK and SCC-13 cells in the presence and absence of retinoic acid was determined by Western blot analysis, the profiles depicted in Figure 22-10 were obtained. What is the interpretation of these data?

d. Several of the keratins migrate together in the blots shown in Figure 22-10. How could these comigrating keratins be distinguished?

e. Keratins K6 and K16 are often associated with hyperproliferation. Design an experimental protocol to determine if the actively dividing cells in the NHEK and SCC-13 cultures are the only ones that express K6 and K16.

70. When various types of animal cells are viewed in the phase microscope, several types of movement are observed. These include cytoplasmic streaming, ruffling, axonal transport, chromosome movement, cytokinesis, and movement of granules. Since 1985 it has been clear

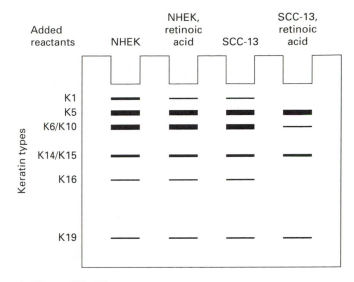

▲ **Figure 22-10**

that many of these movements involve microtubules and kinesin, but recent studies have indicated that actin filaments may be involved in some cellular movements. More specifically, Theresa Hegmann and coworkers at the University of Iowa have suggested that tropomyosin may play a key role in a postulated actin-based granule movement.

The Iowa team isolated five isoforms of tropomyosin (a, b, 1, 2, and 3) from chicken embryo fibroblast (CEF) cells and produced monoclonal antibodies against them. Each of the antibody preparations was used in a Western blot analysis of a total cytosolic extract from CEF cells containing all five tropomyosin isoforms. The resulting autoradiogram depicted in Figure 22-11 reveals the specificity of each of the antibodies (CH291, CG1, CG3, and CGβ6).

a. After each of these antibodies was injected in CEF cells, the speed of granule movement was deter-

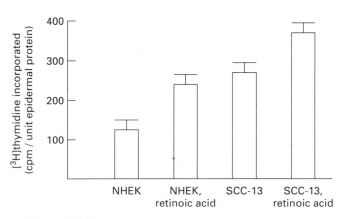

▲ **Figure 22-9**

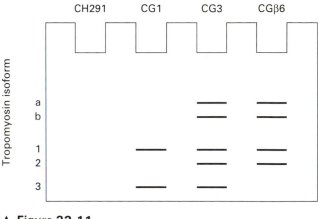

▲ **Figure 22-11**

mined. The observed speeds were as follows (S.E.M. in each case is less than 1.5 μm/min):

Antibody injected	Granule speed (μm/min)
None	20.8
CH291	17.3
CG3	17.3
CGβ6	17.9
CG1	4.9

Calculate the change in the speed of granule movement following injection of CG1.

b. Based on the data presented in part (a), which tropomyosin isoform is necessary for granule movement?

c. How could CG1 be used as its own control so as to demonstrate that the interaction of this antibody with isoforms 1 and 3 causes the change in speed of granule movement?

d. In subsequent experiments, the Iowa team monitored the speed of granule movement in uninjected and CG-1 injected CEF cells at various times after injection. Data similar to their results are shown in Figure 22-12. What do these data indicate?

e. Finally, the Iowa researchers mixed isolated actin filaments and tropomyosin and then incubated the mixture with each of the four antibodies used in the previous experiments. After a 30-min incubation, the samples were separated into a pellet containing actin filaments and a supernatant containing soluble components. Both the pellet (P) and supernatant (S) were analyzed by SDS gel electrophoresis for the presence of actin and tropomyosin isoforms 1, 2, and 3. What is the purpose of this experiment and what do the results depicted in Figure 22-13 indicate? Why is CH291 a good control for this experiment?

71. Genetic experiments have revealed that myosin is a critical component of cytokinesis in nonmuscle cells. Myo-

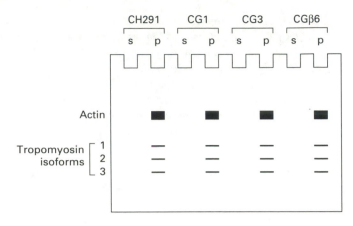

▲ **Figure 22-13**

sin is known to translocate to various regions of nonmuscle cells during cell movement; this process is associated with the phosphorylation of the myosin light and heavy chains. One of the most intriguing cases in which myosin is critical to cell motility is the slime mold *Dictyostelium*. Myosin exists in thick filaments in cells of this mold and moves from the cytoplasm to the cortex in response to the chemoattractant cAMP. This translocation correlates with the phosphorylation of myosin. James Spudich and his colleagues at Stanford have examined the possible role of myosin phosphorylation in the in vitro assembly of thick filaments. The data presented in this problem are representative of their results.

a. Spudich's group prepared myosin isolated from *Dictyostelium* using a rotary shadow technique. When this myosin was examined under low ionic strength conditions, the three forms diagrammed in Figure 22-14 were visible in the electron microscope. Describe the differences in these three forms.

b. When the myosin preparation was treated with high salt, only form 2 in Figure 22-14 was visible. What does this effect of high salt suggest about the nature of forms 1 and 3? Why is this finding significant in terms of what might occur in vivo in these mold cells?

c. As indicated in Figure 22-14, form 3 myosin is staggered. Using an antibody (MY1) specific to a region of the myosin molecule that is 120 nm from the head-tail junction, Spudich's group determined that the two myosin molecules in form 3 are staggered by 16 nm. How could this determination be accomplished?

d. When partially purified myosin heavy-chain (HC) kinase was used to phosphorylate myosin isolated from *Dictyostelium* at low salt, there was a phosphorylation-dependent increase in the number of bent monomers (form 1). The data are summarized in

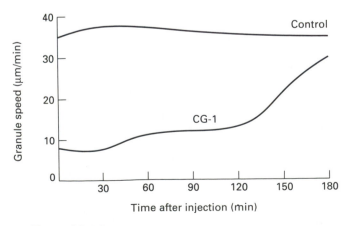

▲ **Figure 22-12**

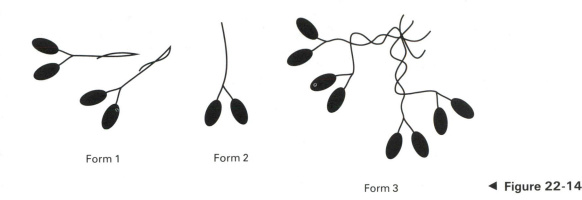

Form 1

Form 2

Form 3

◀ **Figure 22-14**

Figure 22-15. Can you conclude from these data that phosphorylation of monomers causes them to bend their tails?

e. What type of controls should be performed for the experiment described in part (d)?

f. Design an experiment involving differential centrifugation to further test the hypothesis that phosphorylation of myosin monomers causes bending of their tails to give form 1 myosin.

72. In the past few years, it has become apparent that several types of myosin occur in nonmuscle cells. Two types have been identified based on their abilities to form filaments. Myosin II, which has two heavy chains and two pairs of light chains, is the conventional myosin of muscle; it also is found in certain nonmuscle cells. Myosin I, which consists of several isotypes including 1A and 1B, is a monomeric form that lacks a fibrous tail; this myosin is found only in nonmuscle cells. Isozymes of myosin I occur in *Dictyostelium,* intestinal brush borders, and amebas. Both myosin I and II have an actin-binding site in the head region; myosin I also has an actin-binding site in its nonhelical carboxyl-terminal domain.

a. Hidetake Miyata, Blair Bowers, and Edward Korn of the National Institutes of Health have recently determined the location of myosin I in *Acanthamoeba* in order to help reveal the functions of myosin I in nonmuscle cells. They prepared plasma membrane preparations of these cells and did immunofluorescence microscopy using an antibody against myosin I. Punctate patches indicating the presence of myosin I were evident in these preparations. What type of experiment could be performed to determine whether myosin I is associated with actin filaments in these preparations?

b. When lysed *Acanthamoeba* cells were centrifuged and the supernatant and pellet fractions analyzed for myosin I and actin filaments, a substantial amount of myosin I was found in the supernatant. Is this observation consistent with the localization of myosin I in the plasma membrane described in part (a)?

c. When whole *Acanthamoeba* cells were analyzed by fluorescent immunocytochemistry, all of the anti-myosin I staining was found in a ring coincident with the protoplasmic surface of the plasma membrane. How does this finding affect your answer to part (b)?

d. The amounts of actin and myosin I in isolated *Acanthamoeba* plasma membranes were determined before and after salt extraction. The results are shown in Table 22-2. Do these data support the hypothesis that the association of myosin with the plasma membrane is mediated by actin filaments?

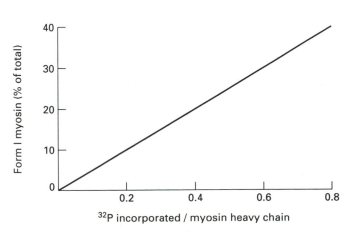

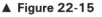

▲ **Figure 22-15**

Table 22-2

Plasma membrane treatment	Relative concentration (%)	
	Actin	Myosin I
No salt extraction (native membranes)	100	100
KI extracted	13	58
KC1 extracted	39	45

e. The amounts of myosin I that binds to the KI-extracted plasma membrane described in part (d) are determined at increasing concentrations of free myosin I. Predict the results of this experiment in view of the data in Table 22-2. (Assume that bovine serum albumin is used to block nonspecific binding.)

f. Would preincubation of isolated plasma membranes with actin filaments accelerate the association of myosin I with the plasma membranes? (Assume that all nonspecifically associated F actin is washed from the plasma membranes before the addition of myosin I.)

C H A P T E R # 23

Multicellularity: Cell-Cell and Cell-Matrix Interactions

PART A: *Reviewing Basic Concepts*

Fill in the blanks in statements 1–24 using the most appropriate terms from the following list:

I	integrin
II	lignin(s)
III	lysyl oxidase
VI	meristem
agrin	N-cadherin
basal lamina	N-terminal
C-terminal	pectin(s)
Ca²⁺	peroxidase
collagen	phloem
desmosomes	plasmodesmata
duplication	K⁺
extensin	propeptides
fibronectin	proteoglycan
glycoprotein	rough endoplasmic reticulum
glycosylation	splicing
homotypic	uvomorulin
hyaluronic acid	
hydrolysis	

1. _____ interactions are those that occur between cells of the same type not between cells of two different tissues.

2. The most abundant extracellular matrix component in animals is _____.

3. The extracellular matrix substance _____ gives tissues a gel-like consistency because of its hydrated characteristic.

4. The present-day collagen genes arose through _____.

5. The _____ of collagen are globular domains at the N- and C-termini that are cleaved at the time collagen is released from cells.

6. Disulfide bonds in collagen molecules form first at the _____ end.

7. Glycosylation of collagen begins in the _____.

8. Oxidation of lysine and hydroxylysine residues in collagen by _____ in the extracellular space is partially responsible for formation of collagen fibrils.

9. Type _____ collagen is the major form of collagen found in cartilage.

10. Type _____ collagen is different from other collagen subtypes in that much of the molecule does not form triple helices.

11. The _____ is a 60- to 100-nm-thick sheet of extracellular matrix material, usually containing both collagen type IV and laminin, that underlies epithelial and endothelial cells.

12. _____ is an extracellular matrix component made up of repeating units of the disaccharide glucoronic acid $\beta(1 \rightarrow 3)N$-acetylglucosamine.

13. Entactin, an extracellular matrix component present in basal laminae, is a _____.

14. A group of cell-surface receptors collectively called the _____ receptor family can bind laminin and fibronectin and are characterized by α and β transmembrane polypeptide chains.

15. Cell-adhesion proteins can be classified into two major classes based on their requirement for _____ to effect adhesion.

16. If the epithelial-specific cell-adhesion protein _____ is removed from or blocked in cultured kidney cells, tight junctions and gap junctions do not form.

17. During morphogenesis ectodermal cells destined to differentiate into neuronal tissue lose their uvomorulin and develop the cell-adhesion protein _____.

18. The diversity of nerve-cell adhesion molecule (N-CAMs) arises through _____ and _____.

19. _____, a protein found in the basal lamina of muscle cells, causes the clustering of both acetylcholinesterase and acetylcholine receptors.

20. Cell division in plants is restricted to specific regions called the _____.

21. _____ are major constituents of plant cell walls and can be highly hydrated, thus forming a gel.

22. _____ is a cell-wall glycoprotein that is characterized by an abundance of glycosylated hydroxyprolines.

23. _____, a cell-wall component, is very prevalent in woody plants and helps guard against infection by predators.

24. Fluid movement through the phloem is effected by enlargement of the _____.

PART B: *Linking Concepts and Facts*

Circle the letters corresponding to the most appropriate terms/phrases that complete or answer items 25 – 36; more than one of the choices provided may be correct in some cases.

25. The shape of tissues is maintained by which of the following extracellular matrix components?

    a. laminin   d. elastin

    b. fibronectin   e. proteoglycans

    c. collagen

26. The functions of the extracellular matrix include

    a. supporting differentiation.

    b. inducing morphogenesis.

    c. binding growth hormones.

    d. filtering.

    e. providing a dense framework for some structures and tissues.

27. Which of the following extracellular matrix components is a right-handed triple helix 30 nm in length?

    a. fibronectin   d. proteoglycans

    b. collagen   e. microtubules

    c. laminin

28. Which feature(s) of the collagen molecule is (are) most important in determining the varied structures and properties of the different native collagens?

    a. the C-terminal propeptide

    b. the N-terminal propeptide

    c. the nonhelical domains

    d. the substitution of glycine by hydroxyproline

    e. the length of the α chains

29. Which of the following extracellular matrix components contains three different polypeptide chains that form a cross-shaped molecule?

a. collagen type I   d. laminin

b. collagen type IV   e. hyaluronic acid

c. fibronectin

30. Proteoglycans are a group of cell-surface and extracellular matrix substances that

     a. are variable in molecular weight.

     b. are highly positively charged.

     c. have a molecular weight less than 1000.

     d. may contain heparan sulfate as a constituent.

     e. bind to collagens and fibronectin.

31. Fibronectin

     a. is found on the surface of transformed cells.

     b. promotes adhesion of cells.

     c. is evolutionarily related to the collagens.

     d. is a glycoprotein

32. Cadherins

     a. are soluble proteins found in the cytosol.

     b. are calcium-independent.

     c. are cell-adhesion molecules.

     d. can have their function blocked by antibodies directed against them.

     e. consist of at least three subclasses.

33. The major difference between N-CAM in adult tissues and N-CAM in embryonic tissues is that

     a. N-CAM from embryonic tissue has more sialic acid.

     b. N-CAM from adults has a much greater molecular weight.

     c. N-CAM is not expressed in embryonic tissues.

     d. N-CAM is more rapidly turned over in adult tissues.

     e. N-CAM promotes better cell-to-cell adhesion in embryonic versus adult cells.

34. Experiments with developing neurons indicate that growth cones follow the correct path and associate with the correct synaptic partner. In this process,

     a. laminin provides all the specificity necessary.

     b. glial cells provide some of the specificity necessary.

     c. the point of exit from the ganglion provides all the necessary information.

     d. neurons follow an original ("pioneer") neuron to the right partner.

     e. specific cell-adhesion molecules on the growth cone guide it to the proper partner.

35. Regenerating axons follow chemical cues when reinnervating the neuromuscular junction. These receptor molecules are present

     a. on glial cells.

     b. on the soma of the regenerating neuron.

     c. on the myofibril.

     d. on the basal lamina.

     e. in the interstitial fluid.

36. The pectins and hyaluronic acid are both

     a. secreted by animal cells.

     b. very negatively charged.

     c. proteins.

     d. intracellular substances

     e. highly hydrated.

PART C: *Putting Concepts to Work*

37. An unusual characteristic of type I collagen is that it contains a glycine residue every third amino acid. Why is this critical to the tertiary structure of this protein?

38. What is the functional significance of the proline and hydroxyproline residues in collagen?

39. Which structural features of collagen I fibrils are responsible for their high tensile strength?

40. In vitro studies have shown that the renaturation of thermally denatured procollagen is substantially accelerated once the proper interchain disulfide bonds have

formed. What does this finding imply about the biosynthesis of type I collagen?

41. Which properties of hyaluronic acid contribute to its ability to resist compression forces?

42. The proteoglycan aggregate found in cartilage contains a central molecule from which radiate many core protein molecules in a brushlike array. What chemical compound constitutes this central component of the proteoglycan aggregate?

43. Why is the basal lamina often called the type IV matrix?

44. In what sense is laminin a bifunctional matrix protein?

45. The fibronectin receptor has been shown to bind not only fibronectin but also vinculin, talin, and actin. What does this finding suggest about the nature and function of this receptor?

46. Some researchers have transplanted presumptive epidermal or dermal cells from one area of an early chick embryo to another area and examined the resulting effects on skin differentiation. Such experiments provide information about what type of tissue-tissue interaction? What is the result of such interactions in vivo?

47. Basic fibroblast growth factor (FGF) and acidic FGF are tightly bound to heparin sulfate in the basal lamina. How does this association enhance the activities of these hormones?

48. Experiments with developing grasshopper neurons indicate that these neurons (and probably others as well) have specific surface markers that can recognize other surface markers on target cells. These interactions are specific and can account for synaptic specificity. Remarkably, however, only a few cell-adhesion molecules have been identified in nerves, most notably N-cadherin and N-CAMs. How could just a restricted number of such adhesion molecules account for the cell-to-cell specificity observed during formation of neuronal networks?

49. Explain the statement "cellulose is to plant cells as collagen is to tendons."

50. The synthesis of phytoalexins in plants exposed to pathogenic fungi involves a positive-feedback mechanism. Describe this induction process.

51. How are pectins and hemicelluloses involved in maintaining the multicellular integrity of a plant?

52. Collagen in the cornea of the eye and cellulose in the cell wall of plants form layers of parallel fibers. In both cases, the fibers in adjacent layers are oriented at right angles to each other. What purpose is served by this repeating, perpendicularly oriented organization of the fibers in these structures.

53. What evidence suggests that microtubules may be involved in the organization of cellulose microfibrils?

PART D: *Developing Problem-solving Skills*

54. You want to apply a solution of type I collagen to plastic petri dishes in order to observe its effects on the morphology of liver cells maintained in vitro. However, after isolating type I collagen from adult rat tails, you note that it is not soluble in water. Why is this collagen preparation insoluble in water? How could you prepare a collagen solution to apply to the surface of the petri dish, thus forming a uniform layer of type I collagen?

55. Many connective tissue diseases result from synthesis of aberrant collagens. Some of these collagens can be distinguished from native collagens by their different migratory pattern on a denaturing (SDS) gel. SDS gel electrophoresis of collagen from a child who exhibits a possible collagen-deficiency disease produces a profile indicative of a type I collagen deficiency. When the separated chains were dialyzed from the gel and recombined in an appropriate buffer, they *failed* to reassociate into a native triple helix. Do these data indicate that the molecular defect in this person involves the inability of collagen to form its triple helix once secreted in the extracellular matrix?

56. Cultured fibroblasts were treated with a drug known to disrupt the RER → Golgi transport pathway of collagen. The extracellular matrix and cell contents from

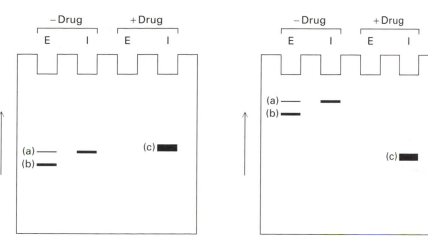

(a) Denaturing gel

(b) Nondenaturing gel

E = extracellular matrix
I = intracellular matrix

◄ **Figure 23-1**

treated and nontreated cells were subjected to Western blot analysis on an SDS gel using a monoclonal antibody that recognizes the internal domain of one of the α chains in type I collagen. The resulting profiles are shown in Figure 23-1a.

a. Which polypeptides are represented by the indicated bands (a, b, and c) in Figure 23-1a? What do these data suggest about the possible mechanism(s) of action of this drug and its ability to inhibit the depositon of type I collagen in the extracellular matrix?

b. When this experiment was repeated with a native (nondenaturing) gel, the profiles shown in Figure 23-1b were obtained. Which polypeptides are represented by the indicated bands (a, b, and c) in this profile? What do these data suggest about the action of the drug on type I collagen synthesis and secretion?

57. You discover a collagen-related disease clinically manifested by an unusual deformity of the dermis of the skin. Western blot analyses indicate that collagen in the extracellular matrix of affected individuals has a slightly greater molecular weight than does collagen from normal individuals under both denaturing and nondenaturing conditions. What is the probable molecular defect causing this particular collagen disease?

58. MDCK cells deposit a basement membrane containing type IV collagen when grown on microporous membranes but not when maintained on conventional plastic cell culture dishes. How could you determine whether the difference between these two cell forms results from the ability of the cells on microporous membranes to permit/induce synthesis of type IV col-

lagen or from the ability of the microporous membranes to facilitate release of collagen to the basal portion of the cell compared with plastic-grown cells?

59. Neurons can develop and extend axonlike processes when grown on laminin films. When hippocampal neurons from the fetal brain are grown on laminin-coated surfaces, a similar phenomenon is seen. However, when these cells are grown on non-laminin-coated surfaces but soluble laminin is added to the medium, the same phenomenon is seen. How could you determine whether the cells are responding to soluble laminin in the medium or to laminin that is probably adhering to the surface of the cell culture dish?

60. The adhesion of the cell line CL5 is greatly enhanced by the addition of a dried film of fibronectin to the culture dish. When radioactively tagged proteolytic fragments of purified fibronectin are incubated with plasma membranes, one particular fragment binds to the membranes and can be specifically displaced by native fibronectin. What is the probable nature of this fragment? How would CL5 cells pretreated with this fragment behave when plated on a fibronectin film overlay in vitro?

61. You wish to examine the effects of the basal lamina on differentiation of a mammary epithelial cell line. First, you grow endothelial cells on a substrate because these cells deposit a basal membrane. Next, you strip off the endothelial cells using a deoxycholate method, leaving an intact basal membrane behind. When the mammary epithelial cells are plated on this basal lamina previously deposited by endothelial cells, the former aggregate into alveolarlike aggregates. On fibronectin-coated dishes, however, the mammary epithelial cells do not form simi-

lar aggregates. What criteria can be used to indicate that the cells on the endothelial basal laminia are more differentiated than those on the fibronectin-coated surface?

62. Most epithelial cells grown in culture require the addition of the calcium chelator EDTA or EGTA to dissociate confluent monolayers into single cells. Often trypsin must be added as well to generate a suspension of single cells. What is the probable action of these agents? What enzymes would you choose if you wanted to dissociate cultured plant-cell monolayers?

PART E: *Working with Research Data*

63. The integrin superfamily of adhesive receptors are currently under intensive investigation. These cytoadhesion receptors are 140-kDa heterodimeric molecules. They interact with extracellular matrix components such as laminin, collagen, vitronectin, and fibronectin and are involved in a vast variety of cell-to-cell and cell-to-matrix interactions. Many of the integrin receptors are characterized by the ability to bind specifically to an Arg-Gly-Asp (RGD) sequence. This tripeptide has been useful in affinity chromatographic techniques to isolate and characterize the RGD-binding integrin receptors. The mechanism(s) by which cells can (a) increase the number of integrin receptors and/or (b) modulate the activity of individual integrin receptors is of extreme interest. Recent investigations have shown that the cytoplasmic domains of some integrin receptors can be phosphorylated by oncoviral tryosine kinases. This, in turn, may lead to an altered cell morphology and adhesion characteristics of transformed cells.

Dr. Danilov and R. L. Juliano at the University of North Carolina have examined the possible role of protein kinase C in modulating the interaction of integrin receptors with fibronectin. They plated ^{35}S-labeled Chinese Hamster Ovary (CHO) cells on surfaces pretreated with fibronectin and then added the phorbol ester PMA, which is known to stimulate protein kinase C. The percentage of cells adhering to the substrate at various times after addition of PMA was determined. The data shown in Figure 23-2 are representative of their results.

a. Was the use of ^{35}S as a probe critical in these experiments?

b. Based on the data in Figure 23-2, what effect does PMA have in this system?

c. Danilov and Juliano found that increasing concentrations of PMA enhanced adhesion of cells up to a concentration of 100 nM PMA. What does this dose-dependent response indicate about how PMA affects cell adhesion?

d. What types of experiments could be done to further test the hypothesis that the PMA effect is modulated through an increase in protein kinase C

e. When the concentration of fibronectin on the surface of the culture dish was varied, it was noted that PMA had an effect at lower, but not at higher, concentrations of fibronectin; that is, adhesion of both PMA-treated and control cells saturated at about

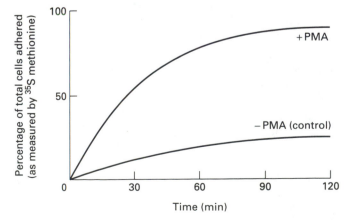

▲ **Figure 23-2**

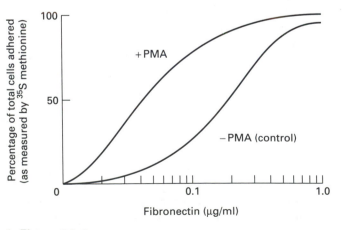

▲ **Figure 23-3**

1.0 μg/ml fibronectin, as shown in Figure 23-3. Do these data suggest that PMA acts by increasing the number of integrin receptors or their affinity for fibronectin?

f. When the peptide Gly-Arg-Gly-Asp-Ser-Pro (GRGDSP) was added to the CHO cell suspension prior to plating the cells on the fibronectin surface, adhesion of both control and PMA-treated cells to the substrate was inhibited. Explain these results.

g. In similar experiments, an anti-fibronectin receptor antibody, PB1, was substituted for GRGDSP. Would you expect PB1 to have the same effect on adhesion as the inhibitory peptide?

h. When culture dishes were pretreated with polylysine rather than fibronectin, PMA exerted no stimulatory effect on the adhesion of CHO cells. What addi-

tional information does this finding provide about the effect of PMA?

i. The results described in part (e) demonstrate that PMA does not increase the number of fibronectin receptors but rather increases the receptor affinity for fibronectin. Propose a mechanism by which PMA inreases receptor affinity. How could you test this hypothesis?

64. Elizabeth Hay at Harvard Medical School has been a leader in examining the effects of the extracellular matrix on cell differentiation. In recent studies with Anna Zuk and Karl Matlin of Harvard, Hay demonstrated that the Madin-Darby canine kidney (MDCK) cell line can become fusiform in shape when migrating on a type I collagen gel (Figure 23-4a). However, when these cells become confluent on a synthetic basement-membrane

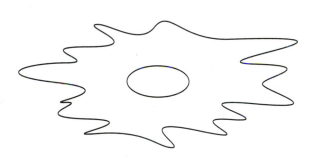

(a) Fusiform cell in collagen gel

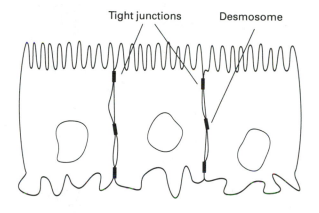

(b) Polarized monolayer on basement-membrane gel

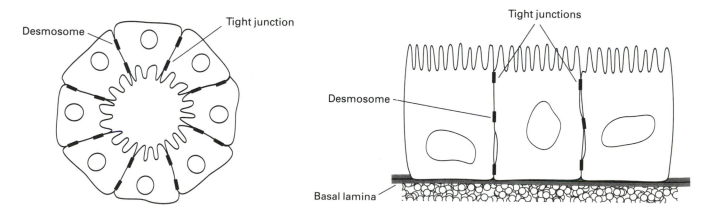

(c) Tubule in collagen gel

(d) Polarized monolayer with basal lamina on microporous membrane

▲ **Figure 23-4**

gel, they organize themselves into a highly polarized monolayer with the same intercellular junctions evident in renal tubules in vivo (Figure 23-4b). When MDCK cells explanted from this basement-membrane gel are grown within a hydrated type I collagen gel, they give rise to tubulelike structures with microvilli pointing inward (Figure 23-4c). Finally, other investigators have shown that MDCK cells grown on a type I or laminin-coated microporous membrane polarize and secrete a basement membrane (Figure 23-4d), a feature not seen in Hay's experiments. Because fusiform MDCK cells resemble mesenchymal cells in their morphology, Zuk and her colleagues have performed numerous experiments to determine whether these cells in their various morphological forms contain components typical of true mesenchyme cells. Some, but not all, of the data below are representative of their results.

a. When fusiform MDCK cells were stained with an antibody for type I procollagen, they exhibited no intracellular immunofluorescence, indicating that these fusiform cells may not be true mesenchymal cells. Why was it critical that the antibody be able to recognize procollagen instead of only mature collagen?

b. What control could be done for the experiment described in part (a) to demonstrate that the antibody would have recognized type I procollagen in the MDCK cells if it had been present?

c. When another group performed a similar immunofluorescent analysis of MDCK cells growing on microporous membranes, they noted no staining of the basal lamina. Assuming that the antibody used in this experiment can recognize type I procollagen, what is the most plausible reason for this result?

d. When fusiform MDCK cells were exposed to an antibody against ZO-1 (a marker of tight junctions), no staining was detected. What does this finding suggest about the nature of these cells? Would you expect this antibody to stain MDCK cells grown on a basement-membrane gel?

e. In preliminary experiments performed by another group, the expression of type IV collagen was analyzed in fusiform MDCK cells and in MDCK cells grown on microporous membranes. An antibody that recognizes both procollagen and collagen was used. A punctate staining pattern was seen in cells grown on the microporous membrane, but no staining was seen in fusiform cells. Why might these results be consistent with the morphologies of these cells as depicted in Figure 23-4?

f. Which morphological cell type sketched in Figure 23-4 would be expected to have the greatest number of integrin receptors?

g. Intracellular laminin can be demonstrated immunocytochemically in fusiform MDCK cells but not in cells grown on a microporous membrane. Propose a negative-feedback system that would account for these findings.

65. Nerve-cell adhesion molecules (N-CAMs) can affect a variety of cellular processes including cell adhesion, cell division, junctional communication, neurotransmitter synthesis, and guidance of axons via glial cells to their target cells. These substances contain long chains of polysialic acid (PSA); the greater their PSA content, the weaker the cell adhesion mediated by N-CAMs. The N-CAMs in embryonic tissues usually have a high PSA content, whereas the N-CAMs in the same tissues in adults have a relatively lower PSA content. It is thought that embryonic tissue is more plastic than adult tissue and that the high PSA content of its N-CAMs prevents adhesion during morphogenesis and development.

Schwann cells are specialized glial cells that wrap around neurons, thus forming a myelinated nerve. Both Schwann cells and myelinated neurons contain two types of cell-adhesion molecules, N-CAMs and L1. Bernd Seilheimer, Elke Persohn, and Melitta Schachner at the University of Heidelberg in West Germany have recently examined the expression of these two types of cell-adhesion molecules in Schwann cells and dorsal root ganglion cells. Some of the data below are representative of their results.

a. Dorsal root ganglion cells and Schwann cells grown in pure cultures were subjected to quantitative immunogold labeling with antibodies against L1 and N-CAMs to determine the density of these cell-adhesion molecules. The density of gold particles in pure cultures exposed to the L1 antibody was 1 particle per 690 ± 30 nm in Schwann cells and 1 particle per 2010 ± 51 nm in dorsal root ganglion cells. The density in pure cultures exposed to antibodies to N-CAMs was 1 particle per 1720 ± 34 nm in Schwann cells and 1 particle per 1668 ± 35 nm in ganglion cells. Why are these data expressed in terms of a linear measure (nm) rather than an area measure (μm^2)?

b. When the experiment described in part (a) was repeated with cocultures of Schwann cells and dorsal root neurons, a decrease in the expression of L1 and N-CAMs (as indicated by the density of gold particles) was noted. The percentage decreases in cocultures compared with pure cultures are shown in Table 23-1. Why were fibroblasts used in this study?

c. How could one determine whether the coculture-induced decrease in L1 and N-CAM expression resulted from Schwann cell–neuron *contact* or from a factor produced by one cell type affecting the other?

Table 23-1

Coculture components	Decrease in expression (%)	
	L1	N-CAM
Schwann cell	90 ± 3	41 ± 3
Dorsal root neurons	36 ± 2	9 ± 2
Fibroblasts	—	6 ± 3
Dorsal root neurons	4 ± 3	4 ± 3

d. The West German researchers exposed cocultures of Schwann cells and dorsal root neurons to an antibody against the dorsal root neurons. This treatment produced immunocytolysis of the neurons but left the Schwann cells intact. What feature of the coculture-induced decrease in the expression of L1 and N-CAMs could be investigated in this experiment?

e. Another research group has used an antibody to N-CAMs to examine embryonic chick brain cells for the presence of N-CAMs. In cultures treated with this antibody, the apposition of cell membranes was less than in untreated cultures. What is the most probable reason for the effect of this antibody?

f. If binding of N-CAM antibody was found to be less in embryonic chick neurons than in adult neurons, what would be the most likely reason for this difference? (Assume that the number of receptor molecules expressed in both cases is the same.)

g. When embryonic chick neurons were treated with endoneurominidase N, an enzyme that modifies N-CAMs, the apposition of the cell membranes of the treated embryonic neurons appeared similar to that of untreated adult neurons. What effect of endoneurominidase N treatment would explain these findings?

66. A critical question concerning cell development is the spatial and temporal expression of the extracellular matrix and its respective receptors during differentiation. For instance, does the appearance of laminin precede the expression of the laminin receptor? Similar questions can be posed concerning fibronectin, the collagens, etc. To answer such questions, a variety of techniques can be used including Northern blot analysis, Western blot analysis, and immunoprecipitation. These techniques require specific cDNAs and antibodies corresponding to the various matrix proteins. Dr. Yamada and coworkers at the National Institute of Dental Research have recently used these techniques to investigate the temporal and spatial expression of laminin and the laminin receptor in developing mice kidney. The data presented in this problem are representative of their results.

a. Messenger RNA from developing mice kidneys was subjected to Northern blot analysis using cDNA probes encoding the laminin receptor (LMR), the laminin B1 chain (LMB1), and the laminin A chain (LMA). The resulting blots are depicted in Figure 23-5 for kidney mRNA from mice at various developmental times from 16 days of gestation to 3 weeks after birth (lanes 1–5). Messenger mRNA from the F9 cell line was also analyzed with the three probes (lane 6). What conclusions can be drawn from these blots concerning the temporal expression of LMR, LMB1, and LMA mRNAs in developing mice kidneys?

b. The F9 cells have nothing to do with the events under study in part (a). Why was it critical that they be analyzed nonetheless?

c. Based on the Northern blots in Figure 23-5, what protein bands would you expect to be revealed by Western blot analysis of the same cells?

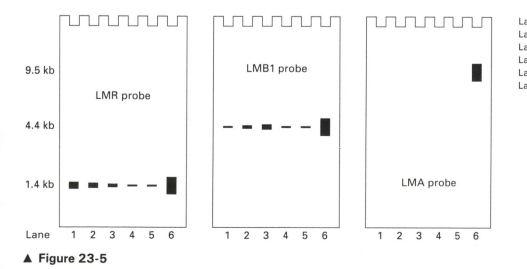

Lane 1 = kidney mRNA at 16 days gestation
Lane 2 = kidney mRNA at birth
Lane 3 = kidney mRNA 1 week postnatal
Lane 4 = kidney mRNA 2 weeks postnatal
Lane 5 = kidney mRNA 3 weeks postnatal
Lane 6 = F9 cell mRNA

▲ **Figure 23-5**

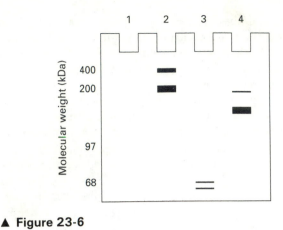

▲ **Figure 23-6**

d. Figure 23-6 shows the results of immunoprecipitaton of ^{35}S-labeled primary cultures of newborn mouse kidney epithelial cells with preimmune serum (lane 1), anti-laminin antibodies (lane 2), anti-laminin receptor antibodies (lane 3), and anti–type IV collagen antibodies (lane 4). Do these data demonstrate that all three laminin chains are present in these cells?

e. In Figure 23-6, two closely spaced bands appear in the laminin receptor lane (lane 3). What species might this doublet represent?

f. Lane 4 in Figure 23-6 contains one larger band corresponding to one of the type IV collagen chains. However, there also is a second, smaller band representing a contaminating protein with a molecular weight similar to that of one of the laminin chains in lane 2. Assuming that the small band in lane 4 in fact is a laminin chain, why might it be revealed in this case?

g. What type of experiment could be performed to determine whether the small band in lane 4 of Figure 23-6 is indeed a laminin chain?

Cancer

PART A: *Reviewing Basic Concepts*

Fill in the blanks in statements 1–21 using the most appropriate terms from the following list:

Ames

antioncogene

basal lamina

benign

bone marrow

coda

collagen

cytochrome P-450s

electrophile(s)

enhancer insertion

epigenetic

Epstein-Barr

fibronectin

genetic

G protein

inactivated carcinogens

infrared radiation

initiators

line

hepatitis B

HIV

inactive gene

long terminal repeat (LTR)

malignant

mutant

nucleophile(s)

oncogene

plasminogen activator

promoter(s)

promoter insertion

quinones

retrovirus

serine

strain

threonine

transduction

transformation

transforming growth factor (TGF)

tyrosine

ultimate carcinogens

UV radiation

visible light

x-rays

1. _____ tumors are localized and contain cells with a normal appearance, whereas _____ tumors do not remain localized and often are composed of cells with less differentiated characteristics.

2. The main physical barrier that keeps body tissues separated is the proteinaceous _____.

3. A culture of cells with an indefinite life span is considered immortal; such a culture is called a cell _____ to distinguish it from an impermanent cell _____.

4. The changes in the growth properties of cultured cells that confer on them the ability to form tumors when injected into susceptible animals are collectively called _____.

5. A(n) _____ is a protein secreted by transformed cells that can stimulate the growth of normal cells.

6. Transformed cells often secrete a protease that is called _____.

7. Normal quiescent cells in monolayer culture are covered with a dense fibrillar network whose major protein component is _____; transformed cells either lack or have greatly reduced amounts of this protein.

8. A(n) _____ is a gene whose product can function either in transformation of cells in culture or in cancer induction in animals.

9. Each virion of a(n) _____ contains about 50 molecules of a DNA polymerase that is capable of copying its genomic RNA into DNA.

10. When a retroviral RNA is copied to produce a double-stranded DNA, the short repeat at either end of the viral RNA is extended to form a(n) _____.

11. When the 5′ end of a c-*myc* gene transcript contains a sequence from a retroviral LTR, the mechanism by which transcription of this gene was activated is likely to be _____.

12. _____ virus probably plays a role in at least two human tumors: Burkitt's lymphoma and nasopharyngeal carcinoma.

13. The virus that causes acquired immunodeficiency syndrome is called _____.

14. The direct-acting carcinogens, of which there are only a few, are reactive _____.

15. Indirect carcinogens are converted to _____ by introduction of electrophilic centers; these oxidation reactions are catalyzed by a set of proteins that include _____.

16. The _____ test measures the ability of potential carcinogens to mutagenize bacteria.

17. Two types of radiation, _____ and _____, are especially dangerous because they modify DNA.

18. Phosphorylation of _____ residues in proteins may occur at 10-fold higher levels in transformed cells than in normal cells.

19. A(n) _____ condition can be passed from a cell to its progeny without any alteration in the coding sequence of the DNA.

20. Phorbol esters, which act on the cellular enzyme protein kinase C, are a class of tumor _____.

21. A(n) _____ is a gene whose loss leads to malignant transformation. One example of this type of gene is the *RB* gene.

PART B: *Linking Concepts and Facts*

Circle the letters corresponding to the most appropriate terms/phrases that complete or answer items 22–32; more than one of the choices provided may be correct in some cases.

22. 3T3 cells, which are derived from embryonic rodent tissue,

   a. constitute a cell strain.

   b. contain an oncogene that produces a product found in the nucleus.

   c. do not respond to growth factors in serum.

   d. can be transformed with SV40 virus.

   e. can grow indefinitely.

23. Compared with normal cells, transformed cells in culture

   a. have decreased levels of glucose transport.

   b. have fewer actin microfilaments.

   c. form gap junctions more frequently.

   d. have more mobile surface proteins.

   e. can grow in medium with a lower serum concentration.

24. Papovaviruses

   a. are RNA viruses.

   b. include SV40.

   c. cause nonpermissive cells to replicate continuously.

   d. integrate into the host chromosome at a specific site in the viral but not in the host-cell DNA.

   e. produce late proteins that cause transformation.

25. Proteins that can confer immortality to cells

    a. are usually found in the cytoplasm.

    b. include polyoma T protein.

    c. include *myc* gene products.

    d. include *ras* gene products.

    e. can act without producing morphological changes in cells.

26. The reverse transcripts of retroviruses

    a. are shorter than the viral RNA.

    b. integrate at random sites in the viral and host DNA.

    c. include promoter and enhancer sequences.

    d. include a polyadenylation signal sequence.

    e. may contain sequences derived from cellular genes.

27. The conversion of a proto-oncogene to an oncogene

    a. may occur when the gene is transcribed at a greater rate.

    b. may be caused by a mistake in DNA repair.

    c. could be caused by deletion of a DNA sequence.

    d. probably occurs whenever a retrovirus acquires cellular DNA.

    e. may be caused by insertion of a virus near the gene.

28. Human immunodeficiency virus (HIV)

    a. causes an infection that makes the patient prone to other infections and cancers.

    b. primarily attacks cells of the nervous system.

    c. contains fewer genes than most retroviruses.

    d. grows less vigorously than most retroviruses in infected cells.

    e. has an RNA genome.

29. DNA-repair systems

    a. are sometimes error-prone.

    b. may cause activation of oncogenes.

    c. are found only in eukaryotic cells.

    d. that are defective may be lethal.

    e. that are defective are associated with increased probability of developing certain cancers.

30. The product of a *ras* gene

    a. is located in the nucleus.

    b. binds guanine nucleotides.

    c. can transform 3T3 cells.

    d. has tyrosine kinase activity.

    e. can act synergistically with the product of a *myc* gene to produce transformation.

31. The 3T3 cell assay for oncogenes

    a. involves the uptake of DNA by the 3T3 cells.

    b. detects all known oncogenes with equal sensitivity.

    c. relies on detection of transformed cell foci to determine the presence of an oncogene.

    d. involves isolation of the oncogene product.

    e. can detect oncogenes in approximately 20 percent of human tumors.

32. A tumor promoter

    a. often causes a tumor to be produced when it is applied alone.

    b. must be metabolized before it promotes tumor formation.

    c. is probably an electrophile.

    d. must be present for weeks or months to promote a tumor.

    e. may lead to an irreversible alteration in cellular metabolism when applied following the application of an initiator.

33. Oncogenes can be classified according to the nature and cellular location of their protein products. The general types of oncogene products and the possible cellular locations are as follows:

Product type	Cellular location
nuclear transcription factor (NTF)	Secreted (S)
growth factor (GF)	cytoplasm (C)
protein kinase that phosphorylates tyrosine (PTK)	plasma membrane (PM)
protein kinase that phosphorylates serine or threonine (PK)	nucleus (N)
guanine nucleotide-binding protein with GTPase activity (GTPase)	
protein related to a phospholipase C and to the src gene product (PLC)	
thyroid hormone receptor (THR)	
receptor with protein-kinase activity (RPK)	

Using the abbreviations given in the list above, indicate the type of protein product, and its cellular location, produced by each of the following groups of oncogenes.

	Product	Location
a. *jun, myc, fos, ski*	——	——
b. *mos, raf(mil)*	——	——
c. *sis*	——	——
d. Ha-*ras*, N-*ras*	——	——
e. *erb*A	——	——
f. *src, abl, met, fps*	——	——
g. *crk*	——	——
h. *ros*, *erb*B, *fms*	——	——

PART C: *Putting Concepts to Work*

34. What features define the G_0 phase? How does it relate to the G_1 phase? How does the length of time spent in the G_0 phase in normal cells compare with the length of time spent in the G_0 phase in cancer cells?

35. Describe the course of an SV40 infection in permissive and nonpermissive cells.

36. What is the usual effect of a retroviral infection on a somatic cell?

37. What are the two different ways in which retroviral infections can produce cell transformation?

38. Explain why a double-strand break in chromosomal DNA is difficult to repair. What type of chromosomal change can this type of damage cause? How can such a chromosomal rearrangement lead to transformation of a cell?

39. Describe the types of changes in DNA that may be responsible for conversion of a proto-oncogene to an oncogene.

40. Describe the "two-step" model of transformation induced by oncogenes. What evidence supports this model? What evidence argues against this model?

41. Explain the observation that teratocarcinoma cells injected into adult mice form lethal tumors, whereas the same cells behave like normal cells when they are injected into an early embryo. What type of mechanism is thought to be involved in formation of teratocarcinomas?

42. Assuming that cigarette smoke contains a tumor promoter, what factors might affect whether or not a particular smoker gets lung cancer? Which factors do you think are most important?

43. Describe the probable function of the *RB* gene. Why is this gene considered an antioncogene?

PART D: *Developing Problem-solving Skills*

44. Researchers investigating "retrovirus Q" observed that infection of rat cells with the virus resulted in transformation but no production of virions, whereas infection of chicken cells with the virus led to transformation and production of virions. Fusion of retrovirus Q–transformed rat cells and noninfected chicken cells produced hybrid cells that were capable of producing new virions. Explain these observations. How could these researchers obtain production of retrovirus Q virions in rat cells?

45. In the Ames test, *Salmonella* cells that are unable to produce histidine are mixed with a rat liver extract and a suspected carcinogen. The cells are then plated on a medium without histidine. The plates are incubated to allow any revertant bacteria (those able to produce histidine) to grow. The number of colonies is a measure of the mutagenicity of the suspected carcinogen.

   a. Why is the rat liver extract included?

   b. What would happen if the strain of *Salmonella* used in the Ames test had a defective *recA* gene?

   c. Would UV radiation as well as chemical carcinogens induce a reversion of the mutation in the histidine gene?

46. The development of teratocarcinoma cells into normal cells or tumor cells is dependent on their environment. When placed in early mouse embryos, mouse teratocarcinoma cells develop into normal cells; when injected into adult animals or grown in tissue culture of early embryo cells, the cells develop into tumors.

   a. Which observation would lead you to conclude that the early embryonic environment contains a factor that directs normal development of teratocarcinoma cells?

   b. How would you go about identifying the factor that directs the normal development of teratocarcinoma cells in the early embryo?

47. Plasminogen activators, which are secreted by many tumors, can catalyze the conversion of the serum protease precursor plasminogen to the broad-spectrum protease plasmin, which in turn can activate other proteases that degrade collagen and other components of the basal lamina. However, serum contains high levels of plasmin inhibitors, which theoretically can render plasmin inactive. Recently, workers have shown that fibrosarcoma cells have cell-surface receptors for both plasminogen activators and plasminogen/plasmin. Furthermore, plasmin bound to these cells is catalytically active and is known not to be inhibited by at least one of the serum plasmin inhibitors.

   a. Discounting possible effects of serum plasmin inhibitors, suggest a role for these proteases in tumor growth and progression.

   b. What do the observations concerning serum plasmin inhibitors and cell-surface receptors on fibrosarcoma cells suggest about the localization of the relevant proteolytic processes?

48. A431 cells, which are derived from a human squamous-cell carcinoma of the vulva, express very high levels of epidermal growth factor (EGF) receptor on their sur-

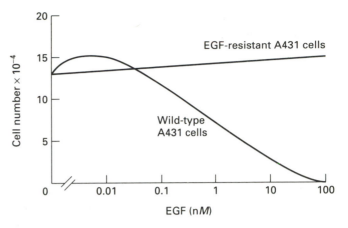

▲ **Figure 24-1**

faces. Kawamoto and coworkers found that very low levels of EGF stimulate the growth of these cells in culture, whereas higher levels of EGF inhibit their growth. King and Sartorelli were able to select clones of A431 cells that are resistant to growth inhibition by EGF; such clones have fewer EGF receptors per cell than do wild-type A431 cells. Typical data showing the number of wild-type and EGF-resistant A431 cells after a 4-day incubation in media containing various concentrations of EGF are depicted in Figure 24-1. King and Sartorelli also examined the ability of wild-type and resistant A431 cells to undergo differentiation, defined as the ability of the cells to form cornified envelopes. Only 18 percent of the wild-type cells differentiated in serum-free medium, whereas 58 percent of the EGF-resistant cells underwent differentiation.

   a. Would wild-type or EGF-resistant A431 cells be more likely to form tumors when injected into experimental animals? Why?

   b. Estimate the EGF concentration to which A431 tumor cells are exposed in vivo.

49. The oncogene *fgr* from a feline transforming virus encodes a protein that consists of a 128-aa peptide of actin fused to the active site of a tyrosine-specific protein kinase. Bearing in mind that actin and vinculin are often found in close association, suggest a role for this protein in producing some of the observed effects of transformation on the cytoskeleton.

50. When diethylnitrosamine is administered to rats in a single dose a few days after birth, a few enzyme-altered foci are detected in the liver at 32 weeks of age. If, in addition to the single dose of diethylnitrosamine, the rats are placed on a diet containing phenobarbital for 4 weeks (beginning at 8 weeks of age), a few enzyme-altered foci are again found at 32 weeks. If, in addition to the single dose of diethylnitrosamine, the rats are placed on a diet containing phenobarbital from the age

of 8 weeks onward, then many enzyme-altered foci and some carcinomas are detected at 32 weeks of age.

a. What do these data indicate about how diethylnitrosamine and phenobarbital act in the induction of carcinomas?

b. What pathology would you expect at 32 weeks if the rats were not treated with diethylnitrosamine and consumed the phenobarbital diet from 8 weeks onward?

51. HIV contains at least six genes in addition to *gag*, *pol*, and *env*. These are *tat*, *rev*, *vif*, *nef*, *vpr*, and *vpu*. The *tat* gene product is a transactivator, which increases viral gene expression by acting on a sequence in the viral LTR immediately downstream from the mRNA start site. Gilbert Jay and his colleagues introduced the *tat* gene under the control of the viral LTR into the germline of mice and subsequently detected *tat* mRNA in the skin of both male and female animals but not in brain, thymus, liver, heart, lung, intestine, kidney, pancreas, spleen, muscle, or testes. At 4 months of age, 33/37 of the male transgenic mice, but none of the 15 female transgenic mice or the 10 nontransgenic littermates, had developed progressive dermal lesions. At 12–18 months of age, approximately 15 percent of the male mice (no females) developed skin tumors that resemble Kaposi's sarcoma.

a. Suggest a mechanism by which the *tat* gene might induce the Kaposi's sarcoma–type lesions.

b. How would you interpret these observations in relation to Kaposi's sarcoma induced by HIV in humans?

52. 3T3 cells, given adequate nutrients, will grow to confluency (i.e., form a monolayer covering the plate) and then stop. When these cells have been transformed with SV40, however, they will continue to grow after reaching confluency, piling on top of each other. Working in Luis Glaser's laboratory, Dan Raben, Brock Whittenberger, and Mike Lieberman extracted the membranes of mouse 3T3 cells with the detergent octylglucoside and then fractionated the octylglucoside extract. One fraction, which contained less than one-tenth of the cell-membrane protein, was called S_4. Addition of fraction S_4 to 3T3 cells that had not reached confluency caused the cells to decrease their rate of DNA synthesis by almost 50 percent. In contrast, addition of fraction S_4 to SV40-transformed 3T3 cells did not cause a change in the rate of DNA synthesis. When fraction S_4 was heated to 80°C for 10 min or subjected to a pH of 2 for 30 min, the ability of the fraction to decrease the rate of DNA synthesis in sparse 3T3 cells was substantially reduced.

a. Suggest a possible function for fraction S_4 in controlling the growth of normal, untransformed 3T3 cells.

b. What do the data suggest about the type of molecule that constitutes the active factor in fraction S_4?

c. What do the data suggest about the mechanism of growth control in the SV40-transformed cells?

53. Both epidermal growth factor (EGF) and transforming growth factor α (TGF-α) are soluble proteins. Both of these factors bind to the EGF receptor, stimulating its intrinsic tyrosine kinase activity. Both of these growth factors are synthesized as larger precursor proteins that contain a membrane-spanning segment. Studies on TGF-α have shown that the mature form of TGF-α is released from the cell by protease cleavage of its transmembrane precursor. One interesting possibility is that these growth factors are active in their membrane-bound, as well as their soluble, forms. Such membrane-bound growth factors might be important in promoting the growth of specific neighboring cells during development, for example. Design an experiment to determine whether the transmembrane forms of the growth factors are active as growth factors or are simply inactive intermediates in the production of the soluble growth factors.

54. A 66-year-old woman was diagnosed as having a malignant stomach tumor. She needed surgery, but a compatible blood donor could not be found. A-B-O blood-typing indicated that her blood was type O. However, in another blood-group system, the P-system, she had the very rare *p* blood type. The P-system comprises two immunological markers, P and P_1; the absence of these markers is designated as the *p* blood group. In addition to having *p* blood, the woman also had serum antibodies against the P and P_1 antigens. Because *p* blood could not be found, the woman was given a 25-ml transfusion of blood containing the P and P_1 antigens. Because the woman had a serious reaction to the trial transfusion, the surgeons decided to remove only part of the tumor, so that a blood transfusion would not be required. Surprisingly, the tumor disappeared completely after the surgery. How might the "cure" be explained?

PART E: *Working with Research Data*

55. Sung-Hou Kim, Jasper Rine, and their coworkers have examined the function of the protein encoded by the human *ras* gene c-H-*ras*[val12]. The Ras protein used in their experiments was synthesized by expression of the human gene in *E. coli*. The ability of this protein to promote cell division was determined by injecting it into frog oocytes and monitoring the subsequent breakdown of the germinal vesicle, which is correlated with meiosis in these oocytes. The effect of compactin and mevalonate on the ability of the Ras protein to promote cell division were also determined in this system. Compactin is a potent inhibitor of the enzyme HMG CoA reductase, which converts 3-hydroxy-3-methylglutaryl CoA (HMG CoA) to mevalonate. Mevalonate is a precursor of cholesterol and other isopentenoid compounds. The results of these experiments are shown in Table 24-1.

  a. Which experiments suggest that mevalonate is necessary for the Ras protein to promote cell division?

  b. Although Ras proteins generally are located in the plasma membrane, altered Ras proteins have been identified in the cell cytoplasm; these altered proteins do not promote cell division. Considering that many of the known metabolites of mevalonate are highly lipophilic compounds, can you suggest a role for mevalonate in the function of the Ras protein?

  c. Based on the data in Table 24-1, suggest an approach to treating tumors in which the *ras* gene is activated.

56. Acoustic neuroma is a human tumor derived from the Schwann cells that surround the vestibular branch of the vestibulocochlear nerve. Although most cases of acoustic neuroma occur as unilateral, apparently noninherited tumors, bilateral tumors are characteristic of an inherited form of this cancer. In order to understand the basis for the inherited disorder, James Gusella and colleagues examined the DNA of 21 patients with acoustic neuroma. They made restriction digests of the DNA from leukocytes (normal cells) and from the tumor cells of these patients. The digests were fractionated by agarose gel electrophoresis, transferred to nylon membranes, and hybridized individually to 23 different ^{32}P-labeled DNA probes with known locations on human chromosomes. Hybridization of the labeled probes with two bands in the patient's DNA digests indicates heterozygosity in the patient's restriction enzyme fragments, whereas a single band indicates that only one restriction fragment species hybridized to the probe DNA. The patients' leukocyte DNA was first hybridized to each of the 23 probes. In cases where two leukocyte restriction fragments hybridized to a particular DNA probe, the tumor-cell DNA from the patient also was analyzed with the same probe. The data from this study are shown in Table 24-2.

  a. Which chromosomal probes reveal differences in the leukocyte and tumor DNA from some patients? What is the most likely cause of these differences?

  b. What other cancer involves a similar type of alteration in the tumor-cell DNA compared with normal DNA?

  c. What do these data suggest about the mechanism of tumorigenesis in familial acoustic neuroma?

  d. In the tumor DNA from nine patients, no loss of heterozygosity was detected with the chromosome 22 probes. How might these data be explained?

57. DNA from human papillomaviruses (HPV) types 16 and 18 has been shown to be present in 70–80 percent of cervical carcinomas. However, there have been many suggestions in the literature that other factors, such as cigarette smoking and oral contraceptive use, are signifi-

Table 24-1

Experiment no.	Material injected			Germinal vesicle breakdown
	Ras protein	Compactin	Mevalonate	
1	0	0	0	0
2	+	0	0	+
3	0	+	0	0
4	0	0	+	0
5	+	+	0	0
6	+	+	+	+

Table 24-2

Probe	Chromosomal location of probe	Number of patients with two leukocyte fragments that hybridize with probe	Number of patients with two leukocyte fragments that also have two tumor fragments that hybridize with probe
A	1	5	5
B	1	5	5
C	4	9	9
D	4	3	3
E	10	3	3
F	11	10	10
G	11	6	6
H	11	8	8
I	11	4	4
J	12	5	5
K	13	6	6
L	13	9	9
M	13	6	6
N	13	2	2
O	14	11	11
P	17	10	10
Q	18	9	9
R	19	2	2
S	19	7	7
T	21	11	11
U	22	6	3
V	22	12	8
W	22	4	3

cant factors in the development of cervical carcinoma. Pater and coworkers examined the transformation of primary rat kidney cells transfected with c-Ha-*ras*-1 and/or HPV-16 DNA in the presence and absence of dexamethasone or progesterone. The resulting data are shown in Table 24-3.

a. Which factors are necessary for transformation of these primary cells?

b. What type of gene might be present in the HPV-16 DNA that would affect transformation? Where would you expect the product of this gene to be located?

c. How might the steroid hormones affect the transformation process in these cells?

58. After receptor binding, certain growth factors, including thrombin, vasopressin, and bradykinin, cause the activation of a phosphatidylinositide-specific phospholipase C through a GTP-binding (G) protein. This protein can be inactivated by ADP-ribosylation catalyzed by pertussis toxin. Action of the phospholipase C generates mediators, such as inositol trisphosphate and diacylglycerol, leading to a partially unknown series of events that presumably culminate in cell division. The pertussis-sensitive protein is the inhibitory G protein G_i, which simultaneously inhibits adenylate cyclase when this thrombin-stimulated pathway is activated.

The product of the *ras* oncogene is also a GTP-binding protein. David Kelvin and coworkers investigated the effect of pertussis toxin on the action of the *ras* gene product in order to determine whether the Ras protein acts in the same pathway as thrombin. Using BC3H1 muscle cells, these workers demonstrated that [^{3}H]thymidine incorporation stimulated by thrombin was decreased by 80 percent in the presence of pertussis toxin. Pertussis toxin also induced differentiation, as measured by the production of the muscle-specific enzyme creatine kinase. Differentiation was inhibited by thrombin, but pertussis toxin reversed the inhibition. When BC3H1 cells were transfected with the Ha-*ras* gene, differentiation did not occur, and addition of pertussis toxin did not induce differentiation. Also, addition of

Table 24-3

Treatment	Addition of dexamethasone[1]	Addition of progesterone[2]	Transformation
Transfection with c-Ha-*ras*-1	− +	− −	0 0
Transfection with HPV-16 DNA	− +	− −	0 0
Transfection with c-Ha-*ras*-1 and HPV-16 DNA	− + −	− − +	0 + +

[1] Dexamethasone is a glucocorticoid, a steroid hormone.
[2] Progesterone is another steroid hormone. An analog of this hormone is used in oral contraceptives.

pertussis toxin to the Ha-*ras*–transfected cells did not significantly affect [³H] thymidine incorporation. Do these data support the hypothesis that the *ras* gene product acts in the same pathway as thrombin to stimulate cell division and inhibit differentiation?

59. Gabriele Mugrauer, Fred Alt, and Peter Ekblom examined the expression of the N-*myc* proto-oncogene during mouse development. *In situ* hybridization of N-*myc* antisense mRNA was used to detect N-*myc* mRNA. Examination of mouse brain, kidney, heart, lung, and liver on the 12th day of development showed that the N-*myc* gene was expressed to some extent in each of these tissues. In the developing kidney, there are at least four cell lineages: the ureter epithelium, the endothelium, a mesenchyme that will become stroma, and a mesenchyme that will become the epithelium of the kidney tubules.

Hybridization of 16-day mouse embryo kidney sections revealed that the N-*myc* gene was expressed only in the periphery of the developing kidney in the mesenchymal cells destined to become the epithelium of the kidney tubules. As the tubules develop, these differentiating mesenchymal cells are gradually displaced toward the inner part of the developing kidney by proliferating mesenchymal cells in an earlier stage of differentiation to epithelium. Detailed analysis of autoradiograms of the developing tubules showed that the mesenchymal cells that were developing into epithelium expressed N-*myc* product only when the cells were located near the periphery of the kidney and did not express N-*myc* product after they were displaced to the inner part of the kidney, where the cells had stopped proliferating.

Examination of N-*myc* expression in other developing tissues suggested that its expression did not correlate with cellular proliferation; in the lung and liver, N-*myc* was not expressed in cells in the late stages of differentiation, although the cells were proliferating; in brain, N-*myc* expression continued after proliferation had concluded, while differentiation continued. A hypothesis consistent with these data is that the N-*myc* proto-oncogene is expressed during early differentiation and can serve as a marker for early differentiation. Suggest an experiment that can test this hypothesis.

CHAPTER 25

Immunity

PART A: *Reviewing Basic Concepts*

Fill in the blanks in statements 1–20 using the most appropriate terms from the following list:

allelic exclusion

antigen-dependent

antigen-independent

B cells

binding

class switching

clonal selection

complement

cytotoxic T lymphocytes (CTLs)

decrement

determinant

effector

G-A transition

heavy

IgT

IgM

immunoglobulins

immunopathology

interferons

light

lymphokines

macrophages

matrix

memory cells

major histocompatibility complex (MHC)

myelomas

pathogens

pheochromocytomas

plasma cells

somatic mutation

T$_H$ cells

tolerance

T-cell receptor

X-inactivation

1. The effector molecules of humoral immunity are called _____.

2. The site on an antigen at which an antibody binds is called an antigenic _____.

3. Antibody molecules can be divided into two functional domains, called the _____ domain and the _____.

4. The antigen-binding site on the surface of a T cell is called the _____.

5. Activated B cells can become either _____ or _____.

6. _____ is the name of the process whereby individual bases in a joined VDJ segment are replaced with alternative bases.

7. Animals generally do not make antibodies against their own proteins; this phenomenon is called _____.

8. The immune system protects organisms against _____.

9. The theory of _____ can be used to explain the production of large numbers of active B cells from only a few reactive virgin B lymphocytes.

10. Both the heavy and light chains of an IgM molecule bind to antigens, but only the _____ chain can bind to membranes or to soluble components of the blood.

11. The class of T cells known as _____ directly kill target cells; the class of T cells known as _____ assist B cells in their reaction to antigen.

12. The process whereby a specific B cell shifts from producing IgG to producing IgA is called _____ .

13. Antigens can be processed and re-expressed on the surface of _____ and _____ .

14. Self-recognition molecules found on the surface of all cells are encoded by genes in the region called the _____ .

15. B-lymphoid tumors called _____ have been useful sources of purified immunoglobulin chains for structural studies.

16. Inactivation of immunoglobulin genes on one chromosome of a diploid lymphocyte is called _____ .

17. The _____ phase of B-cell development takes place in the bone marrow.

18. Cytotoxic proteins that are found in the blood and that are activated by IgM and IgG are collectively called the _____ system.

19. Class I MHC proteins are found on all cells; class II MHC proteins are found mainly on _____ and _____ .

20. Factors secreted by some immune system cells that regulate the activity of other immune system cells are collectively called _____ .

PART B: *Linking Concepts and Facts*

Circle the letters corresponding to the most appropriate terms/phrases that complete or answer items 21–30; more than one of the choices provided may be correct in some cases.

21. Binding of antigen to a B-cell surface protein
    a. results in division of the B cell.
    b. is required in order for B cells to become plasma cells.
    c. causes changes in proteins synthesized by the B cell.
    d. activates some B cells to become cytotoxic T cells.
    e. causes immediate secretion of IgM.

22. The mammalian immune system exhibits which of the following properties?
    a. regulation
    b. memory
    c. specificity
    d. adaptability
    e. ability to distinguish self from nonself

23. Antibodies
    a. contain an antigen-binding site.
    b. have a carbohydrate component.
    c. contain effector domains.
    d. have two or more light chains.
    e. can be membrane bound or soluble.

24. Lymphokines are directly involved in which of the following processes?
    a. B-cell mitogenesis
    b. cell killing by cytotoxic T cells
    c. helper T-cell mitogenesis
    d. platelet activation
    e. inhibition of macrophage migration

25. Which of the following processes or molecules are important in the production and secretion of IgG by a B cell?
    a. DNA rearrangements
    b. membrane fusion
    c. polyadenylation
    d. signal peptide
    e. cell-surface receptors

26. Specific antibodies that recognize which of the following molecules can be produced.
    a. actin
    b. bacterial cell-wall components
    c. dinitrophenol

d. nucleic acids

e. class I MHC gene products

27. Interactions directly involved in antigen binding to IgG include

a. electrostatic interactions.

b. disulfide bonds.

c. hydrogen bonds.

d. hydrophobic interactions.

e. van der Waals interactions.

28. Virgin B lymphocytes can differentiate into

a. natural killer cells.

b. plasma cells.

c. memory cells.

d. dendritic cells.

e. granulocytes.

29. Macrophages, also called monocytes, have the ability to

a. process and present antigens to T cells.

b. produce antibodies.

c. produce peptides that are mitogenic for B cells.

d. phagocytose bacterial cells.

e. express IgM on their surface membranes.

30. Helper T cells (T_H cells)

a. stimulate division of B cells.

b. stimulate division of cytotoxic T cells (CTLs).

c. can kill other cells.

d. stimulate migration of macrophages.

e. are inactivated by HIV infection.

31. Describe the cellular source(s) and the function(s) of each component of the vertebrate immune system listed below.

a. IgG

b. IL-2

c. Class II MHC proteins

d. CD4

e. IgA

f. IgM

g. IgD

h. IgE

32. Listed below are some immunopathologies, all of which are due to an excess or a lack of some component of the human immune system. Indicate (1) the component responsible for each characteristic pathology (can be a cell or a molecule) and (2) whether is is lacking or in excess.

a. Severe combined immunodeficiency disease (SCID).

b. Myeloma.

c. Acquired immune deficiency syndrome (AIDS).

d. Allergies.

33. The major classes of immunoglobulin differ in the type of heavy chain they contain and in their relative concentrations in serum. In the table below indicate the heavy-chain type and the relative serum concentration for each immunoglobulin class (1 = lowest and 5 = highest serum level).

Immunoglobin class	Heavy-chain type	Relative serum concentration
IgA	_____	_____
IgM	_____	_____
IgG	_____	_____
IgD	_____	_____
IgE	_____	_____

34. The development of a specific B lymphocyte occurs in two phases, an antigen-independent and an antigen-dependent phase. Each phase of this maturation process is associated with some of the specific sites or molecular processes listed below. In the space provided, write I if the site or process is confined to the antigen-independent phase, write D if it is confined to the antigen-dependent phase, and ID if it occurs during both phases of B-cell maturation.

a. DNA rearrangements ____

b. Allelic exclusion ____

c. Cell located in blood and lymph ____

d. Interaction with T_H cell ____

e. Secretion of IgM into extracellular fluid ____

f. Binding of IL-4 to specific receptors ____

g. Production of IgA ____

h. Polyadenylation of immunoglobulin gene transcripts ____

i. Cell located in bone marrow ____

j. Production of memory B cells ____

PART C: *Putting Concepts to Work*

35. Describe two observations that argue against an *instructive* theory and are consistent with a *selective* theory of antibody generation.

36. What are two roles for macrophages in protecting vertebrates from infection by bacteria?

37. What molecular mechanisms are known to operate in the generation of antibody diversity?

38. Why are children generally much more susceptible to infection than are adults?

39. List the major structural and functional differences between T cells and B cells.

40. What is the evidence that tolerance (inability to react

with self-antigens) develops early in the life cycle of vertebrates?

41. A B cell that produces IgG cannot switch to production of IgM, but an IgM-producing cell can switch to production of IgG. Why?

42. Why does only one of every three V_κ-J_κ joints result in production of a light chain?

43. What are three possible fates for a virgin B lymphocyte?

44. What is the evidence that cytotoxic T lymphocytes (CTLs) must recognize both a foreign antigen and a class I MHC gene product before initiation of a cytolytic response?

PART D: *Developing Problem-solving Skills*

45. Cancers of the hematopoietic cells (leukemias, myelomas, lymphomas, etc.) are among the most common cancers of human beings. Based on what you know about growth control and gene regulation mechanisms in these cells, why do you think that these cells commonly become cancerous?

46. Erythroblastosis fetalis is a condition found in newborn infants who are Rh$^+$ and whose mothers are both Rh$^-$ and have previously given birth to another Rh$^+$ infant. Infants with this condition have very few red blood cells. Their mothers, who were presumably exposed to Rh$^+$ blood when giving birth to the first infant, make antibodies to the Rh antigen.

   a. Why do these infants have low red blood cell counts?

   b. What class of immunoglobin is responsible for this condition, and why?

   c. Serum from most people contains antibodies that react with the highly antigenic ABO blood-group determinants. These antibodies, termed *isohemagglutinins,* are thought to be synthesized in response to bacterial cell-wall components that are similar to

the ABO carbohydrate structures. Although there are many cases of mothers with type A blood giving birth to two or more infants with type B blood, none of these infants develop the symptoms typical of erythroblastosis fetalis. How can you explain this?

47. Germ-line DNA from the organism *Incredulosa dubium* contains genes for the constant regions of five different immunoglobulin heavy chains, which are designated H, I, J, K and L. Unstimulated lymphocytes (virgin B cells) from this organism can make membrane-bound IgH. Virgin B cells, as well as IgL-secreting plasma cells and IgI-secreting plasma cells, have been analyzed for the presence of the various heavy-chain constant regions. The resulting data are shown in Table 25-1.

   a. How do you think scientists determined that specific cells lacked certain heavy-chain sequences.

   b. Assuming that mechanisms known to operate in the generation of different human and mouse immunoglobulins also operate in this organism, suggest the order in which the five heavy-chain constant regions occur in the genome, based on the data in Table 25-1. (*Note:* As in mouse and human systems, the

Table 25-1

Type of cell	Heavy-chain constant region				
	H	I	J	K	L
Virgin B cell	+	+	+	+	+
Plasma cell secreting IgL	−	+	−	+	+
Plasma cell secreting IgI	−	+	−	−	−

name of an immunoglobulin is derived from the name of its heavy chain; for example, IgL contains L heavy chains, and IgI contains I heavy chains.)

48. Assume that an organism has 400 V_κ and 10 J_κ light-chain regions and 400 V, 10 D, and 10 J heavy-chain regions in its haploid genome.

   a. If no nucleotides are lost or gained during recombination within the heavy- and light-chain DNA, how many functional immunoglobulin light-chain genes and heavy-chain genes theoretically could be generated in this organism?

   b. How many different immunoglobulin molecules could be generated from these heavy and light chains?

   c. If the joining of V_κ and J_κ light-chain gene segments to produce the light-chain gene is imprecise, but always occurs within the same codon, how many different immunoglobulins could be generated in this organism? Assume that no out-of-phase stop codons exist in any of the reading frames.

49. Two inbred strains of mice, A and B, which differ in the genes in the major histocompatibility complex (MHC), can be infected with viruses called Q and Z. Strain A mice were infected with virus Z. After 7 days, CTLs were isolated from these mice and cultured with various types of mouse fibroblasts. Based on your knowledge of the role of MHC molecules in cell-mediated immunity, predict whether the fibroblasts in the following cultures would be killed by the isolated CTLs. In each case state your reasoning.

   a. Normal fibroblasts from strain A mice

   b. Z-infected fibroblasts from strain A mice

   c. Normal fibroblasts from strain B mice

   d. Z-infected fibroblasts from strain B mice

   e. Q-infected fibroblasts from strain A mice

50. Immunoglobulins are often injected into patients who have been bitten by a venomous snake or spider. This is known as passive immunization. Why is IgG the antibody of choice in such situations?

51. Tolerance to a specific antigen can be induced in adult organisms; studies of such induced tolerance have aided in our understanding of the development of tolerance to self-antigens, which occurs early in life. Several factors have been shown to contribute to the ability of an antigen to be *tolerogenic* rather than *immunogenic*. The form of the antigen is important. Adult mice will become tolerant of monomeric bovine IgG, but polymeric bovine IgG or monomeric IgG in an adjuvant is highly immunogenic in these animals. The metabolic fate of antigens also seems to be a factor. Synthetic polypeptides made of D-amino acids are tolerogenic when administered to adult rabbits or mice at doses of 10 μg/animal. These polypeptides are very resistant to enzymatic digestion. What do these observations suggest about the mechanism(s) of induction of tolerance, or about the mechanism(s) of induction of immunity?

52. In humans, some alleles of MHC genes are associated with a higher risk of contracting various diseases. The best known of these is the association of an allele of the B locus, known as B27, with ankylosing spondylitis, an inflammatory disease that leads to stiffening of the vertebral joints. Although well over 90 percent of individuals affected with this disease carry the B27 allele, not everyone with this allele contracts the disease, indicating that some other factor also contributes to the cause of this disease. Propose a hypothesis to account for these observations.

53. IgM is the first immunoglobulin produced in response to foreign proteins. Why is this polymeric immunoglobulin particularly well suited for this early response?

54. In order for T-cell recognition and activation to occur, a T cell needs to encounter a foreign antigen in association with a class I MHC protein. It would seem feasible for a single type of class I MHC protein to suffice for this purpose; however, as transplant surgeons have found to their dismay, many different MHC proteins exist in humans (and in mice). Why are MHC proteins so polymorphic; that is, why are there so many different class I MHC alleles?

PART E: *Working with Research Data*

55. Human immunodeficiency virus (HIV) binds to a membrane glycoprotein known as CD4 on the surface of T_H cells (and other cells). This binding is mediated through the HIV glycoprotein known as gp120. CD4 is important in the binding between T_H cells and antigen-presenting cells; it is thought to bind to invariant regions of class II MHC gene products. CD4 is an extremely potent immunoregulatory molecule; antibodies that react with CD4 can suppress T-cell–dependent immune responses both *in vitro* and *in vivo*.

    An indirect test of T-cell activation can be performed by measuring Ca^{2+} influx into T cells after stimulation with a mitogen or specific antigen. Internal Ca^{2+} concentration, measured by fluorescence of a Ca^{2+}-specific dye inside the cells, increases dramatically when the cells are treated at time 0 with a specific agent known to stimulate resting T cells (Figure 25-1). Mittler and Hoffmann found that this activation is not altered when the T cells are pretreated with either gp120 (solid triangles) or serum from a patient infected with HIV (solid squares); this serum, like that from almost all HIV-infected patients, contained antibody to gp120. However, when the T cells are pretreated with both gp120 and the serum from the HIV-infected patient, Ca^{2+} influx is almost completely inhibited (open circles). Data obtained from similar experiments, using serum from patients infected with HIV or with other viruses, are shown in Table 25-2.

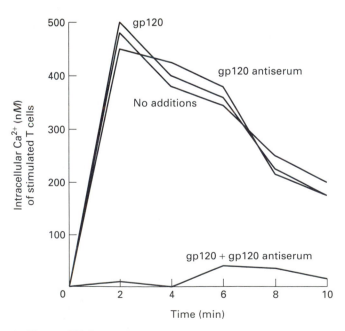

▲ **Figure 25-1**

a. Propose a hypothesis that explains these data.

b. What other experiments can you propose in order to test your hypothesis?

56. The syndrome known as severe combined immunodeficiency disease (SCID) can result from the absence of the enzyme adenosine deaminase (ADA), which functions in the purine (deoxy)ribonucleoside salvage pathway. The organs with the highest specific activity of ADA in normal humans are the spleen and the thymus. Several molecular forms of ADA are known in humans; erythrocytes and leukocytes contain primarily a low-molecular-weight form; other tissues have a higher-molecular-weight form. Both forms are missing in patients with SCID, implying that both forms are the product of a common gene. This gene is on chromosome 20, and thus is not linked with the genes of the MHC complex on chromosome 6. Patients with ADA deficiency are usually diagnosed in the first few months of life; these individuals often have multiple infections of the respiratory tract, the skin, and gastrointestinal organs. Cytologically, few white cells are present; B cells are moderately decreased, and T cells are profoundly decreased. Immunoglobulin levels are low to absent; without therapy, SCID patients die within the first year of life.

    Another, even rarer immunodeficiency syndrome is now known to result from the absence of purine nucleoside phosphorylase (PNP), another enzyme in the purine salvage pathway. The symptoms typical of PNP deficiency, although somewhat milder than those typical of ADA deficiency, also include lowered T-cell populations and death within the first few years of life due to viral or bacterial infections.

    The reactions catalyzed by ADA and PNP are shown in Figure 25-2. The question naturally arises as to why a deficiency in one or the other of these purine salvage–pathway enzymes results in a seemingly specific depletion of T cells. Pathogenesis in such metabolic disorders can result from four general causes:

    - deficiency of the product of the absent enzyme

    - accumulation of toxic levels of the substrate of the absent enzyme

    - accumulation of a toxic metabolite that is the product of an alternative reaction because of the accumulated substrate

    - accumulation of a toxic metabolite that is normally detoxified by the absent enzyme but that is not the classical substrate of the enzyme

Table 25-2

Virus in patient serum	Number of sera tested	Number with gp120 reactivity	Suppression of Ca²⁺ influx (% of control)	
			− gp120	+ gp120
HIV	14	14	12	75
Cytomegalovirus	4	0	5	7
Hepatitis B	2	0	5	8

a. Based on the information presented, which of these pathogenic mechanisms can be eliminated as a cause of ADA-deficiency SCID? Which of the remaining mechanisms is (are) most likely to be important in the pathogenesis of SCID?

b. What experiments could be performed to determine whether the likely mechanism(s) is(are) correct?

57. The existence of a third type of T cell, the suppressor T (T_S) cell, has been postulated and is currently the subject of some debate; many investigators doubt that such a cell exists. The existence of T_S cells was first postulated when it was shown that T cells from mice that had been made tolerant to sheep red blood cell antigens could induce tolerance when injected into another mouse. Subsequent genetic experiments indicated that a region within the MHC, named I-J, governs the function of T_S cells and soluble T suppressor factors. Many investigators have adapted the notion of antigen-specific T_S cells to explain tolerance, immunosuppression by tumors, and a wide variety of other immune phenomena.

Several recent findings, however, cast doubt on the existence of specific T_S cells. These include the inability of any investigator to define a surface antigen that is specific for T_S cells, thus precluding the isolation of appreciable quantities of normal T_S cells. Furthermore, recombinant DNA studies indicate that the I region of the MHC does not contain a gene with the predicted properties of I-J. Finally, studies of putative T_S hybridomas indicate that these cells have rearranged the genes for the T-cell receptor to render them nonfunctional. It is difficult to explain the existence of antigen-specific T_S cells if these cells do not have a functional T-cell receptor. Thus, although the existence of antigen-specific suppressor cells has been a useful theoretical construct, it is unclear at this time if these cells will turn out to be a physiological reality.

Many workers, however, feel that these arguments against the existence of T_S cells are not convincing. Would you argue that this evidence is conclusive proof that these cells do not exist? Assuming that T_S cells do exist, how might these apparently negative observations be explained?

58. Allelic exclusion is the name given to the phenomenon of inactivation of one chromosomal set of immunoglobulin genes when the other chromosomal set is expressed. This means that an individual B lymphocyte expresses only one allele for the heavy chain and one allele for the light chain, to the exclusion of the other allele. Models to account for this phenomenon have focused almost exclusively on DNA rearrangements as the signal for allelic exclusion.

The working hypothesis for heavy-chain allelic exclusion, based on a variety of approaches, is that feedback control mechanisms operate to ensure that only one allele is functionally rearranged. If rearrangement of heavy-chain genes is initiated in both alleles, it is possible that one of these rearrangements will yield a nontranslatable heavy-chain gene. Successful rearrangement of one allele, resulting in a translatable mRNA for the heavy chain, inhibits further rearrangement in the other allele. Subsequent versions of this hypothesis predict that production of a complete immunoglobulin molecule inhibits further rearrangements of the light-chain gene.

A prediction of this model is that nonexpressed (excluded) heavy-chain genes would have nonfunctional rearrangements. Early studies, in tranformed B lineage cells, seemed to confirm this prediction. Three types of nonfunctional heavy chain gene rearrangements were detected: incomplete rearrangements (only D-J joining had been completed), complete rearrangements with in-frame termination codons or frameshifts, and trans-

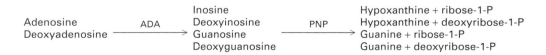

▲ **Figure 25-2**

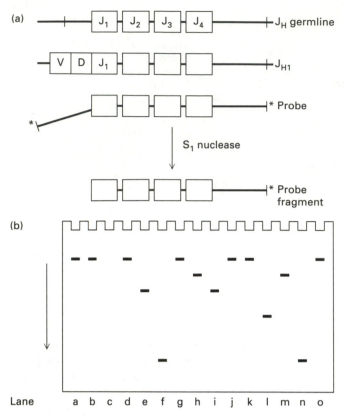

▲ **Figure 25-3**

locations of the heavy-chain genes to another chromosome.

Recently, Weissman's group at Stanford analyzed a series of normal B lymphocytes that expressed IgD at their cell surface. DNA restriction fragment patterns and sequences of seven rearranged clones from the nonexpressed allele were determined. In their experimental protocol, outlined in Figure 25-3a, allelic DNA from each B-cell clone was hybridized with a ^{32}P end-labeled *Xba*I restriction fragment (*). This probe included the four J segments of heavy-chain DNA ($J_1–J_4$). The hybrids were digested with S_1 nuclease, leaving a duplex DNA whose protected length is indicative of the rearrangement breakpoint. The digests were then electrophoresed on agarose gels and analyzed by autoradiography. The gel profiles of 14 such digests are depicted in Figure 25-3b (lanes a–n); the sample in lane o is from unrearranged (germ-line) DNA.

a. How many of the nonexpressed alleles shown in Figure 25-3b have been rearranged?

b. Sequence analysis of the rearranged but nonexpressed alleles indicated that several classes of rearrangements were present in the seven clones. These included four clones with VDJ rearrangements with no obvious impediments to translocation (i.e., they

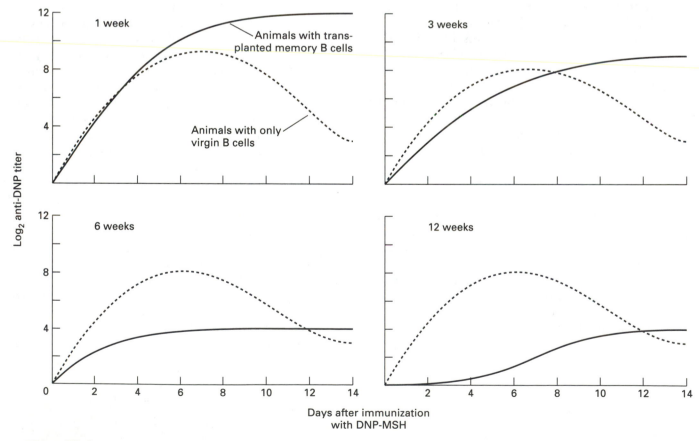

▲ **Figure 25-4**

were in-frame, contained no termination codons, and were translationally viable). What do these data indicate about the feedback model of allelic exclusion described above?

c. What other mechanisms can you postulate to explain allelic exclusion of heavy-chain genes in B lymphocytes?

59. After stimulation with appropriate antigen and interactions with appropriate T_H cells, B cells divide repeatedly; the progeny of these divisions can be either antibody-secreting plasma cells or so-called memory B cells. These latter cells are thought to be very long-lived; indeed, memory B cells are thought to persist for the life of the organism even without further antigenic stimulation.

Gray and Skarvall of the Basal Insitiue of Immunology in Switzerland recently published some observations that bear directly on the fate and activity of memory B cells in the absence of antigenic stimulation. These workers injected rats of a certain allotype with DNP conjugated to a protein called haemocyanin from the marine organism *Maia squinada* (DNP-MSH). Thoracic duct lymphocytes from these DNP-MSH–primed rats were isolated 2–3 months later and injected into sublethally irradiated congenic rats (adoptive transfer). The adoptive hosts also were injected with DNP-MSH at 1, 3, 6, and 12 weeks after lymphocyte transfer and the serum anti-DNP titer was measured at various times after immunization. The results, shown in Figure 25-4, compare the responses of host animals containing donor memory B cells (open circles) with the responses

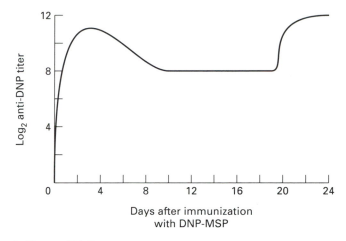

▲ **Figure 25-5**

of host animals that had not received lymphocytes and thus contained only virgin B cells. The results of an additional experiment in Figure 25-5 show the serum anti-DNP titer in host animals that received an injection of DNP-MSH at the same time as the lymphocyte transfer.

a. What do these data suggest about the longevity of memory B cells in the absence of antigen stimulation?

b. There is good evidence for maintenance of memory B cells in humans for as long as 60–70 years. How can these observations be reconciled with the data presented in Figures 25-4 and 25-5?

CHAPTER 26

Evolution of Cells

PART A: *Reviewing Basic Concepts*

Fill in the blanks in statements 1–14 using the most appropriate terms from the following list:

archaebacteria	mitochondria
chloroplasts	nuclei
cross-joining	oxygen
cytosine	precell
DNA(s)	progenote
endosymbiont	prokaryotes
eocytes	protein(s)
eubacteria	reverse transcriptases
eukaryotes	RNA(s)
exon shuffling	RNA polymerases
flexible	self-splicing
guanosine	stable
hydrogen	stromatolites
hydrogenation	transplicing
mammals	unstable
methane	

1. A gas that was probably not present in the early atmosphere of the earth was _____.

2. Attempts are currently being made to determine whether peptide synthesis can be carried out by _____ molecules alone, a result that would add more weight to the strong case that interactions among these molecules were basic to the evolution of translation.

3. Reassociated bacterial ribosomes can take part in translation even if some of the _____ are omitted.

4. The enzyme that cleaves *E. coli* tyrosyl-tRNA, called RNase P, is composed of a catalytically active _____ and a catalytically inactive _____.

5. Many RNAs with group I introns can perform _____ reactions using a(n) _____ cofactor.

6. The joining of two different RNA molecules is called _____.

7. Enzymes that can produce DNA from an RNA template, called _____, have been discovered in many diverse organisms.

8. Ancient microfossils resemble structures called _____, which are being laid down today where the ocean sediment precipitates around colonies of cyanobacteria and other bacteria.

9. The three ancient lineages revealed by sequencing rRNAs are _____, _____, and _____.

10. The cell from which the three ancient lineages were derived is termed the _____.

11. Based on rRNA sequencing, James Lake has proposed that a group of sulfur-metabolizing, thermophilic pro-karyotes, generally considered to be _____,

should be distinguished as a separate group, called _____.

12. According to the _____ hypothesis, eukary-otic _____ were derived from a purple pho-tosynthetic bacterium and _____ were de-rived from a cyanobacterial type of organism.

13. The intron-exon structure of eukaryotic nuclear genes is very _____ over time.

14. The reassortment of different functional domains, such as is likely to have occurred in the evolution of the low-density lipoprotein receptor, has been termed _____.

PART B: *Linking Concepts and Facts*

Circle the letters corresponding to the most appropriate terms/phrases that complete or answer items 15 – 22; more than one of the choices provided may be correct in some cases.

15. When hydrogen, methane, and ammonia are mixed with water and exposed to electric discharges or UV light,

    a. amino acids are produced.

    b. purines are produced.

    c. more pyrimidines than purines are produced.

    d. peptides including insulin are produced.

    e. ATP may be produced in the presence of calcium phosphate.

16. Ribonucleotides

    a. are obtained in high yield when the products of the formose reaction are coupled with a purine.

    b. may have been preceded in evolutionary time by nonchiral nucleotide analogs.

    c. formed from heating mixtures of urea, NH_4Cl, phosphate, and hydroxyapatite are a combination of the D- and L-enantiomers.

    d. have phosphorimidazolide analogs that can undergo nonenzymatic, template-directed polymerization.

    e. containing cytosine can be made abiotically by a well-known reaction mechanism.

17. The original translation system

    a. is likely to have arisen because of the special chemi-cal affinity that exists between particular amino acids and their three-base codons.

    b. is likely to have used a coding system with only two bases per codon.

    c. may have produced proteins that were only slightly ordered.

    d. may have involved a triplet code in which only the first two of the three bases per codon were used to specify an amino acid.

    e. may have used codons capable of specifying only that a hydrophobic or hydrophilic amino acid be added to a growing peptide chain.

18. Group II introns

    a. are similar to group I introns in sequence and struc-ture.

    b. require a cofactor for self-splicing.

    c. require energy input for self-splicing.

    d. self-splice using the same pattern of reactions as snRNP-assisted splicing in the nucleus.

    e. end up in a lariat structure after being spliced out of an RNA transcript.

19. Properties of the L-19 RNA molecule, which is derived from the *Tetrahymena thermophila* group I intron, in-clude

    a. the ability to carry out polynucleotide synthesis in the absence of protein.

b. the ability to self-splice.

c. the ability to act as a site-specific protease.

d. the ability to act as a site-specific nuclease.

e. the ability to regenerate itself after accelerating the rate of a chemical reaction.

20. The most promising approach to discovering the relationships among organisms that existed at the beginning of evolution is

   a. examination of the earliest preserved specimens of life.

   b. simulation of the postulated conditions on the early earth.

   c. morphological comparison of present-day organisms.

   d. the study of microfossils.

   e. comparison of nucleic acid sequences of different present-day oganisms.

21. Archaebacteria

   a. often live in habitats that might be considered harsh.

   b. are not known to have introns in their genes.

   c. probably were separated from other prokaryotes early in evolutionary time.

   d. include cyanobacteria.

   e. have the same basic domains in their 16S-rRNA structure as yeast do.

22. Introns

   a. may have been lost from nuclear genes during evolution.

   b. are known to have been inserted into nuclear protein-coding genes during evolution.

   c. are known to have been both lost and gained by eukaryotic organelle DNA.

   d. may be exons for a reverse transcriptase in yeast mitochondria.

   e. in plant nuclear genes are known to encode site-specific nucleases.

PART C: *Putting Concepts to Work*

23. How might you determine whether the amino acids found in a meteorite were synthesized abiotically or resulted from contamination by biological materials produced by living organisms on earth?

24. Comparison of the properties of proteins, DNA, and RNA suggest that RNA is the most likely candidate for the primordial "molecule of life." Provide support for this statement.

25. Some group I mitochondrial "introns" actually code for proteins, called maturases, that promote excision of the RNA "introns" containing their coding sequences. It has been suggested that maturases may simply cause the RNA to fold in such a way that self-splicing occurs, rather than function by actively taking part in the catalysis. What is the basis for this suggestion?

26. How might RNA transplicing have functioned to advance the abilities of precellular molecules?

27. The mitochondrial DNA of the flagellated protozoans *Leishmania, Trypanosoma,* and *Crithidia* does not contain sequences corresponding exactly to the mitochondrial mRNAs encoding several proteins (e.g., cytochrome *c* oxidase, NADH dehydrogenase). This difference has been attributed to a process called RNA editing.

   a. What evidence supports the existence of RNA editing?

   b. What changes occur in the structure of an RNA during RNA editing and what is the functional effect of these changes?

28. Explain how some present-day eukaryotes, such as *Giardia lamblia,* might have evolved so as to lack mitochondria.

29. What may have been the structural significance of exons in the evolution of present-day proteins?

30. What change in the progenote probably has contributed to the ability of present-day eubacteria to grow rapidly?

PART D: *Developing Problem-solving Skills*

31. What is the selective pressure for evolution of a molecule with reverse transcriptase activity? In other words, why is DNA preferable to RNA as the genomic material?

32. The great evolutionary biologist J.B.S. Haldane proposed that "the critical event that may best be called the origin of life was the enclosure of several different self-reproducing polymers within a semipermeable membrane." This event isolated the replicating structures from the vagaries of existence in the primordial sea, and defined the origin of the first cell. This event would require the association of relatively high concentrations of self-replicating polymers with organized semipermeable membranes. Assuming that amphipathic molecules similar in properties to modern-day membrane lipids existed in the primordial soup, suggest a scenario by which tides and tidal pools might have played a role in the origin of these hypothetical first cells.

33. The endosymbiont hypothesis cannot explain the observation that many (in fact most) mitochondrial proteins are coded for by nuclear genes. If mitochondria arose from endosymbiotic visitors to the primitive eukaryotic cell, but are now relegated to the status of organelles, what is the origin of nuclear genes encoding mitochondrial proteins? Put another way, why aren't all mitochondrial proteins coded for by nuclear genes? Recently, Van den Boogaart and coworkers have discovered that the common mold *Neurospora crassa* has both a nuclear and a mitochondrial gene for a subunit of the mitochondrial ATPase. Only the nuclear gene is transcriptionally active, however.

    a. Describe two hypotheses to explain the presence of redundant genes in mitochondrial and nuclear genomes.

    b. What experimental evidence could distinguish between these two hypotheses?

34. The theory that protein, rather than RNA, was the first biotic molecule has been proposed and defended by several investigators, including Sidney Fox. This theory attempts to bridge the gap between prebiotic chemical evolution and biotic evolution with proteins rather than with RNA. Experimental evidence for this theory rests on observations using protein "microspheres," which can be generated in high yield from reaction conditions mimicking the conditions on prebiotic earth. In these experiments, thermal polycondensation of mixed amino acids in the absence of excess water can yield polymers that are remarkably like protein. These compounds can catalyze reactions (including the polymerization of amino acids or nucleotides) and can associate into cell-like structures called, variously, "proteinoids," "thermal proteins," or "proteinoid microspheres." How can these observations be integrated with the "RNA-first" theory of the evolution of life?

35. The terrestrial phylogeny proposed by C. R. Woese is based on rRNA sequence comparisons that indicate there are at least three main branches to the phylogenetic tree (archaebacteria, eubacteria, and eukaryotes) and is consistent with the observation that most eukaryotic genes (and some prokaryotic genes) contain noncoding regions called introns. This phylogeny is depicted in simplified form in Figure 26-1a. Recent discoveries regarding mitochondrial nucleic acids, however, may force a modification of this phylogenetic tree. These discoveries are as follows:

    • The respiratory enzymes of bacteria and mitochondria are very similar, but their genetic systems are very different.

    • Codon usage by mitochondrial tRNAs varies considerably from that of prokaryotic genomes or eukaryotic nuclear genomes.

    • Adenine-uracil content is greater than 60 percent in animal and fungal mitochondrial tRNA; less than 40 percent in prokaryotic tRNA; less than 40 percent in nuclear-coded tRNA from fungi, plants, and animals, and 40–50 percent in chloroplast tRNA. (The abundance of adenine in successful prebiotic simulations

Table 26-1

Source of mitochondrial DNA	Number of base pairs	Type of rRNA genes	Number of tRNA genes	Number of protein genes	Presence of introns
Mammals	15–18,000	12S, 16S	22	13	None
Fungi	18–78,000	15S, 21S	25	> 20	Some
Plants	> 100,000	5S, 18S, 26S	> 25(?)	> 25(?)	Many

(a) Woese phylogenetic tree

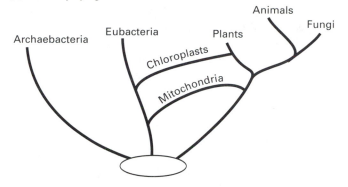

(b) Modified phylogenetic trees

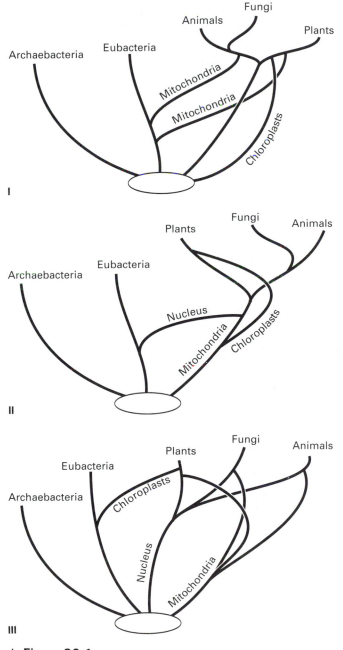

▲ **Figure 26-1**

suggest that adenine content was quite high in primitive replicating systems.)

- Comparative analysis of tRNA molecules indicates that mammalian mitochondrial tRNAs are significantly different from prokaryotic and eukaryotic tRNAs, are significantly different from chloroplast tRNAs, and are different from fungal mitochondrial tRNAs. Considerable homology is found only among mammalian mitochondrial tRNAs themselves; insufficient data about plant cellular tRNAs are available for comparison.

- Differences in the size, structure, and complexity of mitochondrial DNA, summarized in Table 26-1, strongly suggest that mammalian, fungal, and plant mitochondria differ considerably from each other.

a. Which of the modified phylogenetic trees (I, II, or III) shown in Figure 26-1b is most consistent with the rRNA sequence comparisons that formed the basis of the Woese phylogeny and with the comparative tRNA data presented above? State the reasons for your conclusion.

b. According to scheme III in Figure 26-1b, which organelle(s) or cell type(s) represent the most likely living representative of primitive living systems?

36. The major membrane lipids in eukaryotes and eubacteria are phosphoglycerides, in which fatty acids are linked to glycerol through an ester bond. In contrast, the known membrane lipids of archaebacteria have phytanol chains ether-linked to glycerol; in phytanol chains, every fourth carbon has a methyl group. In addition to these differences, the stereochemistry of these lipids is reversed; that is, the phosphoryl groups of archaebacterial lipids are on the *sn*-1 position of glycerol, whereas the phosphoryl groups of eukaryotic and eubacterial phosphoglycerides are on the *sn*-3 position.

a. Are these data consistent with the Woese phylogenetic tree shown in Figure 26-1a? Explain your answer.

b. What selection pressures might have caused archaebacteria to evolve ether-linked, rather than ester-linked, membrane lipids?

37. DNA-DNA hybridization has been used to compare DNA sequences from various avian species in order to determine evolutionary relationships. In such studies, single-copy DNA is first isolated from (nucleated) avian red blood cells. The single-copy DNA from one species is labeled with ^{125}I and mixed with unlabeled single-copy DNA from another species. The mixture is then heated to 100°C to dissociate the DNA into single strands and cooled to 60°C, where it is held for 120 h to allow the strands to reassociate. The mixture is then

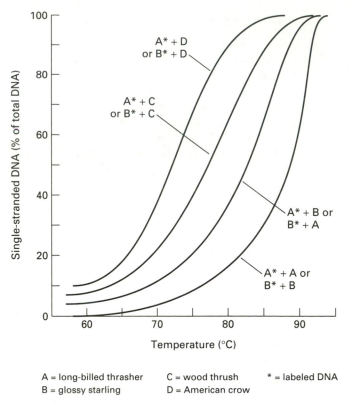

A = long-billed thrasher   C = wood thrush   * = labeled DNA
B = glossy starling   D = American crow

▲ **Figure 26-2**

placed on a hydroxyapatite column, which binds double-stranded, but not single-stranded, DNA. The temperature of the column is raised from 55°C to 95°C in 2.5°C increments. At each of the intermediate temperatures, single-stranded DNA is washed from the column into a scintillation vial and quantitated by scintillation counting. The results are plotted as a graph showing the percentage of the DNA that has melted at each temperature; from such a graph, the $T_{\frac{1}{2}}$—the temperature at which 50 percent of the DNA has melted—can be calculated. The difference in the $T_{\frac{1}{2}}$ values of a homoduplex DNA (labeled and unlabeled DNA from the same species) and of a heteroduplex DNA containing one strand from another species is a measure of the genetic difference and thus the time of evolutionary divergence of the species being compared.

Results similar to those obtained by Charles Sibley and Jon Ahlquist using this experimental approach are shown in Figure 26-2. From these data, draw an evolutionary tree showing the relative times at which the four species diverged.

PART E: *Working with Research Data*

38. The DNA sequences encoding part of a particular protein from five different species are shown in Figure 26-3a.

    a. Assuming that the divergence in these sequences is representative of the other genes in these species, draw an evolutionary tree that fits these sequence data. Explain how you deduced this tree. (*Hint*: Examine the sequences pairwise and note the total number of differences in the DNA sequences.)

    b. Which of the evolutionary trees shown in Figure 26-3b best fits the data? Does this tree match yours?

39. Chen and Roufa, in investigating the structure of genes for ribosomal proteins, isolated several γ phage cones, containing human genomic DNA inserts, that hybridized to a cDNA from the mRNA for ribosomal protein S17. Southern blot analysis of these clones, probed with radioactive S17 cDNA, indicated the presence of at least six different genomic sequences complementary to the cDNA. Although ribosomal RNA genes occur in multiple copies in vertebrate genomes, previous analysis of ribosomal protein genes had indicated that these genes occur only once per haploid genome.

Partial nucleotide sequences of the human genomic DNA inserts in three of these γ clones (HGS17-1, 17-2, and 17-6) are shown in Figure 26-4 along with the corresponding partial sequence for the cDNA for human ribosomal protein S17. The numbers 472 and 1469 indicate positions downstream from the transcription start site determined from the cDNA sequence. Coding sequences are shown in capital letters; noncoding sequences are shown in lower-case letters. One insert (HGS17-6) contains DNA that is complementary to the cDNA sequence but also contains four intervening sequences spanning almost 4kb of DNA. The other inserts (HGS17-1 and 17-2) contain DNA complementary to that of the cDNA only; these clones include only the exons and differ slightly in the coding sequence from the cDNA.

Species A: ATA ACC ATG TAT ACT ACC ATA ACC ACC TTA ACC CTA ACT CCC TTA ATT CTC CCC ATC CTC
Species B: ACA GCC ATG TTT ACC ACC ATA ACT GCC CTC ACC TTA ACT TCC CTA ATC CCC CCC ATT ACC
Species C: ATA ACT ATG TAC GCT ACC ATA ACC ACC TTA GCC CTA ACT TCC TTA ATT CCC CCT ATC CTT
Species D: ATA GCA ATG TAC ACC ACC ATA GCC ATT CTA ACG CTA ACC TCC CTA ATT CCC CCC ATT ACA
(a) Species E: ATA ACC ATG CAC ACT ACT ATA ACC ACC CTA ACC CTG ACT TCC CTA ATT CCC CCC ATC CTT

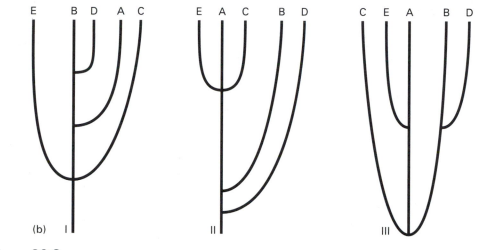

(b)

▲ Figure 26-3

a. Do you think that clone HGS17-1 represents the functional gene for this ribosomal protein? Why? If not, what does it represent?

b. How would you determine which of the DNA inserts represents the transcriptionally active gene?

40. Protein synthesis remains as a significant roadblock in all evolutionary schemes leading from RNA or RNA-like polymers to primitive cells. The exquisitely complicated machinery that catalyzes protein synthesis in modern cells involves two large RNA molecules, one or two smaller RNAs, 50–80 polypeptides, mRNA templates, many initiation, elongation, and termination factors, dozens of specifically aminoacylated tRNAs, and at last 20 different tRNA synthetases. None of these components appears to be very useful in isolation; yet the entire apparatus must have evolved in a stepwise manner, implying the usurpation of molecules previously used for another purpose. In particular, the evolution of tRNA molecules, which are very ancient, has elicited considerable headscratching because these molecules seemingly have few alternative uses, nor do they seem to be related to other useful molecules that might be their evolutionary precursors.

Strategies for elucidating the steps in the evolution of tRNA must involve searches through a sort of "molecular fossil record," as the original conditions for this evolutionary pathway have disappeared long ago. This molecular fossil record probably includes RNA catalysts (like RNase P and *Tetrahymena* introns) and enzymes involved in RNA metabolism. One of these enzymes is the so-called Qβ replicase, an RNA polymerase that functions to replicate the single-stranded RNA genome of the coliphage known as Qβ. This protein enzyme is a tetramer consisting of a phage-encoded subunit (subunit II) complexed to three host-derived proteins—the elongation factors Tu and Ts and the ribosomal protein S1.

The involvement of host proteins in Qβ replicase led Weiner and Maizels to speculate on the relationship of replication and tRNA evolution. Tu binds to amino-acylated tRNA, and Ts binds to Tu during protein synthesis. Both the plus and minus strands of the Qβ genome contain the 3′-terminal CCA motif found in tRNA. Indeed, many other bacterial and plant RNA viruses have 3′-terminal CCA sequences. One interpretation of these observations is that the virus has simply appropriated translation factors to function in RNA replication. However, the presence of tRNA-like ter-

S17 cDNA: G A T A G C A G G T T A T G T C

HGS17-1: G A C A G C A G G T T A T G T C

HGS17-2: G A A A G C A G G C T A T G T C

HGS17-6: G A T A G C A G G g t g a g t c g g g . . . Intron 2 . . . T T A T G T C

472   1469   ◀ Figure 26-4

mini in many otherwise unrelated nucleic acids, assuming that these did not evolve independently, led these workers to speculate that the 3′-terminal tRNA-like structures represent a molecular fossil of the ancient "RNA world."

These speculations also might help solve two other problems faced by early replicating systems: (1) identification of a specific site on the template for initiation of replication and (2) distinguishing the specific template from the myriad of other RNA molecules that are postulated to have coexisted with the early replicating systems. If the templates had 3′ identifying sequences (or "tags"), replication could have proceeded completely from the 3′ to 5′ end (analogous to modern-day replication) and would only proceed on specific template molecules. Thus the evolution of a 3′ tag would have been highly favored, and, indeed, essential for the evolution of more sophisticated living systems.

Other information relevant to this discussion includes the demonstration by Cech that the self-splicing intron of *Tetrahymena* can be experimentally transformed into a poly C polymerase. This finding indicates that early RNA replication might have been performed by an RNA, rather than by a proteinaceous catalyst. If this "RNA replicase" recognized a specific 3′ terminus, then it could proceed in template-directed polymerization of RNA from the same pool of prebiotically synthesized ribonucleotides that must have given rise to the RNA replicase itself. This primitive RNA thus functioned both as genome and as enzyme. Evolution of this RNA might have resulted in generation of variants, but each would have had the same 3′-terminal structure to ensure that it could be recognized and replicated. Weiner and Maizels postulate that one possible variant could be "charged" by a mononucleoside (N), exactly as documented for the *Tetrahymena* intron. The labile bond between the RNA replicase and the N monophosphate would be discharged by attack of an amino acid carboxyl group, rather than the 3′ hydroxyl of an RNA, thus generating an aminoacyl-NMP intermediate. Another RNA, with the recognized CCA tag, then attacks the aminoacyl-NMP intermediate, and is "charged" with the amino acid. This hypothetical reaction scheme is diagrammed in Figure 26-5.

a. Propose a model for the evolution of tRNA molecules assuming that the scheme in Figure 26-5 was operating.

b. Does your model explain the known association of specific amino acids with specific tRNAs (i.e., specific anticodons)?

c. What is the fundamental difference between the functions of RNA before these postulated events and afterward?

d. What observations concerning tRNA, tRNA genes,

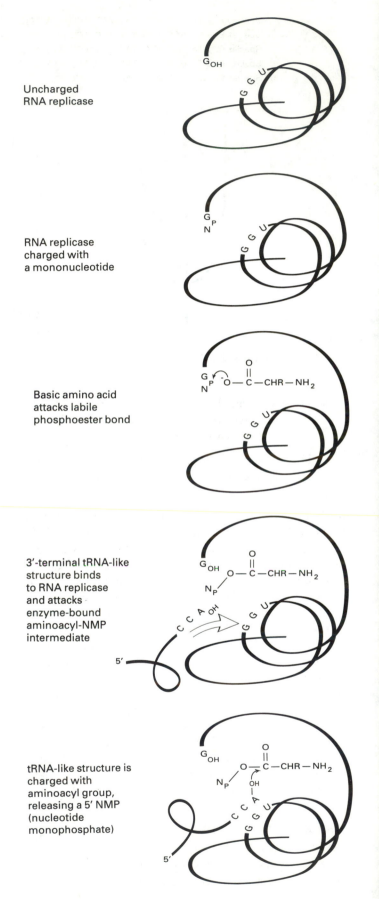

Uncharged
RNA replicase

RNA replicase
charged with
a mononucleotide

Basic amino acid
attacks labile
phosphoester bond

3′-terminal tRNA-like
structure binds
to RNA replicase
and attacks
enzyme-bound
aminoacyl-NMP
intermediate

tRNA-like structure is
charged with
aminoacyl group,
releasing a 5′ NMP
(nucleotide
monophosphate)

▲ **Figure 26-5**

and viral RNA replicases would be partially explained or made more plausible if the Weiner and Maizels hypothesis is true?

41. Before the advent of sequencing of entire RNA molecules, short segments of RNA could be sequenced. In 1969, in order to compare the sequences of 16S rRNAs from various species, Carl Woese used T_1 ribonuclease to cut the RNAs to be compared into small pieces. This enzyme cuts RNA on the 3′ side of guanine, producing many short oligonucleotides, which Woese sequenced. Nucleotide sequences less than six bases long are usually repeated in a 16S RNA, these redundant oligonucleotides were not analyzed. In contrast, when oligonucleotides containing six or more bases are found to be identical in different 16S RNAs, this identity usually reflects truly homologous sequences in the rRNAs.

Woese analyzed data from T_1 of 16S rRNAs by calculating an association coefficient, S_{AB}, which is defined as twice the number of nucleotides in the oligonucleotides common to the two RNAs being compared divided by the number of nucleotides in all the oligonucleotides with six or more nucleotides in both RNAs. S_{AB} can vary from 1 when the RNAs being compared are identical to less than about 0.1 when they are unrelated. (S_{AB} is usually greater than 0 because of chance similarities in the sequences of unrelated molecules.) When the 16S rRNAs from seven species were compared by this method, the S_{AB} values in Figure 26-6 were determined.

a. From these data, construct an evolutionary tree

	A	B	C	D	E	F	G
A	1.0	.05	.08	.11	.11	.10	.07
B	.05	1.0	.21	.11	.12	.07	.07
C	.08	.21	1.0	.14	.12	.12	.06
D	:11	.11	.14	1.0	.51	.34	.17
E	.11	.12	.12	.51	1.0	.31	.15
F	.10	.07	.12	.34	.31	1.0	.19
G	.07	.07	.06	.17	.15	.19	1.0

Species key:
A = *Saccharomyces cerevesiae*
B = *Escherichia coli*
C = *Lemna* chloroplast
D = *Methanobacterium thermoautotrophicum*
E = *Methanobrevibacter ruminantium*
F = *Halobacterium halobium*
G = *Sulfolobus acidocaldarius*

▲ **Figure 26-6**

showing the relative times of divergence for each of these species.

b. What do these data indicate about the relatedness of the archaebacterium *Sulfolobus acidocaldarius*, which James Lake classifies as belonging to a separate group called eocytes, to other archaebacteria (e.g., *Methanobacterium thermoautotrophicum*, *Methanobrevibacter ruminantium*, and *Halobacterium halobium*) and to eubacteria and eukaryotes?

Glossary

abscisic acid A plant **hormone** that has growth-inhibitory, senescence-promoting activities and also prevents water loss from leaves during dry periods.

absorption spectrum Relative intensity of light absorbed by a compound (e.g., a photosynthetic pigment) at various wavelengths of incident light. Comparison of the absorption spectra of chlorophylls with the action spectrum of **photosynthesis** implicates the **chlorophylls** as the primary pigments in photosynthesis.

acetyl CoA Metabolite found in the matrix of **mitochondria** that is a common intermediate in many biosynthetic reactions. Pyruvate from glycolysis, fatty acids, and amino acids are all converted to acetyl CoA, which then transfers its two-carbon acetyl group to citrate at the beginning of the **citric acid cycle**.

acid Any molecule or ion that can release a hydrogen ion.

actin Major protein of most animal cells, which can exist in a monomeric globular form (G actin) and polymeric filamentous form (F actin). Actin has diverse roles in both the structure and function of cells. In muscle cells, F actin interacts with **myosin** during contraction.

action potential The all-or-none electric impulse that propagates down the length of an **axon** as the result, in part, of the opening and closing of voltage-sensitive ion channels and the influx of Na^+ ions.

action spectrum Relative rates of **photosynthesis** (or other light-dependent process) at various wavelengths of incident light. See also **absorption spectrum.**

activation energy The input of energy required to initiate a reaction.

activator A compound that increases the ability of an **enzyme** to catalyze a reaction.

active ion transport Movement across a membrane of ions against a concentration gradient by a process directly coupled to ATP hydrolysis. Active ion transport proteins are often termed ATPases, for their ATP hydrolysis activity, or ion pumps.

active site The site in an **enzyme** where the substrate binds and the reaction is catalyzed.

adenylate cyclase Enzyme that catalyzes formation of the **second messenger** cyclic AMP (cAMP) and plays a key role in transducing many hormonal signals. Binding of several different **hormones** (e.g., epinephrine, glucagon, FSH, vasopressin) to their cell-surface receptors leads to activation of adenylate cyclase and a rise in intracellular cAMP.

adhesion plaques Area of attachment of cells to a substrate, which is thought to contain actin and actin-binding proteins.

aerobe, obligate A microorganism or eukaryotic cell that needs O_2 to grow. Obligate aerobes metabolize glucose to CO_2 and use O_2 as the final electron acceptor.

affinity chromatography A technique for purification of macromolecules based on the binding of a macromolecule with a specific **ligand** (e.g. an antibody or a substrate analog). Usually the ligand is coupled to an insoluble matrix such as a plastic bead.

agrin A protein found in the **basal lamina** that can cause aggregation of both acetylcholinesterase and acetylcholine receptors.

allelic exclusion The ability of heterozygous T and B **lymphocytes** to produce only one allelic form of antigen-specific receptors (T-cell receptor and **immunoglobulin**, respectively) when they have the genes to produce two forms.

allosteric site The site in an **enzyme** to which **effectors** of enzymatic activity bind; also called the regulatory site. The allosteric site is distinct from the **active site.**

α helix A spiral-shaped secondary structural element of proteins in which the carboxyl oxygen of each peptide bond is hydrogen-bonded to the amino group of a second amino acid that is separated from it in the primary structure by three residues. See also **β-pleated sheet.**

α-actinin   A coiled-coil protein, containing two identical polypeptides, that bundles the ends of **actin** filaments in striated and smooth muscle cells.

Alu **sequence**   A very common, short (150- to 300-bp) interspersed repetitive element (SINE) in mammalian DNA containing a recognition site for the **restriction endonuclease** *Alu*I; an example of a **retroposon.** *Alu* sequences have considerable sequence similarity with mammalian 7SL RNA, which is found in the cytoplasm and is a component of the **signal recognition particle.**

amino acid   One of the monomers that make up proteins. Amino acids contain an amino group linked to a carbon atom that is linked to a variable "side chain" and an acidic carboxyl group.

aminoacyl-tRNA synthetase   Any enzyme that couples an **amino acid** to the end of a tRNA.

amphipathic   Refers to a molecule or structure that has both a **hydrophobic** part and a **hydrophilic** part. Both proteins and lipids may be amphipathic.

amyloplast   A nonpigmented, starch-containing **plastid** found in plant cells.

anabolism   Cellular processes whereby energy is used to synthesize complex molecules from simpler ones. See also **catabolism.**

anaerobe, facultative   A microorganism or eukaryotic cell (e.g., yeast cell) that can grow either in the presence or absence of O_2.

anaerobe, obligate   A microorganism that can grow only in the absence of O_2. No eukaryotic cells are obligate anaerobes.

aneuploid   Having an inappropriate number of **chromosomes.**

anion   A negatively charged ion.

anterograde   Movement of cytoplasmic material from the cell body of a neuron down the **axon** to the growth cone or **synapse.** Material moving in this direction is usually associated with the growth or development of the synaptic junction. See also **retrograde.**

antibody   See **immunoglobulin.**

anticodon   Sequence of three nucleotides in a transfer RNA that is complementary to a **codon** in messenger RNA.

antigen   Any foreign material that is specifically bound by an **antibody.** See also **immunogen.**

antigenic determinant   The part of an **antigen** molecule that binds to the antigen-combining region of an **antibody;** also called an epitope.

antioncogene   A gene whose loss is the cause of a malignant **transformation.**

antiport   Cotransport process in which the movement of molecules or ions against their concentration gradient is driven by the movement in the opposite direction of a second ion with its concentration gradient. See also **symport.**

apocytochrome *c*   The cytoplasmically synthesized precursor of cytochrome *c* (a major mitochondrial protein), which is translocated to mitochondria. Apocytochrome *c* does not contain a heme group, whereas cytochrome *c* does. The presence of the heme group probably induces a conformational change in cytochrome *c*, which prevents its translocation to mitochondria.

archaebacteria   Anaerobic prokaryotic organisms, generally found in extreme environments such as brine ponds or hot sulfur springs, which are related to but distinct from the so-called true bacteria (**eubacteria**).

ascus   In yeast and fungi, a sac that encloses **haploid** cells called ascospores.

astral fibers   **Microtubules** that radiate outward from the mitotic poles of a dividing cell to the periphery.

asymmetric carbon   A carbon atom with four different substituents.

ATP (adenosine triphosphate)   A nucleotide that is an important cellular molecule for capturing and transferring free energy.

ATP synthase   Multimeric protein particle bound to inner mitochondrial membranes that catalyzes synthesis of ATP coupled to proton movement down the electrochemical gradient; also called the F_0F_1 particle. This bipartite enzyme contains two oligomeric complexes: the F_0 particle, through which protons are believed to flow, and the F_1 particle, which contains the catalytic site for phosphorylation of ADP. A similar enzyme is found in bacteria and the thylakoid membranes of chloroplasts. See also **oligomycin; CF_0CF_1.**

attenuation   Transcriptional control mechanism that regulates the frequency of completion of a mRNA chain.

autogenous regulation   Control by a regulatory protein of its own synthesis.

autonomously replicating sequence (ARS)   Any sequence that allows for autonomous replication of a chromosome.

autoradiography   Technique for visualizing radioactive molecules in a sample (e.g., a tissue section or electrophoretic gel) by exposure of a photographic emulsion.

auxin   A plant **hormone** that can induce cell elongation. This effect may be mediated by a proton pump system that softens part of the cell wall, thus allowing growth.

auxotroph   A cell that will proliferate only when the culture medium is supplemented with a specific nutrient not normally required by a wild-type cell.

avidity of antibody binding   The apparent binding affinity exhibited between an **antibody** and **antigen,** one or both of which have multiple binding sites that bind independently. Binding of a multivalent antibody to a multivalent antigen is a common example. In such cases, it is not possible to define a single affinity constant for the interaction from an antigen saturation curve; instead, the half-saturation concentration reflects the avidity.

axon   The portion of a neuron that conducts electric impulses away from the cell body. Axons can be more than 1 m long in humans.

axon hillock   The junction of a nerve cell body and axon; site where an **action potential** originates.

axon terminal   Small branch at the distal end of an **axon** that forms a **synapse** with a postsynaptic cell.

axoneme   The bundle of **microtubules** present in **cilia** and **flagella.** It is enveloped by an extension of the plasma membrane and connects with the **basal body.**

B cell   See **lymphocyte, B.**

bacteriophage (phage)   A virus that infects bacterial cells.

Barr body   Dark-staining, peripheral nuclear structure consisting of an inactive, condensed X chromosome in female mammals, who are XX. See also **lyonization.**

basal body   A structure at the base of **flagella** and **cilia** from which **microtubules** radiate. The (−) ends of microtubules are associated with a basal body, and the (+) ends point away from it.

basal lamina   A thin sheetlike extracellular network underlying most animal epithelial and endothelial cells. It usually contains type IV **collagen** and **laminin.**

base   Any molecule or ion that can combine with a hydrogen ion.

base composition   Frequency of adenine (A), guanosine (G), cytosine (C), thymine (T), or uracil (U) in a **nucleic acid.**

benign   Refers to a **tumor** containing cells that closely resemble normal cells. Benign tumors stay in the tissue where they originated. See also **malignant.**

β-pleated sheet   A secondary structural element of proteins in which there is a series of **hydrogen bonds** between the backbone atoms of the peptide bonds in different polypeptide chains or between the peptide bonds in different sections of a folded polypeptide. See also α **helix.**

bilayer   A symmetrical two-layer structure formed by **phospholipids** in water; the basic structure of all biomembranes. In a bilayer, the **polar** lipid head groups are exposed to the water, while the hydrocarbon chains are not.

binding-change mechanism hypothesis   The hypothesis that relates conformational shifts of the F_0F_1 **particle** to **oxidative phosphorylation.** A shift in one of the β subunits is responsible for the release of newly formed ATP, whereas another shift in a different β subunit facilitates binding of ADP and P_i.

biological membrane   Permeability barrier, surrounding cells or organellar compartments, that consists of a phospholipid **bilayer** composed of two lipid leaflets in which proteins are embedded. Other lipids including **cholesterol** and glycolipids may be present. See also **integral membrane protein** and **peripheral membrane protein.**

brittle bone disease   See **osteogenesis imperfecta.**

brush border   See **microvilli.**

buffer   A solution of the acid (HA) and base (A⁻) form of a compound that undergoes little change in pH when small quantities of strong acid or base are added.

C₄ pathway   Two-step pathway of CO_2 fixation that involves shuttling CO_2 from mesophyll cells to bundle sheath cells, where it is used in the **Calvin cycle.** This pathway, which is common in plants growing in hot, dry environments, increases the CO_2 level in the bundle sheath cells and consequently reduces the rate of **photorespiration.**

C value   Total amount of DNA per **haploid** cell in an organism.

cadherins   A family of Ca^{2+}-dependent cell-adhesion molecules, which play roles in tissue differentiation and structure. See also **uvomorulin.**

calorie   Unit of energy equal to the amount of heat required to raise the temperature of 1 ml of water at 14°C by 1°C.

Calvin cycle   See **dark reactions.**

cap-binding complex   The complex of four proteins that recognizes the 5′-5′ methylated guanylate residue at the 5′ end of eukaryotic mRNA.

capping, 5′   Addition of 7-methylguanosine via a 5′ to 5′ triphosphate bridge to the 5′ end of hnRNA. The 5′ cap is retained in almost all eukaryotic mRNAs.

capsid   The outer proteinaceous coat of a virus, which encloses the nucleic acid. The capsid plus enclosed nucleic acid is called a *nucleocapsid.*

carbohydrate   A polyhydroxyaldehyde, polyhydroxyketone, or a compound derived from these compounds. Examples include **monosaccharides, oligosaccharides,** and **polysaccharides.**

carcinogen, chemical   A substance that can cause **tumors** when animals or humans are exposed to it.

carcinogenesis   The induction of a **tumor.**

carcinoma   A malignant **tumor** derived from endoderm or ectoderm.

cardiolipin   A common **lipid** in the mitochondrial inner membrane that is thought to enhance the permeability of this membrane to protons; also called diphosphatidyl glycerol.

catabolism   Cellular processes whereby energy is extracted from the breakdown of complex molecules into simpler ones. See also **anabolism.**

catabolite repression   Depression by glucose or a glucose metabolite of **transcription** of genes coding for a wide variety of sugar-metabolizing enzymes. Catabolite repression is mediated by catabolite activator protein (CAP).

catalyst   A substance that increases the rate of a chemical reaction without undergoing a permanent change in its structure.

catenated DNA   Covalently linked circles of DNA.

cation   A positively charged ion.

cDNA (complementary DNA)   DNA copied from an mRNA molecule.

cell cycle   The ordered cyclic phases that follow division of eukaryotic cells and include G_1 (gap 1) before DNA synthesis occurs, S (the DNA-synthesis period), G_2 after DNA synthesis occurs, and M (**mitosis**). Some workers also distinguish the G_0 state, a diversion from G_1, in which **quiescent** (nongrowing, nondividing) cells are stopped. Appropriate stimulation of such cells can induce them to return to G_1 and resume growth and division.

cell line   A culture of cells that has undergone a change allowing the cells to grow indefinitely.

cell strain   A culture of cells that does not have the capacity to grow indefinitely.

cellulose   A polysaccharide made of glucose units linked together by $\beta(1 \longrightarrow 4)$ glycosidic bonds. It is the major constituent of plant cell walls.

central nervous system   The part of the nervous system comprising the brain and the spinal cord.

centriole   The central structure within the **microtubule organizing center (MTOC)** from which **microtubules** radiate. The (−) ends of microtubules are associated with a centriole, and the (+) ends point away from it.

centromere   Region of the chromosome, containing conserved CEN sequences, where sister chromatids are attached.

CF₀CF₁ complex   Integral **thylakoid membrane** protein complex that catalyzes synthesis of ATP coupled to proton movement down the electrochemical gradient. The catalytic site for ADP phosphorylation is within the CF_1 portion, which is located on the stromal side of the membrane. The CF_0CF_1 complex is functionally equivalent to mitochondrial **ATP synthase.**

chemiosmotic hypothesis   The hypothesis that ATP synthesis in both **mitochondria** and **chloroplasts** is driven by a proton (pH) gradient and an electric potential (200 mV).

chiasma (pl. chiasmata)   Point at which nonsister chromatids of recombinant chromosomes cross each other.

chlorophylls   A group of porphyrin pigments, critical in photosynthesis, that contain a central Mg^{2+} ion surrounded by five rings, one of which has a long hydrocarbon tail. See also P_{680} and P_{700}.

chloroplast   An organelle in plant cells that is surrounded by a double membrane and contains **chlorophyll** and the other photosynthetic machinery.

cholera toxin   A peptide produced by the bacterium **Vibrio cholerae** that causes severe diarrhea through irreversible activation of **adenylate cyclase.**

cholesterol   A four-ringed hydrocarbon with a hydroxyl group; a major component of many eukaryotic membranes.

cholinergic   Refers to a neuron-to-neuron or neuron-to-muscle synapse that releases acetylcholine from the presynaptic cell.

chromatin   Complex of DNA and associated protein isolated from eukaryotic cells that constitutes the genetic material.

chromatosome   **Chromatin** structural unit comprising one **nucleosome** with one bound H1 histone molecule.

chromoplast   A pigmented **plastid** found in plants cells, which contains yellow, orange, and red pigments. **Chloroplasts** are one example.

chromosome   The structural unit of the genetic material of eukaryotes, consisting of DNA in association with basic proteins called **histones.**

cilium (pl. cilia)   Membrane-enclosed locomotory structure extending from the surface of eukaryotic cells and composed of a specific arrangement of **microtubules** (called an **axoneme**). Cilia usually occur in groups, and their beating causes movement of particles or liquid along the apical surface of cells. See also **flagellum.**

cis-active   Refers to DNA elements that can control genes only on the same piece of DNA. In bacteria, cis-active elements must be adjacent or proximal to gene(s) to exert control, whereas in eukaryotes they may be distal. An **operator** or **promoter** is a cis-active controlling element. See also **trans-active.**

cistron   Smallest genetic unit that encodes one polypeptide: independent **trans-active** complementation unit.

citric acid cycle   A set of nine reactions that occur in the matrix of **mitochondria,** generating NADH, $FADH_2$, and GTP, all of which can be used to produce ATP; also called Krebs cycle.

class switching   The shift of a B cell or its progeny from production of one class of **immunoglobulin** to the production of another class by a change in the type of heavy chain. See also **lymphocyte, B.**

clathrin   A fibrous protein of high molecular weight and multiple subunits that coats in a basketlike manner small portions of the plasma membrane and Golgi apparatus. The polymerization and association of clathrin with endocytic receptor proteins is thought to be anchored by assembly particles. See also **endocytosis.**

clone   A strain of cells descended in culture or in vivo from a single cell.

clone, cDNA   A selected **host cell** with a vector containing a **cDNA** molecule from a different organism.

clone, genomic   A selected **host cell** with a vector containing a fragment of **genomic DNA** from another organism.

codon   Sequence of three nucleotides in messenger RNA that specifies a particular amino acid during protein synthesis; also called triplet.

coenzyme   A tightly bound small molecule or **prosthetic group** essential to enzymatic activity.

colchicine   A plant alkaloid that binds to free tubulin and is incorporated into polymerizing microtubules, inhibiting the addition of more tubulin. The drug and its derivatives are useful in genotyping as well as in anticancer therapy.

collagen   A triple-helical protein present in bone, skin, cartilage, tendon, and other structures. Collagen forms fibrils that have great tensile strength. The subtypes of collagen are distinguished from each other by the cells that synthesize them and the substances that facilitate association of collagen with the cell.

complement   A group of cytotoxic serum proteins involved in the mediation of immune responses. The complement cascade is initiated by binding of **antibody** to **antigens** on cell surfaces.

condensation   A chemical reaction in which two molecules are linked and water is eliminated. An example is the formation of a **peptide bond.**

consensus sequence   Conserved nucleotide sequence present in a number of genes, which often is a DNA or RNA element involved in control of transcriptional or processing events.

constitutive mutant   (1) A mutant in which a structural gene is expressed at constant levels, as if continuously induced; (2) also a bacterial regulatory mutant in which operon products are produced in the absence of inducer.

contact inhibition of growth   The cessation of cell proliferation when a cell comes in contact with another cell or cells.

contact inhibition of movement   The cessation of fibroblast movement in a given direction when the fibroblast comes in contact with another cell.

cooperative interaction   In a protein with more than one binding site, an alteration in the conformation of the protein that occurs after binding of one or more molecules and that affects the binding affinity of the remaining unbound site(s).

$C_0t_{\frac{1}{2}}$ The product of the molar concentration of a DNA sample (C_0) and the reaction time (t) in seconds at which one-half of the sample renatures. This is a convenient parameter for comparing the reassociation rates of different fractions of DNA.

cotransport   Protein-mediated transport of ions, amino acids, or sugars across a membrane against a concentration gradient, which is driven by coupling to movement of a second ion down its concentration gradient. See also **antiport** and **symport.**

coupled transcription-translation   In bacteria, translation of an mRNA molecule while the DNA encoding it is still being transcribed. In eukaryotes, **transcription** and **translation** occur in the nucleus and cytoplasm, respectively; thus they are not coupled.

covalent bond   A chemical force that holds the atoms in molecules together by sharing of electrons through the overlap of the electronic orbitals of two atoms; also called an electron-pair bond. Such a bond has a strength of 50–200 kcal/mol.

critical micelle concentration   Minimum detergent concentration at which **micelle** formation occurs.

CURL   Smooth endocytic compartment (named after its function as the *c*ompartment of *u*ncoupling of *r*eceptor and *l*igand) in which acid-dependent uncoupling occurs.

cyclic AMP (cAMP)   A **second messenger,** produced from ATP by **adenylate cyclase** and degraded by cAMP phosphodiesterase, that amplifies the signal of many cell-surface–active **hormones.**

cyclic AMP (cAMP) phosphodiesterase   Any enzyme that hydrolyzes **cyclic AMP (cAMP).**

cytokinin   A plant **hormone** that can initiate cell division.

cytoplasm   Viscous material lying outside the **nucleus** of a eukaryotic cell. The part of the cytoplasm not contained in any **organelle** is called the *cytosol.*

cytoplasmic face   The face of a membrane directed toward the **cytoplasm.**

cytoplasmic plaque   A structure present in smooth muscle that is analogous to the Z disk in striated muscle and to which **actin** filaments are anchored.

cytoskeleton   Network of fibrous elements (including **microtubules, microfilaments,** and **intermediate filaments**) found in the cytoplasm of eukaryotic cells. The functions of the cytoskeleton include structural support for the cytoplasm, organelle and chromosome motion, and cell motion.

cytosol   See **cytoplasm.**

dark reactions   The sum total of the reactions and processes that fix CO_2 into sugar during **photosynthesis;** also called Calvin cycle. These processes are indirectly dependent on the **light reactions** and occur both in the dark as well as under light conditions.

denature   For proteins, to disrupt **noncovalent bonds** and cause the polypeptide chain to unfold; for nucleic acids, to disrupt **hydrogen bonds** and cause a double-stranded molecule to become single-stranded.

de novo pathway   Sequence of reactions resulting in synthesis of required compounds (e.g., nucleotides) from simple carbon and nitrogen compounds.

dendrite   Process extending from a nerve cell body that receives signals from **axons** or other neurons.

depolarization   Change in membrane potential that results in a new membrane potential closer to 0 mV. This effect is typical of an excitatory neurotransmitter.

desmin   A 55-kDa intermediate-filament protein located at the edge of Z disks in striated muscle. Desmin fibers anchor and orient Z disks, so that the myofibrils are in proper register with one another.

diphosphatidylglycerol   See **cardiolipin.**

diploid   An organism or cell having two full sets of homologous **chromosomes.** Somatic cells contain the diploid number of chromosomes ($2n$) characteristic of a species. See also **haploid** and **merozygote.**

dipolar bond   A bond between two atoms of differing **electronegativity.** One end of a dipolar bond is slightly negatively charged, and the other end is slightly positively charged.

dipole   Polar molecules that tend to line up in an electric field because one part of the molecule carries a partial negative charge and another a partial positive charge. Dipoles occur when one atom in a molecule has a greater **electronegativity** than another.

disulfide bond   A covalent link between two cysteine residues in the same or different polypeptide chains, which is formed by an **oxidation** reaction.

DNA fingerprint   Southern blot pattern of minisatellite DNA distribution, which is unique for each human.

DNA loop   Bend in a DNA **double helix,** which can result in

two linearly separated DNA sequences being close to each other.

DNA polymerase   Any enzyme that catalyzes synthesis of double-stranded DNA in the $5' \longrightarrow 3'$ direction using a DNA **template.** *E. coli* has three DNA polymerases (I, II, III) and mammalian cells have four ($\alpha, \beta, \gamma, \delta$). *E. coli* polymerase III is a large molecule with two active sites that catalyzes synthesis of both the **leading strand** and **lagging strand.** In mammals, polymerase α is responsible for lagging-strand synthesis, and polymerase δ for leading-strand synthesis. See also **primase; processivity;** and **proofreading, DNA.**

DNA repair   Correction of DNA lesions, which frequently are caused by environmental insult.

DNA replication   A general term encompassing the initiation and termination of DNA synthesis and chromosome separation.

double helix   The most common structure for cellular DNA in which the two sugar-phosphate strands are wound around each other with the base pairs stacked between the strands. Most double helices are "right-handed." See also **Z-DNA.**

downstream   In the 3' direction within DNA. The +1 position of a gene is the first transcribed base pair; nucleotides downstream from the +1 position are numbered +2, +3, etc. See also **upstream.**

duplicated gene   A protein-coding gene present in the genome in multiple copies. About one-half to three-fourths of all protein-coding genes in multicellular organisms are duplicated, whereas only a few bacterial protein-coding genes are duplicated. See also **solitary gene.**

dynein   An ATPase in **cilia** and **flagella** that accounts for the energy necessary to drive the movement of these organelles. ATP binding to dynein and its hydrolysis causes the dynein bridges to sequentially break and form new bonds with adjacent microtubular doublets.

dystrophin   A 400-kDa muscle protein that is missing in patients with Duchenne muscular dystrophy. The physiological function of dystrophin is unknown.

editing, DNA   See **proofreading, DNA.**

Edman degradation   Technique for obtaining the amino acid sequence of a polypeptide by repeated removal and analysis of the amino terminus.

effector   A compound that affects the ability of an **enzyme** to catalyze a reaction. See also **activator** and **inhibitor.**

elaioplast   A lipid-containing **plastid** found in plant cells.

electron-pair bond   See **covalent bond.**

electronegativity   The power of a molecule to attract electrons to itself, measured on a scale of 0 to 4 with 4 representing the most electronegative atom, fluorine.

electrophoresis   Technique for separation of molecules (DNA, RNA, or protein) based on their migration in a strong electric field.

electrophoresis, pulse-field   Technique for separation of large DNA molecules (in the range of 1 million base pairs) using pulsed electric fields rather than the more standard continuous electric field.

electroporation   Technique for introducing macromolecules such as DNA into eukaryotic cells by using brief electric shocks to open transient holes in the cell membrane.

elongation factor   One of a group of helper proteins involved in **translation** whose functions include regulation of binding of charged tRNA to the A site or release of uncharged tRNA, among others. See also **initiation factor.**

Emerson effect The increased photosynthetic rate that occurs in light of two different wavelengths (e.g., 680 and 700 nm) compared with the sum of the rates at each wavelength. The original discovery of this effect suggested that leaves contain two different photosystems, now known as **PSI** and **PSII.**

endergonic Refers to a chemical reaction that absorbs energy (has a positive **free-energy change**).

endocytosis The uptake of extracellular materials by invagination of the plasma membrane to form a small membrane-bound vesicle (**endosome**) ~0.05–0.1 μm in diameter. Receptor-mediated endocytosis involves the specific uptake of a receptor-bound **ligand** typically by invagination of **clathrin**-coated regions of the plasma membrane of animal cells. See also **exocytosis** and **phagocytosis.**

endocytotic vesicle See **endosome.**

endoplasmic reticulum, rough Network of interconnected membranous structures within the **cytoplasm** of eukaryotic cells that function in protein synthesis and processing. See also **endoplasmic reticulum, smooth.**

endoplasmic reticulum, smooth Network of interconnecting membranous structures within the **cytoplasm** of eukaryotic cells, whose functions include lipid biosynthesis and detoxification of hydrophobic compounds. See also **endoplasmic reticulum, rough.**

endosome Small smooth vesicular intermediate in **endocytosis;** also called endocytotic vesicle.

endosymbiont (endosymbiotic) hypothesis The hypothesis that **mitochondria** and **chloroplasts** arose from the fusion of primitive prokaryotes with a precursor to present-day eukaryotic cells.

endothermic Refers to a chemical reaction that absorbs heat (has a positive change in **enthalpy**).

energy The ability to do work.

enhancer A DNA sequence that modulates the rate of transcription catalyzed by eukaryotic **RNA polymerase** II. Enhancers may be located at a great distance either upstream or downstream from the RNA start site.

enthalpy (*H*) Heat; in a chemical reaction the enthalpy of the reactants or products is equal to their total bond energies.

entropy (*S*) A measure of the degree of disorder or randomness. It can be expressed in the unit calories/degree.

enzyme A biological macromolecule that acts as a **catalyst.** Most enzymes are proteins, but RNA is also known to act as a catalyst (these molecules are called *ribozymes*).

enzyme regulatory site See **allosteric site.**

epigenetic Refers to a condition or process that can be passed from a cell to its progeny without any alteration in the coding sequence of the DNA.

epitope See **antigenic determinant.**

equilibrium The state of a chemical reaction in which the concentration of all products and reactants is constant and at which the rates of the forward and reverse reactions are equal.

eubacteria Prokaryotic organisms including the so-called true bacteria (e.g., the common bacteria *Escherichia coli* and *Salmonella typhimurium*). See also **archaebacteria.**

euchromatin Light-staining, less condensed portions of **chromatin.**

eukaryotes The class of organisms, including all plants, animals, fungi, yeast, protozoa, and most algae, that are characterized by the presence of a true **nucleus** and **organelles.** See also **prokaryotes.**

exergonic Refers to a chemical reaction that releases energy (has a negative **free-energy change**).

exocytosis Fusion of intracellular membrane-bound structures with the plasma membrane of eukaryotic cells, resulting in release of material to the extracellular fluid. See also **endocytosis.**

exon Part of a **primary transcript** (or the DNA encoding it) that *exits* the nucleus and reaches the cytoplasm as part of a mRNA molecule.

exon shuffling The reassortment by genetic recombination of nucleic acid segments that correspond to various functional protein domains.

exoplasmic face The external face of a membrane directed away from the **cytoplasm.**

exothermic Refers to a chemical reaction that releases heat (has a negative change in **enthalpy**).

extensin A glycoprotein present in plant cell walls that contains many glycosylated hydroxyprolines. It is secreted as a soluble molecule but becomes insoluble once incorporated into the cell wall.

extracellular matrix A usually insoluble matrix consisting of **collagens, laminin, fibronectin, heparan sulfate,** and other substances, which are secreted by mammalian and other animal cells and can affect the differentiation, morphogenesis, and biochemical functions of cells.

extrinsic protein See **peripheral membrane protein.**

F_0F_1 particle See **ATP synthase.**

facilitated diffusion Protein-aided diffusion of molecules across a membrane, which is distinguished by three kinetic properties: (1) a rate faster than that predicted by simple diffusion according to Fick's law; (2) ligand specificity (expressed by K_M); and (3) saturation (indicated by V_{max}).

fatty acid Any long hydrocarbon chain linked to a carboxyl group.

feedback inhibition Decrease in the catalytic activity of one of the enzymes in a metabolic pathway caused by the ultimate product of that pathway. Usually, the inhibited step is the first step in the pathway that does not lead to other products.

fibroblast A type of connective tissue cell, found in almost all vertebrate organs, that functions in mammals as stem cells for adipocytes and in wound healing.

fibronectin A glycoprotein that promotes cell attachment and migration of vertebrate cells. Its receptor may also be linked to **actin**-binding proteins inside cells.

fibronectin A major protein of the dense fibrillar network that covers normal quiescent cells in culture; also called LETS protein.

Fick's law Mathematical expression that relates the movement of an uncharged molecule across a membrane by simple diffusion (the diffusion rate) to its concentration gradient, the membrane area, and the permeability coefficient, which is proportional to the partition coefficient.

flagellum (pl. flagella) Membrane-enclosed locomotory structure extending from the surface of eukaryotic cells and composed of a specific arrangement of **microtubules** (called an **axoneme**). Usually there is only one flagellum per cell (as in sperm cells), and its bending propels the cell forwards or backwards. See also **cilia.**

flip-flop The movement of a phospholipid from one leaflet to the other in **bilayer** membranes. This process is thermodynamically unfavorable.

fluid mosaic model Currently popular model in which the membrane is seen as a two-dimensional mosaic of lipid and protein molecules that can move laterally.

follicle-stimulating hormone (FSH) A polypeptide **hormone** released by the anterior pituitary that influences the size of the ovarian follicle.

footprint Feature of a sequencing gel pattern indicating the presence of a protected nucleic acid segment(s); by extension, the nucleic acid segment(s) protected from chemical or enzymatic attack by binding of a specific protein (e.g., RNA polymerase).

FRAP (*fluorescence recovery after photobleaching*) Method for quantitating the diffusion constant of membrane molecules after localized bleaching of an area.

free-energy change (ΔG) A measure of change in potential energy, which can be used to predict the direction of a chemical reaction.

fusicoccins Fungal compounds that act in a similar way as **auxin** by lowering the pH of cell walls, thus facilitating cell extension.

ganglioside A glycolipid containing *N*-acetylneuraminic acid.

gap junctions Patches of particles, containing connexin, that form tiny channels directly linking the cytoplasm of two animal cells. These junctions support metabolic coupling or metabolic cooperation between cells.

gelsolin A protein found in amebas that can cleave **actin** in vitro, thus enhancing the sol form, and that may be critical to the ability of these cells to extend and retract pseudopods.

gene Physical and functional unit of heredity, which carries information from one generation to the next. Genes usually are composed of a DNA sequence containing both a transcribed region and a regulatory region, which modulates **transcription**. See also **cistron** and **transcription unit.**

gene control Regulation of the formation and use of mRNA, which can be effected at several molecular steps (levels of control) including **transcription, RNA processing,** RNA stabilization, and **translation.** In a few cases, amplification or rearrangement of chromosomal segments affect gene expression.

gene conversion Phenomenon in which alleles appear to be lost during meiotic recombination.

genetic drift Spontaneous variation in the sequence of **duplicated genes;** also called sequence drift.

genomic DNA All DNA sequences from a given organism.

ghost Hemoglobin-depleted erythrocyte, which retains the overall size of the intact cell.

gibberellic acid A plant **hormone** that induces germination of seeds through induction of specific mRNAs.

glucagon A peptide **hormone** synthesized in the A cells of the islets of Langerhans that elicits the degradation of glycogen. Blood levels of glucagon rise when glucose levels drop in the blood.

glycocalyx Membrane coat formed by the **oligosaccharide** side chains of lipids and proteins.

glycogenolysis Breakdown of glycogen to glucose. This process can be enhanced by **hormones.** A key site of this process is in the liver and fat cells.

glycosidic bond The **covalent bond** between two **monosaccharides.**

glycosyl transferase Any enzyme that can catalyze the addition of a sugar residue to proteins and phospholipids in the **endoplasmic reticulum** and the **Golgi complex.** All are **integral membrane proteins.**

glyoxysome An **organelle** in plant cells, similar but not identical to a **peroxisome** in animal cells, that contains enzymes involved in conversion of stored lipid into carbohydrate.

Golgi complex Stacks of membranous structures in eukaryotic cells that function in processing and sorting of proteins and lipids destined for other membranous compartments or for secretion to the extracellular fluid; also called Golgi apparatus. See also **trans Golgi reticulum (TGR).**

gram-negative Refers to any prokaryote that cannot be stained by the Gram technique (a specific histological stain for bacteria). Gram-negative bacteria have two lipid-containing membranes (inner and outer membrane) as well as a peptidoglycan structure called the cell wall.

gram-positive Refers to any prokaryote that can be stained by the Gram technique (a specific histological stain for bacteria). Gram-positive bacteria have a single lipid-containing surface membrane as well as a peptidoglycan structure called the cell wall.

group translocation Bacterial transport process in which the transported molecule is chemically modified (e.g., phosphorylated) during its movement across the membrane.

growing fork Site in double-stranded DNA at which the **template** strands are separated and addition of **nucleotides** to each newly synthesized chain occurs. As DNA synthesis proceeds, the growing fork continuously moves in the direction of leading-strand synthesis. See also **leading strand.**

gyrase See **topoisomerase II (topo II).**

haploid An organism or cell having only one member of each pair of homologous **chromosomes.** Gametes contain the haploid number of chromosomes *(n)* characteristic of a species. See also **diploid** and **merozygote.**

heat-shock response Increased expression of a specific group of genes in response to elevated temperature. This response is very widespread among both prokaryotic and eukaryotic species.

helix-turn-helix motif Conserved protein structural element common to one class of DNA-binding proteins. One of the two α helices is termed the recognition helix. See also α **(alpha) helix.**

hemidesmosome A structure similar to a **spot desmosome** that serves to attach cells to the **basal lamina.**

heparan sulfate A sulfated glycosaminoglycan that is chemically linked to matrix proteins such as **fibronectin** and **laminin** to form **proteoglycans.**

heterochromatin Dark-staining regions of condensed **chromatin.**

heteroduplex A DNA **double helix** formed by annealing single DNA strands from different sources.

heterokaryon A cell with more than one functional **nucleus.**

hexose A six-carbon **monosaccharide.**

histones A family of small highly conserved basic proteins found in the **chromatin** from all eukaryotic cells as part of the **nucleosomes.** The five major types are H1, H2A, H2B, H3, and H4 histone. Histonelike proteins have also been found in some prokaryotes.

hnRNA (heterogeneous nuclear RNA) **Primary transcripts** present in the **nucleoplasm** that are formed from eukaryotic protein-coding genes; also called pre-mRNA, hnRNA is processed to form mature mRNA. See also **RNA processing.**

Holliday structure Intermediate structure in DNA recombination whose resolution can result in **gene conversion.** Isomeric Holliday structures appear as chi forms in electron micrographs.

homeobox A conserved 60-aa region in DNA-binding proteins encoded by **homeotic genes.** The homeobox sequence has been found in genes from frogs to mammals.

homeotic gene A gene whose product is involved in controlling early development in multicellular organisms and in which a mutation can cause substitution of one developmental pathway for another. See also **homeobox.**

hormone Any substance that functions as an endocrine signal by inducing specific responses in target cells distant from its site of synthesis and release. Hormones coordinate the growth, differentiation, and metabolic activities of various cells, tissues, and organs in metazoans.

host cell A cell (usually bacterial) in which a vector can be propagated.

hyaluronic acid A rigid polymer consisting of repeating units of the disaccharide glucoronic acid $\beta(1 \rightarrow 3)N$-acetylglucosamine. Because of its **hydrophilic** residues, hyaluronic acid can bind a great deal of water and thus provides lubrication to joints and connective tissue.

hybrid dysgenesis Production of nonfertile hybrid progeny in a cross.

hybridization, in situ Technique for determining the location of a nucleic acid sequence (DNA or RNA) within a tissue or cell based on molecular hybridization of radioactive complementary sequences followed by **autoradiography.**

hybridization, molecular Association of complementary nucleic acid sequences to form double-stranded molecules. Hybrids can contain two DNA strands, two RNA strands, or one DNA and one RNA strand.

hydrogen bond A relatively weak association between an electronegative atom (the acceptor atom and a hydrogen atom covalently bonded to another electronegative atom (the donor atom).

hydrophilic **Polar** and soluble in water.

hydrophobic **Nonpolar** and insoluble in water.

hydrophobic interaction The force that causes the aggregation of nonpolar molecules or parts of molecules in an aqueous solution. This "interaction" is based on the positive change in **entropy** when a **nonpolar** molecule is removed from water and, to a lesser degree, on the **van der Waals interactions** between closely packed nonpolar molecules.

hyperpolarization Change in membrane potential that results in a new membrane potential farther from 0 mV. A hyperpolarization can be either negative or positive. This effect is sometimes characteristic of an inhibitory neurotransmitter.

hypertonic Refers to a solution with an osmotic strength greater than that of a cell, which is typically about 300 mOsm. See also **hypotonic.**

hypotonic Refers to a solution with an osmotic strength less than that of a cell. See also **hypertonic.**

I-cell disease An inherited disease resulting from a defect in the ability of cells to phosphorylate lysosomal enzymes on mannose moieties.

immunogen A substance capable of eliciting an immune response. See also **antigen.**

immunoglobulin (Ig) Collective term for proteins that function as antibodies. An immunoglobulin molecule contains two or more identical heavy chains and two or more identical light chains. The five major classes of immunoglobulin (IgA, IgD, IgE, IgG, and IgM) differ in the heavy chains they contain, their average serum levels, and their specific functions in the immune response. See also **lymphocyte, B.**

inhibitor A compound that decreases the ability of an **enzyme** to catalyze a reaction.

initiation factor One of a group of helper proteins involved in initiation of **translation** whose functions include modulation of interactions between ribosomal subunits and mRNA, among others. See also **elongation factor.**

initiator, tumor A chemical that damages DNA and that can induce **tumor** formation. See also **promoter, tumor.**

inositol 1,4,5-trisphosphate A water-soluble carbohydrate, derived from a phospholipid, that acts as a **second messenger,** releasing Ca^{2+} ions from stores in the **endoplasmic reticulum** and thus affecting various cellular activities.

insertion sequence See **IS element.**

integral membrane protein Any membrane-bound protein that can be removed from the membrane only by extraction with detergent; also called intrinsic protein. Some integral membrane proteins contain **hydrophobic** amino acid residues whose side chains interact with fatty acyl groups of membrane **phospholipids;** others are bound to the membrane by covalently attached **lipid.**

integration Recombinational event whereby foreign (e.g., viral) DNA is stably incorporated into a chromosome of a **host cell.**

integrin receptors A superfamily of cell-surface receptors that can bind **extracellular matrix** components such as **laminin** and **fibronectin** and contain two different transmembrane polypeptide chains.

interferons A group of 20 to 25 low-molecular-weight proteins that cause cells to become resistant to the growth of a wide variety of viruses.

intermediate filaments Cytoskeletal fibers, 8–12 nm in diameter, that are formed by polymerization of several classes of cell-specific subunit proteins (e.g., cytokeratins, **desmin,** vimentin). They connect the spot desmosomes of epithelial cells; form the major structural proteins of skin and hair; form the scaffold that holds Z disks and myofibrils in place in muscle cells; and generally function as important structural determinants in many metazoan cells and tissues.

intermediate-repeat DNA The eukaryotic DNA fraction that reassociates at an intermediate rate and consists of short interspersed elements (SINES) and long interspersed elements (LINES). This fraction constitutes 25–40 percent of total cellular DNA. See also **simple-sequence DNA** and **single-copy DNA.**

interneuron Any neuron that connects two neurons within a neuronal circuit and that is not a primary sensory or final motor neuron.

intervening sequence See **intron.**

intrinsic protein See **integral membrane protein.**

intron Part of an eukaryotic **primary transcript** (or the DNA encoding it) that is not included in the finished mRNA, rRNA, or tRNA; also called intervening sequence. In **hnRNA,** introns are bounded by 5′-GU and 3′-AG sequences.

inverted repeat Self-complementary region of a nucleic acid containing an axis of symmetry and capable of base-pairing about this axis of symmetry (e.g., CGCAT. . . . ATGCG), which can fold back on itself.

ion channel protein Any transport protein that allows selective ion transport down an ion concentration gradient.

ionic bond An electrostatic force holding an **anion** and a **cation** together, which does not involve sharing of electrons.

ionophore Any compound that increases the permeability of membranes to a specific ion.

IS element Any bacterial **mobile genetic element** that ends in **inverted repeats** and generally is less than 1500 bp in length; also called insertion sequence.

isoelectric focusing Technique for separation of molecules in a pH gradient subjected to an electric field. This technique separates molecules on the basis of charge.

K_M A parameter that describes the affinity of an **enzyme** for its **substrate,** the affinity of a transport protein for the transported molecule, or the affinity of a receptor for its **ligand.**

karyotype Number, sizes, and shapes of the metaphase chromosomes of a eukaryotic cell.

kinesin A 380-kDa protein that binds to **microtubules** and transports vesicles from the ($-$) to the ($+$) end. Its function is dependent on ATP hydrolysis.

kinetic energy The energy of movement.

kinetochore fibers **Microtubules** that extend from the centromere of each mitotic chromosome and extend toward the poles of the cell.

Krebs cycle See **citric acid cycle.**

lagging strand Newly synthesized DNA strand that is formed by $5' \longrightarrow 3'$ copying of the $5' \longrightarrow 3'$ **template** strand into short, discontinuous segments with the aid of RNA primers and ligation of these segments, called **Okazaki fragments.** The direction of lagging-strand synthesis is opposite to the movement of the **growing fork.** See also **leading strand.**

lamellipodia Thin cellular extensions that do not contain **organelles** but do contain **actin** filaments and are important to the motility of nonmuscle cells.

laminin A component of the **extracellular matrix** that contains three polypeptide chains and is commonly found in the **basal lamina.** One portion of the laminin molecule has a **heparan sulfate** binding site, and another portion binds to surface receptors.

lamins A group of proteins, about 10 nm in diameter, that are partly responsible for the cytoarchitecture of the **nucleus.** The phosphorylation and dephosphorylation of lamins A, B, and C play an important role in governing nuclear integrity during **mitosis.**

leading strand Newly synthesized strand of DNA that grows continuously in the $5' \longrightarrow 3'$ direction by copying of the $3' \longrightarrow 5'$ **template** strand. The direction of leading-strand synthesis is the same as the movement of the **growing fork.** See also **lagging strand.**

lectin A plant protein with multiple binding sites for specific sugars.

leucine-zipper protein Any DNA-binding protein that in dimeric form functions as a **transcription factor** and contains a DNA-binding domain and leucine-rich region in each subunit. Interaction of the leucine-rich regions leads to dimer formation.

leukemia A **sarcoma** that grows as individual cells in the blood, rather than as a solid mass.

library (1) A complete set of genomic clones from an organism (genomic library); (2) a complete set of cDNA clones from one cell type (cDNA library).

ligand Any chemical other than an enzyme substrate that binds to a specific macromolecule forming a macromolecule-ligand complex. Naturally occurring **hormones** (e.g., epinephrine) and certain synthetic analogs (e.g., isoproteronol) form active receptor-ligand complexes, which can initiate changes in the metabolic behavior of the cell containing the receptor.

light reactions The sum total of the light-dependent reactions and processes that generate ATP and NADPH during **photosynthesis.** These reactions are not dependent on the **dark reactions,** but the dark reactions are indirectly dependent on the light reactions. See also **CF$_0$CF$_1$ complex.**

light-harvesting complex A large complex consisting of several polypeptides with no catalytic activity that serves to capture light energy and transfer it to **PSI** or **PSII.** This complex may help to maintain **chlorophyll** molecules in the proper orientation for energy capture and transfer.

lignin A complex, insoluble polymer of phenolic residues, which strengthens cell walls and also serves to guard against infection.

linking number An integer equal to the number of times one strand of a DNA **double helix** crosses the other within the boundaries being considered.

lipid A biological molecule that is soluble in organic solvents. Lipids are commonly found in membranes or fat droplets, and frequently are derived from the two-carbon compound acetate.

liposome Spherical lipid bilayer structure that has an aqueous interior and that forms in vitro. Liposomes may contain lipid only or lipid and protein.

low-density lipoprotein (LDL) A major **cholesterol**-carrying lipoprotein in mammals, which is internalized by receptor-mediated **endocytosis.** In this process, the receptor is recycled and the LDL delivered to **lysosomes** in which hydrolytic reactions result in the release of the cholesterol and degradation of the LDL.

lymphocyte, B A small cell that arises from the bone marrow and has IgM or IgM plus IgD on its surface; also called B cell. When B cells are exposed to **antigen,** they give rise to plasma cells, which produce and secrete **antibody** specific for the stimulating antigen, and long-lived memory cells. See also **class switching; immunoglobulin;** and **lymphocyte, T.**

lymphocyte, T A small cell that originates in the bone marrow and passes through the thymus during maturation; also called T cell. Cytotoxic T lymphocytes (CTLs) mediate cellular immunity; helper T lymphocytes (T$_H$ cells), after exposure to antigen degradation products, release **lymphokines,** which have a variety of specific and nonspecific effects on other cells of the immune system. See also **lymphocyte, B.**

lymphokine General term for soluble protein factors secreted by h T$_H$ cells, that can promote antibody production by B cells or stimulate other cells (e.g., CTLs and macrophages) that participate in the immune response. See also **lymphocyte, T.**

lyonization The random inactivation and condensation of one of the two female sex chromosomes (the X chromosomes) in virtually all somatic cells of female animal mammals. The inactive X chromosome is visible during interphase as a dark-staining, peripheral nuclear structure called the **Barr body.**

lysogeny Process of bacteriophage DNA **integration** into a host chromosome.

lysosome Membrane-bound **organelle** containing acid hydrolases and having an internal pH of 4–5.

lysyl oxidase Enzyme that oxidizes certain lysine and hydroxylysine residues into reactive aldehydes, which in turn cross-link **collagen** helices into fibrils.

major histocompatability complex (MHC) A cluster of genes encoding polymorphic cell-surface **antigens** that are recognized by receptors on T lymphocytes. These antigens lead to rapid graft rejection between individuals of a species that differ at an MHC locus.

malignant Refers to a **tumor** that is less well-differentiated than a **benign** tumor, and that contains cells with an altered genotype as compared to normal cells. Malignant tumors are characterized by their ability to invade surrounding tissues and to metastasize. See also **metastasis.**

mannose 6-phosphate The "address tag" added to lysosomal enzymes that permits them to sort into the proper vesicles and accumulate in the **lysosomes.**

MAP1C A 1000-kDa protein that binds to **microtubules** and transports vesicles from the (+) to (−) end. Its structure is similar to cytoplasmic dynein.

matrix-targeting sequence A sequence, usually at the N-terminal end, in a cytoplasmically synthesized protein that is sufficient to direct its translocation to the mitochondrial matrix or chloroplast stroma. See also **uptake-targeting sequence.**

meiosis In eukaryotes, process whereby two successive nuclear and cellular divisions produce genetically nonequivalent, **haploid** gametic cells. See also **mitosis.**

merodiploid See **merozygote.**

merozygote Partially **diploid** bacterial cell containing a part of a second chromosome, which may be carried on a **plasmid;** also called merodiploid.

metastasis The spread of **tumor** cells and establishment of areas of secondary growth.

micelle A spherical, single-layer structure assumed by some **amphipathic** compounds, such as detergents, when dispersed in water. The **hydrophilic** part of the molecules faces outward and the **hydrophobic** part cluster inward.

microfilaments, actin Cytoskeletal fibers, 7 nm in diameter, that are formed by polymerization of monomeric **actin** (G actin) and can associate with a variety of actin-binding proteins, including **myosin.** Actin filaments play an important role in muscle contraction, cytoplasmic streaming, **microvilli,** cytokinesis, cell movement, platelet activation and other cellular functions and structures.

microtubule-associated proteins (MAPs) A diverse group of proteins that bind to **microtubules** in a constant ratio and act to accelerate the polymerization of **tubulin dimers** and to stabilize microtubules.

microtubule-organizing center (MTOC) Origin of the radial array of **microtubules** seen in nondividing, confluent cells in culture. Although this array seems quite stationary, a perturbation to one edge of the cell can result in rapid assembly of new microtubules in this region.

microtubules Cytoskeletal fibers, 24 nm in diameter, that are formed by polymeriation of **tubulin dimers** and exhibit

structural and functional polarity. **Cilia, flagella,** the mitotic spindle, and other cellular structures contain microtubules. Intracellular transport of vesicles and protein particles often occurs along microtubules.

microvilli Extensively folded region of an epithelial cell layer whose function is to increase surface area for transport of metabolites or ions; also called brush border.

missense mutation Any mutation that alters a **codon** so that it codes for a different amino acid. See also **nonsense mutation.**

mitochondrion (pl. mitochondria) Large **organelle,** surrounded by two **bilayer** membranes and containing DNA, that consumes oxygen and produces **ATP** for cellular use.

mitosis-promoting factor (MPF) A protein present in mitotic cells, but not in nonmitotic cells, that triggers **chromatin** condensation and the disassembly of the interphase **nucleus.**

mitosis In eukaryotes, process whereby nuclear and cellular division produces two genetically equivalent daughter cells. See also **meiosis.**

mobile genetic element Any **repetitious DNA** sequence that is not located in identical places in the chromosomes of individuals of the same species. Examples of mobile genetic elements include Tn elements in bacteria, *copia* and P elements in *Drosophila,* Ty elements in yeast, and Ac and Ds in maize.

monoclonal antibody **Immunoglobulin** produced by the progeny of a single B-lymphocyte clone.

monomer Any small molecule that can serve as a component of a polymer. Examples include **amino acids, nucleotides,** and **monosaccharides.**

monosaccharide Any simple sugar with the formula $(CH_2O)_n$, where $n = 3, 4, 5, 6,$ or 7.

multimeric For proteins, containing several polypeptide chains.

muscle cell See **myofiber.**

mutagenesis The induction of mutations (changes) in a DNA sequence.

myofiber A cell that contains **actin** and **myosin,** is capable of contraction, and generally is longer than other cells, measuring 1–40 nm in length; also called muscle cell.

myosin A protein containing both a head and a tail region, each of which consists of two polypeptides, that is prevalent in muscle and nonmuscle cells and that interacts with **actin** to effect muscle contraction and cell movement.

myosin light-chain (LC) kinase Enzyme that stimulates contraction in vertebrate smooth muscle cells by phosphorylating myosin light chains. Ca^{2+} activates this enzyme via the Ca^{2+}-calmodulin complex.

N-CAMs (nerve-cell adhesion molecules) A group of **integral membrane proteins** that are responsible for cell-to-cell adhesion and commonly are found in neural cells. The degree to which an N-CAM promotes adhesion is affected by the amount of extracellularly attached sialic acid and sulfate **proteoglycan** that it contains.

***N*-linked oligosaccharides** Sugar residues that are covalently attached to the amide nitrogen of asparagine residues in proteins. They include mannose and N-acetylglucosamine.

Nernst equation Mathematical expression that defines the electric potential E across a membrane as directly proportional to the concentration ratio of the ions and inversely proportional to the valency of the ions.

neurotransmitter A chemical that is released by a presynaptic neuron and usually elicits an electric response in a postsynaptic cell. A neurotransmitter can elicit either an excitatory or inhibitory response, but the response is specific to the receptor system activated by the neurotransmitter, not to the chemical nature of the neurotransmitter itself.

nexin   A protein that appears to connect adjacent **microtubule** pairs in **flagella** and **cilia**. It is extremely elastic and is one of the components that helps maintain the shape of cilia and flagella.

nick   A break in one strand of a double-stranded **nucleic acid.**

noncovalent bond   A weak chemical bond between atoms that are not covalently bonded to each other. Examples include the **hydrogen bond, ionic bond, van der Waals interaction,** and **hydrophobic interaction.**

nonpermissive   Refers to cells that are incapable of hosting a lytic viral infection. Viral infection may occur in these cells without production of new **virions.**

nonpolar   Containing neither ions nor **dipolar bonds.** See also **hydrophobic.**

nonsense mutation   Any mutation that alters a **codon** so that it codes for a stop or termination codon. See also **missense mutation.**

Northern blot analysis   Technique for detecting the presence of specific mRNAs. A mRNA sample is separated by gel **electrophoresis** under denaturing conditions, and the separated species are transferred to a nitrocellulose sheet, which is then exposed to a radioactive **cDNA** complementary to the mRNA in question. Any mRNA-cDNA hybrids that form are revealed by **autoradiography.**

nuclear envelope   Two-membrane structure that defines the perinuclear space. The nuclear envelope is continuous with the **endoplasmic reticulum** and contains special points of fusion between the inner and outer membranes where **nuclear pores** are found.

nuclear pore   A passageway in the **nuclear envelope,** surrounded by an array of proteins, that permits the passage of proteins from the **cytoplasm** to the interior of the **nucleus** and the passage of RNA from inside the nucleus to the cytoplasm.

nucleic acid   A polymer of **nucleotides** linked by phosphodiester bonds.

nucleocapsid   See **capsid.**

nucleolus   Large structure, in the **nucleus** of eukaryotic cells, that functions as the site of ribosomal RNA synthesis and processing, and assembly of ribosome subunits.

nucleoplasm   Semifluid matrix in the interior of the **nucleus,** which is distinct from the **nucleolus.**

nucleoplasmin   A 165-kDa pentameric protein that is synthesized in the cytoplasm of frog eggs and is translocated through **nuclear pores** to the interior of the nucleus. The tail portion of each monomer contains the targeting sequence that guides nucleoplasmin into the nucleus.

nucleoside   A small molecule composed of a **pentose** and an organic base (a **purine** or **pyrimidine**).

nucleosome   Small 10-nm-diameter structural unit of **chromatin** consisting of a disk-shaped **histone** core around which an ~140-bp segment of DNA is wrapped. In intact chromatin, adjacent nucleosomes are linked by DNA. When chromatin is extracted in a solution of low ionic strength, the nucleosomes are visualized as "beads on a string." In condensed chromatin, the nucleosomes are packed into a spiral or solenoid arrangement, with six nucleosomes per turn.

nucleotide   A small molecule composed of a phosphate group, a **pentose,** and an organic base (a **purine** or **pyrimidine**).

nucleus   Large structure, surrounded by two **bilayer** membranes, that contains DNA and functions as the site of RNA synthesis and processing and ribosome assembly in eukaryotes.

Okazaki fragments   Short (<1000 bases), RNA-primed DNA segments formed as short-lived intermediates during discontinuous synthesis of the **lagging strand.** Okazaki fragments are linked by DNA ligase to complete lagging-strand synthesis.

oligodendrocytes   Cells that produce a myelin sheath around neurons in the peripheral nervous system. Myelin increases the speed of impulse conduction.

oligomycin   An antibiotic that binds to the F_0 portion of the F_0F_1 particle (**ATP synthase**) and is thought to block protons from flowing through the F_0 portion, thus inhibiting **oxidative phosphorylation.**

oligosaccharide   A short, sometimes branched polymer of **monosaccharides** containing up to about 15 monosaccharides.

O-linked oligosaccharides   Sugar residues that are covalently attached to serine or threonine residues in proteins. They include galactose and N-acetylgalactosamine.

oncogene   A gene whose product is involved either in transforming cells in culture or in inducing cancer in animals. See also **transformation.**

operator   DNA site at which a **repressor** binds.

operon   Cluster of genes transcribed from one **promoter** site; a single multigene transcription unit that produces polycistronic mRNA.

organelle   Any membrane-limited structure found in the cytoplasm of eukaryotic cells.

osmotic pressure   The hydrostatic pressure required to stop the net flow of water across a membrane separating solutions of different concentrations.

osteogenesis imperfecta   A genetic disease characterized by a mutation in **collagen,** which may involve just a single amino acid substitution; also called brittle bone disease.

oxidation   Loss of electrons from an atom or molecule.

oxidative phosphorylation   The phosphorylation of ADP to ATP driven by the **proton-motive force** in bacteria, **chloroplasts,** and **mitochondria.** See also **substrate-level phosphorylation.**

oxygen-evolving complex   A manganese-containing protein complex that is involved in the splitting of H_2O in **PSII.**

P_{680} and P_{700}   Specialized **chlorophyll** molecules present in the reaction centers of **PSII** and **PSI,** respectively. They are named based on their absorption optima. Other pigments shuttle their energy to P_{680} and P_{700}.

pectins   A group of galacturonic acid-containing **polysaccharides** that are major constituents of plant cell walls. Pectins are highly hydrated and can form gels. They are partially responsible for linking plant cells together.

pentose   A five-carbon **monosaccharide.**

peptide   A polymer of **amino acids,** connected by peptide bonds, usually containing fewer than 30 amino acids.

peptide bond   The amide bond that connects two **amino acids** in a polymer.

peripheral membrane protein Any membrane protein that does not interact directly with the **hydrophobic** core of the phospholipid **bilayer** and is attached to the membrane indirectly by interactions with **integral membrane proteins** or with **lipids;** also called extrinsic protein. Peripheral membrane proteins can be extracted by solutions of high ionic strength, chemicals that bind divalent cations, or high pH; most are water soluble.

periplasmic space The space between the plasma membrane and the cell wall of a bacterium.

permease Any transport protein that speeds the rate of movement of a molecule across a membrane by **facilitated diffusion;** also called a transporter.

permissive Refers to cells that are capable of hosting a lytic viral infection.

peroxisome An **organelle** in animal cells, similar but not identical to a **glyoxysome** in plant cells, whose functions include degradation of fatty acids and amino acids by means of reactions that generate hydrogen peroxide.

petite mutant A type of yeast mutant that grows more slowly than wild-type cells and produces ATP through fermentation only. Petite yeast have **mitochondria** that are structurally aberrant and cannot perform **oxidative phosphorylation.** For this reason, they have been very useful in elucidating mitochondrial inheritance.

pH A measure of the acidity or alkalinity of a solution; $pH = -\log [H^+]$.

phagocytosis Process that occurs in eukaryotic cells in which relatively large particles (e.g., bacteria) are bound to the cell surface and then internalized. See also **endocytosis.**

phase transition "Melting" of the fatty acyl side chains of lipids in a membrane; change from a highly ordered, gel-like state to a more mobile state or vice versa.

phosphodiester bond The **covalent bond** between **nucleotides** in a nucleic acid.

phospholipase Any enzyme that hydrolyzes certain ester or phosphodiester bonds in **phospholipids.**

phospholipid The major component of most biomembranes; contains phosphate and usually two hydrocarbon chains.

photorespiration A light-dependent process, catalyzed in part by **ribulose 1,5-bisphosphate carboxylase,** that consumes O_2 and releases CO_2.

photosynthesis Complex series of **light reactions** and **dark reactions** that generate carbohydrates from CO_2 and H_2O with evolution of O_2 in bacteria and **thylakoid membranes** of plants.

phylogeny The evolutionary development of species.

phytoalexins Small organic molecules that are synthesized by plant cells following wounding or exposure to ultraviolet light. These compounds are toxic to fungi and thus confer disease resistance to plants.

picosecond absorption spectroscopy Technique of optical absorption spectroscopy that can monitor changes in the conformations of molecules during time intervals as small as 1×10^{-12} s.

pinocytosis The nonspecific uptake of small droplets of extracellular fluid by vesicles.

plaque assay A technique for quantitation of infectious viral particles, which are identified with a clear area (plaque) on an otherwise confluent layer of prokaryotic or eukaryotic cells.

plasma membrane The **bilayer** membrane that separates a cell from its external environment.

plasmadesmata Tubelike continuities of cytoplasm that interconnect adjacent plant cells.

plasmid A DNA molecule capable of autonomous replication in bacteria.

plastids Group of specialized **organelles** in plant cells that originate from **proplastids** and contain pigments or reserve substances. They include **chromoplasts, amyloplasts,** and **elaioplasts.**

pleiotropic mutation An individual mutation that affects the production of multiple gene products.

pluripotent Refers to stem cells capable of producing multiple types of differentiated cells. See also **unipotent.**

polar Containing ions or **dipolar bonds.** See also **hydrophilic.**

polar fibers **Microtubules** that extend from the two poles of the mitotic spindle toward the equator of a dividing cell.

polarized cell Any cell whose plasma membrane is organized into at least two discrete functional regions (e.g., apical and basolateral).

poly A tail The 3'-terminal sequence of adenylate residues found at the end of most eukaryotic mRNAs. The poly A tail is added post-transcriptionally, and is not encoded in DNA. The site of poly A addition is signaled by an AAUAAA sequence in mRNA.

polymerase Any enzyme that **catalyzes** the synthesis of a macromolecule from similar or identical monomeric subunits.

polypeptide A polymer of **amino acids** connected by peptide bonds.

polysaccharide A large polymer of **monosaccharides,** linked by glycosidic bonds, usually having more than 15 monosaccharide residues.

polytenization An amplification process resulting in parallel arrays of DNA in a chromosome. One common example of this process is the polytene chromosomes of the *Drosophila* salivary gland in which as many as 1000 identical DNA molecules may be aligned in one chromosome.

pool In **pulse-chase** experiments, the total amount of precursor molecules available to the cell.

porin An outer mitochondrial membrane protein that forms channels through the phospholipid **bilayer.** Porin is an unusual cytoplasmically synthesized mitocondrial protein in that its **matrix-targeting sequence** is not cleaved off and its topological organization in the outer mitchondrial membrane is due to hydrophobic amino acids that span the membrane.

potential energy Stored energy.

pre-mRNA See **hnRNA.**

Pribnow box Bacterial **promoter** consensus sequence located at -10 from the RNA start. The Pribnow box is analogous in function to the eukaryotic **TATA box** located at about -30.

primary structure, protein The linear arrangement of **amino acids** and the location of covalent (mainly disulfide) bonds.

primase RNA-synthesizing enzyme whose specific role is to catalyze the synthesis of RNA **primers** required for DNA synthesis by **DNA polymerase.**

primer A nucleic acid sequence, hydrogen-bonded to a complementary strand, that contains a free 3'-hydroxyl end.

processed gene A DNA sequence corresponding to a retroviral-copied RNA that has been inserted into an eukaryotic chromosome.

processivity Ability of an enzyme to catalyze successive polymerization steps without release of the enzyme from the **template.** This term is applied particularly to replicative **DNA polymerases,** which are high processivity enzymes.

profilin A 15-kDa protein that binds to monomeric **actin** and can inhibit the incorporation of this actin monomer into an actin filament.

progenote The postulated precellular ancestor to all cells.

prokaryotes The class of organisms, including the **eubacteria** and **archaebacteria,** that are characterized by lack of a true **nucleus** and other membrane-bound **organelles.** See also **eukaryotes.**

promoter Segment of DNA to which **RNA polymerase** binds selectively to begin **transcription.** Promoters often are located **upstream** of the start site; however, elements within a gene that function as promoters are known.

promoter The regulatory region of a **gene** or group of genes that acts as the **RNA polymerase**-binding site.

promoter, tumor A chemical that, when applied after a tumor initiator and for a continual period of time, will enhance the probability of **tumor** formation. See also **initiator, tumor.**

proofreading, DNA Nucleotide error-correction mechanism in which the $3' \longrightarrow 5'$ exonucleolytic activity of **DNA polymerases** removes incorrect bases inserted during DNA synthesis.

proplastids A group of precursor **organelles** that are present in plant embryonic tissue and contain DNA but are devoid of major plastid proteins and complexes until induced to develop into **chloroplasts** and other **plastids.**

prostaglandins Group of lipid-soluble, hormone-like substances that contain a cyclopentane ring and are synthesized from arachidonic acid in many different cell types. The prostaglandins modulate the responses of other **hormones** and can have profound effects on many cellular processes, including platelet aggregation and contraction of smooth muscle.

prosthetic group A small nonpeptide molecule that binds tightly to a protein (either covalently or noncovalently) and plays a crucial role in its function.

protease See **proteolytic enzyme.**

protein A polymer of **amino acids,** linked by peptide bonds, usually having more than 50 amino acids.

protein disulfide isomerase (PDI) An enzyme located in the **endoplasmic reticulum** that catalyzes the rearrangement of **disulfide bonds,** thus permitting a thermodynamically stable conformation.

protein domain An independent structural-functional region in a protein.

proteoglycans A group of glycoproteins, found in nearly all extracellular matrices, that contain a core protein to which is covalently attached one or more glycosaminoglycans (e.g., **heparan sulfate**). They are generally negatively charged.

proteolytic enzyme Any enzyme that digests proteins or peptides by hydrolyzing the **peptide bonds;** also called a protease.

proto-oncogene A normal cellular gene from which an **oncogene** is derived.

proton-motive force The force responsible for **oxidative phosphorylation,** which consists of the transmembrane electrical gradient and transmembrane proton (pH) gradient. In **mitochondria,** these gradients are generated by the electron transport system and maintained by the inner mitochondrial membrane. See also **chemiosmotic hypothesis** and **Q cycle.**

pseudogene A duplicated gene copy that has become nonfunctional because of **genetic drift,** which may introduce a stop codon or an alteration that prevents processing of any RNA transcript that is formed.

PSI (photosystem I) Complex of many **chlorophyll** molecules, carotenoid pigments, and electron-carrier proteins in **thylakoid membrane** that is used in both linear and cyclic electron flow and functions to transfer electrons to NADP, forming NADPH. PSI is optimally stimulated by light of 700 nm.

PSII (photosystem II) Complex of many **chlorophyll** molecules, carotenoid pigments, and electron-carrier proteins in the **thylakoid membrane** that functions to split H_2O releasing O_2 and in conjunction with **PSI** carries out noncyclic electron flow. PSII is optimally stimulated by light of 680 nm.

pulse-chase Addition of a precursor molecule (usually radioactive) for a brief period (pulse), followed by its removal (chase). This technique is often used for studying precursor-product relationships.

purine A basic compound, containing two fused heterocyclic rings, that occurs in nucleic acids. The purines commonly found in DNA and RNA are adenine and guanine.

pyrimidine A basic compound, containing one heterocyclic ring, that occurs in nucleic acids. The pyrimidines commonly found in DNA are cytosine and thymine; the pyrimidines commonly found in RNA are cytosine and uracil.

Q cycle The ability of coenzyme Q (ubiquinone) to pump four protons for every two electrons transported. This cycle, along with the contributions of other electron transport complexes, produces a proton gradient across the inner mitochondrial membrane, which, in turn, can generate ATP.

quarternary structure, protein The organization of several polypeptide chains into a single protein molecule, such as hemoglobin.

quiescent Refers to a cell that is not increasing its mass or passing through the **cell cycle.**

radioisotope An unstable form of an element in which the nucleus undergoes random disintegration, accompanied by the emission of radiation and yielding another atom.

reading frame The **codon** sequence that is determined by reading a nucleotide sequence in groups of three from a specific start codon to a termination codon.

receptor-mediated endocytosis See **endocytosis.**

reduction Gain of electrons by an atom or molecule.

reduction potential Measurement of the readiness with which an atom or molecule takes up an electron. It is expressed in volts (V) from an arbitrary zero point set at the reduction potential of the reaction, $H^+ + e^- \rightleftharpoons 1/2 H_2$, at 25°C, 1 atm, and 1 M reactants.

regulon Scattered group of coordinated regulated genes, which all have **operators** that are recognized by the same repressor.

repetitious DNA Repeated sequences in eukaryotic chromosomes, including **simple-sequence DNA,** and **intermediate-repeat DNA.** The latter fraction contains **retroposons** and **transposons.**

replicon   Region of DNA served by a single DNA replication origin.

replisome   Entire collection of accessory proteins involved in DNA synthesis plus **DNA polymerase.**

resting potential   The electric potential in an excitable cell when no **action potential** is being propagated and little **neurotransmitter** is being released at the axon terminal. The resting potential ranges from -30 to -70 mV (inside negative).

resident endoplasmic reticulum (ER) protein   Any protein that remains in the **endoplasmic reticulum** following its synthesis.

residue   A **monomer.**

resolution, limit of   Minimum distance that can be distinguished by an optical apparatus; also called resolving power. For example, an apparatus with a resolving power of 0.5 μm cannot distinguish between two objects that are positioned less than 0.5 μm apart.

resolving power   See **resolution, limit of.**

restriction endonuclease   An enzyme (usually bacterial) that recognizes and cleaves a specific short sequence (4–8 bases) in double-stranded DNA molecules.

restriction fragment length polymorphism (RFLP)   Inherited differences in the length of DNA fragments resulting from digestion of chromosomal DNA with **restriction endonucleases.** These differences reflect polymorphism in the presence or absence of specific **restriction sites.** RFLPs have been widely used in genetic studies of humans.

restriction site   A short sequence (rarely more than 4–8 base pairs) in double-stranded DNA that is recognized and cut by a **restriction endonuclease.**

retrograde   Movement of cytoplasmic material from the growth cone or synaptic end of a nerve cell to the cell body. Material moving in this direction is usually degraded by **lysosomes.** See also **anterograde.**

retroposon   A **mobile genetic element** whose movement is mediated by a RNA intermediate. Retroposons are divided into two classes—nonviral and viral; the latter contain retroviral-like sequences. *Drosophilia copia* elements and yeast Ty elements act as viral retroposons. See also **transposon.**

retrovirus   An RNA virus in which the plus-strand RNA of the genome directs the synthesis of a double-stranded DNA, which ultimately acts as the template for making mRNA.

reverse transcriptase   An enzyme that catalyzes the synthesis of DNA from an RNA strand. This enzyme was first identified in retroviral particles.

rho protein   A protein factor that regulates termination of **transcription** in some bacteria.

rhodopsin   A transmembrane protein in photoreceptors that interacts with transducing G proteins and eventually causes the closing of a Na^+ channel.

ribosome   A complex comprising several different ribosomal RNA molecules and more than 50 proteins, organized into a large subunit and small subunit, that is the site of protein synthesis.

ribulose 1,5-bisphosphate carboxylase (RBPase)   The Calvin-cycle enzyme that catalyzes fixation of CO_2 to the five-carbon sugar ribulose 1,5-bisphosphate, forming a hexose that quickly splits to form two molecules of 3-phosphoglycerate. It is the most abundant enzyme on earth. See also **dark reactions.**

rigor   State in muscle contraction characterized by the lack of ATP or creatine phosphate. Permanent rigor is induced by death but transient rigor is part of the myosin-ATPase cycle in muscle contraction.

RNA, messenger (mRNA)   Any RNA molecule, transcribed from **genomic DNA,** from which a protein is translated by the actions of **ribosomes.**

RNA polymerase   Any enzyme catalyzing **transcription** of DNA. In prokaryotes, there is only one RNA polymerase; in eukaryotes, there are three RNA polymerases numbered I (products—5.8S, 18S, and 28S rRNA; location—**nucleolus**), II (products—mRNAs; location—**nucleoplasm**), and III (products—tRNA, 5S rRNA, U1 RNA, etc.; location—nucleoplasm).

RNA processing   Co- and post-transcriptional modifications of RNA, including methylation, nucleolytic cleavage, **splicing, capping,** and polyadenylation. Many but not all RNAs are processed. RNA processing is more complex in eukaryotes than prokaryotes.

RNA, ribosomal (rRNA)   One of a class of large RNA molecules, coded in the nucleolar organizer region in eukaryotes, that are structural and functional components of **ribosomes.**

RNA, transfer (tRNA)   One of a class of small RNA molecules that contain covalently bound amino acids and that function as amino acid donors during **translation.**

run-on assay   Procedure for determining the fraction of total RNA copied from a particular gene by incubating isolated nuclei with a radioactive nucleoside triphosphate and measuring the amount of label incorporated into gene-specific nascent RNA chains. Little, if any, initiation of new transcripts occurs in this assay.

S1 fragment   Head portion of the **myosin** molecule, which can be isolated by papain digestion. This fragment can be used to determine the polarity of actin filaments and to monitor **actin** polymerization in vitro.

salvage pathway   Sequence of reactions that yield required compounds (e.g., nucleotides) from similar preformed compounds present in the extracellular fluid.

sarcoma   A malignant **tumor** derived from mesoderm.

sarcomere   The portion of a striated muscle cell that extends from one Z disk to the next and shortens in length during contraction.

sarcoplasmic reticulum   Network of membranes derived from the smooth **endoplasmic reticulum,** which release Ca^{2+}, thus eliciting muscle contraction, and serve to coordinate the contraction of the entire **myofiber.**

satellite DNA   See **simple-sequence DNA.**

saturated   For a fatty acid, containing no double bonds.

Schwann cells   Cells that produce a myelin sheath around neurons in the central nervous system. Myelin increases the speed of impulse conduction.

second messenger   An intracellular signaling compound whose concentration increases (or decreases) in response to binding of a **ligand** to cell-surface receptors and that functions to mediate the effects of the ligand on cellular activities.

secondary structure, protein   The folding of polypeptide chains into regular structures, such as α helices and β-pleated sheets.

self-splicing of RNA   See **splicing of RNA.**

semiconservative replication   Replication mechanism whereby each old chain becomes paired with a new chain copied from it.

sequence drift See **genetic drift.**

sigma factor Any of several accessory proteins to bacterial **RNA polymerase** responsible for selective **promoter** choice. Different sigma factors recognize different promoters.

signal peptidase A **proteolytic enzyme** located in the **endoplasmic reticulum** that cleaves the **signal sequence.**

signal peptide See **signal sequence.**

signal recognition particle (SRP) A complex consisting of six polypeptides and a 300-nucleotide RNA that is part of the protein initiation complex in the cytoplasm and binds to the **signal recognition particle receptor.**

signal recognition particle (SRP) receptor A protein consisting of an α and β subunit that binds the **signal recognition particle** and the growing nascent polypeptide chain.

signal sequence A sequence of 30–40 amino acids at the N-terminal end of a nascent protein that causes the growing polypeptide to cross the **endoplasmic reticulum** membrane; also called signal peptide.

signal sequence receptor A protein in the rough **endoplasmic reticulum** (ER) that binds the **signal sequence** after its release from the **signal recognition particle** and aids in the insertion of the protein into the lumen of the ER membrane.

silencer sequence A distal DNA sequence to which binding of a silencer protein results in total shut-off of gene transcription.

simple diffusion A type of passive transport in which movement of a molecule occurs without the aid of a protein. For example, movement of a molecule across a membrane without the aid of a transport protein occurs by simple diffusion. See also **Fick's law.**

simple-sequence DNA The most rapidly reassociating eukaryotic DNA fraction, which generally consists of short 5- to 10-bp segments that are tandemly repeated thousands of times in the genome; also called satellite DNA. Most simple-sequence DNA is located in **centromeres** and **telomeres.** This fraction constitutes 10–15 percent of total cellular DNA. See also **intermediate-repeat DNA** and **single-copy DNA.**

single-copy DNA The most slowly reassociating eukaryotic DNA fraction, which contains most of the protein-coding genes. This fraction constitutes 50–60 percent of total cellular DNA. See also **intermediate-repeat DNA** and **simple-sequence DNA.**

snRNP Small nuclear ribonucleoprotein particle that functions in nuclear **RNA processing,** especially **splicing.** Several different snRNPs have been identified; they are named based on the small RNA species they contain (e.g., U1 . . . U5 snRNP).

solitary gene A protein-coding gene present in one copy per **haploid** genome. About one-quarter to one-half of all protein-coding genes in multicellular organisms are solitary, whereas almost all bacterial protein-coding genes are solitary. See also **duplicated gene.**

SOS repair system An inducible DNA-repair system in bacteria, which makes errors in the DNA sequence as it repairs.

Southern blot analysis Technique for detecting specific DNA sequences. A sample is digested with one or more **restriction endonucleases;** the fragments are separated by electrophoresis on a denaturing gel and then transferred by blotting to a nitrocellulose sheet. The blot is exposed to a radioactive probe with a known nucleotide sequence. Any hybrids that form between the probe and fragments containing complementary sequences are revealed by **autoradiography.**

specific activity Amount of radioactivity per unit of material, usually expressed in disintegrations per minute (dpm) per mole.

spliceosome Large multicomponent complex of snRNPs and proteins involved in the **splicing** of eukaryotic pre-mRNA.

splicing of RNA An excision-ligation process that results in joining of RNA segments separated by an intervening sequence (**intron**). Splicing is very common for higher eukaryotic pre-mRNA. It generally requires snRNPs, but RNA-mediated self-splicing can occur. See also **spliceosome** and **transplicing of RNA.**

spot desmosome Buttonlike attachment between cells (usually epithelial or endothelial), that enhance adhesion of cells and often are connected to other cytoplasmic filaments such as **intermediate filaments.**

stereoisomers (D and L) Compounds that have identical molecular formulas and atoms linked in the same order, but which differ from one another because their atoms have different spatial arrangements. D and L refer to particular configurations related to the configurations of D- and L-glyceraldehyde. Biological amino acids are L-isomers, and biological sugars are generally D-isomers.

stomata Pores through which CO_2 enters a leaf. Stomata are surrounded by guard cells whose shape controls stomatal opening and closing.

stop-transfer anchor sequence A domain within a protein that anchors it within a membrane.

stringent response Decreased rRNA and tRNA synthesis in bacteria transferred from rich to poor media, initiated by the accumulation of the guanine polyphosphates ppGpp and pppGpp.

stromatolite Layered structure in sedimentary rock formed by the accumulation of mineral grains (ocean sediments) around colonies of microorganisms.

substrate-level phosphorylation Phosphorylation of ADP to ATP in the **cytoplasm.** This type of phosphorylation involves the chemical transformation of ADP to ATP by cytoplasmic enzymes and does not depend on the **proton-motive force** responsible for **oxidative phosphorylation** in the mitochondria.

substrate The chemical that undergoes change in a reaction catalyzed by an **enzyme.**

supercoil Twisting of DNA double helices around each other.

suppressor mutation Any mutation that counteracts the effects of an independent mutation. Classical suppressor mutations code for altered tRNA species that read a termination codon as an amino acid **codon.**

symport Cotransport process in which movement of ions or molecules is in the same direction. See also **antiport.**

synapse Region between the **axon terminal** of one neuron and the beginning of a dendrite of an adjacent neuron or other excitable cell across which electric impulses are transmitted from the presynaptic to the postsynaptic cell. At chemical synapses, the impulse is conducted by release of a **neurotransmitter,** which diffuses across the synaptic cleft and then activates (usually depolarizes) the postsynaptic cell. At electrical synapses, impulse conduction occurs directly via **gap junctions** and does not involve a neurotransmitter.

synapsin I A fibrous phosphoprotein located in the **axon terminal,** which is a substrate for cAMP-dependent protein kinases and Ca^{2+}-calmodulin–dependent protein kinases and may function to position the **synaptic vesicles** in the postsynaptic process.

synaptic vesicle A membrane bound vesicle in the presynaptic axon terminal of a neuron. It houses neurotransmitter(s), proteins, and sometimes ATP. Vesicles can be released upon stimulation, usually by an **action potential.**

synaptophysin An **integral membrane protein** present in **synaptic vesicles,** which may function in the uptake of **neurotransmitters** from the cytosol.

T cell See **lymphocyte, T.**

target-site repeat Short **inverted repeat** found at the site of **transposon** insertion.

targeting The ability of a protein containing certain domains or groups to direct its destination within a cell (e.g., its ability to be exocytosed, to be retained within a cell, and to associate with a specific organelle). Amino acid sequences, glycosylation, the conformation of a protein, and the presence of terminal phosphates are important in targeting.

TATA box A **promoter** sequence that is common to eukaryotic protein-coding genes transcribed at high rates and that is functionally analogous to the prokaryotic **Pribnow box.** The TATA box is typically located 25–35 bp **upstream** from the RNA start site.

telomere End region of the chromosome containing the special telomeric (TEL) sequences, which are added by telomerases after **DNA replication.**

temperate phage A bacterial virus that can recombine with the host DNA and be replicated along with the host chromosome without lysing the host cell.

template A molecular "mold" that dictates the structure of another molecule. For example, one strand of DNA acts as a template for synthesis of a complementary strand during replication.

terminal web A structure, immediately basal to microvilli in many polarized epithelial cells such as intestinal epithelial cells, that is connected to desmosomes and is thought to span the entire cell.

termination factor One of a group of proteins that recognize a termination **codon** and regulate the release of newly synthesized proteins from the **ribosome.**

tertiary structure, protein The folding of regions of a polypeptide chain between α helices and β-pleated sheets and the combination of these secondary features into domains.

thermogenin A 33-kDa protein found in the mitochondria in brown fat that acts as a natural **uncoupler** of **oxidative phosphorylation,** thus converting much of the energy of the **proton-motive force** to the generation of heat rather than the phosphorylation of ADP to ATP.

thick filament A bipolar aggregate containing several hundred individual myosin molecules all oriented in a specific fashion, which is the chief myosin component of **myofibers.**

thin filament Actin microfilament in striated muscle.

threshold potential Magnitude of the electric-potential change necessary to elicit an **action potential** in a neuron. This potential is the summation of both excitatory and inhibitory signals impinging on a postsynaptic neuron.

thylakoid membrane Membrane structure within **chloroplasts** that contains **chlorophyll** and is the primary site of **photosynthesis.** See also CF_0CF_1 **complex.**

thymine dimer Covalent cyclobutyl linkage of adjacent thymine residues in DNA, which is produced by absorption of UV radiation.

thyroid hormones Collective term referring to three **hormones** —thyroxine or tetraiodothyronine (T_4), triiodothyronine (T_3), and calcitonin—that are synthesized and released by the thyroid gland.

tight junction Ribbonlike junctional complex composed of thin bands that completely encircle a cell and are in contact with similar thin bands on adjacent cells. A common feature of epithelial cell layers, tight junctions form seals that limit diffusion of most molecules through the epithelium.

titin A 1000-kDa protein that may function to connect myosin **thick filaments** in striated muscle to the Z disks and center these filaments during contraction.

tolerance Diminished or nonexistent capacity to make a specific antibody to an antigen. Tolerance is produced by contact with an antigen during development of the immune system (self-tolerance) or under nonimmunizing conditions (induced tolerance).

topogenic sequence Amino acid sequence in an **integral membrane protein** that determines the proper orientation, or topology, of the protein within the membrane.

topoisomerase I (topo I) An enzyme that can change the **linking number** of DNA by 1. A topo I molecule binds to DNA and cuts one strand, forming a covalent complex with the free phosphate on the DNA molecule; after rotation, the cut strands are resealed. See also **topoisomerase II.**

topoisomerase II (topo II) An enzyme that can change the **linking number** of DNA by 2; also called gyrase. In a topo II-catalyzed reaction, a double-strand cut is made in a DNA duplex, another portion of the duplex (or another duplex molecule) is passed through the cut, and the cut is resealed. ATP is consumed in this reaction. Topoisomerase II is involved in the release of final products after chromosome replication. See also **topoisomerase I.**

trans Golgi reticulum (TGR) The last processing compartment in the **Golgi complex** where proteins are sorted into vesicles; also called trans Golgi network (TGN).

trans-active Refers to DNA elements that can control genes on the same or different chromosomes. Trans-active control is mediated via a diffusible factor. See also **cis-active.**

transcript, primary Initial RNA product, containing **introns** and **exons,** formed during DNA **transcription.** Many primary transcripts undergo **RNA processing** to form the final physiologically active RNA species.

transcription activator Any DNA sequence that can accelerate transcription, including **upstream activating sequences** (UASs) and **enhancers.** Such sequences are present in many, if not all, eukaryotic genes.

transcription factor (TF) Any of several accessory transcription proteins in eukaryotes required for initiation, elongation, or formation of a preinitiation complex. General factors are required for **transcription** of all genes transcribed by a given **RNA polymerase;** specific factors regulate transcription of particular genes.

transcription unit A region in DNA, bounded by an initiation and termination site, that is transcribed to produce a **primary transcript.** Complex transcription units in eukaryotes can produce two or more different mRNAs.

transcription Synthesis of RNA using a DNA template in a reaction catalyzed by **RNA polymerase.**

transduction The process whereby genes are moved from a bacterial donor to a bacterial recipient using a bacterial virus as the vector.

transformation (1) Process whereby bacterial cells are stably modified by application of DNA from a cell of a different genotype; (2) process of conversion of a "normal" mammalian cell in tissue culture to a cell characterized by immortality and/or uncontrolled cell division.

transforming growth factor A protein secreted by transformed cells that can stimulate the growth of normal cells.

translation The **ribosome**-mediated production of a **polypeptide** whose primary amino acid sequence is derived from the triplet **codon** sequence of an mRNA.

transplicing of RNA Joining of two separate RNA species. Transplicing between certain RNAs can occur in the absence of a protein catalyst.

transporter See **permease.**

transposon A relatively long **mobile genetic element** that contains protein-coding genes, ends in **inverted repeats,** and moves in the genome by a mechanism involving DNA synthesis and transposition. Transposons are found in prokaryotes and eukaryotes and may encode a transposase. Examples include Ac and Ds elements in maize and P elements in *Drosophilia*. See also **retroposon.**

triacylglycerol Any lipid formed from three **fatty acids** esterified to one glycerol molecule.

tropomyosin A coiled-coil protein containing two parallel α-helical polypeptides that is associated with **actin** filaments in striated muscle and binds **troponin.**

troponin Complex of three proteins—named T, I, C—that are attached to **actin** filaments in striated muscle via **tropomyosin.** Troponin C binds Ca^{2+}, and troponin I inhibits the actin-stimulated myosin ATPase, and troponin T binds the complex to tropomyosin.

tubulin dimer Dimeric subunit of **microtubules** consisting of α- and β-tubulin monomers, each of which has a molecular weight of approximately 50,000; also called $\alpha\beta$-tubulin.

tumor A mass of cells, generally derived from a single cell, that is not controlled by normal regulators of cell growth.

turgor Hydrostatic pressure inside plant cells which is generated by entry of water into the cell **vacuole.**

uncoupler Any compound that dissipates the **proton-motive force** generated across inner mitochondrial and chloroplast **thylakoid membranes** but does not affect either the electron transport system or the **ATP synthase.** Common uncouplers include dinitrophenol (DNP), valinomycin, and FCCP.

unequal crossing over Crossover event in which unequal segments are transferred between chromatids of the same or different chromosomes.

unipotent Refers to stem cells capable of producing only a single type of differentiated cells. See also **pluripotent.**

unsaturated For a fatty acid, containing one or more double bonds.

upstream activating sequence (UAS) A 15- to 20-bp protein-binding site in DNA that is necessary for maximal gene expression in eukaryotes. Yeast genes generally have only one associated UAS, whereas mammalian genes normally have multiple UASs.

upstream In the 5' direction within DNA. The +1 position of a gene is the first transcribed base; nucleotides upstream of the +1 position are numbered −1, −2, etc. See also **downstream.**

uptake-targeting sequence Sequence of 20–60 amino acids at the N-terminus of most cytoplasmically synthesized precursor proteins that directs these proteins to the proper compartment of the **mitochondria** or **chloroplasts.** The uptake-targeting sequence of mitochondrial precursor proteins contains a **matrix-targeting sequence,** which first directs translocation to the matrix, and an additional sequence(s) that targets proteins to the intermembrane space or membranes. In most, but not all, mitochondrial proteins, the uptake-targeting sequence is removed once translocation has occurred.

uvomorulin An **integral membrane protein** present in belt desmosomes and responsible for the cell-to-cell attachment of epithelial cells; also called E-cadherin. See also **cadherins.**

V_{max} A parameter used to describe the maximal velocity of enzymatic catalysis or other processes such as transport of small molecules across membranes.

vacuole In plant cells, an **organelle** surrounded by a single **bilayer** membrane (the tonoplast) that functions as a reservoir for both food materials and waste products and also as a source of **turgor** pressure necessary for cell elongation during growth.

van der Waals interaction A weak attraction, between atoms that are not covalently bonded, due to the formation of transient **dipoles** in each atom when the atoms are close to one another.

vector, expression Plasmid containing a DNA sequence sufficient to direct host-mediated protein synthesis of a gene on the plasmid itself.

virion An infectious viral particle.

vinculin An **actin**-associated protein found in muscle that binds to α-**actinin** in cell-free experiments but whose in vivo function is unknown.

Western blot analysis Technique for detecting the presence of specific proteins. A protein sample is separated by one- or two-dimensional **electrophoresis,** and the separated proteins are transferred to a nitrocellulose sheet, which is exposed to a radioactive antibody specific for the protein in question. Any antibody-protein complexes that form are revealed by **autoradiography.**

wobble The ability of certain bases (particularly inosine) at the first position of an **anticodon** to form **hydrogen bonds** in several ways, enabling alignment of the anticodon with more than one mRNA **codon.**

Z-DNA A "left-handed" helical configuration of double-stranded DNA (alternative to the "right-handed" **double helix**), which may occur in regions of DNA rich in guanine and cytosine.

zinc-finger motif A region of protruding protein folds ("fingers") marked by zinc bridges between cysteine- and/or histidine-rich amino acid clusters that characterizes one class of DNA-binding proteins (e.g., $TF_{III}A$) and enables recognition of specific sites in DNA.

zwitterion A doubly ionized molecule containing both a positively and negatively charged atom.

zymogen An inactive precursor of a **proteolytic enzyme.**

ANSWERS

CHAPTER 1

1. Covalent bonds
2. dipolar
3. donor, acceptor
4. Hydrogen bonds
5. water
6. van der Waals interactions
7. hydroxyl
8. base
9. zwitterion
10. exothermic
11. Entropy
12. oxidation, reduction
13. ATP
14. catabolism

15. endergonic
16. activation energy
17. catalyst, enzymes
18. a b c d
19. b d e
20. b c d
21. b d
22. c d
23. a b
24. a b c e
25. a b d e
26. a b c d e
27. a d
28a. H, I, V, Hy
28b. I

28c. V, Hy (also H if interactions between water molecules are considered.)

28d. C, H, I, V, Hy   28e. C

29.

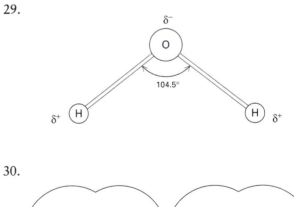

30.

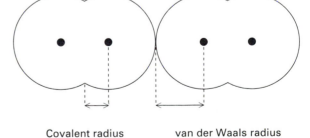

Covalent radius   van der Waals radius

31. $pK_a = 6.75$; this is the concentration at which half of the HA is dissociated. In other words, this is the concentration at which $[HA] = [A^-]$.

32. phosphoanhydride bonds

33.

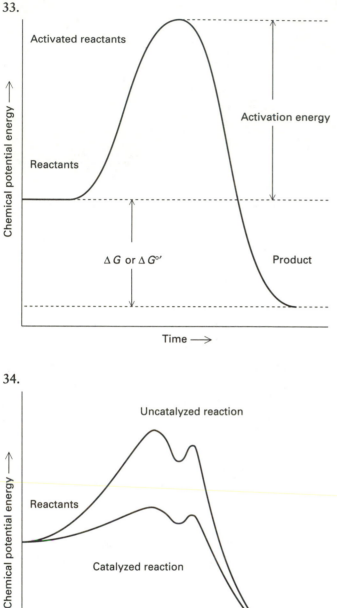

34.

35a. The interaction between fat molecules in aqueous solution is called the hydrophobic interaction.

35b. When a polar molecule such as ATP is dissolved in water, the disorder or entropy in the system is increased.

35c. If a nonpolar molecule such as a triacylglycerol were surrounded by water, the disorder or entropy in the

system would be decreased because the freedom of motion of nearby water molecules, unable to hydrogen bond with the nonpolar molecule, would be restricted.

35d. ΔH for fat droplet formation is probably negative for two reasons: first, because the number of hydrogen bonds in the water is maximized and, secondly, because van der Waals interactions are formed between the fat molecules themselves.

35e. When nonpolar molecules exclude water and form into a fat droplet, S is increased and H is decreased. Since $\Delta G = \Delta H - T \Delta S$, the change in free energy is negative. Entropy probably is a more important factor than enthalpy in driving the formation of hydrophobic structures.

36a. The fact that the solution becomes cold means that heat is absorbed; that is, the reaction is endothermic, and ΔH for the reaction is positive. Since urea in fact dissolves under these conditions, ΔG must be negative. In order for ΔG to be negative when ΔH is positive, ΔS must be positive. Indeed, the increase in the degree of disorder when urea is dissolved in water is the driving force of the dissolution reaction.

36b. A decrease in temperature will decrease the value of the term $T \Delta S$, increasing the value of ΔG, since $\Delta G = \Delta H - T \Delta S$. (The values of ΔH and ΔS are relatively independent of temperature.)

37. Catabolic reactions, such as the breakdown of glucose, produce ATP for the cell. Anabolic reactions, such as the synthesis of proteins or nucleic acids, generally consume ATP.

38a. pH $= 4$. First calculate the H^+ concentration of the diluted solution and then take the negative log of that value.

$$\frac{(10^{-2}\ M\ H^+)(1\ ml)}{10^2\ ml} = 10^{-4}\ M\ H^+$$

$$pH = -\log [H^+] = -\log [10^{-4}] = 4$$

38b. pH $= 9$

$$\frac{(1\ M\ OH^-)(1\ ml)}{10^5\ ml} = 10^{-5}\ M\ OH^-$$

$$[H^+][OH^-] = 10^{-14}\,M^2 \quad \text{or} \quad [H^+] = \frac{10^{-14}\,M^2}{10^{-5}\,M\ OH^-}$$
$$= 10^{-9}\,M\ H^+$$

$$pH = -\log[H^+] = -\log[10^{-9}] = 9$$

39a. [sodium acetate] = 0.0645 M; [acetic acid] = 0.0355 M. The solution is calculated using the Henderson-Hasselbach equation.

$$pH = pK_a + \log\frac{[A^-]}{[HA]} \quad \text{or} \quad pH - pK_a = \log\frac{[Ac^-]}{[HAc]}$$

(see MCB, p. 30)

Since $[Ac^-] + [HAc] = 0.1\,M$, the following substitutions can be made:

$$5 - 4.74 = \log\frac{[Ac^-]}{0.1\,M - [Ac^-]}$$

Taking the antilog of both sides and solving for $[Ac^-]$ gives $[Ac^-] = 0.0645\,M$; thus $[HAc] = 0.0355\,M$.

39b. pH = 4.82

When H^+ is added to the solution, the following reaction occurs: $H^+ + Ac^- \rightarrow HAc$. Thus the concentration of Ac^- is reduced by 0.01 M to 0.0545 M and the concentration of HAc is increased to 0.0455 M. Again,

$$pH = pK_a + \log\frac{[Ac^-]}{[HAc]}$$

So,

$$pH = 4.74 + \log\frac{0.0545\,M}{0.0455\,M} = 4.82$$

40a. The fraction of an acid in the undissociated form (HA) and dissociated form (A^-) at any pH can be calculated using the following form of the Henderson-Hasselbach equation:

$$pH - pK_a = \log\frac{[A^-]}{[HA]}$$

If the pH is one unit below the pK_a, then $pH - pK_a = -1$ or

$$-1 = \log\frac{[A^-]}{[HA]}$$

Taking the antilog of both sides gives $0.1 = [A^-]/[HA]$. Since $[A^-] + [HA] = 100\%$,

$$0.1 = \frac{100\% - [HA]}{[HA]}$$

Solving this expression gives [HA] = 91%. Similarly, if the pH is one unit above the pK_a, then $pH - pK_a = 1$ or

$$1 = \log\frac{[A^-]}{[HA]}$$

Taking the antilog of both sides and substituting $[HA] = 100\% - [A^-]$ gives

$$10 = \frac{[A^-]}{100\% - [A^-]}.$$

Solving this expression gives $[A^-] = 91\%$.

40b. [HA] = 24%. At 0.5 pH units above the pK_a, $pH - pK_a = 0.5$. Thus

$$0.5 = \log\frac{[A^-]}{[HA]} = \log\frac{100\% - [HA]}{[HA]}$$

Taking the antilog of both sides (antilog 0.5 = 3.16) and solving gives [HA] = 24%.

40c. [HA] = 99%. At 2 pH units below the pK_a, $pH - pK_a = -2$. Thus

$$-2 = \log\frac{[A^-]}{[HA]} = \log\frac{100\% - [HA]}{[HA]}$$

Taking the antilog of both sides and solving gives [HA] = 99%.

41a. The negative value for $\Delta G^{\circ\prime}$ indicates that the reaction would be spontaneous under standard conditions in the direction it is written. In other words, the net direction of the reaction would be $A + B \rightarrow C + D$.

41b. $\Delta G = +5.7$ kcal/mol; the net direction of the reaction is $C + D \rightarrow A + B$.

$$\Delta G = \Delta G^{\circ\prime} + RT\ln\frac{[C][D]}{[A][B]} \quad \text{(see MCB, p. 32)}$$

Substituting and solving for ΔG:

$$\Delta G = -2400\ \text{cal/mol} + (1.987\ \text{cal/degree mol})(310\ \text{K})\ln\frac{(5\ mM)(10\ mM)}{(0.01\ mM)(0.01\ mM)}$$
$$= +5700\ \text{cal/mol} = +5.7\ \text{kcal/mol}$$

The positive ΔG value indicates that the reaction, as originally written, is spontaneous in the reverse direction.

42. At equilibrium, [3-phosphoglycerate]/[1,3-bisphosphoglycerate] = 200. From the definition of the equilibrium constant (see MCB, p. 27)

$$K_{eq} = \frac{[\text{3-phosphoglycerate}][\text{ATP}]}{[\text{1,3-bisphosphoglycerate}][\text{ADP}]}$$

where brackets indicate equilibrium concentrations. Rearranging the expression $\Delta G^{\circ\prime} = -2.3\ RT \log K_{eq}$ (see MCB, p. 33) gives

$$\log K_{eq} = -\frac{\Delta G^{\circ\prime}}{2.3\ RT}$$

Thus

$$\log K_{eq} = -\frac{-4500\ \text{cal/mol}}{(2.3)(1.987\ \text{cal/degree mol})(298\ \text{K})} = 3.30$$

Taking the antilog of both sides gives $K_{eq} = 2000$. Rearranging the definition of the equilibrium constant and substituting gives

$$\frac{[\text{3-phosphoglycerate}]}{[\text{1,3-bisphosphoglycerate}]} = K_{eq}\frac{[\text{ADP}]}{[\text{ATP}]} = (2000)\frac{1}{10} = 200$$

43. The concentration outside could be as high as 22.6 M. The energy available to power the transport of substance A is the 7.3 kcal/mol available from the ATP hydrolysis. The transport process in this question is reversed with regard to inside and outside the cell from the example given in the text (MCB, p. 33). To answer the question, you need to calculate the value of C_2 (concentration outside the cell) for $\Delta G = +7.3$ kcal/mol and $C_1 = 100\ \mu M$. If the outside concentration were any greater than the value calculated in this manner, the ΔG value for the transport process would be $> +7.3$ kcal/mol and, coupled with ATP hydrolysis at -7.3 kcal/mol, ΔG for the overall process would be positive; in this case, the transport of A from inside to outside could not occur.

The ΔG associated with a concentration gradient is given by the expression

$$\Delta G = RT \ln \frac{C_2}{C_1} \qquad \text{(see MCB, p. 33)}$$

where a molecule is being transported from C_1 to C_2; in this case, C_2 = outside concentration and C_1 = inside concentration. Rearranging and substituting into this expression gives

$$\ln \frac{C_2}{C_1} = \frac{\Delta G}{RT} = \frac{7300\ \text{cal/mol}}{(1.987\ \text{cal/degree mol})(298\ \text{K})} = 12.3$$

Taking the natural antilog of both sides and solving for C_2 when $C_1 = 100\ \mu M$,

$$\frac{C_2}{C_1} = 2.26 \times 10^5$$

$$C_2 = (100 \times 10^{-6}\ M)(2.26 \times 10^5) = 22.6\ M$$

Thus, under standard conditions, the hydrolysis of one mole of ATP theoretically would provide energy for one mole of substance A (with an intracellular concentration of 100 μM) to be exported from the cell as long as the concentration of A outside the cell remained below 22.6 M!

44. $\Delta E_0' = 0.72$ V; $\Delta G^{\circ\prime} = -33.2$ kcal/mol. The standard electric-potential change of an oxidation-reduction reaction is the sum of the E_0' values of the partial reactions. For the oxidation of $CoQH_2$, the partial reactions are as follows:

$CoQH_2 \rightleftharpoons CoQ + 2e^- + 2H^+$	$E_0' =$	-0.10 V
$\frac{1}{2}O_2 + 2e^- + 2H^+ \rightleftharpoons H_2O$	$E_0' =$	0.82 V
Sum: $CoQH_2 + \frac{1}{2}O_2 \rightleftharpoons CoQ + H_2O$	$\Delta E_0' =$	0.72 V

The relationship between ΔG and ΔE for an oxidation-reduction reaction is given by the expression

$$\Delta G^{\circ\prime}\ (\text{cal/mol}) = -n\mathscr{F}\Delta E_0'$$

$$= -n\left[\frac{96,500\ \text{joules/(V)(mol)}}{4.18\ \text{joules/cal}}\right]E_0'\ (\text{volts})$$

$$\text{(see MCB, p. 35)}$$

where n is the number of electrons transferred, $\mathscr{F}$ is the Faraday constant, and 4.18 is the factor for converting joules to calories. Substituting $n = 2$ and $\Delta E_0' = 0.72$ V into this expression and solving $\Delta G^{\circ\prime} = -33,200$ cal/mol $= -33.2$ kcal/mol.

CHAPTER 2

1. positively

2. hydrocarbons, hydrophobic

3. glycine, L

4. condensation, peptide

5. fibrous, globular

6. Secondary

7. amphipathic

8. x-ray crystallography

9. hydrogen

10. conservative

11. prosthetic group

12. disulfide bridges

13. phosphorylation

14. denatured

15. substrates

16. active site

17. zymogens

18. K_M

19. allosteric site

20. Ligand

21. antibody

22. hapten

23. nucleotide

24. purines

25. phosphodiester

26. antiparallel

27. nicked

28. saturated, unsaturated

29. phosphoglyceride

30. Liposomes

31. asymmetric

32. glycosidic

33. glycogen

34. gangliosides

35. glycolipids and glycoproteins

36. a c e

37. b c d e

38. a d

39. a b c d e

40. b d

41. a c

42. b

43. a e

44. a b d e

45. b e

46. a c d e

47. b c d e

48. a b c d e

49. b e

50. b d

51. P

52. P

53. C

54. L

55. C

56. N

57. P

58. O

59. P

60. C

61. O

62. C

63. L

64. P, or O and P (heme + globin)

65. The *trans* form is favored. In this form the amino acid side chains linked to the α carbons are farther apart and thus are less sterically hindered.

66. Amino acids 1–14 and 19–28 are likely to form helices. Proline generally is not found in helices because it is unable to rotate appropriately about one of its peptide bonds. Glycine also is rarely found in helices. See MCB, pp. 48–49.

67. Phospholipids and proteins cover the surface of lipo-proteins in contact with the blood plasma, while triacyl-glycerols and cholesterol esters form the cores of lipo-protein particles. The proteins may have amphipathic helices; the hydrophobic amino acids face the nonpolar lipid core and the hydrophilic amino acids interact with the aqueous blood plasma. See MCB, pp. 75–77.

68. A short peptide would more easily form an antiparallel β pleated sheet because this structure can result from the chain simply folding back on itself. See MCB, Figure 2–9, p. 50. In a parallel β pleated sheet, the hydrogen-bonded chains run in the same direction; that is, they are oriented with the N-terminal ends of their primary sequences at the same end of the β pleated sheet. The two segments in the parallel structure must have an intervening region in the primary structure; this intervening region would not be involved in the same parallel β pleated sheet. Thus a parallel β pleated sheet structure would require a longer peptide chain than an antiparallel structure.

69. The native form of some proteins is the conformation with the lowest free energy, whereas the native form of other proteins is not the form with the lowest free energy. Proteins in the first class renature and reform the proper disulfide bridges spontaneously; those in the second class do not. See MCB, p. 55.

70. The residues recognized by chymotrypsin are all hydrophobic amino acids, which often are buried in the interior of proteins, away from the aqueous environment. Thus it is likely that these residues are unavailable for recognition by chymotrypsin when the protein is in its native state. In the body, chymotrypsin works in the intestine to further hydrolyze proteins that have been denatured and partially hydrolyzed in the stomach.

71a. Accumulation of product E is likely to inhibit the conversion of C to D, the first step in the pathway that does not lead to the formation of other products. Often this step requires an input of energy (often in the form of ATP). Thus inhibition of this reaction, when its product is not needed for the synthesis of the endproduct, conserves energy. Inhibition of a reaction (in a series of reactions) by the ultimate product of that series is termed *feedback inhibition*.

71b. Accumulation of product H inhibits the conversion of F to G by feedback inhibition.

71c. E and H are likely to inhibit the enzymes catalyzing the conversion of C to D and F to G, respectively, by binding to their target enzyme at an allosteric (regulatory) site. Binding at an allosteric site causes the enzyme to assume an inactive conformation, thus inhibiting catalysis at the active site. See MCB, p. 63.

72. One reasonable approach would be to use antibody affinity chromatography. To use this technique, you could produce an antibody able to bind specifically to your purified protein by introducing this protein into an animal, collecting the animal's blood, and isolating the antibody molecules that bind to your protein. If this antibody is then coupled to plastic beads, it can be used to isolate your protein from other proteins. See Figure 2-28 in MCB (p. 66).

73. Fragment 1 would denature (separate into single strands) at a lower temperature than fragment 2 because 1 has more A-T pairs than 2. A-T pairs, which have two hydrogen bonds, are less stable than G-C pairs, which have three hydrogen bonds (see Figure 2-35 in MCB, p. 69). The relationship between the G-C content of DNA and denaturation temperature T_m is shown in Figure 2-41 in MCB (p. 72).

74a. Assuming that the average amino acid residue has a molecular weight of about 120 and the molecular weight of the peptide is 3000, there are about 25 amino acid residues present.

74b. The data suggest that the C-terminal residue may be lysine. If there are 25 amino acid residues in the protein, about 4 percent of the total amino acid composition would be equivalent to 1 residue. Thus the compositional data suggest that there are two chymotrypsin sites (1 Phe + 0 Trp + 1 Tyr) and two trypsin sites (0 Arg + 2 Lys). There are three chymotrypsic fragments, as would be expected from two cleavages, but only two tryptic fragments. Unless there is a fragment(s) that was not detected, this implies that one lysine residue did not act as a site for trypsin, suggesting that this residue is the C-terminal amino acid.

74c. A reasonable experiment would be to hydrolyze the tryptic fragments with chymotrypsin and the chymotryptic fragments with trypsin. Sizing the resulting pieces should provide the information necessary for constructing a map of the enzyme hydrolysis sites.

75. TCPK is an analog of the chymotrypsin substrate phenylalanine in a peptide linkage. However, instead of the hydrolyzable amide linkage on the C-terminal side of phenylalanine, TCPK has a nonhydrolyzable O=C-CH_2-Cl group. Thus TCPK competes with peptide substrates to bind to chymotrypsin in the active site, but it cannot be hydrolyzed by the enzyme. In fact, the O=C-CH_2-Cl group reacts to form a covalent complex with histidine 57 of the chymotrypsin active site, permanently inactivating the protein. Obviously, TCPK does not react with trypsin because phenylalanine does not bind to trypsin's active site. See MCB, p. 57–59.

76. You might hypothesize that the bacteria provide something necessary for the health of the rats, which is absent in treated rats. The low activity of acetyl CoA carboxylase in bacteria-depleted rats suggests that the material provided by the bacteria is necessary for the function of this enzyme. In fact, intestinal bacteria synthesize biotin, a vitamin that as a prosthetic group and coenzyme of acetyl CoA carboxylase is necessary for catalytic activity of this enzyme.

77a. When the HMGCoA reductase activity is measured in the cell homogenate, the V_{max} would be greater for the cells grown with compactin than for the cells grown without compactin, but the K_M would be the same for the cells grown with or without compactin. V_{max} increases when there is more enzyme present; in this case there is more HMGCoA reductase per mg cell protein (see Figure 2–24 in MCB, p. 62). On the other hand, a change in K_M would represent a change in the enzyme's affinity for its substrate. There is no evidence presented to indicate that any such change occurs in response to compactin; the number of molecules of HMGCoA reductase per cell is simply increased without any alteration in the properties of individual enzyme molecules.

77b. Measured as the enzyme rate per mg purified HMG CoA reductase, there would be no difference in the K_M or V_{max} for the preparations from cells grown with or without compactin. After purification, the two enzyme preparations would be identical; there would just be more of the purified enzyme from the cells grown with compactin.

78a. The data suggest that valine 57 is part of or related to the allosteric site of enzyme M. Alteration of this residue has little effect on the "basal" activity of enzyme M (the activity in the absence of the activator). The basal activity represents the activity of the catalytic (active) site when the protein's conformation has not been affected by the binding of activator. In contrast, alteration of valine 57 has a large effect on the ability of compound C to act as an activator by binding to the allosteric site. This suggests that replacement of valine 57 either affects binding of the activator or affects the ability of the enzyme to undergo the conformational change that activates catalysis.

78b. Both valine and alanine are nonpolar, uncharged amino acids; thus substitution of alanine for valine is conservative and does not lead to a large functional change. Serine is polar but uncharged, and glutamine is both polar and negatively charged; substitution of each of these amino acids for valine 57 reduces enzyme activity considerably. Thus the more the chemical properties of an amino acid differ from the residue it replaces, the more drastic its effect on a protein's functional properties.

79. The problem with this approach is that it will produce not only 1-stearoyl,2-oleoylphosphatidylcholine but also 1-oleoyl,2-stearoylphosphatidylcholine,1,2-dioleoylphosphatidylcholine and 1,2-distearolyphosphatidylcholine because the chemical addition of acyl groups occurs nonspecifically at the available positions.

Cells utilize the specificity of enzymes to acylate each position on the glycerol with the desired fatty acyl group. Such specificity causes cells to place more unsaturated fatty acyl species in the 2-position than in the 1-position of phospholipids. In the laboratory, particular phosphatidylcholine molecular species can be synthesized most easily by first acylating the glycerol phosphorylcholine with the fatty acyl group that is desired in the 1-position. In the case above, such acylation would produce 1,2-distearoylphosphatidylcholine. The fatty acyl group in the 2-position of this compound then is removed by an enzyme that is specific for this reaction, called phospholipase A_2. (Phospholipases A_2 are easily obtained from snake venoms.) Lastly, the compound is acylated again at the 2-position with oleic anhydride to produce the desired compound.

80. Branched polysaccharides have numerous free ends available for the formation and hydrolysis of glycosidic bonds. Therefore such compounds can incorporate large amounts of glucose when it is in excess and, conversely, rapidly release glucose by hydrolysis when it is in short supply.

81a. The presence of a sugar residue in a substance derived from a membrane suggests that the substance is either a glycoprotein or a glycolipid. The isolated material, however, contains no protein, implying it is a glycolipid. Glycolipids that contain N-acetylneuraminic acid are called gangliosides. In fact, the cell-surface receptor for cholera toxin is a ganglioside.

81b. Gangliosides contain other monosaccharides as well as ceramide and a fatty acyl group. Chemical analyses for any of these constituents would be appropriate. Another approach would be to compare the chromatographic and/or toxin-binding characteristics of the isolated substance with those of known gangliosides. Other approaches are possible.

CHAPTER 3

1. aminoacyl-tRNA synthetase
2. Okazaki fragments
3. transcription, translation
4. polymerases   5. anticodon, codon
6. initiation
7. initiation
8. suppressor
9. wobble
10. ribosome
11. methionine
12. nonsense
13. introns

14. template, primers

15. tRNA (also rRNA)

16. amino, carboxyl

17. initiation, elongation, and termination

18. termination

19. toposiomerase

20. protein and nucleic acid

21. b c d e	34. P
22. b c	35. E P
23. d	36. E P
24. a c d e	37. E P
25. a b d e	38. E P
26. c d e	39. E P
27. a b c d e	40. E
28. b c	41. E P
29. b c e	42. E P
30. c d	43. E P
31. E P	44. P
32. E	45. E P
33. E P	

46. The highly specific chemical reactions of translation take place at a much higher rate if the individual components (mRNA, tRNA, and the appropriate enzymes) are confined by mutual binding to another component, the ribosome. This interaction limits diffusion of one component away from the rest and enables protein synthesis to proceed at the rate of nearly 1 million peptide bonds/s in the average mammalian cell. Similarly, electron donors and acceptors, in a highly organized array, such as that found in the inner membrane of the mitochondrion or the plasma membrane of a bacterium, can operate much more efficiently than they would if diffusion in three dimensions occurred.

47. The amber suppressor is a tRNA with an altered anticodon (5′)UAG(3′); its function is to act as an amino acid donor for a codon that would normally function as a stop codon (3′)AUC(5′). Since these components function during protein synthesis, and not during RNA synthesis, rescue of a mutation (growth of the cell containing the mutation) by placing it in a genetic background containing an amber suppressor implies that the mutation must be in a gene that codes for a protein. Because genes coding for tRNA or rRNA are not translated, mutations in these genes cannot be alleviated by components of the protein translation machinery (such as an amber suppressor). See MCB, p. 101.

48. Wobble may speed up protein synthesis by allowing the use of alternative tRNAs. If only one codon-anticodon pair was permitted for each amino acid insertion, protein synthesis might be temporarily halted until a reasonable level of that particular charged tRNA was regenerated. In fact, as discussed in Chapter 7 of MCB, a slowdown of protein synthesis, due to the lack of a particular aminoacyl tRNA, is used to regulate the levels of enzymes involved in synthesis of some amino acids. This process is called *attenuation*.

49. (1) Both proteins and nucleic acids are made up of a limited number of subunits, which are added one at a time. (2) The synthesis of both is directed by a template (mRNA in the case of protein; a complementary strand in the case of DNA). (3) Both are synthesized in one direction only, starting and stopping at specific sites in the template. (4) The primary synthesis product is usually modified; these modifications include cutting, splicing, and addition of chemical groups.

50. RNA is less stable chemically than DNA because of the presence of a 2′-hydroxyl group on the ribose backbone. Additionally, chemical deamination of cytosine (found in both RNA and DNA) to give uracil is recognized and repaired by cellular enzymes if found in DNA (where uracil is not normally found) but not in RNA. Thus the presence of thymine and deoxyribose make DNA more stable than RNA and might help explain the use of DNA as a long-term information-storage molecule.

51. A strong conclusion from this observation is that life on earth evolved only once.

52. Eukaryotic mRNAs are highly modified between transcription and translation. These modifications include cutting, slicing, and modification of the 5′ end (capping) and the 3′ end (polyadenylation). These modifications occur in the nucleus before mRNA is transported to the cytoplasm for translation. Prokaryotic mRNAs are not modified and often begin to be translated even before transcription is complete. In addition, prokaryotic mRNAs are often polycistronic (contain sequences for translation into more than one protein), whereas eukaryotic mRNAs are monocistronic (contain a translatable sequence for one protein).

53. The genetic code was broken by determining which amino acids were present in polypeptides formed by translation of synthetic polynucleotides in bacterial ex-

tracts. For example, it was shown that polyuridylate (UUU_n) was translated into polyphenylalanine. Similar analyses of other synthetic nucleotides allowed scientists to discover the triplet codons for all of the standard 20 amino acids.

54. DNA synthesis is discontinuous because DNA polymerase can synthesize DNA only in the $5' \rightarrow 3'$ direction. Thus one strand is synthesized continuously at the growing fork, but the other strand is synthesized in fragments that are joined together by DNA ligase.

55. Modification of the CAU anticodon of a tRNA^{Met} to give UAC (the anticodon for tRNA^{Val}) resulted in recognition by valine aminoacyl-tRNA synthetase and addition of valine rather than methionine to the altered tRNA molecule. Additional evidence was obtained when the converse modification was made: the altered tRNA^{Val} with a CAU anticodon was recognized and charged by the methionine aminoacyl-tRNA synthetase.

56. Proline, XYX; leucine, YXY; threonine, YXX; alanine, XXY; and tryptophan, XXX.

57. Since $A = T$, AT content $= 72$ percent and GC content $= 28$ percent. Since $G = C$, the content of G is 14 percent.

58. Table 3-1 is arranged so that codons in the same column in the same "box" differ only in the third (3′) base. Thus most of the codons for an individual amino acid appear in the same column and the same box because different codons for an amino acid usually differ only in the third (3′) base. This occurs because single tRNA species can bind to multiple codons, which differ only in the nucleotide found in the third (3′) position.

The formation of "nonstandard," in addition to standard, base pairs between the third base of the codon and the first (5′) base of the anticodon was hypothesized in 1965 by Francis Crick, and is called the wobble hypothesis. Base pairs that can form between codon and anticodon are as follows:

Codon — third (3′) base	Anticodon — first (5′) base
U	G or I
C	G or I
A	U or I
G	C or U

Note that A never appears as the first (5′) base of the anticodon. An adenosine in this position is deaminated by the enzyme *anticodon deaminase* to form hypoxanthine (the base for the nucleoside inosine) whenever it appears in a newly transcribed tRNA. Thus U in the third (3′) position of a codon pairs with inosine (I) or forms the so-called "wobble pair" U-G.

The pairings shown above have several implications. First, the codons ending in C-3′ and U-3′ are always synonymous, since both would respond to either G or I in the first (5′) position of the anticodon. Second, a codon ending in A-3′ would respond to anticodons beginning with 5′-I or 5′-U. If the anticodon starts with 5′-I, then codon 5′-(XY)A-3′ will be synonymous with both 5′-(XY)U-3′ and 5′-(XY)C-3′. If the anticodon starts with 5′-U, then 5′-(XY)A-3′ will be synonymous with 5′-(XY)G-3′. Thus, because of synonymous codons, an amino acid with four codons (e.g., Ala), requires only two tRNAs. One will recognize only codons ending in G-3′; the other will recognize 5′-(XY)A-3′, 5′-(XY)C-3′, or 5′-(XY)U-3′. Still, amino acids with only one codon, such as methionine, encoded by (5′)AUG(3′), may be translated unambiguously because anticodons starting with 5′-C will only pair with codons ending in G-3′. Obviously, these pairing schemes imply that some anticodons are not used; for example, the anticodon (5′)UAU(3′) is not used because it would pair with (5′)AUG(3′), encoding methionine or start, as well as with (5′)AUA(3′), which encodes isoleucine. Likewise, there are no anticodons that pair with the stop codons.

In conclusion, those anticodons that are used pair with either one, two, or three codons, depending on the base in the 3′ (wobble) position of the anticodon. If this base is C, only one codon is read. If it is G or U, two codons are read. If it is the modified base inosine, then three codons are read.

59. Prokaryotes, but not eukaryotes, have an N-formylmethionyl residue at the amino terminus of most proteins. In fact, peptides containing an N-formylmethionyl residue at the amino end are potent activators of mammalian immune system cells such as neutrophils and macrophages.

60. Stop codons were identified by genetic techniques, which were the only techniques available for this task at the time. By the time that Nirenberg, Khorana, and Ochoa had deciphered the code, it was clear that 61 of the 64 possible codons could be associated with a particular amino acid. The remaining three codons (UAA, UAG, and UGA), however, could not be identified as stop codons merely on the basis of evidence that no polypeptides were synthesized from artificial RNA

templates containing a substantial amount of any of these three codons.

The genetic evidence for the identity of stop codons came from sequence analyses of proteins in certain mutant and revertant strains of bacteriophage and *E coli*. This approach is illustrated by A. Garen's studies of an amber mutant of *E. coli* alkaline phosphatase. This amber mutation had a stop codon that arose at a site containing a tryptophan in the wild-type protein. Revertants of this mutant (presumably resulting from single-base changes) contained proteins with tryptophan lysine, serine, tyrosine, leucine, glutamic acid, or glutamine at the site corresponding to the amber codon in the mutant protein. From Table 3-1, it is clear that the only codon related to the codons for all these amino acids by a single-base change is UAG (now called the amber codon) as shown below:

$$(5')UAG(3') \rightarrow (5')UGG(3')\text{-Trp}$$
$$(5')UAG(3') \rightarrow (5')AAG(3')\text{-Lys}$$
$$(5')UAG(3') \rightarrow (5')UCG(3')\text{-Ser}$$
$$(5')UAG(3') \rightarrow (5')UAU(3')\text{-Tyr}$$
$$(5')UAG(3') \rightarrow (5')UUG(3')\text{-Leu}$$
$$(5')UAG(3') \rightarrow (5')GAG(3')\text{-Glu}$$
$$(5')UAG(3') \rightarrow (5')CAG(3')\text{-Gin}$$

61. It is possible that the primitive genetic code, although a triplet code, used only the first two bases of a triplet and that 16 (or fewer) amino acids were actually coded for by primitive replicating entities. The addition of additional amino acids (evolutionary latecomers) would have necessitated the refinement of the code such that the third base of the triplet could be used as part of an unambiguous code. Thus the number of codons for a particular amino acid might be related to the time when that amino acid first was incorporated into the metabolic machinery of living systems; methionine and tryptophan could be evolutionary latecomers.

62a. The analytical technique described, called a *shift assay,* separates the 30S and 50S ribosomal subunits from each other and from fully associated 70S prokaryotic ribosomes. At the higher Mg^{2+} concentration, the 30S and 50S subunits form a ternary 70S complex with the synthetic polyribonucleotide, whereas at the lower concentration, they do not (see Figure 3-8). Because this association must occur before protein synthesis is initiated, translation occurs only at the higher Mg^{2+} concentration.

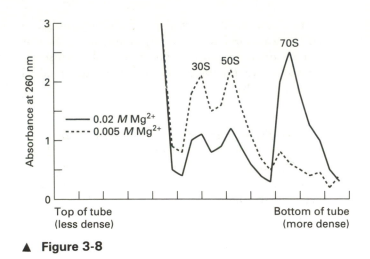

▲ **Figure 3-8**

62b. The fractionation profile for a mixture containing a biological mRNA should resemble the profile depicted with the solid line in Figure 3-8; that is, the ribosomal subunits and the mRNA are associated, and protein synthesis can proceed on the 70S ribosomes. This productive association forms only after the 30S subunit finds an AUG codon on the biological mRNA; GTP, *N*-formylmethionyl tRNA, and initiation factors are then bound. This *initiation complex* can form at low Mg^{2+} concentrations if an AUG triplet is present on the ribonucleotide to be translated.

63a. If a single "hit" was enough to kill a cell, the survival curve would be log-linear (i.e., linear when graphed on a semilogarithmic scale). This is clearly not true for the wild-type and the resistant bacteria, whose survival curves are quite flat at low radiation doses. This could mean that more than one hit is needed to kill an individual cell; alternatively, it could mean that the cells are capable of repairing damage incurred at low doses but cannot repair all the damage done at higher doses. The killing curve for the sensitive bacteria is nearly log-linear. This indicates that either the cells have changed so that a single hit is enough to be lethal (highly unlikely) or that repair mechanisms used to restore the cells to their original condition have been inactivated (much more likely).

63b. Enzymes likely to be altered include DNA polymerase (particularly the error-repair function of that enzyme), DNA (or polynucleotide) ligase, and enzymes involved in inactivating reactive peroxides or other free radicals. Other types of enzymes likely to be altered include those involved in DNA repair (discussed in Chapter 12 of MCB). In addition, mutations that result in higher O_2

consumption (hence lower O_2 concentrations intracellularly) would also be manifested as a radiation-resistant phenotype.

63c. Same as 63b.

63d. The radiation-sensitive mutant will be more sensitive to mutagens if the radiation-sensitive phenotype is due to defects in DNA-repair mechanisms. It is thought that most mistakes in DNA synthesis (due to mismatched bases or thymine dimers, for example) are repaired before the next round of DNA replication; thus the variant base sequence (mutation) is not transmitted to the next generation. Defects in this repair system would enable more modified bases to persist until the next round of replication, thus more mutations would be observed in succeeding generations. However, if the radiation sensitivity is due to depletion of peroxide-scavenging mechanisms, base-modifying mutagens should not be any more potent in this cell than in the parental cell.

64a. AZT is an analog of thymidine, which is a component of DNA. When phosphorylated and converted to AZT triphosphate, it may act as a competitive inhibitor of thymidine triphosphate incorporation during DNA synthesis. An alternative (and more likely) mode of action is as a chain terminator, since AZT does not contain the 3'-OH group needed to form the bond for the addition of the next nucleotide triphosphate. Once incorporated at the end of a growing DNA strand, it cannot be removed, and DNA synthesis ceases. The unique specificity of AZT for retroviral infections results from the greater preference of the reverse transcriptase for AZT triphosphate than for thymidine triphosphate; normal human DNA polymerases do not prefer this analog.

64b. As noted in 64a, the active form of the drug is the phosphorylated derivative, not AZT itself.

64c. A likely explanation is that there is a change (mutation) in the reverse transcriptase enzyme lowering its affinity for AZT. If this polymerase is mutated in such a way, then the AZT doses required for viral inhibition will also be inhibitory for the patient's DNA polymerases.

65. Components that are directly involved in the step inhibited by the antibiotic can be identified by growing bacteria in the presence of lethal concentrations of the antibiotic and obtaining mutants that are resistant to its

effects. Translational components (tRNA, rRNA, initiation and elongation factors, etc.) in the resistant and sensitive cells can be examined and compared in order to determine which component of the translational machinery is altered in the resistant mutant. Other, nongenetic, approaches might include examination of binding of radiolabeled antibiotic to cell components, which can then be separated chromatographically or electrophoretically for further analysis.

66. Since DNA is the molecule of inheritance, replication errors must be scrupulously avoided. Base-pairing without a 3'-OH primer, as necessarily performed by any RNA polymerase, is very error-prone. If a DNA polymerase performed this function, errors in the DNA sequence (mutations) would be introduced during replication and then transmitted to future generations. An error rate of one base in 10^5, which is not unusual for RNA polymerases, would result in an enormous increase in the mutation rate. However, the RNA primers made during DNA replication are erased and replaced with high-fidelity DNA copies; any mismatched bases will be replaced before being passed on to the next generation.

67a. These data are consistent with each other and do not rule out either hypothesis. Two pieces of evidence are consistent with hypothesis 1 ("masked" RNA): the observation that foreign mRNA can be translated by egg lysates, and the minimal stimulation of the reticulocyte lysate by 10 percent urchin egg lysate (panel A), even though the amount of egg mRNA added would be expected to produce a large stimulation of protein synthesis in the reticulocyte lysate assay system. In contrast, the rate of protein synthesis was much higher when reticulocyte lysate and a 10 percent sea urchin larval lysate were mixed (panel C). Evidence consistent with hypothesis 2 (missing translational factor) includes the ability of at least some of the egg mRNA to be translated at high ratios of reticulocyte lysate to egg lysate (right side of panel A). Stimulation of translation by addition of reticulocyte lysate (panel A) is due to addition of sea urchin egg mRNA; clearly this mRNA is not completely "masked," since it can be translated under these conditions. However, the decline in synthetic rate (left side of panel A) at low reticulocyte to egg lysate ratios indicates that translation is limited by some component of the reticulocyte lysate. This decline is not seen with the zygote lysates (panel B). The missing component cannot be mRNA, as the reticulocyte lysate had been treated to remove mRNA. The data are consistent with the hypothesis that some initiation or elongation factor supplied by the reticulocyte lysate becomes rate-limiting in mixtures containing high levels of egg lysate. It seems likely that both mRNA availability and translational factors are limited in the sea urchin egg; that is, both hypotheses are probably true.

67b. You could add purified initiation factors, elongation factors, tRNAs, or ribosomal subunits to the egg lysate and determine the effect of these components on translation of the egg mRNA species. Alternatively, you could fractionate the rabbit reticulocyte lysate using various chromatographic techniques, determine the effects of various fractions on the protein synthetic rate of the egg lysate, and attempt to identify the components in the most active fraction(s). Application of the former approach by Hershey and coworkers implicated the elongation factor eIF_{4F} as the component that limits translation in the egg lysate but not in the zygote lysate.

67c. RNA from eggs should not shift the 40S *preinitiation complex* into the 80S initiation complex; RNA from zygotes should shift the 40S complex to the 80S complex in this assay.

68a. The damaged DNA in the photoreactivated cells is repaired. If it were not repaired, the cells would exhibit the same survival curve as the UV-irradiated cells kept in the dark.

68b. The damaged DNA in the cells held in the non-nutritive medium is repaired as well. Since this repair process occurs in the dark, it is probably not exactly the same process that occurs in the photoreactivated cells. The data imply the existence of at least two separate DNA-repair mechanisms.

68c. Mutant 1 is incapable of photoreactivation but can repair damaged DNA if held in non-nutritive medium for several hours. Mutant 2 is capable of photoreactivation but is incapable of repairing DNA in the dark in non-nutritive medium. These data further support the hypothesis that at least two DNA-repair systems are present in these bacteria, since components of each can be mutated independently of the other. Enhanced survival of cells maintained in non-nutritive medium after UV irradiation is called *dark repair* to distinguish it from photoreactivation or light-dependent repair.

68d. The ability to repair UV damage to DNA probably had great survival value in the eons before the earth had an ozone layer (before the appearance of photosynthetic organisms). These repair systems thus reflect the history of the organism. Repair systems, once in place, probably evolved to become capable of repairing other types of damage to DNA. This is true for dark repair, at least; the enzymes involved in this pathway are capable of repairing DNA damaged by other agents. However, the persistence of photoreactivation in bacteria that live in the dark is currently unexplained, since it seems to repair only thymine dimers, which are characteristic of UV-irradiated DNA.

69a. Synthesis of the 13 ribosomal proteins examined seems to be coordinated, since all of these proteins show an increase in the relative rate of synthesis after shift-up. Furthermore, the time course of this increased synthesis is similar for all the proteins examined.

69b. Synthesis of ribosomal proteins and ribosomal RNA seems to be coordinated, since both show an increased rate of synthesis after shift-up. Furthermore, the time courses for the increased synthetic activities are similar for rRNA and ribosomal proteins.

69c. These data imply that at least two of the components of the translation machinery, ribosomal proteins and rRNA, are coordinately regulated. It seems likely that tRNA (and aminoacyl-tRNA synthetases), the other major component of the protein synthetic machinery, would also show a concomitant increase in synthetic rate after shift-up.

CHAPTER 4

1. nucleolar organizer
2. chromosomes
3. resolving power
4. organelle
5. plasma membrane
6. gram-negative
7. cytoskeleton

8. nucleus and mitochondria

9. plasma membrane, rough endoplasmic reticulum

10. smooth endoplasmic reticulum

11. exocytosis   13. turgor

12. lysosomes

14. peroxisomes, glyoxysomes

15. mitochondria and chloroplasts

16. plasma membrane   18. Golgi complex

17. transmission electron   19. coated vesicles

20. photosynthesis, chloroplasts

21. a d	31a. P E
22. a b c e	31b. P E
23. b c e	31c. E
24. c d	31d. E
25. a	31e. P E
26. a c d	31f. P E
27. a c d	31g. P E
28. a b e	31h. E
29. a b c d e	31i. P E
30. b c d e	31j. P E

32. See table below.

Process	Eukaryotes	Prokaryotes
Photosynthesis	Chloroplasts	Plasma membrane, mesosome, and thylakoid vesicles
DNA synthesis	Nucleus, chloroplasts, and mitochondria	Cytoplasm (DNA is attached to plasma membrane in most prokaryotes)
Amino acid transport	Plasma membrane	Plasma membrane
Protein synthesis	Cytoplasm, endoplasmic reticulum, chloroplasts, and mitochondria	Cytoplasm and plasma membrane

33. Gram-negative bacteria are surrounded by two membranes. The inner membrane is analogous in many ways (especially as a permeability barrier) to the plasma or surface membrane of eukaryotic cells. The outer membrane is much more porous and forms an additional permeability barrier for some larger molecules. In between these membranes is the bacterial cell wall. Gram-positive bacteria, on the other hand, have only a cell wall and a surface membrane. The cell wall (peptidoglycan) in gram-positive bacteria is considerably thicker than that in gram-negative bacteria.

34a. Motion of chloroplasts in plant cells is best visualized by light microscopy; electron microscopy (EM) is not suitable because EM specimens need to be killed and fixed before they can be viewed. Phase-contrast microscopy differential interference (Nomarski interference) microscopy, and perhaps standard transmission light microscopy would suffice for this purpose. Another suitable technique utilizes the natural fluorescence of the chlorophyll in chloroplasts as a visual marker; in a flourescence microscope under the proper conditions, the chloroplasts would have a red fluorescence and could be easily visualized. Phase-contrast microscopy is the preferred technique because objects the size of chloroplasts can be easily distinguished using this technique and because the difference in refractive index of the cytosol and the organelles is sufficient to give good contrast.

34b. Viral particles are best visualized by transmission electron microscopy; they are usually too small to be seen with the light microscope.

34c. Motion of bacterial cells is best observed with the light microscopic technique called dark-field microscopy, although phase-contrast microscopy would also suffice for this purpose. Electron microscopy could not be used because it requires fixed dead cells. Dark-field microscopy allows the use of live, unstained cells and is sufficiently sensitive so that objects the size of bacterial cells are easily resolved.

34d. The localization of a specific protein in a tissue is best accomplished with the light microscopic technique known as immunofluorescence microscopy, although the use of a specific antibody attached to an enzyme that generates a colored product would also allow the use of standard bright-field microscopic techniques. Electron microscopy could not be used because the size of the subject (a whole tissue) is prohibitively large. Immunofluorescence (or immunocytochemistry) are the only techniques that allow the investigator to localize specific macromolecules in tissue sections.

35. The lysosomal membrane keeps the lysosomal enzymes separated from the cytosol and nucleus. In addition, the pH of the cytosol (~ 7.0) is inhibitory for most lysosomal enzymes, which function best at pH 4–5.

36. The contractile vacuole present in many protozoans acts as a site for accumulation of water, which would normally enter the cell from the extracellular fluid. The water in this vacuole is periodically returned to the extracellular space by fusion of the vacuole with the cell plasma membrane. The vacuole in plant cells contains many small molecules and ions, which are present in sufficient concentrations to result in the movement of water from the cytoplasm into the vacuole. This movement expands the vacuole and thus causes the internal volume of the cell to expand; this expansion is critical to plant cell growth. Unlike the contractile vacuole, the plant cell vacuole does not periodically fuse with the plasma membrane to release its contents into the extracellular space.

37. The purpose of the microvilli in intestinal epithelial cells is to increase the surface area of these cells. Since a major function of intestinal cells is to absorb food from the lumen of the intestine, an increased surface area allows this process to proceed more rapidly. In addition, enzymes found at the cell surface act to break down complex molecules in the intestinal lumen; increased surface area results in exposure of more of these enzymes to the intestinal contents. See MCB, p. 146.

38. The high protein concentration of the cytosol enhances weak protein-protein interactions in the cell. These weak interactions, which are important to cellular metabolism, may never be detected by biochemists because of the difficulty in isolating sufficient protein so that the high cytosolic protein concentration can be mimicked in the test tube. For the electron microscopist, dehydration, fixing, and staining of cytosol with high protein concentrations could cause aggregation or precipitation of proteins, forming structures (called *artifacts*) that may not exist in the living cell.

39. The presence of organellar structures in cells permits the following advantageous phenomena to occur: (1) compartmentalization of antagonistic processes (e.g., protein synthesis in the endoplasmic reticulum and protein degradation in lysosomes); (2) allocation of membrane-dependent processes to increased intracellular membrane surfaces; and (3) confinement of diffusion-limited processes to a small area, thus increasing the rate of these processes.

40. The cytoskeleton functions to maintain cell shape, to aid in cell motility, and to anchor and/or move specific cellular structures (e.g., chromosomes during cell division).

41. The cell walls of eukaryotes such as plants and yeast function as determinants of cell shape, as permeability barriers, and as rigid structural elements, which maintain cell integrity in the face of osmotic stress. The cell walls of bacteria also have the same functions.

42. Differential-velocity centrifugation separates particles on the basis of their size. Equilibrium density-gradient centrifugation separates particles on the basis of their density.

43. Genes that code for antibody proteins are generated by rearrangement of DNA sequences coding for part of the antibody. In addition, mutations in these rearranged genes are quite frequent. This combination of DNA mutation and reorganization vastly increases the number of possible coding sequences for antibody proteins. These DNA rearrangements and high mutation rates occur in a specific subset of cells in mammals, the so-called B-lymphocytes. For more detail, see Chapter 25 of MCB.

44. The "magnification" scale in this problem is about 10^5, since a 25-nm (25×10^{-9} m) particle is now the size of a 2.5-mm (2.5×10^{-3} m) particle: $2.5 \times 10^{-3} \div 25 \times 10^{-9} = 1 \times 10^5$. The magnified dimensions and common objects of equivalent size are listed in the table at the top of the facing page.

45a. $D = \dfrac{(0.61 \times 600 \text{ nm})}{1 \times \sin 70°} = \dfrac{366 \text{ nm}}{0.94}$
$= 390 \text{ nm or } 0.39 \text{ } \mu\text{m}$

45b. 260 nm or 0.26 μm

45c. 190 nm or 0.19 μm

45d. A mitochondrion should be visible under all of the conditions specified in 45a–c, since its dimensions are considerably than the value of D in all cases.

Cellular structure	Actual size	"Magnified" size	Familiar object
Bacterial cell	1 μm diameter	0.1 m (4 in.)	Softball
Mitochondrion	1 × 1 × 2 μm	0.1 × 0.1 × 0.2 m (4 × 4 × 8 in.)	Telephone
Muscle actin filament	0.007 μm thick × 1 μm long	0.7 mm × 0.1 m (0.027 in. × 4 in.)	Extra-long mechanical pencil lead
Nucleus	5 μm diameter	0.5 m (19.7 in.)	Beach ball
Intestinal epithelial cell	7.5 × 7.5 × 15 μm	0.75 × 0.75 × 1.5 m (30 × 30 × 60 in.)	Desk
Human egg cell	70 μm diameter	7.0 m (23 ft)	Hot air balloon

46. Both bacteria and mitochondria contain DNA, perform oxidative phosphorylation in their inner membrane, and have an outer membrane that is porous to large molecules. In addition, ribosomal structure, antibiotic sensitivity, and many other biochemical attributes are very similar in bacteria and mitochondria. (See Chapter 26 of MCB for details).

47a. See table below.

Organelle	Marker molecule	Enriched fraction (no.)
Lysosomes	Acid phosphatase	11
Peroxisomes	Catalase	7
Mitochondria	Cytochrome oxidase	3
Plasma membrane	Amino acid permease	15
Rough endoplasmic reticulum	Ribosomal RNA	5
Smooth endoplasmic reticulum	Cytidylyl transferase	12

47b. Rough endoplasmic reticulum is more dense than smooth endoplasmic reticulum, since it is found at a higher sucrose concentration (more dense solution) on the gradient.

48a. Osteoclasts contain an above-average amount of lysosomal enzymes. In a strict sense, these cells do not have more lysosomes, as the degradative activity occurs outside the cell; osteoclasts could be considered to have a large external lysosome.

48b. Anterior pituitary cells contain an above-average amount of rough endoplasmic reticulum (ER) and Golgi complex; these two organelles are involved in proteins synthesis and processing for secretion.

48c. Palisade cells of the leaf contain an above-average amount of chloroplasts which are solely responsible for photosynthesis in eukaryotes.

48d. Brown adipocytes contain an above-average amount of peroxisomes, which are involved in fatty acid degradation and heat production. In addition, these cells contain above-average amounts of the other oxygen-utilizing, fatty acid–catabolizing organelle, the mitochondrion.

48e. Ceruminous gland cells contain above-average amounts of smooth ER, which is the site of lipid biosynthesis.

48f. Schwann cells contain above-average amounts of both smooth and rough ER, these organelles are the site of biosynthesis of the proteins and lipids that compose the myelin sheath.

48g. Intestinal brush border cells contain an above-average amount of plasma membrane (the microvilli), which is the site of nutrient uptake.

48h. Leydig cells of the testis contain above-average amounts of both smooth ER and mitochondria. The smooth ER is the site of cholesterol biosynthesis; the mitochondria contain the oxidative enzymes involved in transforming cholesterol into the male sex hormones.

49a. See Figure 4-5, panels A and B.

49b. Comparison of panels A and B indicates that the cells used in analysis A were dividing more slowly; that is, a smaller proportion were engaged in synthesizing DNA (channel numbers 120–180) and a larger proportion were in the G_1 and G_2 phases of the cell cycle.

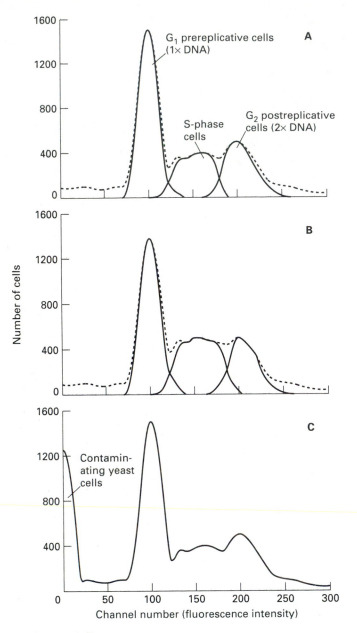

▲ **Figure 4-5**

occurs in most cells containing more than one Golgi apparatus, might be much more difficult for the chloroplast, which has a substantial amount of internal membranous structure. Thus the single chloroplast in these algal cells probably divides, without fragmentation, before or during cell division. In fact, electron microscopic evidence indicates that the chloroplast actually divides at the time the cell divides.

50b. Like the Golgi membranes, the nuclear membrane breaks into fragments (so that the nucleus disappears during cell division); these fragments are then distributed approximately equally between the daughter cells.

51a. Yes. The disappearance of the sucrosomes suggests that the invertase is taken up by the cells and becomes localized in the lysosomes, where it catalyzes breakdown of the sucrose. Additional experiments to test this hypothesis might include labeling the invertase with a fluorescent or radioactive marker, adding it to cells, isolating the lysosomes, and determining if the invertase activity co-purified with acid phosphatase or some other lysosomal marker molecule. In fact, if you prepared a fluorescent invertase preparation, you might even be able to detect lysosomal fluorescence using a fluorescence microscope.

51b. Invertase in the fused cells is also located in a lysosomal compartment; this conclusion is consistent with the hypothesis discussed above. The observations may or may not be consistent with other hypotheses that you might have formulated. These data indicate that lysosomes inside cells mix rapidly, probably by fusion of their membranes, since invertase in one lysosome obviously comes into contact with sucrose in another lysosome where it breaks down the sucrose.

52a. Because O_2 was probably toxic to most cells when it first appeared in large quantities, an organelle that could remove O_2 at all O_2 concentrations would have been most beneficial; thus the peroxisome was most likely the first O_2 utilizing organelle to appear in evolutionary time. A corollary of this hypothesis is that adaptation of the peroxisomal reactions, to perform useful metabolism, occurred later in the evolutionary development of eukaryotic cells.

49c. Because yeast cells contain much less DNA per cell than mammalian cells, the yeast cells should exhibit much less fluorescence after exposure to chromomycin A_3. If yeast cells contaminated the cell culture in analysis A, then there would be a large peak in cell number (the yeast cells) at a low channel number. See Figure 4-5, panel C.

50a. The single Golgi apparatus in these cells forms a multitude of fragments, which are then divided approximately equally between the daughter cells. This process of membrane fragmentation during mitosis, which also

52b. Peroxisomes might function to keep cellular O_2 levels low, thus eliminating toxic oxidative side-reactions, particularly when mitochondrial respiration (O_2 consumption) is already operating at maximal capacity.

CHAPTER 5

1. plasmids
2. cell cycle
3. clone
4. auxotroph
5. transformation
6. fibroblasts
7. blastocyst
8. erythroleukemia
9. somatic
10. heterokaryons
11. nucleic acid
12. salvage
13. monoclonal antibody
14. bacteriophages
15. retroviruses
16. virion
17. undifferentiated cell
18. meiosis
19. *Saccharomyces cerevisiae*
20. plaque
21. a c d e
22. c d e
23. a b d
24. b c e
25. a d
26. a b c
27. a b d e
28. b
29. a b c d
30. a b c d
31. See table below.
32. See Figure 5-6.

33. One could use the technique of replica plating and search for mutants that grow on medium containing leucine but not on medium deficient in leucine. The procedure would be analogous to that shown in Figure 5-10 of MCB (p. 161), which depicts the isolation of bacteria defective in arginine synthesis.

34. 1 in 2. In the first meiotic division, each daughter cell would receive one of two homologous chromosomes (one of which was descended from your father and one from your mother). The division of homologous chromosomes into two daughter cells is called *segregation*. See Figure 5-7 in MCB (p. 159).

35. Firstly, since cellular enzymes synthesize many viral macromolecules, viral infections afford the opportunity to study the actions of cellular machinery. Secondly, this opportunity is enhanced because many viruses shut down cellular macromolecule synthesis, allowing researchers to analyze a simplified synthetic pattern. Finally, since many viral genomes are integrated into the DNA of the host and passed to subsequent generations, the heritability, expression, and recombination of the viral genes can be analyzed in infected cells.

36. Cells can be maintained in selective media to eliminate those organisms that revert to a wild-type phenotype, or cells can be recloned periodically in order to assure genetic purity of the clonal isolate.

37. Glutamine, in quantities sufficient to supply the entire organism, is synthesized in human liver and kidney cells but not in most other cells, including fibroblasts. Thus this amino acid must be supplied to cultured fibroblasts.

38. Prokaryotes synthesize DNA for virtually the entire time between one cell division and the next. Eukaryotes

Cell type	Type of organism (e.g., bird, yeast)	Generation time (approx.)	Advantages for molecular cell biology research
Escherichia coli	Bacteria	30 min	(1) Has short generation time. (2) Has well-understood genetic system. (3) Requires only simple growth medium.
Saccharomyces cerevisiae	Yeast	2 h	(1) Has short generation time. (2) Has well-understood genetic system. (3) Can be grown as haploid or diploid.
Cultured human cells	Mammal	24 h	(1) Is a higher eukaryote with short generation time. (2) Results often applicable to medical problems. (3) Is suitable for genetic studies (somatic-cell and DNA manipulation methods) not feasible with whole organism because of its long generation time.

6

Early anaphase

3

Middle prophase

1

Interphase

2

Early prophase

8

Telophase

7

Late anaphase

5

Metaphase

4

Late prophase

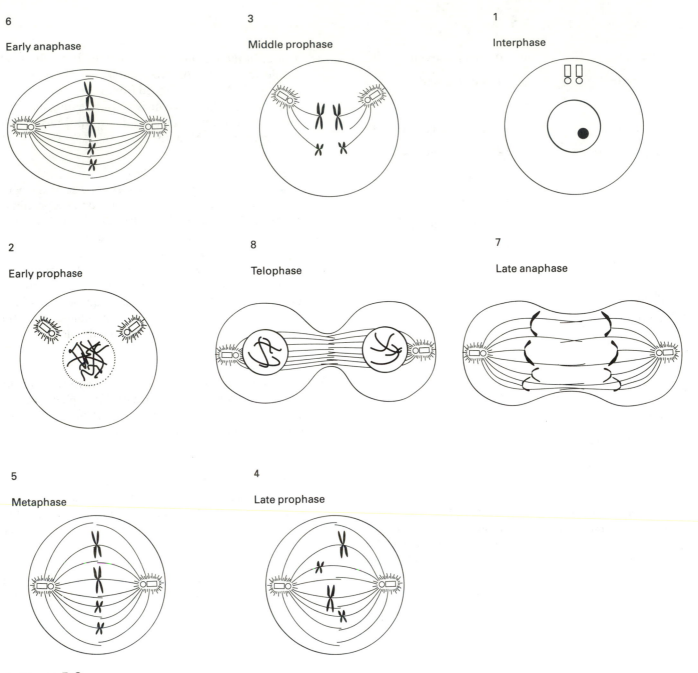

▲ **Figure 5-6**

synthesize DNA during a discrete portion of the cell cycle, the S phase, which averages approximately 6–8 h in rapidly dividing cultures of mammalian cells. See Figure 5-2 in MCB (p. 153).

39. Cultured cells are preferable to whole organs because they consist of a single cell type and can be derived from organisms that are not routinely used as experimental animals (e.g., human). In addition, environmental and genetic variables can be more closely monitored (if not controlled) with cultured cells than with organs. A dis-

advantage of cultured cells is that cell-cell interactions, which are present in an organ and which may be important determinants of the process under study, are abnormal in cell cultures. Also, synthesis of materials, especially tissue-specific macromolecules (e.g., glutamine synthetase in the liver), may be low or nonexistent in cultured cells and very high in the intact organ.

40. Foreign DNA is integrated into yeast by homologous recombination, usually at the site of the appropriate yeast gene if yeast DNA is introduced. Other eukaryotic

cells integrate foreign DNA in nonspecific locations, and sometimes at multiple sites in the chromatin of individual cells. The particular advantages of the homologous recombination seen in yeast are (1) that the site of integration is known and (2) that mutant forms of a gene can be delivered to the diploid and analyzed in the haploid (after ascus formation), enabling the investigator to determine if that particular gene is necessary for yeast-cell viability.

41. The characteristics used in this classification scheme are type of nucleic acid (DNA or RNA); number of strands in nucleic acid; size of genome; site of viral replication; presence of lipid envelope; infectivity of viral nucleic acid; and enzymes used in replication of viral nucleic acid. See MCB, pp. 182–185.

42. Because isolation of the foreign gene is not required for somatic-cell hybridization, genes for vaguely quantifiable properties such as "tumorigenesis" can be introduced and analyzed by this technique. This technique also allows the relatively rapid assignment of genes to whole chromosomes or parts of chromosomes. Finally, genes with proper regulatory elements are easily introduced, making it possible to engineer cells that synthesize differentiated gene products such as monoclonal antibodies.

43. The gene for thymidine kinase (TK) is on human chromosome 17, the only one present in all TK^+ clones and absent from TK^- clones. This type of analysis, called *concordance analysis,* correlates the presence of retained human chromosomes with the presence of the enzyme and takes advantage of the fact that the hybrid cells lose human chromosomes in a seemingly random manner. In order to confirm this conclusion, one could test whether the observed TK activity is due to the human enzyme by use of specific antibodies, electrophoretic analysis, kinetic characteristics, etc.; such experiments could eliminate the possibility that the activity is due to some other enzyme or to reversion of the mouse TK^- mutation, rather than to the presence of the human enzyme.

44. The genes *a* and *b* are on the same chromosome. The homozygous father produces only one type of gamete: a_2b_2. If the genes were on separate chromosomes, the mother would produce four types of gametes: a_1b_1, a_1b_2, a_2b_2, and a_2b_2, yielding offspring with four different phenotypes. However, as evidenced by the lack of offspring who were either homozygous or heterozygous for both genes, the mother produced only two types of

gametes: a_1b_2 and a_2b_1. The *a* and *b* genes did not assort independently; thus they must be "linked," that is, on the same chromosome.

45a. The haploid cell that when mated with each of its parents shows the same phenotype as the parent is the double recessive, A^-B^-. (The diploid cell from the first mating is A^-A^+/B^-B^+. After meiosis the haploid progeny are A^-B^-, A^-B^+, A^+B^-, and A^+B^+.)

45b. The *B* gene codes for the earlier step in the pathway. Because the haploid cell is defective in both genes but accumulates product b, one can infer that the reaction that uses this product is an earlier step in the pathway.

46a. The assembly initiation site must reside between nucleotide 1 and nucleotide 1200 on the 3' end of the RNA.

46b. Assembly starts with interaction of a nucleotide sequence (so-called "initiation loop") with a disk ("lock washer") of protein subunits. This sequence is located about 1000 nucleotides from the 3' end of the RNA; as assembly proceeds, the 5' tail is drawn up through the central hole of the growing rod by subsequent additions of new protein disks. Therefore, assembly could be initiated normally if 1200 bases were deleted from the 5' end; however, assembly could not be initiated normally if the initiation sequence were deleted from the 3' end.

47a. The total concentration of viral particles is calculated from the hemagglutinin assay results (HU = hemmagglutinin unit): $(400 \text{ HU/ml}) \times (1 \times 10^6 \text{ particles/HU}) = 4 \times 10^8$ particles/ml

47b. The concentration of infectious viral particles is calculated from the results of the plaque assay. Averaging the number of plaques that formed with dilution 2 and adjusting for the dilution factor gives the following:

$$\frac{368 \text{ plaques/ml diluted soln}}{1 \times 10^{-6} \text{ ml orig. soln/ml diluted soln}}$$
$$= 3.68 \times 10^8 \text{ infections particles/ml}$$

47c. 92 percent of the viral particles are infectious, as follows:

$$\frac{3.68 \times 10^8 \text{ infectious particles/ml}}{4 \times 10^8 \text{particles/ml}} = 0.92$$

48a. Enrichment for *cdc* mutants, which are almost always temperature-sensitive (ts) mutants, is easily achieved by looking for clones that grow and divide at low temperatures (20–24°C) but not at elevated temperatures (36°C). However, most of the ts mutants obtained will be defective in some critical metabolic reaction, such as ATP production or DNA synthesis, and will not be true *cdc* mutants.

48b. True *cdc* mutants can be distinguished from nonspecific ts mutants because the *cdc* mutants will stop at a defined point in the cell cycle, whereas the nonspecific mutants will stop at random points in the cell cycle. A corollary of this observation is that *cdc* mutants should show synchronous entry into the cell cycle when shifted to the permissive conditions (lower temperature, in this case). If the mutant clone in question shows synchronous entry into the S phase (as measured by [³H]thymidine incorporation into DNA) when shifted from the high temperature to the low temperature, it is probably a *cdc* mutant.

49. In the nucleotide salvage pathway, the products (nucleotides) are directly required for DNA synthesis (and thus for cell division). Since all cells make DNA, this salvage pathway is found in all eukaryotic cells examined so far. Cells that lack the pathway and that are blocked in de novo synthesis by inhibitors will not progress through the S phase and will be rapidly outnumbered by salvage-competent cells. Additionally, the precursors and products of this pathway are cheap, soluble, and readily available, and specific inhibitors for various parts of the pathway are known and well characterized. Finally, the products do not readily cross cell membranes, thus reducing the chances for metabolic cooperation between a competent cell and a neighboring incompetent cell. See MCB, pp. 171–172.

50. First you need to modify the cloned gene so that it will not code for a functional protein. The easiest way to do this is to introduce a small deletion, or you might introduce an insertion by site-directed mutagenesis (see Chapter 6 of MCB for details). Then you can incorporate the inactivate cloned gene into diploid yeast spheroplasts and screen the resultant cells to ensure that the modified DNA is integrated into the genome. Because yeast incorporate DNA by homologous recombination, the inactivate gene should replace an active gene in one of the two chromosomes in these diploid cells. After induction of sporulation and generation of haploid ascospores, you can dissociate the spores and perform *tetrad analysis* to determine if all or only half of the spores from a given meiotic division give rise to viable

colonies. If half (or a statistical analog of half) of the ascospores fail to give rise to viable colonies, you can reasonably conclude that the cell-wall protein you have isolated is essential for life as yeast know it.

51. Since the primary human cells have a finite life span, you can perform what is called a "half-selection." Simply grow the hybridization mixtures in HAT medium to kill the mouse parental cell (see MCB, p. 172); only the hybrids and the human parental cell will grow under these conditions. After 20–40 generations, however, the human cells will stop dividing and will be overgrown by the vigorously growing hybrid cells. Cloning of this population should then ensure that you have selected for hybrids and eliminated both types of parental cells.

52a. G_2 is approximately 6 h long (the time from cessation of thymidine incorporation into DNA until cell division; that is 30 − 24 h).

52b. S is approximately 8 h long (the time during which thymidine is incorporated; that is, 24 − 16 h).

52c. The cells are trapped in early G_1.

52d. Mutagenize the cells, grow them in a hormone-deficient medium, and look for variants that grow under these conditions.

53a. The factor is probably not a protein or nucleic acid because both of these would be denatured by boiling and/or protease treatment. One could hypothesize that the factor consists of one or more fatty acids, which are stable to these treatments.

53b. To test this hypothesis, you could analyze the adipocyte-conditioned medium for fatty acid content. If fatty acids are found in this medium, then you could supplement the minimal medium with fatty acids (of the type found in the conditioned medium) in order to determine if one or more of these compounds can stimulate the growth of the epithelial cells.

54a. The cells that survive the amp-str plates are recombinants after conjugation; they have an ampicillin-resistance gene from their F⁻ parent and a streptomycin-resistance gene from their Hfr parent.

54b. Cells that grew on plates without y must be y^+; cells that grew without x must be x^+; and cells that grew without z must be z^+.

54c. The genes are brought into the F⁻ cells in the order y, x, z.

54d. Only a fraction of the recombinant cells ever receive the x, y, or z genes because conjugation is often interrupted spontaneously before the cells are placed in the blender.

55a. Peak 1 represents the cells' first DNA replication after the radioactive thymidine pulse; thus it corresponds to cells that were in the S phase during the pulse. Peak 2 represents the second DNA replication. The distance from peak 1 to peak 2 thus corresponds to one complete cell cycle.

55b. Total cell-cycle time = 20 h; G_1 period = 1.5 h; G_2 period = 7.5 h; and S phase = 10 h. The total cell-cycle time can be determined from the distance between the two mitotic peaks (or any two analagous points in peak 1 and peak 2), as indicated in Figure 5-7. The average length of the S phase corresponds to the distance between the two points in peak 1 at which 50 percent of the cells are labeled, as shown in Figure 5-7. The time after the pulse required to reach the first of these 50 percent points is $G_2 + 1/2M$; $1/2M$ is used because cells usually are scored in metaphase, which is approximately halfway through the M period. As indicated in Figure 5-7, $G_2 + 1/2M = 8$ h; since the problem states that M = 1 h, then $G_2 = 7.5$ h.

The value of G_1 can be calculated from the expression

for the total cell-cycle time (C_t) as follows:

$$C_t = G_1 + G_2 + S + M$$

Rearranging this expression and substituting the known values gives

$$G_1 = C_t - G_2 - S - M = 20\text{ h} - 7.5\text{ h} - 10\text{ h} - 1\text{ h} = 1.5\text{ h}$$

55c. The length of the M period can be determining by examining any of the autoradiograms and determining the ratio of mitotic cells to total cells (the mitotic index). For example, if this ratio is 0.05 and the total cell-cycle time is 20 h, the M = 0.05 × 20 H = 1 h.

56a. These observations are incompatible with the hypothesis that two classes of fibroblasts exist because genetically one of these classes does not "breed true."

56b. An alternative hypothesis is that these cells are predisposed to become adipocytes but that some compound that causes the cells to differentiate is present at a level too low to act on every cell. Experiments designed to identify this compound, which might be a hormone, a nutrient, a cell-surface component, or a combination of the above, would test this hypothesis. These experiments could be as simple as analyzing and manipulating the hormonal composition of the medium, or as complex as coculturing these cells with other cells known to secrete specific cell-surface or extracellular compounds.

57. One strategy would be to clone multiple copies of the EPSPS gene into the Ti plasmid and use the plasmid to introduce the gene into tomato cells in culture. Since glyphosate is a competitive inhibitor, it should be possible for a plant cell to become resistant simply by making excess enzyme. In this way the variant plant cell would always have some enzyme that is not inhibited. These cells would have a selective advantage over normal wild-type cells and should be easily selected from a mixed culture. An alternative strategy would be to find resistant plants (not necessarily a tomato) and clone the EPSPS gene from these strains into the Ti plasmid. If resistance is due to an alteration of the enzyme itself such that glyphosate no longer is a competitive inhibitor, than integration of this gene into the genome of tomato cells should generate resistant tomato cells as well. After resistant tomato cells have been isolated by either of these techniques, the cells can be used to produce entire tomato plants, using standard culture techniques. Theoretically (and in practice) these plants should be resistant to the herbicide.

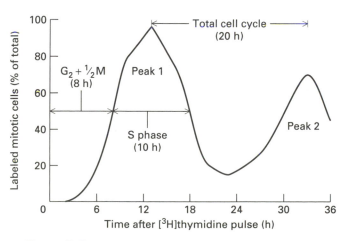

 Figure 5-7

CHAPTER 6

1. autoradiogram
2. Edman degradation
3. restriction endonucleases
4. library
5. isoelectric focusing
6. pool
7. rate-zonal
8. polymerase chain reaction
9. cDNA, reverse transcriptase
10. pulse-field
11. vector
12. in situ hybridization
13. proteins, antibodies
14. specific activity
15. single-stranded, double-stranded
16. carboxyl, amino
17. site-directed mutagenesis
18. exonucleases
19. sodium dodecyl sulfate (SDS)
20. isopycnic
21. d e
22. b
23. d
24. d
25. a b c d e
26. a b c d e
27. a b d e
28. a b d
29. a d e
30. b c d e
31. See table below.

Molecule or particle	S constant	Order of sedimentation (fastest to slowest)
Human ribosomal RNA (large)	28	3
Cytochrome c	1.7	5
Bacterium	5000	1
Fibrinogen	7.6	4
Poliomyelitis virus	150	2

32. See table on the facing page.

33. High specific activities are preferable when it is important to incorporate large amounts of radioactivity in a short time and when pool sizes are small; in the latter situation, addition of a small amount of very radioactive compound would not alter the pool size appreciably.

34. Binding of cesium ions to nucleic acids via electrostatic interactions appreciably increases the density of nucleic acids. RNA density is increased more dramatically than DNA density because cesium binds to both the phosphate groups and the hydroxyl groups of RNA, whereas it only binds to the phosphate groups of DNA.

35. SDS binds to a polypeptide at a ratio of approximately one SDS molecule per amino acid residue. This binding denatures or unfolds a protein so that chain length (i.e., molecular weight) becomes the most important determinant of motion through the pores of the polyacrylamide gel. In other words, the effects of variations in protein shape and charge are minimized.

36. One method is to examine the contents of a cell extract for newly synthesized protein. Cells are pulsed with a radioactive amino acid (usually [^{35}S]methionine) and then extracted; the extract is analyzed by electrophoresis and autoradiography. Another approach is to label proteins in an in vitro cell-free translation system from wheat germ or rabbit reticulocytes. In this method, endogenous mRNA in the wheat germ extract is first destroyed with a ribonuclease, the ribonuclease is then inhibited by a specific inhibitor, and extracts from the cells to be tested are added along with radioactive amino acids. Products of this cell-free translation system are detected after electrophoresis and autoradiography.

37a. Only cDNA clones can be used as expression vectors in bacteria because prokaryotes generally lack the enzymes necessary for splicing the primary transcript made from genomic DNA.

37b. A cDNA clone would probably be best for in situ hybridization because it more closely resembles the mRNA being detected.

37c. If the purpose is to make a DNA library in the lambda vector, clearly the genomic DNA clone is the only alternative. However, if one merely wants to generate multiple copies of a known gene sequence, the cDNA clone, which would always be shorter than the genomic DNA clone, would be better for packaging into a lambda phage because only 20–25 kb can be inserted into this vector.

Enzyme	Source	Substrate	Use
*Eco*RI	*E. coli*	DNA	DNA digestion for sequence analysis of fragments; comparison of restriction fragment length polymorphisms
Reverse transcriptase	Retroviruses	RNA	Synthesis of DNA from mRNA in preparation of cDNA clones
S1 nuclease	*Aspergillus oryzae*	Single-stranded DNA or RNA	Digestion of unhybridized (single-stranded) nucleic acid in order to determine how much of a particular DNA restriction fragment is complementary to an mRNA region
Taq polymerase	*Thermus aquaticus*	DNA	Catalysis of the polymerase chain reaction, which is useful for synthesis of a DNA sequence from genomic DNA when only a partial sequence is known
Trypsin	Mammalian pancreas	Protein	Preparation of peptide fragments for sequence analysis, fingerprinting, etc.

37d. Either type of clone would be suitable for this purpose; however, smaller fragments (e.g., only a small portion of the gene) are usually used. Although the great majority of base changes that cause restriction fragment length polymorphisms occur in the intervening sequences (introns) and not in the exons, detection of these changes does not depend upon the type of DNA used to probe the electrophoresed and blotted DNA digests.

38a. Because only the exon portions of a gene are retained in the corresponding mRNA, hybridization occurs between the β-globin mRNA and the exons of the β-globin gene only, as diagrammed in Figure 6-5.

38b. R-loops.

39a. Electrophoresis on polyacrylamide (for smallest molecules) or agarose gels.

39b. Isoelectric focusing on polyacrylamide gels.

39c. Pulse-field electrophoresis on agarose gels.

39d. Rate-zonal centrifugation on sucrose density gradients.

39e. Electrophoresis on polyacrylamide gels in the presence of SDS.

40. Synthesis of oligonucleotides based on codons predicted from a protein sequence is greatly simplified if the amino acids in that sequence have only a limited number of codons. Since there is only one codon each for tryptophan and methionine, the number of possible nucleotide sequences for peptides enriched in these amino acids will be relatively low.

41. The discovery and characterization of restriction enzymes, which function as mechanisms to protect prokaryotes from viral or other foreign DNA, was a vital prerequisite for the development of recombinant DNA technology. In a similar manner, the discovery and characterization of bacterial viruses and their replication was an equally important prerequisite. Certainly no one predicted that basic studies of such esoteric topics as bacterial viruses and mechanisms by which bacteria protect themselves from these agents would ever be useful in exactly this way.

42. Radioactive amino acids can be effectively chased because they rapidly enter and exit cells. Thus, even though the intracellular pools may be large, added amino acids are easily equilibrated with the extracellular fluid, facilitating the removal of radioactive material by a large nonradioactive chase. Pulses of thymidine can be chased effectively because the intracellular thymidine

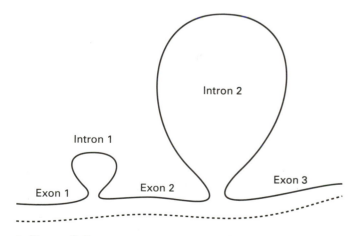

▲ **Figure 6-5**

pool is quite small and is used up in only a few minutes of DNA synthesis. See Figure 6-3 in MCB (p. 193).

43. In the examples given (cystic fibrosis and retinoblastoma), the genetics (how the diseases are inherited) are understood, but the function(s) of the proteins that are coded for by the defective genes, and ways in which expression of these genes are regulated, are still not completely understood. Thus we do not know what genetic changes to make in order to correct many genetic defects in humans. In other words, the structural information gained from these recombinant DNA techniques, although very useful in many ways, must be accompanied by other basic functional information before human maladies such as these can be corrected or cured.

44. See Figure 6-6. Experiments a and d would have qualitatively similar results; that is, in both further incorporation of labeled thymidine would be effectively inhibited after 20 min. Experiment b, in which the radiolabeled thymidine is diluted to one-half its original concentration, would have the same kinetics as the original experiment, but the amount of label incorporated into DNA would be lower. In experiment c, incorporation of exogenous labeled thymidine would be effectively inhibited because the precursor must be phosphorylated before it can be incorporated into DNA.

45. See Figure 6-7. Refer to MCB (pp. 206–209) for the details of restriction-site mapping.

46. Presumably both embryonic and sperm DNA contain all the DNA sequences found in an organism. If differentiation is accompanied by loss of specific DNA sequences (as can occur in white blood cells), a genomic

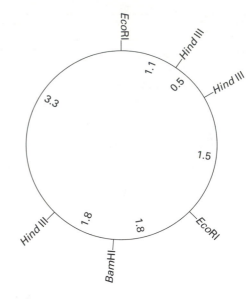

▲ **Figure 6-7**

library prepared from adult organs or cells might be incomplete.

47a. Since a λ vector can safely accommodate about 20 kb of DNA, you theoretically would need about 7500 clones to constitute the 1.5×10^8 bp in the *Drosophila* genome. However, statistical considerations (Poisson distribution) indicate that in order to ensure that every sequence has a 95 percent chance of being represented at least once, you should prepare a library containing about five times the theoretical minimum. The actual library, then, would contain 3.75×10^4 clones for the *Drosophila* genome. See Figure 6-33 in MCB (p. 217).

47b. Based on the same reasoning, about 500 yeast clones theoretically could represent the entire *Drosophila* genome, and 2500 clones would form a useful library.

48a. The number of codons for each of the amino acids in this sequence is as follows: Met, 1; Ala, 4; Cys, 2; His, 2; Trp, 1; and Asn, 2. Since the number of oligonucleotide probes needed is the product of the possible codons, you will need to prepare $1 \times 4 \times 2 \times 2 \times 1 \times 2 = 32$ oligonucleotide probes.

48b. If a leucine were substituted for the tryptophan residue, you would need $1 \times 4 \times 2 \times 2 \times 6 \times 2 = 192$ different oligonucleotide probes, as leucine has six codons. In practice, however, fewer synthetic oligonucleotide probes are required because, as discussed in later chapters, hybridization can occur if one base is not paired correctly. In addition, most organisms do not use

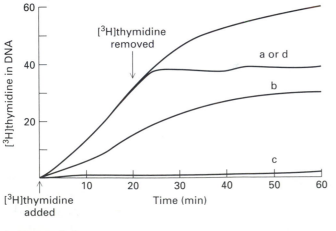

▲ **Figure 6-6**

all the codons randomly; preferred codon usages are known and can be used to design the most probable sequence for the oligonucleotide probes.

49a. The disappearance of the 9-kb band coincides with the appearance of the 5.4- and 3.6-kb bands; these bands must therefore be contained in the 9-kb fragment. Similarly, the disappearance of the 3-kb band coincides with the appearance of the 2.5- and 0.5-kb fragments; these bands must therefore be contained in the 3-kb fragment. There is no way that the actual relationship of the fragments can be unequivocally determined from these data; the two possible maps are shown in Figure 6-8.

49b. You could end-label the once-cut, linear 12-kb fragment that was formed after a brief (5-min) digestion. This end-labeled DNA is then digested completely, electrophoresed, blotted, and analyzed by autoradiography. If the 2.5- and 3.6-kb fragments or the 0.5- and 5.4-kb fragments are labeled, then map A in Figure 6-8 is correct. If the 2.5- and 5.4-kb fragments or the 0.5- and 3.6-kb fragments are labeled, then map B is correct.

50. You could determine the partial sequence of the amino terminus of the purified protein by *Edman degradation*. This sequence could be used to *prepare oligonucleotide probes* with which to *screen a cDNA library* of the organism; alternatively, the oligonucleotide probes could be used in the *polymerase chain reaction* to prepare DNA complementary to the gene for the protein. Either of these techniques will yield a DNA fragment complementary to all or part of the mRNA for the protein. Such fragments can be labeled with [3H]nucleotides and used as probes for *in situ hybridization* on brain tissue slices in order to locate the cells that produce large quantities of mRNA for the protein of interest.

Alternatively, after obtaining the partial peptide sequence, you could synthesize a synthetic peptide with an identical sequence. This peptide, if injected into an experimental animal such as a rabbit or goat, would cause the animal to produce specific antibodies. These specific antibodies, appropriately purified and labeled with fluorescent or radioactive markers, could be used to locate the appropriate cells using immunocytochemical techniques such as those described in Chapter 5 of MCB.

51. One possible attribute of a DNA-protein complex is that the protein could protect the DNA from degradation by nucleases. This hypothesis turns out to be true if the DNA-protein complex has a low dissociation constant. A popular technique that utilizes this property of DNA-protein complexes is called *DNA footprinting*. This method also takes advantage of the ability of electrophoretic methods that can resolve DNA fragments differing in length by only one nucleotide. This technique is outlined in Figure 6-9. Briefly, one incubates a solution of end-labeled DNA, suspected of containing a specific binding site for a protein, in the presence and

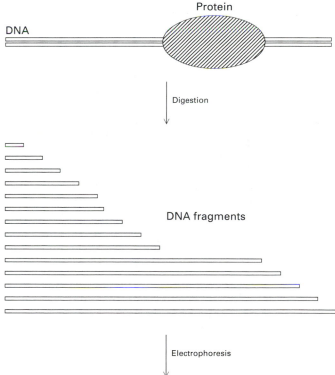

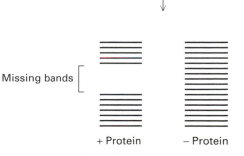

▲ Figure 6-9

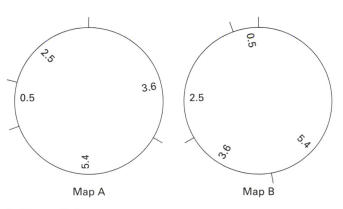

▲ Figure 6-8

absence of the protein. The DNA is then digested briefly with a nonspecific DNase (or chemical reagent that cleaves DNA) and electrophoresed. If the protein interacts with DNA, the (+) protein gel pattern will lack some of the bands present in the (−) protein pattern. The gap (footprint) in the ladderlike array of DNA fragments separated in this way corresponds to the portion of the DNA protected by bound protein.

52. The absorbance at 260 nm decreases as DNA renatures; thus sample B renatures more quickly than sample A. Since the rate of renaturation is inversely proportional to the complexity of a DNA, sample A is more complex than sample B. Thus sample A is the bacterial DNA and sample B is the phage DNA.

53. Restriction enzymes recognize palindromic sequences because they cut both strands of a DNA molecule. If the recognition site is the same on both strands (as it is with a palindromic sequence), then the enzymes (which are also usually dimeric) can recognize and cut both strands simultaneously. The recognition sites in the *Eco*RI sequence shown below are indicated by arrows:

$$(5')G{\downarrow}GATCC(3')$$
$$(3')CCTAG{\uparrow}G(5')$$

54. You could use the antibody to precipitate polyribosomes containing nascent chains of the protein from the cell line that expresses the protein. The mRNA species still attached to the polyribosomes should be greatly enriched for the mRNA of interest, since the antibody should only recognize (and precipitate) polyribosomes containing one type of nascent chain. This mRNA could then be used to prepare cDNA molecules using reverse transcriptase, and the cDNA could be used to screen a genomic library from the animal or plant used to make the cell line. Once specific hybridization to genomic DNA was found, the genomic clones could be investigated for the presence of introns, transcription start sites, and other attributes of an active gene as discussed in later chapters.

55. Three translation start codons (3')TAC(5'), corresponding to the mRNA sequence AUG in the opposite polarity, are present in this cDNA. The actual translation initiation site must be followed by an open reading frame, which by definition lacks the stop codons (3')ATT(5'), (3')ACT(5'), and (3')ATC(5'), corresponding to the mRNA sequences UAA, UGA, and UAG in the opposite polarity. In this cDNA, the only open reading frame coding for a 40-aa protein begins with the

second triplet in the second row and continues until two ATT stop codons are reached. This coding sequence is underlined below; start codons are circled, and the first stop codon after each of the three start codons are enclosed in a box.

(3')CCCTTGTGGATCCACACCC TAC CGGAGG ACT ATTAACTGTCCG
GCA TAC TTTGGCTGCGGTGTGGGGCAAGGTGAAGCTGGATGAA
GTTGGTGGTGAGGCCCTGGGGCAGACGTTGTATCAAGGTTTCA
AGACAGGTTTAAGGCAGACCAATAGAAACTGGGCGGC ATT ATT
GCA TAC ATT GGCCCTCGGAGTGTCAGTTGCAATGCTAGCTAAG(5')

56a. Radioactive oligomeric DNA probes were synthesized that were complementary either to the normal sequence encoding asparagine at position 370 in β-glucosidase or to the mutant sequence encoding serine at position 370. Genomic DNA from normal and affected individuals was digested with restriction enzymes, electrophoresed, and analyzed by Southern blotting with the radioactive probes under so-called "high-stringency" conditions (temperature and buffer conditions such that only absolutely homologous sequences would remain hybridized). DNA from the +/+ individuals hybridized only with the normal sequence; DNA from the −/− individuals hybridized only with the mutant sequence; and DNA from the +/− individuals hybridized with both oligomeric probes.

56b. There are at least four genotypes present among patients with type I Gaucher's disease. These include the three listed in Table 6-2 and a heterozygous genotype with a variant gene encoding proline (rather than leucine) at position 444 of the enzyme as one of the alleles.

56c. Type 2 disease is not due to the Ser-370 allele that is the likely cause of some cases of type 1 disease. Whether other type 1 patients and type 2 patients have the same defective allele cannot be determined from these data.

56d. You could look for *restriction fragment length polymorphisms* (RFLPs) in the DNA of affected individuals (homozygotes) and their parents (heterozygotes) by hybridizing a probe for the β-glucosidase gene on Southern blots of DNA from affected individuals and their parents. The DNA on the Southern blots is prepared by using many different restriction enzymes singly or in pairs. If the mutant gene is **linked** to a section of the chromosome that has lost or gained a DNA sequence (or a restriction endonuclease site), you will see a new restriction fragment that hybridizes to the β-glucosidase gene probe. If this RFLP is also present in the parents (and some of the grandparents, if available) of an affected individual, then the siblings of the affected individual can be analyzed for this RFLP to determine whether they are carriers for the defective gene.

57a. Since the concentration of moose genomic DNA is much greater than that of the labeled cloned DNA, renaturation (reassociation) of the later results in formation of hybrids containing the cloned DNA and genomic DNA fragments. As discussed in Chapter 10 of MCB (see Figure 10-15, p. 362), renaturation at low $C_0t_{1/2}$ values involves repetitive DNA sequences. About half of the ^{32}P-labeled cloned DNA is hybridized at relatively low $C_0t_{1/2}$ values, suggesting that the moose genomic DNA contains many copies (10^3 to 10^4) of part of the cloned sequence (see Figure 6-10). The hybridization seen at high $C_0t_{1/2}$ values involves a sequence that is unique (i.e., repeated only once in the haploid moose genome).

57b. It is possible that the cloned fragment codes for part of an mRNA, since it does contain some unique sequence. However, the presence of repetitive sequences indicates

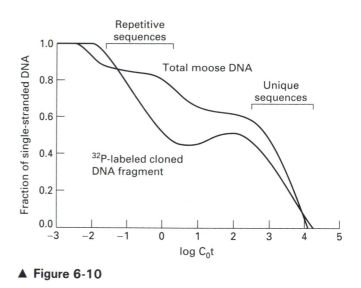

▲ **Figure 6-10**

that approximately half of this fragment does not code for a differentiated gene product, as such products usually are represented by a unique sequence in the haploid genome. The stringency of the conditions used for the molecular hybridization will affect this conclusion.

C H A P T E R 7

1. structural genes, regulatory genes

2. promoter, Pribnow box

3. consensus sequence

4. effector

5. negative control

6. cis-active

7. trans-active

8. autogenous control

9. lysogeny

10. regulon

11. operon, coordinate control, polycistronic, coupled transcription-translation

12. sigma (σ)

13. stringent response

14. attenuation

15. rho

16. repressor

17. catabolite repression

18. positive control

19. constitutive synthesis

20. cI and Cro

21. antiterminators

22. phosphorylation

23. leader sequence

24. dimeric, two

25. a d

26. a c

27. b d e

28. a b d

29. a d e

30. b d e

31. a b c e

32. a b c

33. b c e

34a. N, CR

34b. N, P

34c. N, P, CR

34d. N, CR

35a. *cI, ara, arg*

35b. *gal, ara, arg*

35c. *cI, ara, hut*

35d. *gal*

36a. AN

36b. N

36c. RI

36d. RD

36e. RD, AN

36f. A

36g. AN

37a. I

37b. I, C

37c. N

36d. C

38. Constitutive synthesis of a regulatory protein means that the protein is synthesized at a low level in an unreg-

ulated fashion at all times. Constitutive synthesis as the result of an O^c mutation means that the operator has lost the ability to bind repressor. Thus synthesis of the proteins coded by the structural genes controlled by the operator is not regulated in the usual way, and the mRNA and proteins are produced.

39. Partial diploids that are I^-Z^+/I^+Z^- require inducer for the expression of wild-type β-galactosidase. Thus the repressor coded by one chromosome can act on the *lac* operator present on the other; that is, *lacI* is trans-active. Evidence that O is cis-active comes from study of expression of a structural gene on the same chromosome as an O^c mutation, which causes constitutive synthesis of the *lac* operon in a bacterium of genotype $O^cI^-Z^+$. A partial diploid that is $O^cI^-Z^+/O^+I^+Z^+$ is also constitutive in the absence of inducer. Thus it has the same phenotype as the haploid $O^cI^-Z^+$, indicating that O^c can act only on genes located on the same chromosome. See Table 7-2 in MCB (p. 241).

40. A mutation in the *I* gene that eliminates the ability of repressor to bind to the operator would result in constitutive synthesis of the *lac* operon. A mutation in the *I* gene that eliminates the ability of the repressor to interact with inducer would cause repressor to bind "permanently" to the operator and prevent induction by lactose.

41a. For the *ara* operon, the formation of loops in double-stranded DNA is controlled by the absence or presence of arabinose, which in turn affects the conformation of a dimer of AraC. In the presence of arabinose, a dimer of the AraC protein binds to two sites, $araI_1$ and $araI_2$, and transcription of the *BAD* genes occurs. In addition, a loop forms between dimers bound at two other sites, O_1 and O_2, upstream of the *araI* sites. In the absence of arabinose, a different conformation of the AraC dimer binds to $araI_2$ and to both O_1 and O_2. These dimers interact to form a larger loop of about 190 base pairs between O_2 and $araI_1$. In this loop, AraC does not interact with $araI_1$, and transcription does not occur. See Figure 7-16 in MCB (p. 243).

41b. Loop formation in the leader of the *trp* operon occurs in single-stranded RNA by hydrogen bonding between complementary sequences. Two alternative stem-loop structures control whether transcription continues past nucleotide 140 in the mRNA. In the presence of tryptophan, base pairing occurs and forms two stem-loops between regions 1 and 2 and between 3 and 4, and almost no synthesis occurs past nucleotide 140. When

tryptophan is scarce, an alternative stem-loop forms between regions 2 and 3, allowing the entire operon to be transcribed. See Figure 7-25 in MCB (p. 252).

42. For some operons (e.g., *lac*), the repressor binds to the operator and prevents transcription in the absence of effector. For other operons (e.g., *trp*) the effector binds with repressor and this complex interacts with the operator to inhibit transcription.

43. Since one repressor inhibits transcription of all the genes in a regulon, mutations in the *arg* repressor gene would affect all the *arg* structural genes equally. An O^c mutation in the operator for the first gene would result in constitutive synthesis of this protein and regulated synthesis of the other proteins in the arginine pathway. Transcription of the first *arg* gene could be prevented by a mutation in its operator that caused the repressor to bind so strongly that it could not be removed by arginine or by a mutation in its promoter; in addition, a mutation in the first structural gene could prevent formation of a functional protein. In these latter two cases, the activity of the first enzyme would be low or absent, while that of the other enzymes could be normal, thus mimicking feedback inhibition. However, even if the other enzymes in the *arg* regulon were wild type, the series of reactions controlled by this regulon would not take place because the product of the first reaction would be absent. This effect would not be reversible by low arginine.

44. The interaction of specific sigma factors with RNA polymerase is necessary for initiation of transcription at a promoter. Rho-dependent termination, mediated by the hexameric rho factor, can be overcome by an elongation control particle, comprised of several proteins one of which (NusA) binds to RNA polymerase after the sigma factor is lost. See Figure 7-27 in MCB (p. 253).

45. A plausible conjecture is that the conserved sequences function to bind the precursor nucleoside triphosphates used in the polymerization reactions and that the divergent regions recognize different promoter sequences.

46. The establishment of lysogeny is dependent on the synthesis of sufficient repressor from P_E (also called P_{RE} for repressor establishment) under the influence of the cII protein. The sequence of events leading to lysogeny begins with early transcription from P_R and P_L. A product of P_L is the N protein, which acts as an antiterminator to permit transcription of the *cIII* gene. cIII protein

inhibits the activity of Hfl, an *E. coli* protein that normally depresses the *cII* gene. This inhibition increases cII activity, which in turn increases cI (repressor) synthesis from P_E. cI eventually regulates its own synthesis from P_{RM} once the lysogenic state is established. It accomplishes this by cI dimers occupying the O_R1 and O_R2 sites of P_R, leaving the similar OR_3 site (which together with a portion of O_R2 define P_{RM}) available for RNA polymrase to transcribe the *cI* gene.

Induction of a prophage involves activation of a protease that cleaves the cI protein, so that the dimers cannot form. When O_R1 and $O_R 2$ are not occupied by cI, transcription occurs from P_R, which comprises O_R1 and part of O_R2. This transcription leads to the synthesis of Cro and eventually to the synthesis of phage-coded proteins. See MCB, pp. 255–256.

47. For a tryptophan prototroph, which is capable of normal synthesis of tryptophan, depletion of the amino acid causes release of repressor from the *trp* operator, so that transcription of the operon can occur. Depletion of tryptophan for an auxotroph, which requires the amino acid for growth, causes a much wider, "global" response, called the stringent response. Protein synthesis slows and two novel phosphorylated compounds, ppGpp and pppGpp, are formed by an enzyme called RelA working in conjunction with both a ribosome that cannot continue protein synthesis and an uncharged tRNA. These polyphosphates inhibit the synthesis of rRNA, tRNA, and some, but not all, mRNAs.

48a. If chain growth were in the $5' \rightarrow 3'$ direction, the addition of a large excess of cold ATP would prevent by dilution any further incorporation of $^{32}P_\gamma$ resulting from initiation of new chains, and label would be retained in preexisting chains at their $5'$ end. Hence continued chain growth would have no effect on previous incorporation.

If chain growth were in the $3' \rightarrow 5'$ direction, the addition of a large excess of cold ATP would prevent by dilution any further incorporation of $^{32}P_\gamma$, and previously incorporated label on the $5'$ end of the growing chains would be released as $^{32}P_i$. Hence continued chain growth would depress the level of incorporation to zero.

48b. If chain growth were in the $5' \rightarrow 3'$ direction, ^{32}P incorporation would occur whenever an α-labeled $3'$-deoxyadenosine triphosphate was added to the $3'$ end of a growing chain. However, this addition would cause cessation of further chain growth because no free $3'$-hydroxyl group would be available to react with the next triphosphate.

If chain growth were in the $3' \rightarrow 5'$ direction, no ^{32}P incorporation would occur because $3'$-deoxyadenosine triphosphate lacks a free $3'$-hydroxyl to react with the nascent chain.

49. Because transcription and translation are coupled in bacteria, all mRNA transcription units would be expected to have ribosomes associated with the nascent chains, but rRNA and tRNA transcription units would not. Transfer RNA transcription units would have only very short transcripts, whereas ribosomal RNA transcription units would have long RNAs corresponding to full-length precursors together with shorter RNAs corresponding to incomplete transcripts arranged in an arrowhead type pattern.

50. The addition of glucose would strongly depress the rate of β-galactosidase synthesis because glucose caused catabolite repression of operons encoding sugar-metabolizing enzymes. Since the culture continues to grow, the amount of preexisting β-galactosidase per cell would decrease exponentially with time. The total amount of enzyme in the culture would stay fairly constant, as the enzyme is metabolically stable.

51. Sine the *I* gene codes for repressor and the *Z* gene codes for β-galactosidase, I^-Z^- cells are incapable of synthesizing either repressor or enzyme; these cells do, however, have an intact operator to which repressor can bind. As the donor I^+Z^+ genome entered the I^-Z^- cells, repressor released from the donor operators bound to free operator in recipient cells, allowing synthesis of β-galactosidase to proceed. Gradually, however, new repressor was synthesized, leading to rerepression of the Z^+ gene unless inducer was added.

This explanation implies that repressor is diffusible and limiting in amount with respect to regulation of the *lac* operon. Moreover, it implies that repressor dissociates from the donor operator at sufficient rate to support the observed results.

52. If binding of radioactive CAP to DNA is inhibited by *lac* repressor (or vice versa), you can conclude that their binding sites overlap. Alternatively, the nucleotide sequence protected by each protein can be determined in footprinting experiments, and the sequences compared. This approach would provide more information than binding studies about the extent of overlap between the binding sites.

53. The *arg* regulon consists of a series of paired operator regions and structural genes scattered about the *E. coli* chromosome. All the structural genes have an operator that is recognized by the same repressor. Thus mutations in the repressor can result in the entire regulon being constitutive, similar to the *lac* operon. A mutation in the operator for one of the *arg* structural genes can result in that gene being constitutive, while the other structural genes in the regulon are still subject to repressor regulation, in striking contrast to the *lac* operon.

54. Promoter mutations affect the rate of transcript initiation for the *trp* operon but not the responsiveness of the operon to tryptophan levels. Attentuator mutations affect the incidence of transcript completion under varying tryptophan conditions but have no effect on the rate of transcript initiation. The biochemical phenotype of these mutations can be distinguished by measurements of *trp* leader mRNA versus downstream *trp* RNA for the *trp* operon. Promoter mutants will show differences in the amount of leader RNA relative to that in wild-type or attentuator mutants. Attentuator mutants will show differences in the ratio of full-length RNA to leader RNA relative to that in wild-type or promoter mutants. For any given mutation, the effect may be either "up" or "down."

55a. Comparison of mutant 3 with the wild type indicates that deletion of *araC* gene leads to increased basal expression of *araBAD*, indicating that AraC protein in the absence of inducer functions as a repressor. However, AraC protein also is needed for maximal expression of *araBAD* in the presence of inducer. Therefore, AraC protein acts as both a positive and negative regulator of *araBAD* expression.

55b. The data for mutants 1 and 4 show that constitutive mutations in either the O or I region, neither of which encodes protein, affects the basal level of *araBAD* expression, but not the induced level. This suggests that AraC protein binds to two sites and that binding to both sites in the absence of inducer is necessary for maximal repression. As the O and I regions are separated by 250 base pairs, binding of AraC to both sites likely requires DNA looping. See Figure 7-16 in MCB (p. 243).

56. Lysogenic *E. coli* contain integrated bacteriophage lambda DNA. The integration of this DNA is maintained by the continuous production of lambda repressor, a diffusible protein. Infection by homologous bacteriophage is repressed because the lambda repressor binds to the homologous operator.

57. Cyclic AMP serves as a hunger signal in bacteria and binds to CAP protein to assist in the rapid transcription of sugar-metabolizing operons. An *E. coli* mutant with elevated AMP phosphodiesterase should have low levels of cAMP, and hence transcription of sugar-metabolizing operons should be low. You argue that this is so predictable that it is not worth doing the experiments.

58. Temperature-sensitive mutations must be in protein. Significant changes in protein structure occur over the small temperature range typical of temperature-sensitive mutants. Nucleic acid structure is more stable and is only affected at temperatures that lead to cell death. Likewise, nonsense suppressors only affect proteins. Nonsense suppressors, during protein synthesis, insert at the position of a nonsense mutation an amino acid. This leads to the synthesis of a full-length protein.

59. Mutagenic agents, by causing nucleotide changes, can affect lambda repressor expression or binding and hence lead to lytic production of lambda. Since most carcinogens are mutagens, they would also lead to lytic expression of lambda. Because *E. coli* is cheap to grow, this system might provide a cost-effective assay for many carcinogens.

60. Since the N gene product is an antiterminator for transcription, a defect in N protein would lead to short RNA transcripts. Hence, important gene products would be expressed at a very low rate. Likely, this would lead to a very reduced rate of either lysogeny or lytic viral production.

61. The level of ppGpp should build up. The RelA protein will be produced in the merozygote and lead to ppGpp formation. The $relA^-$ mutation is recessive.

62a. The *araBAD* expression is from the plasmid-introduced genes, as the host *E. coli* is negative for *araBAD*.

62b. The data show that constructs in which the spacing between O_2 and I_1 is unaltered (i.e., O, same as the normal) or is altered by multiples of 10 bp have a low level of expression in the absence of inducer. Constructs in which the spacing is shifted by about 5 bp have a high level of expression; that is, they act like an operator constitutive mutant.

62c. The plasmid *araBAD* genes are subject to cis control through binding of AraC, the repressor, to both O_2 and I_1. Those constructs that exhibit low expression in the absence of inducer have insertions equivalent to one or more helical turns of the DNA, whereas those exhibiting high expression have insertions equivalent to half-helical turns. The likely explanation of the observed results is that half-helical alterations in the O_2-I_1 spacing places one of these sites on the wrong face of the helix, thus preventing binding of AraC protein to both sites. As a result, expression is relieved. The selective binding of protein to one face of DNA, as proposed here, is thought to occur with RNA polymerase binding in the promoter region.

62d. A −11 construct (one helical turn) should be normally repressed, as should a +21 construct (two helical turns).

63a. Addition of coat protein selectively depresses translation of the replicase cistron, as evidenced by the decrease in incorporation of radioactive amino acids into replicase.

63b. Coat protein acts as if it is a translational repressor of replicase production. If so, saturable binding of coat protein of f2 mRNA should be demonstrable. If the binding site is at the beginning of the replicase cistron, as is likely, sequence alterations here ought to affect binding of coat protein and thus its effect on replicase synthesis.

63c. Synthesis of replicase is lower when the N-terminal portion of coat protein is not translated (mutant 1) than when it is (mutant 2).

63d. In analogy with attenuation, the coat nonsense mutation effect on replicase synthesis suggests that ribosome movement is necessary for the unfolding of the mRNA that allows access to the replicase cistron by other ribosomes. Although all three cistrons are present in equimolar amounts, unequal translation could result from the unequal access of ribosomes to each cistron and coat-protein repression of replicase production. Unequal access to each cistron could be caused by RNA folding in a way that buries ribosome binding sites differentially.

The simplest model is one in which RNA folding controls access of ribosomes to the cistrons. Movement of ribosomes down the coat cistron would "promote" ribosome access to the replicase cistron. Production of coat protein would lead to binding of coat protein to the beginning of the replicase cistron and repress further translation of replicase.

64a. Incorporation of label from [$^{32}P_\gamma$]ATP occurs only at the 5′ end of RNA as a result of chain initiation. Since incorporation of $^{32}P_\gamma$ ceased quickly, but $^{32}P_\alpha$ incorporation continued, there probably was little, if any, reinitiation by RNA polymerase. If reinitiation had occurred, $^{32}P_\gamma$ incorporation would have continued for a longer time.

64b. As negative supercoiling increases, the DNA migrates faster in the gel, indicating that it has become more compact and hence occupies less spatial volume.

As indicated in Figure 7-5, active binding of one RNA polymerase per DNA molecule causes melting of about 1.6 helical turns in the DNA. This corresponds to about 17 bp (1.6 helical turns $\times$ 10.5 bp/turn).

65a. Inspection of the gel pattern reveals that LexA protein after incubation in the presence of ATP and RecA protein migrates as two, lower-molecular-weight bands. This indicates that RecA is an ATP-dependent protease. Because the fragments of LexA resulting from proteolysis cannot be reassembled to form active LexA, new protein synthesis is required for repression to be reestablished. Thus chloramphenicol, an inhibitor of protein synthesis, would block rerepression.

65b. The *lexA3* mutation should be dominant. Inspection of the gel pattern indicates that the mutation results in a protease-insensitive form of LexA protein. Hence, an active LexA repressor protein would always be present.

65c. Because RecA protein should always cleave LexA protein, the repressor, any suggested mechanism of rerepression must lead to decreased or inhibited/inactivated RecA protein. For example, a LexA cleavage fragment might act as a repressor for the *recA* gene, leading to depressed production of RecA protein. Also, phosphorylation and dephosphorylation of RecA protein might regulate its activity and hence permit new expression of LexA protein. Other mechanisms are possible.

66a. The amino acids on the outside of the α helix in the recognition sequence should be important for DNA recognition and binding. In Figure 7-7, these are the amino acids at the positions labeled 1, 2, 5, 6, and 9.

66b. Because amino acids on the inside face of the α-helical recognition sequences are unlikely to play a role in DNA interaction, they should be left unchanged. Comparison of the amino acids on the outside face of the recognition helix in the P22 and 434 repressor reveals that they differ in four of the five outside amino acids (those at positions 1, 2, 5, and 9 in Figure 7-7). By site-directed mutagenesis of a plasmid coding for the 434 repressor, the 434 repressor coding sequence can be specifically

altered at these four amino acids to match the P22 amino acid sequence; the sequence modifications can be verifed by DNA sequencing. The 434R repressor protein can then be produced by introducing the redesigned plasmid into a bacterium.

The binding of 434R repressor to the P22 operator can be assessed in vitro by protein-DNA binding assays and by footprinting experiments. Introduction of the redesigned plasmid into a bacterium containing the P22 operon with a defective repressor would lead to normal control of the operon if the 434R repressor binds to the P22 operator in vivo.

C H A P T E R 8

1. histones, chromatin

2. histones

3. nucleolus

4. 5S RNA, 5.8S RNA, 18S RNA, and 28S RNA; 45S RNA

5. poly A tail

6. thymine-thymine dimers

7. spacer RNA

8. TATA box

9. self-splicing

10. enhancers

11. 32S RNA

12. splicing, exons, introns

13. 45S RNA; mRNA; 5S RNA, tRNA, and snRNA

14. nuclear pores

15. leader

16. methylated G cap

17. AAUAAA

18. GU, AG

19. branch point, lariat

20. 5′ intron-exon junction, branch point

21. a c d
22. a b c d
23. b e
24. b d e
25. d
26. a b e
27. a b c d e
28. b e
29. b c e
30. a c e

31. f j
32. f h
33. a c e f h i
34. f g h i
35. g h i
36. j
37. a e i
38. e
39. a e i

40. a. — 1; b. — 2; c. — 4; d. — 3
41. a b c d
42. b c d
43. b c
44. a b c d
45. a d

46. Simple transcription units have one initiation point and one poly A signal. Their primary transcripts contain only one contiguous protein-coding region or only one intron and yield only one mRNA after processing. Although complex transcription units also have a single initiation point, they have multiple poly A sites and/or several introns. The single primary transcript from such units can form multiple mRNAs, encoding different proteins, by selection of alternative poly A sites and/or splice sites, which signal the excision of different introns. See Figure 8-39 in MCB (p. 291).

47. In pulse-chase analysis, the *size* of the species labeled in the pulse is detected by gradient or electrophoretic separation. See Figure 8-4 in MCB (p. 265). In nascent-chain analysis, the *location* of the labeled species on the template is detected by hybridization with specific restriction endonuclease fragments. See Figure 8-9 in MCB (p. 268).

48. The first modification after initiation of transcription is the addition of the methylated G cap, since these caps have been found on incomplete primary transcripts only 20 or 30 nucleotides in length. The poly A tail is added to the 3′ end and therefore cannot be added until the primary transcript has been completed. Full-length

transcripts containing both caps and tails can be found in the nucleus, but cytoplasmic mRNA lacks introns. Therefore, in at least some cases, splicing is the last process to occur.

49. snRNPS are ribonucleoprotein particles containing small RNAs, called U RNAs, (which are unrelated to the ribosomal RNAs) complexed with specific proteins. hnRNPs contain hnRNA complexed with specific proteins, which differ from those of snRNPs. A spliceosome is formed by the joining of hnRNPs and snRNPs.

50. Eukaryotic genes do not form functional polypeptides in bacteria because bacteria are not capable of carrying out splicing reactions. If expression of functional eukaryotic genes in bacteria is desired, the corresponding cDNA rather than genomic DNA must be cloned. Since cDNA is a copy of mRNA in DNA form, the introns are not present and splicing is not required for expression.

51a. TATA boxes are centered at about −30 and promoter-proximal sequences at −60 to −100 relative to the start site; enhancers can be up to several kilobases away from the genes they control. Enhancers, unlike the other two controlling elements, can be oriented in either direction with respect to the gene and still function.

51b. Although none of these control elements is obligatory for initiation of transcription of all genes, they affect the rate of transcription of some genes. Genes transcribed at high rates contain a TATA box, but genes transcribed at low rates often do not. Transcription of genes with TATA boxes may be decreased if their promoter-proximal or enhancer sequences are mutated.

52. The basic mechanism for ensuring one copy each of 5.8S, 18S, and 28S RNA per ribosome is inherent in the structure of the 45S pre-rRNA molecule, in which one copy of each of the mature molecules is encoded. Processing of one pre-RNA molecule thus produces one molecule of each of the structural RNAs of the ribosome. 5S RNA is not made in equimolar amounts to the other RNAs, but is made in excess to assure that under no circumstances will there be a shortage of this component. 5S RNA not included in ribosomes is degraded.

53. 5 S-RNA genes do not have a promoter similar to those present in bacterial genes. The recognition sequence for initiation of transcription of 5S-RNA genes is not upstream of the site at which transcription begins, as it is in bacteria, but lies within the gene. Unlike bacterial pro-

moters, this internal recognition sequence is transcribed into the 5S RNA. Also, the site at which bacterial RNA polymerase binds overlaps the site for binding of controlling factors, whereas polymerase III binds only to the most upstream regions of 5S-RNA genes. However, transcription of both bacterial genes and 5S-RNA genes are controlled by binding of protein to DNA (i.e., repressors and TF_{III} transcription factors).

54. Processing of introns in pre-tRNA involves excision of the intron by endonuclease activity and rejoining of the ends of the exons. Two ATP molecules are cleaved during this reaction. As an intermediate, a 2′,3′-cyclic monophosphate is formed at the downstream side of the 5′ exon. The phosphate in the phosphodiester bond that rejoins the cut ends is derived from ATP. Processing of pre-tRNA does not involve snRNAs. See Figure 8-57 in MCB (p. 305).

Processing of introns in pre-mRNA requires the participation of snRNPs, each of which contains a small U RNA and several proteins, some of which are present in all snRNPs and some of which are unique to each snRNP. A cut at the 5′ intron-exon border is made first, and the free end of the intron is joined to the 2′ hydroxyl of a nucleotide (the branch point) upstream of the 3′ border. (This nucleotide also participates in two normal phosphodiester bonds.) Then a cut is made at the 3′ border to excise the circularized intron. The phosphate used for rejoining the ends of the exons is derived from the 5′ nucleotide at the upstream side of the 3′ exon. ATP may be required for this splicing and for binding of snRNPs to each other or to form the spliceosome. The U1 snRNP interacts with the 5′ border, the U2 snRNP with the branch point, and the U5 snRNP with the 3′ border. The details of the splicing reaction and the function and the time of action of the individual snRNPs are not completely understood. See Figure 8-49 (p. 299) and Figure 8-53 (p. 302) in MCB.

55. The DNA-coded features that remain in functional mRNA are the codons, which define the amino acid sequence, the cap site, and the poly A signal. The methylated G cap and the poly A tail are not encoded. The GU-AG splice signals are encoded in DNA but are not found in mRNA because they are within the intron and are excised.

56. Methylation of A residues in regions of hnRNA destined to be exons and in regions of pre-rRNA that remain in mature rRNA may protect these regions from degradation during processing. For pre-rRNA, many of these methyl groups are added while the chain is being synthesized. Lack of methylation prevents normal processing of primary transcripts.

57. Transcription by both types of polymerase is terminated at sequences downstream of the 3′ terminus found in the mature RNA molecule. For protein-coding genes, transcribed by polymerase II, signals at the site of termination have not been identified, although the poly A signal is required for termination. Loss of antitermination factors may contribute to termination.

For rRNA genes, transcribed by polymerase I, the enzyme continues 2–10 kb past the end of the 28S-RNA genes. Since RNA genes are tandemly arranged, additional mechanisms may allow polymerase I to move from one pre-rRNA gene to the next one in the series. This polymerase may transcribe virtually all of the region between the genes; it is hypothesized that, in some organisms, a loop is formed in the untranscribed spacer, so that the polymerase is positioned close to the initiation site of the next downstream gene.

58. Deletion analysis can identify the DNA sequences within or around a gene that affect the rate of transcription if all other necessary factors are supplied and an assay for the expression of the gene is available. Footprinting analysis of a DNA-binding protein can identify the specific region(s) of DNA to which the protein binds (thus protecting it from chemical or enzymatic degradation) but gives no information about the function of the protein.

59a. Almost all grains will be over the nucleus, following a brief (5-min) uridine pulse, because almost all RNA synthesis in eukaryotic cells occurs in the nucleus.

59b. Almost all grains will be over the nucleus, following a brief (5-min) thymidine pulse, because almost all DNA synthesis in eukaryotic cells occurs in the nucleus.

59c. Most grains will be over the cytoplasm, following a lengthy (2-h) labeling with uridine, because RNA accumulates in the cytoplasm. However, some grains will be over the nucleus because RNA synthesis continues throughout the labeling period; these nuclear grains represent newly formed RNA that has not yet moved to the cytoplasm.

59d. Almost all grains will be over the nucleus, following a lengthy (2-h) labeling with thymidine, because almost all DNA accumulation, as well as synthesis, occurs in the nucleus.

59e. Almost all grains will be over the cytoplasm, with almost none over the nucleus, because nearly all the labeled nuclear RNA formed during and after a short uridine pulse moves to the cytoplasm during a 2-h chase.

59f. Almost all the grains will be over the nucleus because almost all the labeled DNA formed during and after a short thymidine pulse remains in the nucleus during a 2-h chase.

60. The radiolabeled RNA in a pulse-chase experiment is processed over time. This processing produces a change in the sedimentation profile with a general decrease in size being observed. Of course, since some RNA species (e.g., rRNA) are more stable than others, there will also be a pronounced accumulation of radioactivity in certain major RNA species. Because the UV-absorption profile reflects the bulk, steady-state distribution of RNA among various RNA species, this profile is constant with time. See Figure 8-4 in MCB (p. 265).

61a. Let X = numbers of bases in 45S pre-rRNA and set up the following ratio: $45/28 = X^{1/2}/5100^{1/2}$. Solving for X gives 13.2 kb.

61b. The percentage of 45S pre-rRNA in 5.8S RNA = 1.2 percent; in 18S RNA = 14 percent; and in 28S RNA = 39 percent. These values are calculated by dividing the number of bases in each species by the number of bases in pre-rRNA and multiplying the result by 100.

61c. The metabolically stable portion of pre-rRNA forms 5.8S, 18S, and 28S rRNA. Thus you can calculate the percentage that is metabolically stable from the ratio of the total number of bases in the three smaller rRNAs to the number of bases in pre-rRNA:

$$\frac{0.16 \text{ kb} + 1.9 + 5.1 \text{ kb}}{13.2 \text{ kb}} = 0.54, \text{ or } 54 \text{ percent}$$

The metabolically unstable proportion is calculated by difference: $100 - \text{percentage stable} = 100 - 54 = 46$ percent.

62a. Immediately after a 5-min pulse, label will be in a series of RNA species, including many hnRNA species. Hence the sedimentation pattern will appear as a broad "smear," which includes many high-molecular-weight species. After a 15-min actinomycin D chase, most of the label will be in 45S pre-rRNA; after a 1-h actinomycin D chase, most of the label will be in 18S and 28S rRNA with some label in 4S tRNA. During the actinomycin D chase, no new RNA synthesis occurs; hence new molecules of 45S and 32S pre-rRNA are not formed during the later portion of the chase. See Figure 8-4 in MCB (p. 265) for comparison; note that the patterns in MCB are for continuous labeling periods.

The total amount of incorporated label will decrease greatly during the actinomycin D chase, as no new RNA is synthesized and excised introns and spacer RNA are rapidly degraded.

62b. In the beginning, the base composition is DNA-like because the predominant labeled species, hnRNA, consists of a representative sample of the DNA sequences. At later time points, the predominant labeled species is ribosomal RNA, which consists of a small subset of DNA sequences. Although cells make many hnRNA molecules, hnRNA has a high degradation rate and hence rapidly disappears from cells. Ribosomal RNA has a lower rate of synthesis, but its much greater metabolic stability allows it to stay in cells much longer.

63. UV irradiation causes the random formation of thymine dimers, which prevent further elongation of the transcript. For a set of RNAs encoded by a single transcription unit, the most-upstream species (i.e., closest to the initiation site for transcription) will be least affected by UV damage and the most-downstream species will be most affected. Hence 18S rRNA, which is least sensitive, maps most upstream and 28S, which is more sensitive, maps downstream to the 18S. 45S pre-rRNA whose formation requires transcription of almost the entire transcription unit, is most sensitive.

64a. The B factor binds to DNA; hence its name, B for *b*inding. The affinity of B for binding to DNA is increased by association of B with the S factor.

64b. The B factor must have a structural domain that interacts with DNA and a second structural domain that interacts with S. Depending on whether B interacts directly with RNA polymerase I, it may have a third structural domain that interacts with polymerase I. The S factor must have a structural domain that interacts with B factor. The S factor may also have a structural domain that interacts with polymerase I. Hence B factor has a minimum of two or three structural domains, and S factor has a minimum of one or two structural domains.

64c. The B and S factors are species-specific. For example, mouse B recognizes only mouse ribosomal gene promoter sequences and mouse B binds only to mouse S. Hence the DNA-binding domain of B and the B-S inter-

action domains are species-specific. However, during in vitro transcription of mouse rDNA, human RNA polymerase I can substitute for the mouse enzyme. Thus the polymerase-binding domains in B and S are nonspecific.

65. The difference in the amounts of 32S RNA, a precursor to 28S rRNA, and 45S RNA, the precursor to both 18S and 28S rRNA, in the nucleus can be explained on the basis of differential rates of various steps in rRNA precursor processing. The processing of 45S pre-rRNA to generate 32S precursors and 18S rRNA is rapid, but the processing of 32S precursors to 28S rRNA is slow. Therefore, there is a relative accumulation of 32S precursors over 45S precursors, even though the 45S precursor includes both the 18S and 28S rRNAs. See Figure 8-15 in MCB (p. 274).

66a. The data show that roughly 100 bp upstream from the RNA start site can be deleted with no obvious effect on the amount of 5S rRNA or the size of the molecule. Thus upstream sequences seem to be of little importance for the production of 5S RNA.

66b. Deletions of the region between ~+50 and +85 eliminate 5S transcription. Therefore the location of the 5S gene control region must be between +50 and +85.

66c. The deletion analysis suggests that no sequence, other than the internal control region, is very important in determining the production or length of 5S rRNA. Length is probably dictated by the molecular dimensions of RNA polymerase III and the transcription factors that bind to the control region. Thus shifting the control region 60 bp upstream or downstream should lead to the production of a 5S-rRNA product in roughly normal amounts. The nucleotide sequence of this product would, of course, not be normal.

67. Although 5S-rRNA and tRNA genes are both transcribed by RNA polymerase III, 5S genes require $TF_{III}A$, B, and C, while tRNA genes require only $TF_{III}B$ and C. For the 5S genes, $T_{III}A$ and C are DNA-binding factors; for the tRNA genes, $TF_{III}C$ is the DNA-binding protein. $TF_{III}B$ does not bind to DNA. Hence $TF_{III}A$ and C should protect DNA in footprinting experiments, whereas $TF_{III}B$ should not be. See Figure 8-24 in MCB (p. 281).

68a. When mRNA and DNA are hybridized, complementary RNA displaces the noncomplementary strand of a double-stranded DNA molecule to generate a RNA-

DNA hybrid of double-stranded thickness and a displaced single-stranded DNA molecule. The occurrence of an intervening sequence within genomic DNA will result in a double-stranded DNA bubble — the R loop — containing no sequences complementary to the RNA during hybridization. This double-stranded DNA bubble must be of a certain size to be seen. Short introns are not large enough to be seen as a bubble and hence are not detected by the R-loop technique but are detected by direct sequence comparison.

68b. DNA can be cloned readily and cut specifically into defined fragments by restriction endonucleases, but RNA cannot. Therefore, mRNA must be copied into cDNA for both amplification purposes (i.e., cloning) and fragmentation purposes. DNA is sequenced a few hundred nucleotides at a time. Messenger RNA is not used directly because of these problems.

69. You must supply programmers with enough information to develop a set of algorithms capable of recognizing a RNA polymerase II transcription unit. The problem can be simplified by initially considering only the DNA strand that is transcribed and by specifying sequence features of this strand — upstream and downstream from the start site — that are present in some or all protein-coding genes.

   Upstream from the transcription unit may be TATA box, which would lead to the conclusion that a transcription unit was likely to start ~ 30 bases downstream. This is true of high – transcription rate RNA polymerase II genes. In other cases, there might be a GC-rich region (CCAAT or GGGCG), which would suggest a transcription unit 100–200 bases downstream. These sequences are common in low – transcription rate genes. The algorithm should include subroutines to recognize these features first as nucleotide clusters and then to ask if clusters are spaced close to a protein-coding region. Because promoter features are not universal, they can be a helpful, but no sufficient, trait in the recognition of a RNA polymerase II transcription unit. Furthermore, additional sequence features downstream from the start site must be specified in order to identify protein-coding regions.

   Putative protein-coding regions (open reading frames) can be recognized as continuous long stretches of anticodons devoid of termination anticodons. The algorithm must include provisions for recognizing anticodons in any one of the three possible reading frames. As these open reading frames may be interrupted by introns, the alogrithm must be capable of recognizing introns and intron boundaries. Because introns typically are not protein coding, they usually do not contain many anticodons. There should be a CA at the beginning of an intron segment and a TC at the end. There should be a CATTCA consensus sequence towards the beginning of the intron segment; this sequence generates in RNA the U1-binding site (GUAAGU). Introns should also have sequences corresponding to U2- and U5-binding sites; these sequences, however, are poorly conserved. Dinucleotide sequences may be too short to be a useful identification feature.

   Protein-coding genes also include within their transcription units a downstream sequence (TTATTT) that codes for the poly A addition signal (AAUAAA) in primary transcripts. The algorithm must include a subroutine for the recognition of this sequence.

   Because any DNA sequence isolated from the human genome may be from either of the two strands, the algorithm must also include subroutines for recognition of the complementary sequence features. The actual strand sequenced may not be the strand that is transcribed. In summary, a computer program for recognition of RNA polymerase II transcription units (i.e., protein-coding genes) from sequence data must specify a composite of several sequence features. The definition of transcription units from sequence analysis, however, will be imperfect because the sequence features are not universal and because of the existence of simple and complex transcription units.

70a. The CCAAT sequence is a feature of RNA polymerase II transcription units that do not contain TATA boxes. Because such transcription units are transcribed poorly in in vitro systems, the in vitro system is not suitable. Probably the mammalian cell system is better than the *Xenopus* system. The factors needed for RNA polymerase II transcription are complicated and not fully understood. Therefore, results from placing the transcription unit in a homologous mammalian system are apt to best reproduce the physiological situation.

70b. For the mammalian cell system, the plasmid is most easily introduced by transfection into the cell culture. Microinjection could also be done, but this would produce only a limited number of cells containing the plasmid. For the *Xenopus* system, microinjection is the best method because the egg is large and cannot be readily transfected.

70c. The actual data collected would be the amount of transcription product, namely hybridizable RNA, formed in a given time period. By comparing the results from assays of the normal and mutant gene, you could conclude whether the change in the CCAAT sequence had a positive, negative, or no effect on transcription of the thymidine kinase gene.

71. Nucleotides corresponding to the RNA polymerase II termination sequence are not included in cytoplasmic mRNA. Therefore, there would be no effect on the sequence of the corresponding mRNA of DNA sequence alterations near the termination site for the transcription.

72. Two different potential intermediates can be immunoprecipitated by anti-U5 snRNP: (1) a structure in the process of joining two exons together but still containing the intron and (2) a structure that contains the excised intron. The first structure contains U2, U4, U5, and U6 snRNP and is cleaved at the 5' end of the intron. The second structure contains the intron in lariat form and U5 and U6 snRNP. See Figure 8-53 in MCB (p. 302).

73a. Guanosine does not function as an energy source. If it did, only a high-energy phosphate form such as GTP would be effective in self-splicing.

73b. That the reaction is saturable for G (i.e., there is a K_M) and G can be competed against suggest that there must be a folded domain in *Tetrahymena* rRNA capable of specifically binding G. In other words, the rRNA has a cofactor binding site similar to that present in some enzymes.

74. Normally RNA introns have an AG at their 3' end. The G → A substitution results in the generation of a new upstream AG dinucleotide pair, which may function as the 3' end of the intron. If so, splicing would be altered and would be shifted by 19 nucleotides upstream. A 19-nucleotide shift would both insert amino acids into the protein and cause a frame shift for the reading of downstream codons (19/3 = 6.33 amino acids). Since normal β-globin would not be produced if the alternate splice site were used, alternate splicing would lead to a deficiency of β-globin. The observation that some β-globin is produced indicates that some splicing at the original splice site must occur in the patients. The new splice site, however, must be used preferentially, as a large decrease in production of normal β-globin occurs.

75a. For a point mutation to result in a 24-base insertion in the cDNA (i.e., mRNA), the mutation must affect a splice site. Likely, it is in either the 5' GU or 3' AG of an intron. See answer 74.

75b. A single point mutation in the B subunit should affect only one of two alleles in a heterozygote; the other allele should be normal. Because heterozygotes should have

normal expression of one allele, their β-hexosaminidase activity should be at least 50 percent of the normal level. Since all 12 cDNA clones scored had the same mutation, the GM2144 cells are probably homozygous: If the cells were heterozygous, half the mRNA should be normal and half mutant, and 6 of the 12 cDNA clones would be expected to be normal.

75c. Since all 12 of the cDNA clones for B-subunit mRNA were mutant and all exhibited the same mutation, there probably is no cytoplasmic mRNA corresponding to the second B-subunit allele. That is, the second allele, although present, does not support cytoplasmic accumulation of B-subunit mRNA. Remember, if it did, a second set of cDNA clones should have been seen. Any mutation in the gene of the second allele that leads to inability of the allele to be transcribed or results in failure of the transcribed RNA to appear in the cytoplasm as mRNA would explain the observed genotype and phenotype of the GM2144 cells.

76a. As shown in Figure 8-3b, 18S and 28S rRNAs do not compete with each other for hybridization to HeLa cell DNA; therefore, they show no sequence homology. As shown in Figure 8-3c, 32S pre-rRNA does not compete with 18S rRNA in hybridizing to HeLa cell DNA; therefore, they share no nucleotide sequences, and 32S pre-rRNA cannot be a processing intermediate to 18S rRNA. As shown in Figure 8-3d, 32S pre-rRNA does compete with 28S rRNA, suggesting that the 28S rRNA sequence is included within the 32S pre-rRNA; therefore, 32S pre-rRNA can be a precursor to 28S rRNA. The following pathway is consistent with the data in Figure 8-3:

$$45S \text{ pre-rRNA} \rightarrow 32S \text{ pre-rRNA} + 18S \text{ rRNA}$$
$$\downarrow$$
$$28S \text{ RNA}$$

76b. Figure 8-4 shows that at saturation hybridization, there are ~ 5.2×10^{-4} μg 45S pre-rRNA per μg DNA. Knowing the amount of DNA per genome, you can calculate the amount of pre-rRNA/genome:

pre-rRNA/genome = $(5.2 \times 10^{-4}$ μg RNA/μg DNA)
$\times (15 \times 10^{-6}$ μg DNA/genome)
= 7.8×10^{-9} μg

Knowing the molecular weight of 45S pre-rRNA, you can calculate the number of moles of pre-RNA/genome:

moles of pre-RNA/genome = $(7.8 \times 10^{-15}$ g pre-rRNA/genome)
$\times (4.6 \times 10^6$ g/mol pre-rRNA)
= 1.7×10^{-21} mol

Recall that Avogadro's number is the number of molecules (in this case, individual pre-rRNA genes) in 1 mole. Thus,

$$\text{pre-rRNA genes/genome} = (6.023 \times 10^{23} \text{ genes/mol})$$
$$\times (1.7 \times 10^{-21} \text{ mol pre-rRNA/} \text{genome})$$
$$= 10.2 \times 10^{2} \text{ genes/genome}$$

There are about 1000 45S pre-rRNA genes per HeLa cell genome.

77a. The order of the restriction enzyme sites is as follows:

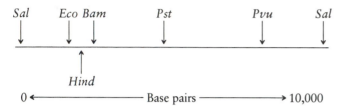

77b. Fragments falling between the *Eco* and *Bam* sites (C and G) are negative for nascent-chain hybridization; therefore the initiation site must be to the right of the *Bam* site. The *Pvu-Sal* fragment (F) is also negative; therefore the transcript does not include the *Pvu-Sal* region. The initiation site has to be somewhere between the *Bam* site and the *Pvu* site. Because the orientation of the transcript with respect to the restriction map is not known, the initiation site could be either towards *Bam* or *Pvu*. By the same reasoning, the termination site must be somewhere between the *Bam* site and the *Pvu* site. Whether it is closer to *Bam* or *Pvu* cannot be determined from the data.

77c. The maximum length of mRNA is 5800 bases, the distance between the *Bam* and *Pvu* sites. This is calculated as the difference between the length of fragment A and B.

77d. Finer mapping of restriction fragments and the use of this finer map in hybridization would improve the determination of mRNA length. Establishment of mRNA orientation relative to the restriction map would allow assignment of the initiation and termination sites relative to left and right on the map.

78a. *Bam* digestion of the DNA results in a 2-kb fragment and a 1-kb fragment. Both fragments are radioactive from the 5′ ^{32}P end-labeling procedure. After strand separation, only the 1-kb radioactive fragment lights up a RNA component in a Northern blot. Thus the 1.0-kb 5′ end-labeled fragment is complementary to a 3′ segment of the mRNA. The mRNA must be oriented from left-to-right on the restriction map. Since the 2-kb end-

labeled fragment does not hybridize, it is a sense rather than an antisense sequence.

78b. Reverse transcriptase will extend the hybridized DNA fragment (primer) from its free 3′ end towards the 5′ end of the mRNA, adding the number of bases equal to the number in the 5′ mRNA sequence missing from the cDNA clone. The 0.2-kb extension indicates that the DNA does not include the first 200 bases at the 5′ end of the mRNA. This procedure of mapping the 5′ end for mRNA is called the primer extension method.

78c. All of the sequences in the single-stranded radiolabeled DNA are included in the mRNA. Therefore, a fully double-stranded hybrid of DNA with mRNA forms and none of the radiolabeled DNA is digested by the S1 nuclease. This results in the 3-kb electrophoretic species. When the single-stranded DNA is hybridized to total-cell RNA, some of the DNA hybridizes to mRNA and the same 3-kb intact DNA species is seen after the S1 digestion. However, some of the DNA also hybridizes to hnRNA (pre-mRNA). If the pre-mRNA has exons and introns, then the introns have no corresponding DNA sequence with which to hybridize, and single-stranded RNA loops and short non-base-paired DNA segments arise. Thus the three bands at 750 bp, 1050 bp, and 1200 bp correspond to three exons. They must be exons because the sequences are found in the cDNA clone.

79a. In this assay, formation of a stable complex between the long gene and any of the transcription factors during the first incubation would prevent expression of the wild-type gene; thus only long rRNA would be observed in the gel. Lanes 3–5, which reveal both products, indicate that no factor alone forms a stable complex. Lanes 2 and 7 suggest that a stable, bound complex of TF$_{III}$A and C is formed when these transcription factors are incubated with 5S-rRNA genes. Formation of this complex does not require RNA polymerase III (lane 9).

79b. If TF$_{III}$A acted catalytically in the formation of a stable complex of TF$_{III}$C with 5S-rRNA genes, then TF$_{III}$A should be available in the incubation mix and would be able to catalyze the formation of a TF$_{III}$C complex with added DNA. No additional A should be needed.

79c. These data indicate that neither TF$_{III}$A or TF$_{III}$C acts catalytically. As complex formation is stable and DNA dependent, the most likely role of the transcription factors is to form a stable, DNA-bound intermediate in the formation of a RNA polymerase III transcription complex.

C H A P T E R 9

1. metaphase, karyotype

2. C value, C-value paradox

3. Barr body   4. centromere

5. heterochromatin, euchromatin

6. nucleosome, solenoid   10. cistron

7. histones   11. complementation

8. telomere   12. lyonization

9. synapsis, chiasma   13. polytenization

14. nuclear scaffold, topoisomerase II

15. b c d e

16. b c d e

17. d e. [Choice (a) is the number of base pairs expected in the genome of a virus that infects eukaryotic cells. Choice (b) is the number of base pairs found in chloroplast DNA. Choice (c) is the number of base pairs found in E. coli DNA.]

18. b c d e   29. i

19. a b d e   30. k

20. a b c d e   31. j

21. b   32a. 6

22. b c d   32b. 8

23. a b c d e   32c. 3

24. a b c e   32d. 1

25. b c d e   32e. 4

26. b c d e g h   32f. 2

27. a b c d e   32g. 5

28. b c d e f

33. Bands in metaphase chromosomes correspond to condensed regions resulting from protein-DNA interactions or to the structures assumed by the repetitive DNA located in these heterochromatic regions. A heterochromatic metaphase band may contain 10^7 base pairs of DNA, but no function has been correlated with the presence of specific bands. The bands in interphase chromosomes result from polytenization, the synthesis of many copies of regions of the chromosome 10^5 base pairs long. These copies are arranged in parallel arrays and correspond to one or a few transcription units.

34. Evidence for the linear arrangement of DNA within a chromosome is derived from recombination experiments, which are interpretable only if linearity is as-

sumed. Also, at the light microscope level, chromosomal regions that display bands retain these morphologic features in the same linear array after tanslocation. These observations, however, do not rule out the possibility that a chromosome is composed of tandemly arranged DNA molecules linked together by protein (or other substance) along the length of the chromosome. The best evidence that there is only one DNA duplex in a chromosome was obtained from yeast and Drosophila. Artificial constructs consisting of a single peice of DNA are capable of behaving as normal chromosomes if they contain specific sequences to provide required functions. In addition, the number of discrete DNA molecules extracted from yeast chromosomes correlates with the number of linkage groups. For the fruit fly, the size of isolated DNA molecules indicates that they contain an amount of DNA equivalent to that measured in situ in the chromosome.

35. Telomeres are replicated using a different mechanism than is used for the rest of the chromosome because of the enzymology of DNA replication. Since DNA replication can proceed only in the $5' \rightarrow 3'$ direction, an RNA primer remains on the lagging strand. Addition of the telomere sequences by telomere tansferase regenerates the telomere, which forms a terminal hairpin. The $3'$ OH at the end of the hairpin then can serve as a "self-primer" so that DNA polymerase can use the $5' \rightarrow 3'$ parental strand as template. See Figure 9-13 (p. 329) and Figure 9-14 (p. 330), in MCB.

36. A major structural difference between transcriptionally active and transcriptionally inactive chromatin is that the DNA-protein and the protein-protein interactions are looser in regions being transcribed. This more "open" structure is shown by the sensitivity of genes undergoing transcription to DNase and the ability of the somatic 5S-rRNA genes in oocytes (which are normally not expressed in this cell type) to be transcribed after H1 histone is removed by salt extraction. The evidence presently available suggests that nucleosomes are probably still present in genes undergoing transcription. However, the interaction between H3 histones in nucleosomes of slime mold rDNA being transcribed is not as close as in the untranscribed spacer region. Methylation is also involved in transcriptional control, since the transcriptionally silent X chromosome is heavily methylated.

37. Complementation can occur between mutations in both the individual gene products of an operon and the individual proteins derived from a polyprotein. In both

cases several proteins are produced, although by different mechanisms — independent initiation of translation in the case of an operon and proteolytic cleavage in the case of a polyprotein.

38. Incomplete complementation would be observed. The diploid can form the product encoded by wild-type gene 4 using the information in the B genome. However, the wild-type product encoded by gene 3 cannot be produced from either genome because the TATA box mutation prevents transcription of this gene in both the A and B genomes.

39. The phrase "one gene, one enzyme" has not stood the test of time, particularly for eukaryotes. The development of more sophisticated techniques for studying the structure of genes and the structure and function of proteins led to the discovery that although some genes code for enzymes, others code for structural proteins without enzymatic activity or for RNA molecules, which are not translated into proteins. These phenomena are true of prokaryotes as well as eukaryotes. In eukaryotes, a region of a genome may code for more than one polypeptide if the codons are included in overlapping genes. That is, two different open reading frames, coding for different proteins, may occur in one region of the genome. In addition, an enzymatic activity (e.g., RNA polymerase) may be coded by more than one gene, and, very recently, experimental evidence has been obtained suggesting that a single polypeptide chain is produced from two mRNAs transcribed from independent genes located on different chromosomes. Finally, some enzymes are active only in a multimeric form containing two or more different polypeptide chains, which are encoded by different genes.

40. If two mutations, one on each chromosome (trans configuration), can complement each other to produce a wild-type phenotype, the two mutations are in different genes. If two mutations introduced in the trans position do not produce a wild-type phenotype, the mutations are in the same gene, since neither chromosome can produce a normal protein (in the absence of recombination). Complementation tests with mutants in which the location of the mutation is known can define the outer limits of a gene.

41. A RFLP can be considered a gene because it is heritable, can occur in the heterozygous state, and has a phenotype (the appearance of a fragment of DNA of a characteristic size upon digestion of donor DNA with a particular restriction endonuclease). Arguments against a RFLP's being considered a gene are that it has no controlling elements and it produces no gene product, either RNA or protein.

42. A straightforward (although not necessarily simple) way to demonstrate that one region of DNA codes for two proteins is to prepare cDNAs from the mRNA produced in the cells. On a Southern blot, two different cDNAs should be detected with a probe prepared from a fragment of the genome containing the coding sequences common to the two proteins. The nucleotide sequence of the cDNAs could be compared to the genomic sequence and some common sequence among the three should be found. Comparison of the cDNA sequences with the genomic sequence will map the introns and exons.

43. Within a given creature, with rare exceptions, the karyotype (i.e., the gross number, shape, and size of chromosomes) of the metaphase cells of all somatic-cell types is identical, as are the G, Q, and R banding patterns. This similarity implies that the overall interaction of DNA with protein is very similar in all cell types within a creature. Because karyotypes are very constant among cell types, chromosome arrangement is unlikely to be a major determinant of cellular differentiation.

   One example of a difference in chromosome size is the polytene chromosomes of *Drosophila* salivary glands. These chromosomes, which are present in interphase cells, can be used to distinguish different stages of the organism's differentiation.

44. In most mammalian females, there are two X chromosomes. One of the two X chromosomes is inactivated and appears as condensed heterochromatin located peripherally in the cell nucleus; this structure is referred to as a Barr body. The presence of a Barr body in a phase-contrast light microscope image of the human cells would be proof that the cell is from a female.

   In rare cases, a female can be XO (i.e., have only one X chromosome). The cells from such an individual do not contain a Barr body. Also, in rare cases, an individual may be an XXY male. The cells from such an individual contain a Barr body. In both these cases, sex assignment based on the presence or absence of Barr bodies would be incorrect. Clearly, the most definitive fetal sex assignment would require karyotyping of amniotic fluid cells.

45. All chromosomes are contained within the nucleus. The diameter of a human cell nucleus is (15 μm cell diameter) $\times$ (0.30 nucleus/cell ratio) = 4.5 μm. Thus each

chromosome must be be folded into segments no longer than 4.5 μm to fit within the nucleus. To calculate the number of folds in a chromosome 10 cm in length, divide this length by the nuclear diameter:

$$\frac{10 \text{ cm}}{4.5 \text{ }\mu m} = \frac{10^{-1} \text{ m}}{4.5 \times 10^{-6} \text{ m}} = 0.22222 \times 10^{+5}$$

Hence, a single human chromosome must be folded approximately 22,222 times to fit within the cell nucleus.

46. Restriction fragment length polymorphisms (RFLPs) are inherited traits. Sufficient heterogeneity of these traits exists within the human population that for each parent a unique set of RFLPs exist. The RFLP pattern for any child will be a composite of that of each of its parents. Hence relatedness can be established by comparison of RFLPs. See Figure 9-7 in MCB (p. 325).

47a. Telomere (TEL) sequences are only required for stable inheritance of linear structures such as chromosomes. Thus a TEL sequence is not required if the gene is introduced as part of a circular DNA molecule.

47b. The terminal sequence of chromosomes is not encoded in the actual chromosomal DNA. Rather it is determined by the enzymes that add the terminal nucleotides. These enzymes, termed telomerases or telomere terminal transferases, are species specific. Thus the observed difference indicates that corn and wheat have different telomerases.

48. After gentle digestion of HeLa cell chromatin, a ladder-like pattern of DNA fragments, spaced about 180 bp apart, should be seen on the gel. During more extensive digestion, trimming of DNA at the ends of the nucleosomes will occur, and the spacing between the fragment bands should decrease to about 140 bp. As digestion becomes extensive, the ladder pattern should gradually collapse to a single band at 140 bp. In the end, all the DNA will be found in 140-bp nucleosome monomers.

49a. The globin gene is sensitive to digestion in the "active" cell type expressing globin but not in the "inactive" cell type in which the globin gene is not transcribed. This difference is apparent from the blot pattern, which shows very little, if any, globin-reactive material (4.6-kb position) in lanes 3–7, but a strong globin band in lane 8.

49b. DNase sensitivity is directly related to the accessibility of the enzyme to the DNA structure. Thus the differential DNase sensitivity of the two cell types implies that the chromatin must be more unfolded, or open, in active, transcribed genes than in inactive, untranscribed genes.

50. Nucleosomes with associated histones occupy about 200 bp of DNA per unit repeat, and each repeat contains nine histone molecules (one H1A and two each of H2A, H2B, H3, and H4). An average human chromosome with 2×10^8 bp thus contains 1×10^6 nucleosomes:

$$\frac{2 \times 10^8 \text{ bp/chromosome}}{2 \times 10^2 \text{ bp/nucleosome}} = 1 \times 10^6 \text{ nucleosomes/chromosome}$$

Multiplying the number of nucleosomes per chromosome by the number of histone molecules per nucleosome gives 9×10^6 histone molecules per chromosome. Since there is 1 DNA molecule per chromosome, there are 9×10^6 histone molecules per DNA molecule. The conversion from number of molecules to moles of molecules requires multiplication of the numerator and denominator by the same constant; hence the molar ratio is equal to the molecular ratio:

$$9.0 \times 10^6 \text{ mol histone/mol DNA}$$

On average, there is one molecule of scaffold protein (topoisomerase II) per 50 kb of DNA in human chromosomes. Thus the molecules of topoisomerase II per human chromosome is 4×10^3:

$$\frac{2 \times 10^8 \text{ bp/chromosome}}{5 \times 10^4 \text{ bp/topoisomerase molecule}}$$
$$= 4 \times 10^3 \text{ topoisomerase II molecules/chromosome}$$

By the same reasoning as above, the molar ratio of topoisomerase II to DNA has the same value:

$$4 \times 10^3 \text{ mol topoisomerase II/mol DNA}$$

51a. Complementation analysis provides the simplest approach to detecting mutations in separate genes. To test for complementation, the eight CHO auxotrophs are fused in a pairwise manner, and the paired fusion products are placed in minimal culture medium. Only those pairs that can complement (i.e., produce adenine) will grow. Fusion of animal cells as a technique is discussed in Chapter 5 of MCB (p. 170).

51b. The distance between the two mutations that appear by complementation to be in the same gene might be mapped by a Benzer-like approach. However, the type of fine mapping done by Benzer requires a large number of individual mutations and only can be done in a reasonable length of time with rapidly growing cells such as bacteria. For this situation, in which the isolation of two mutants in the same gene required one month's work and the cells have a slow generation time (typically 12–15 h), the intragenic recombination appraoch is impractical.

52. Most of the sequence of these two proteins probably is the same. Differential RNA splicing in which additional C-terminal sequences are added to the N-terminal three-quarters of the molecule could result in production of the 95-kDa protein containing C-terminal sequences not included in the 65-kDa protein. Hence antibodies to the C-terminal portion of the 95-kDa protein would not react with the 65-kDa protein.

53a. There are six nucleosomes per helical turn of the solenoid structure, and one helical turn of the solenoid corresponds to slightly less than 30 nm along the length of a chromatin thick fiber (see Figure 9-19 in MCB, p. 333). Assume, for simplicity of calculation, that there is one helical turn per 30 nm. This corresponds to 6 nucleosomes per 30-nm stretch of thick fiber. A 900-nm-long thick fiber thus has 30 solenoid turns (900 nm ÷ 30 nm/turn) and contains 180 nucleosomes (6 nucleosomes/turn × 30 turns). As each nucleosome contains 200 bp of DNA, there are 36,000 bp of DNA within this stretch of thick fiber: (200 bp/nucleosome) × (180 nucleosomes/900-nm thick fiber).

53b. If the chromosome spread was prepared under low-salt and low-magnesium conditions, the chromosome fiber seen would be a thin, 10-nm-diameter fiber with the "beads-on-a-string" appearance. See Figure 9-16 in MCB (p. 332). Each individual nucleosome has a length of 10 nm with 140 bp of DNA wound about it, and there are 60 bp of linker DNA between the individual nucleosome beads. As discussed in Chapter 2 of MCB (p. 70), B-form DNA, the standard double-helical form of DNA, makes a complete helical turn every 3.4 nm, and each turn contains 10 bp. Thus 60 bp of linker DNA corresponds to a linear stretch of 20.4 nm (60 bp × 3.4 nm/10 bp). For each "bead-on-a-string" repeat (nucleosome plus linker DNA), the fiber length should be 30 nm;—10 nm/nucleosome + 20 nm/linker DNA. For a 900-nm-long thick fiber containing 180 (see answer 53a), the beads-on-a-string form should be 5400-nm long—30 nm/repeat × 180 repeats. Thus a beads-on-a-string chromatin fiber is 6 times longer than a thick chromatin fiber (5400 nm ÷ 900 nm).

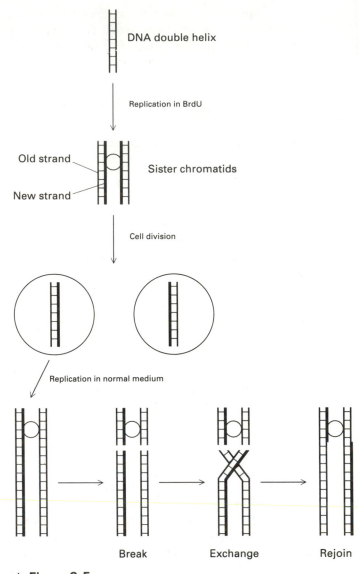

▲ **Figure 9-5**

54a. Figure 9-5 illustrates how normal semiconservative DNA replication followed by sister chromatid exchange can result in the observed staining pattern. In this figure, BrdU-containing chromatids and portions thereof are indicated by thick lines.

54b. In the chromosome spread shown in Figure 9-2, several examples of sister chromatid exchange are apparent. Hence sister chromatid exchange appears to be a frequent event. Because sister chromatid exchange entails a very precise chromosome breaking and rejoining mechanism, few, if any, mutations should result.

55a. The location of the ATGCAAAT sequence at the −70 position suggests that the sequence may be a promoter-proximal element for transcription of immunoglobulin heavy-chain genes.

55b. The gel patterns indicate that there are DNA-binding proteins in the BCL nuclear extract. With the 108-, 56-, and 30-bp fragments, several retarded bands are observed after incubation with nuclear extract, whereas, with the 78-bp fragment, no retarded band is observed. Thus protein binding to DNA is fragment specific.

55c. Maximally, there could be four different DNA-binding proteins, as four separate retarded DNA fragment bands are observed. The consistent position of the retarded bands with the different DNA fragments strongly suggests that the bound protein rather than the size of the DNA fragment is the dominant factor in the migration of the gel-retarded DNA. The four putative DNA-binding proteins appear to be present in unequal amounts, as the intensity of the retarded bands, an indication of relative protein amount, varies. The second band from the top is most intense, suggesting that the protein associated with this band is the most abundant.

55d. If the putative proteins causing retardation are arbitrarily numbered 1–4 in correspondence to the retarded bands, moving downward on the gel, then protein 1 is detected following incubation with the 108-bp fragment. The other three putative proteins are all detected following incubation with the 30-, 56-, or 108-bp fragment. These differences indicate that not all of the proteins bind to the same site in DNA.

The three putative proteins that retard the 30-, 56-, and 108-bp DNA fragments may bind to the same site, which then must be included within the short overlapping 30-bp DNA fragment. The other protein, which causes retardation of the 108-bp fragment must have a different binding site which is located within this fragment but not in the 30- and 56-bp fragments. Thus the minimal number of DNA binding sites is two.

56. β-Galactosidase is a homomeric protein–that is, one containing multiple subunits of the same polypeptide. In a trans-active assay, complementation can occur if two different mutant subunits fit to produce an active enzyme. For example, if normal β-galactosidase is wt/wt in its subunit composition, then activity in a complementation assay might occur for a mut1/mut2 form of the enzyme. Such complementation suggests that the most direct relationship is not between gene and holoenzyme but rather between gene and polypeptide, i.e., protein subunit.

Interallelic complementation may occur for any homoeric protein. However, because some mutations may interfere with multimer formation, not all mutations in genes encoding homomeric proteins will show trans-active complementation.

57a. In this experiment, the nuclear scaffold and any associated DNA is located in the pellet fraction; however, free DNA and DNA fragments are located in the supernatant fraction. Thus detection of a histone-repeat restriction fragment in the pellet indicates that the fragment is bound to the scaffold.

57b. The Southern blot pattern suggests that the 1.35-kb fragment is specifically bound to the scaffold preparation (the pellet fraction), as this fragment is found exclusively in the pellet fraction. The 1.60-kb fragment is completely released into the supernatant fraction by the restriction endonuclease digestion, as is almost all of the 2.10-kb fragment. Hence the 1.35-kb fragment, the only fragment preferentially retained with the scaffold preparation, appears to contain the scaffold-binding site in the histone gene repeat.

57c. The use of additional or alternative restriction endonucleases to produce smaller fragment sizes would increase the precision of this determination.

C H A P T E R *1 0*

1. selfish DNA
2. gene conversion and unequal meiotic crossing over
3. mobile genetic element
4. $C_0t_{1/2}$
5. retroposons
6. deletion
7. rearrangement
8. amplification
9. simple-sequence
10. insertion sequences
11. transposons, transposase

12. SINES, *Alu* family, 7SL

13. Ty, Tn

14. LINES

15. Ac and Ds

16. single-copy

17. LTR

18. genetic drift

19. faster

20. a b c

21. d

22. d

23. a b c d e

24. a b c d e

25. b c d e

26. b d e

27. a b d e

28. a c d e

29. a

30a. 1

30b. 2

30c. 3

30d. 2

31a. TRP

31b. RT3

31c. TRP

31d. UCO

31e. RT2

32a. IR

32b. DR

32c. N

32d. DR

32e. LTR, IR, DR

32f. DR

33. Fingerprinting relies on the ability to demonstrate fragments of different sizes after restriction enzyme digestion of DNA and Southern blotting using radioactive sequences derived from minisatellite DNA as a probe. Among humans, enough differences are present in the lengths of minisatellite DNA so that the Southern blot pattern of each individual's DNA after digestion and probing is virtually unique. For example, if two probes are used, the probability that an individual falsely accused in a paternity suit of being the father of a child will possess, by chance, the paternally derived fragments present in the child is about 4×10^{-8}. See Figure 10-21 in MCB (p. 367).

34. Although many gene products of differentiated cells are coded by somatic-cell DNA that does not differ in organization or copy number from the DNA in the fertilized egg, two exceptions have been discovered. One set of observations points to deletion of DNA during differentiation; a striking example is the alteration that occurs during differentiation of B cells, the cells that produce antibodies. Deletion produces a clone of cells that all synthesize a unique immunoglobulin molecule. The other set of observations points to gene amplification during differentiation. These include gene amplification in cancer cells induced by drugs and amplification of DNA encoding specific proteins required in large amounts in specific tissues (e.g., the chorion protein of *Drosophila*).

35. The critical piece of evidence, which is not yet available, that would confirm the involvement of mobile genetic elements in evolution is detection of reverse transcriptase in gametes or pregametic cells.

36. The $C_0 t$ curve for yeast DNA would be more complex than that for plasmid DNA. The multiple slopes in the yeast curve reflect the different classes of repetitive DNA based on the similarity of sequence. See Figure 10-15 in MCB (p. 362).

37. Even though pseudogenes and processed pseudogenes have protein-coding sequences similar to those of functional protein-coding genes and are both nonfunctional, they are otherwise quite different. Pseudogenes are organized like functional genes but have accumulated enough mutations to prevent production of a functional protein. They are usually arranged in clusters. Processed pseudogenes are chromosomal copies of mRNAs that have also acquired mutations. They lack introns and flanking sequences, often have sequences complementary to poly A tails, and have terminal direct repeats (see Figure 10-27 in MCB, p. 372).

38. The *Alu* transcription unit has a region rich in A residues followed by a direct repeat followed by a stretch of T residues. These Ts are the termination signal for polymerase III transcription (see Chapter 8 of MCB, p. 279). The transcript (from one of the DNA strands) thus would contain the A-rich region, the direct repeat, and a stretch of Us. Loop formation by hydrogen bonding between the Us and the A-rich region would form a hairpinned terminus with a 3′ OH. This can serve as a self-primer for reverse transcriptase, which, like DNA polymerase, requires both a template and a primer. After reverse transcriptase synthesizes a double-stranded DNA from the polymerase III transcript, the cDNA is incorporated by unknown mechanisms into the genome. See Figure 10-29 in MCB (p. 373).

39. The yeast mating-type phenotype (**a** or α) is determined by the presence of peptides termed **a** or α. Only one of these peptides is present in a haploid cell, even though the genes for both are present in the genome, one at a locus called *HML* and the other at a locus called *HMR*. When at these loci, the genes are transcriptionally silent. In order to be expressed, an **a** or α sequence must be present at the *MAT* locus, which is between *HML* and *HMR*, as diagrammed in Figure 10-6. It is this feature that has given rise to the cassette analogy: the mating-type gene inserted into the *MAT* locus is active, just as a tape inserted into a cassette player can be heard.

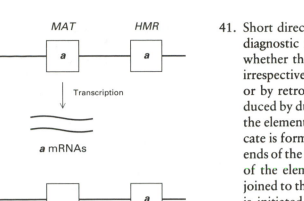

Cell is type **a**

HML   MAT   HMR

α   **a**   **a**

Transcription

a mRNAs

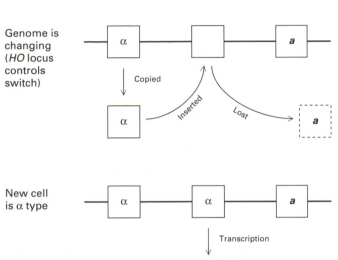

Genome is changing (*HO* locus controls switch)

α

Copied

α

Inserted   Lost

a

New cell is α type

α   α   **a**

Transcription

α mRNAs

▲ **Figure 10-6**

Even when one of the mating-type genes is inserted into the *MAT* locus, it is not lost from *HML* or *HMR* because insertion occurs by gene conversion involving DNA synthesis. In this process, a site-specific eukaryotic endonuclease produces a double-strand cut at the border of *Y* (which contains **a** or α coding sequences) and Z_1 within the *MAT* locus, and the resident mating-type information is degraded. The free termini of Z_1 "invade" an *HM* locus, and then DNA synthesis copies the information in the *HM* locus into *MAT*. Since the sequences in *HM* are transferred to *MAT* by DNA synthesis, they are not lost from these loci and can be passed onto the next generation of cells. See Figure 10-44 in MCB (p. 383).

40. Both *copia* and Ty include long terminal repeats (LTRs) similar to those in retroviruses. Computer comparisons of the nucleotide sequences also reveals that these mobile elements contain protein-coding regions that are similar to retroviral regions encoding reverse transcriptase and integrase. These elements may be transcribed to form "dead-end viruses," since intracellular particles resembling enveloped viruses are found in both yeast and *Drosophila,* and the *Drosophila* particles have reverse transcriptase activity.

41. Short direct repeats on both sides of a transposon are diagnostic for the presence of these mobile elements, whether they are bacterial or eukaryotic in origin and irrespective of whether they move by DNA duplication or by retrotransposition. This repeat sequence is produced by duplication of the target site for integration of the element. For bacterial DNA transposons, the duplicate is formed by generating single-strand breaks at the ends of the target site and the ends of the integration site of the element. The 5′ ends of the target strands are joined to the 3′ ends of the element, and DNA synthesis is initiated at the remaining free 3′-hydroxyl groups. Since there are two single-stranded regions, each with the sequence of the target site surrounding the element, after DNA synthesis the target site is duplicated at the termini of the element. Direct repeats also surround eukaryotic transposons (Ac and Ds of maize; P of *Drosophila*) and retroposons (yeast Ty and *Drosophila copia*). The mechanism for insertion of the elements may be similar to that in bacteria.

42. Repetitious DNA sequences are interspersed throughout the chick genome. The transcription unit for lysozyme contains interspersed repetitious DNA in the flanking regions. When chick DNA is fragmented into 5- to 10-kb pieces, some fragments will contain both lysozyme transcription unit sequences and flanking sequences. If the flanking region includes repetitious DNA, such fragments will contain both solitary protein-coding segments and repetitious DNA. Repetitious sequences are also present in the introns of many solitary protein-coding genes, although the chick lysozyme gene does not have repetitious sequences in its introns.

43. For much of the cell cycle, somatic human cells contain a diploid (2*n*) genome. To simplify the calculation, we will assume a diploid genome. Hence each human cell has 2.5 pg of DNA per haploid genome. If, 1 percent of this codes for protein, then there is 0.025 pg of protein-coding DNA per haploid human genome. For a rapidly growing *E. coli* cell, there is 0.017 pg DNA/4 genomes or 0.00425 pg DNA/haploid genome. Taking the ratio of these two values gives 5.9:

$$\frac{(0.025 \text{ pg DNA/haploid human genome})}{(0.00425 \text{ pg DNA/haploid bacterial genome})} = 5.9$$

Thus somatic human cells contain about sixfold more potential protein-coding capacity than *E. coli* cells. However, because of the greater prevalence of tandemly repeated genes in human cells than in bacterial cells and the presence of duplicated genes, the actual ratio of protein-coding capacity in humans and bacteria probably is less than indicated by this calculation.

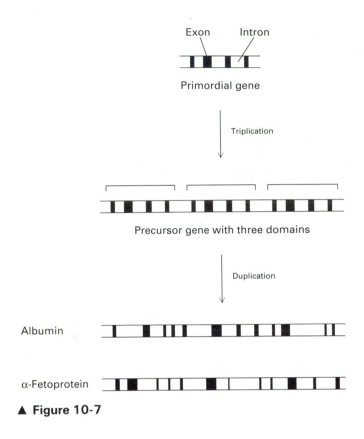

Exon   Intron

Primordial gene

Triplication

Precursor gene with three domains

Duplication

Albumin

α-Fetoprotein

▲ **Figure 10-7**

44a. Comparison of the present-day albumin and α-feto-protein genes reveals a similar repeating pattern of conserved-length exons, consisting of a small exon followed by a large exon and two small exons. The similar exon repeat pattern for the two present-day genes suggests that they arose from gene duplication of a precursor gene. The pattern of exons is repeated three times, suggesting the existence of a primordial gene containing only one set of exons. Triplication of this primordial gene would have given rise to the precursor gene, which subsequently underwent duplication to give the present-day albumin and α-fetoprotein genes. This scheme is diagrammed in Figure 10-7.

44b. Splicing of the transcription product from the primordial gene would give a small protein one-third the size of the present-day proteins. If a simple triplication without sequence loss had occurred, three tandemly arranged repeated genes would have been formed. Thus deletion or mutation of promoter and termination sequences between the repeats must have occurred to generate a single precursor gene. Subsequent sequence drift following duplication of the precursor gene could have given rise to the present-day genes. Since the variation in intron length is greater than the variation in exon length, most of the divergence following duplication must have occurred in the introns rather than in the exons.

45. Because satellite DNA sequences consist of tandemly repeated simple-sequence DNA, many sequence units are localized to one segment of DNA. Therefore, the in situ hybridization of a radioactive simple-sequence probe to the chromosome results in several copies of the probe being clustered in one region of the chromosome (e.g., the centromere). This clustering produces a strong localized signal for autoradiography. *Alu* sequences are interspersed in the genome rather than tandemly repeated. Hence no clustering of the probe occurs and sequence localization is difficult by in situ hybridization.

46. The geographic spread of the P element in the *Drosophila melanogaster* population can be tested by isolating males from various geographic sites and mating these to standard non-P-strain females. If the males are P element positive, hybrid dysgenesis will occur. If the males are P element negative, fertile offspring will be produced. See Figure 10-41 in MCB (p. 381).

47. If the rolling circle mechanism is used for the production of amplified, extrachromosomal rRNA genes, then the new genes are, in essence, reeled out as continuously spun copies of the old genes (see Figure 10-47 in MCB, p. 386). In this case, the nontranscribed spacer regions between individual genes should be the same length as that between the old genes.

48. In the unequal crossing over model, the selective disadvantage of mutant versus wild-type gene copies is the driving force for elimination of mutant gene copies raised to a high frequency by unequal crossing over events. If the crossover events occurred relatively randomly within the spacer DNA, then variability in spacer DNA lengths would be generated at the same time functional gene variability was minimized.

In the gene conversion model, correction is localized by unknown mechanisms to the coding region of the tandem array. Such localized correction would permit greater variability in the nontranscribed portions of the array.

49a. Both mechanisms for methotrexate resistance that have been proposed involve amplification of the DHFR genes. In the first, DHFR genes are amplified as part of the chromosome to generate so-called homogeneous staining regions. In the second, DHFR gene amplification is initially extrachromosomal to generate so-called minute chromosomes, which subsequently are integrated into the chromosomes. By either mechanism, the increased production of DHFR from the amplified genes counteracts the methotrexate inhibition. See Figure 10-49 in MCB (p. 387).

49b. For methotrexate-resistant cells that arise by the extrachromosomal route, removal of the selective pressure of the drug before DHFR gene integration occurs should result in the rapid, random loss of minute chromosomes from the cancer cell population because the minute chromosomes do not segregate normally during mitosis. Thus, after methotrexate is removed, the proportion of resistant cancer cells should decrease, and the remaining daughter cancer cells would once again be susceptible to the next round of methotrexate.

50. Because G-C base pairs have three hydrogen bonds, whereas A-T base pairs have only two, a higher temperature is required to melt a G-C base pair than an A-T base pair. By determining the melting curves of various standard DNAs whose base composition is known, a standard curve of melting temperature versus G-C content can be prepared. This curve can then be used to predict the G-C and A-T content corresponding to the melting temperatue of an isolated satellite DNA, or any other DNA species, of unknown base composition.

51. Transposition allows the introduction of mobile DNA elements into a plasmid. The prescription of multiple antibiotics at the same time or in the same physical location such as a hospital creates selective conditions favoring the propagation of plasmids containing multiple resistance genes. Because transposition provides a mechanism for the insertion of complete genes into a plasmid, it should under these conditions be an important mechanism for the development of bacterial plasmids containing multiple drug-resistance genes. In nature, the R1 plasmid, which confers resistance to five antibiotics, contains within its genome inverted repeats marking multiple transposition events.

52. Most eukaryotic DNA has no known function. In mammals, for example, more than 90 percent of the DNA probably is not functional, suggesting that evolution has not consistently selected against nonfunctional genes.

53. Ds elements have lost, by internal deletion, sequences that are part of Ac elements (see Figure 10-38 in MCB, p. 379). Some of these sequences code for proteins required in transposition. Since these proteins can be trans-active, an Ac element may act as a "helper" for the stable integration of Ds elements into corn. Transposition of a Ds carrier into a chromosome results in a fairly stable integration of new DNA, which may then be inherited through the seed. Plasmids are extrachromosomal and hence less likely to be inherited.

54. "Mutations" resulting from transposition often have a relatively high reversion frequency, whereas true Mendelian mutations do not. Thus genetic analysis to determine the reversion frequency usually can distinguish between apparent and true Mendelian mutations. Mobile genetic elements also have several characteristic sequence features. Detection of these features by sequencing of the affected locus indicates that the mutation resulted from a transposition event. This approach requires either a rapid way to clone the gene of interest or the use of the polymerase chain reaction to amplify the DNA of interest for sequencing purposes. The polymerase chain reaction, provided appropriate primer is available, permits the amplification of the equivalent of the needle in the haystack.

55. Because movement of retroposons occurs through an RNA intermediate, it requires transcription of the retroposon DNA. Mutations in a promoter region might well affect transcription of a retroposon and hence its movement in the genome. In contrast, movement of DNA transposons requires copying of the DNA but does not involve an RNA intermediate. Thus promoter mutations should have no effect on the movement of a DNA transposon.

56a. The absorbance of a DNA sample is inversely proportional to the extent of base pairing; that is, the less extensive the base pairing the higher the absorbance. The higher absorbance value at 37°C for the reassociated intermediate repeat DNA thus indicates that complete base pairing is not occurring along all portions of the DNA. During the renaturation process, rapid intermolecular and intramolecular base pairing of the most repeated DNA may result in a DNA network that prevents complete base pairing.

56b. The denaturation of the reassociated intermediate repeat DNA at the lower temperature is another indication of its less extensive, less complete base pairing. The broader temperature range for the denaturation of this sample is an indication of variable degrees of base pairing in different portions of the DNA. If all the DNA were fully base-paired along its length, then the denaturation would occur over a very narrow temperature range as occurs with the renatured single-copy DNA and native DNA samples.

56c. The broad temperature range over which the reassociated intermediate repeat DNA denatures strongly suggests that the DNA does not consist of exact copies. If the copies were exact, the DNA would be fully base-paired and denature over a narrow temperature range.

57a. You first might incubate the cultured cells with radioactive uridine for 2–3 h to label newly synthesized RNA, extract the RNA, and then pass the extract over the *Alu*-family DNA affinity matrix to pull out any complementary *Alu*-containing RNA. The choice of a 2- to 3-h incorporation period is based on wanting a labeling period long enough to label both nuclear and cytoplasmic RNA. This experiment would tell you if radioactive RNA that hybridizes to the affinity matrix is synthesized by the cells.

To determine if the *Alu* RNA is present in the cytoplasm, you could fractionate the cell extract into a nuclear and cytoplasmic fraction and then analyze each fraction for radioactive RNA binding to the affinity matrix. Hybridization may only be observed with the nuclear preparation.

57b. Experiments to analyze the nature of the hybridized RNA are necessary. The hybridized RNA must first be released from the affinity matrix by denaturing the hydrogen bonding between the RNA and DNA. Probably, the most important analytical step is to determine the size of the hybridized RNA. This can be done by electrophoresis. If all the RNA released from the matrix has a 7S mobility, then likely all the RNA corresponds to 7SL RNA. If species with greater mobility in the gel (i.e., lower molecular weight) are observed, these would be *Alu*-family transcripts. Partial digestion experiments and sequencing of the RNA could be done to confirm the *Alu* nature of the RNA by comparison with the known sequence of the DNA matrix.

58a. Endonuclease activity is detected in this assay by the appearance of higher-mobility bands, representing lower-molecular-weight DNA fragments, following incubation of the DNA with a cell extract. There appear to be two different endonucleases present. One high-mobility band is seen when the pBR or *MAT* DNA was incubated with either the switching (+) or nonswitching (−) extract. This indicates the existence of an endonuclease that is not mating specific. A second high-mobility band is also seen when *MAT* DNA was incubated with a switch-competent yeast strain. This band, which represents a somewhat larger fragment than the other high-mobility band, indicates the existence of a mating-specific endonuclease activity.

58b. An endonuclease is a trans-active gene product. Hence mutations in the endonuclease gene could be trans-active. Mutations in the DNA site recognized or cleaved by the endonuclease would be cis-active, as no diffusible gene product is produced.

59. The sequenced DNA has a direct repeat at its ends, a coding segment that is interrupted by two stop codons, a poly A addition signal, and a brief poly A segment. No introns are present. These are the sequence features of a processed gene. The direct repeats are characteristic of a transposition event. This processed gene is likely to have arisen by reverse transcriptase-dependent copying of β-tubulin mRNA followed by incorporation of the cDNA into the genome. Thus it is a retroposon.

60a. As shown in Figure 10-5, the 207.8 probe hybridizes to a 17-kb piece of DNA from vegetative cells and a 6-kb piece of DNA from heterocysts. These results suggest that an 11-kb deletion event occurs within the 17-kb *nif* region during differentiation and that the probe reacts only with the retained region. However, a base alteration that generates a new *Eco*RI site within this region also could result in the observed blotting pattern with the 207.8 probe.

The fact that the 154.2 probe hybridizes to the same-size (9.5-kb) DNA fragment from both vegetative cells and heterocysts suggests that no sequence change occurs in the 9.5-kb *nif* region during differentiation.

60b. The results in part (a) and (b) are consistent. The 207.3 probe, like the 207.8 probe in part (a), hybridizes to a 17-kb fragment of the vegetative cell DNA. The 11-kb doublet seen with heterocyst DNA and the 207.3 probe represents the 11-kb segment that is deleted from the chromosome during differentiation; the 207.3 probe sequence is included within this deleted segment rather than in the 6-kb retained segment, which hybridizes to the 207.8 probe.

60c. A doublet whose migration is collapsed to a single species of about the same mobility upon digestion with a single-cut restriction enzyme is characteristic of circular DNA. Since a once-cut circle is a linear molecule of the same molecular weight as the complete circular molecule, both have about the same mobility in a gel electrophoretic system. However, the presence of the 11-kb doublet in the *Eco*RI-digested DNA is puzzling and requires further explanation. The doublet perhaps corresponds to nicked and unnicked circular molecules, which would migrate slightly differently.

60d. The doublet DNA should be circular DNA. If the excision event and/or any preceding insertion event were by a transposon-like mechanism, then an inverted repeat would be expected in the 11-kb circle. The inverted repeat should be present in the circle at a position corresponding to the linkage site of the two ends of the 11-kb excised DNA; an inverted repeat gives a means to link the ends together.

C H A P T E R *11*

1. environmental factors, differentiation

2. fat-soluble small molecules and proteins that bind to cell-surface receptors

3. amphiphatic helices (including helix-loop-helix and leucine zippers), zinc fingers, and helix-turn-helix

4. RNA polymerase II and $TF_{II}D$

5. xenobiotics, cytochrome P-450

6. poly A tail, nuclease, ribosome

7. transcription, processing, and translation

8. run-on   9. homeotic

10. chromatin structure and methylation

11. $A(U)_nA$ 13. 100

12. homeobox   14. pluripotent

15. ligand, second messenger

16. inhibitor-chase

17. a b c e   28. d f

18. a b c d   29. a

19. a b c d e   30. c e

20. a b d e   31. b

21. a b c d   32. c

22. a b c e   33. d

23. a b d   34. a

24. b d   35. f

25. a b c e   36. a b

26. c   37. e

27. c f

38. See Figure 11-5.

39. This observation indicates that the cell extract contains several transcription factors that recognize the specific nucleotide sequence attached to the affinity matrix. This result would be expected if the oligonucleotide contained the CCAAT box, for example, because multiple proteins interact with eukaryotic genes containing this element.

40. The idea of Jacob and Monod that gene expression can be regulated in an ordered fashion has been supported by observations in *Drosophila*. For example, during early development, *Drosophila* exhibits an orderly cascade of gene expression, both in time and space (see Figure 11-27 in MCB, p. 421). However, their hypothesis that repressors and inducers are the controlling factors in gene expression has not been supported in *Drosophila*. Instead, the proteins encoded by some of the early-development genes have been shown to activate or suppress other genes in the cascade (see Table 11-4 in MCB, p. 419).

41. Three features of the regulatory circuit of heat-shock genes insure rapid synthesis of heat-shock proteins in eukaryotes. The chromatin encoding these genes is in an "open" configuration, allowing easy access to the polymerase. (It is possible that $TF_{II}D$ and the polymerase are already attached to the chromatin, further shortening the response time.) The factor required for activation of transcription of the heat-shock genes pre-exists in the cells and cycles from an inactive to an active form. Consequently, this factor can be made available without waiting for it to be synthesized. The nucleotide sequence at the 5' end of the mRNAs for the heat-shock proteins is rich in adenine, which reduces secondary structure and permits the ribosomes to attach easily upstream of the translation initiation codon.

42. At least some of these genes encode proteins that normally act as transcription factors. For example, the evidence is very strong that the cellular homolog of the viral gene v-*jun* (called c-*jun*) is the gene that codes for transcription factor AP1. The existence of such homol-

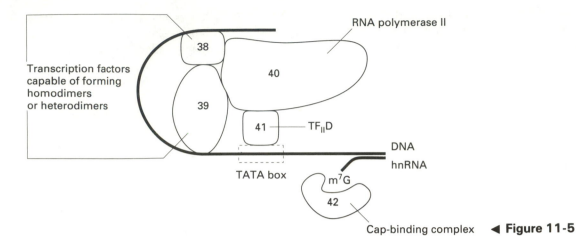

◄ **Figure 11-5**

ogous genes implies that oncogenesis, at some level, may involve an alteration in transcription pattern. If this is confirmed in general, it may suggest strategies for therapy.

43. The cellular DNA sequences are the same or very similar to the SV40 DNA sequences that bind SP1. SP1 may be able to affect the transcription of more than one gene, either alone or in concert with another transcription factor.

44. The specificity of heterodimeric transcription factors (i.e., the DNA sequences with which they interact) may differ depending on which monomers associate to form a dimer. Thus the ability of transcription factors to form heterodimers increases the number of functional transcriptions factors that can form from a limited number of polypeptides.

45. The current model is that one monomer is bound to a proximal controlling element and the other to a more distal enhancer element. Therefore, the bound dimer causes looping of the DNA, bringing the more distal element and its transcription factor close to the RNA polymerase binding site. This model is solely structural and makes no predictions about the order of events. The dimerization might occur first, followed by binding to the two sites on the DNA; or the monomers might bind first, followed by dimerization, which would cause the looping.

46. The basic domains are involved in the binding to the DNA and the acidic domains are involved in the activation of transcription. Acidic domains had been shown to be important in prokaryotic transcription activation,

and this has been confirmed for yeast GAL4 protein. See Figure 11-18 in MCB (p. 411).

47. The activity of transcription factors can be controlled by their differential synthesis, resulting in the presence or absence of specific factors in specific cell types. At least in liver cells, synthesis of factors is controlled at the level of transcription. The activity transcription factors already present within cells can be controlled by the presence or absence of inhibitors (as with NFκB) and by the activation of pre-existing factors. An example of the last mechanism is the protein-protein interaction that forms an active factor following binding of interferon α to cell-surface receptors.

48. The principles illustrated by muscle cell differentiation are that there is a sequence of stages between an undifferentiated cell and a fully differentiated cell, and that differential gene expression is required for passage through these stages. From the investigation of muscle differentiation, conversion from the myoblast stage to a myogenic cell type requires expression of the *MyD-1* gene, and further differentiation to a cell producing muscle proteins requires expression of MyoD-1 and myogenin proteins, which may be transcription factors. Another protein, Myd, has been found to influence muscle differentiation, and undoubtedly additional controlling factors will be identified. Although these proteins are specific for muscle differentiation, proteins with similar properties may regulate differentiation leading to other differentiated cells. See Figure 11-37 in MCB (p. 429).

49. A possible explanation is that the cells may not contain all the transcription factors necessary for expression of all the individual genes necessary for viral replication.

50. Steroid hormone receptors contain one domain responsible for interaction with the effector (hormone) and a second, distinct domain responsible for interaction with specific DNA response sites. Recombinant DNA techniques have been used to produce hybrid receptors containing, for example, the estrogen-binding domain and the glucocorticoid DNA-binding domain. In such hybrids, the hormone specificity of the effector region is maintained, as is the specificity of the DNA-binding domain.

51. Only a partial one. Several mechanisms that control the activity of transcription factors have been identified and the transcriptional control of transcription factors has been demonstrated. However, until the signals that control these regulatory mechanisms are elucidated, the phenomenon of differentiation can only be incompletely understood.

52a. The yeast *HMLα* and *HMRa* genes would be suitable for studies of silencing because both are normally silenced in yeast cells.

52b. The putative silencing sequences upstream of yeast *HMLα* or *HMRa* could be cloned into a plasmid upstream of a "reporter" gene that normally is expressed. If these upstream sequences control silencing, their insertion should cause decreased transcription or inhibition of transcription of the reporter gene. To confirm the silencing function of these sequences, you could determine the effect of mutating specific nucleotides within them. Mutation of the nucleotides that are most important in silencing should cause expression of the reporter gene. To determine if more than one region is involved, the effects of combinations of sequences located upstream of the yeast genes and their relative spacing could be studied. Other useful information could be obtained by searching for similar sequences near other yeast genes and using oligonucleotides containing the identified silencer sequences to isolate proteins that might bind to these sequences.

53. For many viruses that infect eukaryotic cells, it is not too difficult to obtain a reasonable amount of viral DNA, which can then be cloned either as restriction endonuclease–generated fragments or as a complete genome. (The genomes of RNA viruses can be cloned as cDNAs.) The small size of these genomes makes it possible to identify and manipulate individual genes and their controlling sequences much more easily than is possible with chromosomal genes of higher eukaryotes.

54. DNA containing only a protein-coding segment does not constitute a complete gene that could be expressed in a tissue-specific manner in response to cognate DNA-binding factors. Hence additional sequences that can be recognized by liver-specific DNA-binding factors are required. These are referred to generally as upstream activating sequences and may include the TATA box, other promoter-proximal elements, and more distal elements such as enhancer sequences, which may in some cases function in a downstream location. The actual sequences chosen would be identical to those that support liver-specific expression of other genes. The description of such sequences in liver and other tissues is an area of current research. A practical answer in this case is to fuse upstream DNA sequences from a known liver-specific gene (e.g., the metallothionein or albumin gene) to the protein-coding segment.

55a. Two rule-of-thumb approaches can help in quickly analyzing a set of amino acid sequences for conserved features. The first is to scan for whether the same amino acid is found in each position. The second is to scan for similar charge properties and hydrophobicity for the amino acids found at each position. The occurrence of many amino acids conforming to these two rules is evidence for sequence conservation.

Let's examine several positions in the sequences shown in Figure 11-2 as examples. At position 1, Arg (arginine) and Gln (glutamine) are the most common amino acids. Arginine is a basic amino acid (i.e., it carries a positive charge at neutral pH) and is hydrophilic; glutamine, although a neutral amino acid, also is hydrophilic. Lys (lysine) found at position 1 in one of the sequences also is basic and hydrophilic. Thus in 8 of the 10 sequences, the amino acid present at position 1 is hydrophilic; in 7 of 10 sequences it is either arginine or glutamine; and in 6 of 10, it is both basic and hydrophilic. At position 5, Ala (alanine), a hydrophobic amino acid, is present in 8 of the 10 sequences. In the other two sequences, a hydrophobic amino acid also is present. At position 18, all the observed amino acids are hydrophobic, and Trp (tryptophan) is present in 7 of the 10 sequences. Similar scanning of the other positions indicates that the evidence for sequence conservation within these amino acid segments is strong.

55b. The three *Drosophila* proteins have the same amino acid at each position for 15 of 20 positions. At only one position, position 1, is the same amino acid present in CAP and the *Drosophila* proteins. Hence the *Drosophila* proteins are much more related to each other than to *E. coli* CAP. Nonetheless, the *Drosophila* proteins and CAP generally have amino acids of similar charge and hydrophobicity at each position; thus they exhibit sequence conservation.

55c. These homeobox proteins bind to DNA and affect transcription of other early developmental genes in *Drosophila*. Although the sequences of these proteins are very similar in the DNA-binding region, each protein produces a very distinct body segmentation pattern. Thus the apparently small amino acid differences among these proteins must be crucial in determining which genes they activate and hence which proteins are produced in appropriate amounts at different stages of embryonic development.

56. Numerous DNA-binding proteins that affect transcription rates of genes have been identified and characterized. In some cases, the same protein has been found to bind to both distant and proximal sites. In reality, distant sites, such as enhancers, and proximal sites are in three-dimensional space all *proximal* sites. The mechanism by which these sites affect transcription is likely to be the same in many cases; that is, they are the site to which transcription-promoting proteins bind.

57. Nuclear poly A–containing RNA consists of primary transcripts and various processing products. The hybridization of calcitonin cDNA and CGRP cDNA to high-molecular-weight, nuclear poly A–containing RNA from neurons indicates the presence either of a primary transcript or of a processing intermediate containing sequences for both proteins. Detection of an intermediate-molecular-weight, nuclear poly A–containing RNA species in thyroid tissue that is positive for calcitonin cDNA but negative for CGRP cDNA indicates the presence of a calcitonin-specific processing intermediate. Moreover, the presence of this nuclear RNA species suggests that the CGRP-specific sequence must be downstream from the calcitonin-specific sequence in the high-molecular-weight neuron product. Since both the intermediate- and high-molecular-weight nuclear RNAs are poly A positive, the RNAs likely arise from tissue-specific, differential poly A addition to a common poly A–negative primary transcript. The fact that the cytoplasmic poly A–containing RNA from neurons and thyroid tissue hybridizes exclusively to its corresponding cDNA indicates that further tissue-specific RNA processing must occur in the nucleus to generate calcitonin- and CGRP-specific mRNA. These data illustrate tissue-specific mRNA expression resulting from differential RNA processing. This example is depicted in diagrammatic form in Figure 11-41 of MCB (p. 434).

58. Differentiation in plants and animals can be broadly described as a cascade of differential gene expression events, mostly at the level of transcription (see, for example, discussion of *Drosophila* embryogenesis in MCB, pp. 417–422). However, the process is complicated and involves many steps, any one of which may occur through one of several different mechanisms operating at various levels ranging from transcription to translation to molecular turnover. Hence the hope of early biologists that differentiation might be explained by one or a few mechanisms is no longer tenable.

59. RO12347 has a low molecular weight, is soluble in organic solvents but poorly soluble in water, and contains C, H, and O and no S, I, or N. As these are the chemical traits of a steroid hormone, RO12347 probably belongs to the steroid hormone family. Considering its solubility properties, the hormone should diffuse across the cell membrane. Like other steroid hormones, it would bind to a cytoplasmic receptor protein, and this hormone-receptor complex would then be translocated to the cell nucleus, where it would bind to upstream activating sequences, causing the activation of specific genes.

60. Metallothionein is a heavy metal–binding protein, which is preferentially expressed in liver. The protein is induced in animals by exposure to heavy metals such as Cd^{2+}. Livers from fish living in water free of heavy metals should contain little metallothionein, whereas those from fish living in heavy metal–contaminated waters should contain elevated metallothionein levels. Hence metallothionein levels in fish liver could provide a biological assay for heavy metal contamination.

61. Run-on assay systems do not support initiation of new RNA transcripts. ^{14}C-labeled ATP is labeled in the nucleoside portion (base plus ribose) of the nucleotide (base plus ribose plus phosphates), and β-labeled [^{32}P]ATP is labeled in the second or β phosphate position of the nucleotide. The nucleoside is incorporated into RNA, but the β phosphate is released as part of pyrophosphate during RNA synthesis. Hence the ^{14}C label is incorporated during run-on completion of preexisting RNA transcripts, whereas the β ^{32}P label is released.

Because run-on assay systems do not support transcript initiation, the planned experiment will not work. In such a system, the addition of a steroid-receptor complex cannot lead to the production of new transcripts.

62. Cognate DNA-binding proteins typically affect transcription of specific genes. Because differential gene expression is at the heart of embryonic development in *Drosophila,* mutations that affect differential gene expression should be lethal in embryos.

63. If indeed there are several leucine-zipper proteins in hepatocytes and their subunits are structurally similar and interchangeable, then C/EBP might be expected to be isolated as a mixed dimer (i.e., a heterodimer) containing one subunit from another leucine-zipper protein. That C/EBP was isolated as a homodimer suggests either that it is by far the major leucine-zipper protein in hepatocytes or that there is specificity governing subunit interaction. If C/EBP were the major protein, then few subunits from other leucine-zipper proteins would be available to contribute to heterodimer formation. If there were specificity in subunit interaction, then homodimer formation would be favored over heterodimer formation. Any specificity in subunit interaction presumably would reside in the leucine-zipper region. The arrangement of amino acids in the hydrophobic leucine-rich domain of C/EBP probably is unique. See Figure 11-14 in MCB (p. 406).

64. A defect in **a**1 protein would result in the formation of a defective α2-**a**1 protein complex. This protein complex normally blocks α1 gene transcription and haploid-specific gene transcription in diploid yeast. Hence α1 protein production will occur in the mutant. With α1 and α2 protein present and **a**1 protein nonfunctional, the phenotype of the mutant diploid cells will be equivalent to haploid α cells. As the mutant diploid cell is phenotypically equivalent to haploid α cells, mating of the mutant with haploid **a** cells is to be expected. See Figure 11-21 in MCB (p. 415).

65. The heat-shock response is the result of a reversible post-translational modification of a transcription factor termed HSF (heat-shock factor) in *Drosophila.* Because no protein synthesis is required for this reversible modification, transcriptional activation of heat-shock genes would be observed in the presence of cycloheximide. However, in the presence of cycloheximide, translation of the newly made mRNA into protein would be blocked. Thus synthesis of heat-shock proteins would not occur.

66. Binding of interferon to its cell-surface receptor protein causes activation of a specific transcription factor. The inactive form of the factor is found in the cytoplasm. The active form of the factor translocates to the nucleus. Immunofluorescent staining for the transcription factor in cells fixed before exposure to interferon would result in a diffuse pattern of fluorescence throughout the cytoplasm with no fluorescence in the nucleus. Cells fixed and stained after interferon treatment would display transcription factor–specific fluorescence in the nucleus.

67. MyoD-1 and myogenin are muscle-specific cognate DNA-binding proteins, which may belong to the leucine-zipper or helix-loop-helix groups. Microinjection of either protein into 10T1/2 cells causes the cells to cease multiplying and to differentiate into muscle. The expected changes in gene expression include the synthesis of muscle-specific proteins and decreased synthesis of proteins such as histones that are essential for cell multiplicative processes. See Figure 11-37 in MCB (p. 429).

68. A major role of a putative ribosomal nuclease in mRNA turnover has been hypothesized. Since histone mRNA normally turns over rapidly, the greatly prolonged $t_{1/2}$ of histone mRNA observed suggests that the mutation is in this putative ribosomal nuclease. If so, other mRNAs would also show prolonged half-lives.

   Predicting the overall phenotypic consequences to the cell of this mutation is complicated. Messenger RNAs that normally turn over rapidly would turn over slowly, and the production of the corresponding proteins would increase. This might result in unbalanced cell growth and abortive movement of the cells through the cell cycle. In nature, such a mutation might well be lethal.

69a. There are 2.67×10^8 molecules of actin per cell in both the F9 stem cells and retinoic acid–treated cells. Since 10 percent of the protein is actin and there are 200 μg protein per 1×10^6 cells, each cell contains 20 pg of actin:

$$\frac{200 \text{ μg protein}}{1 \times 10^6 \text{ cells}} \times \frac{0.1 \text{ g actin}}{1 \text{ g protein}} = 20 \times 10^{-6} \text{ μg actin/cell}$$

$$= 20 \text{ pg actin/cell}$$

The molecular weight of actin is 45 kDa; thus there are 4.5×10^4 g actin per mole and 4.44×10^{16} moles actin per cell:

$$\frac{20 \times 10^{-12} \text{ g actin/cell}}{4.5 \times 10^4 \text{ g actin/mole}} = 4.44 \times 10^{-16} \text{ moles actin/cell}$$

Multiplying this value by Avogadro's number (6.023×10^{23} molecules/mole) gives 2.67×10^8 molecules of actin per cell.

There is no detectable ERgp76 in the F9 stem cells. The retinoic acid–treated cells contain 1.58×10^5 molecules of ERgp76 per cell. In the retinoic acid–treated cells, 0.01 percent of the protein is ERgp76, which has a molecular weight of 7.6×10^4 g/mole. The solution is obtained by a calculation similar to the one above:

$$\frac{200 \ \mu g \ \text{protein}}{1 \times 10^6 \ \text{cells}} \times \frac{1 \times 10^{-4} \ \text{ERgp76}}{1 \ \text{g protein}} = 2 \times 10^{-8} \ \mu g \ \text{ERgp76/cell}$$

$$= 0.2 \ \text{pg ERgp76/cell}$$

$$\frac{20 \times 10^{-15} \ \text{g ERgp76/cell}}{7.6 \times 10^4 \ \text{g ERgp76/cell}} \times 6.023 \times 10^{23} \ \frac{\text{molecules}}{\text{mole}}$$

$$= 1.58 \times 10^5 \ \text{molecules ERgp76/cell}$$

69b. To determine if either the transcription rate of the ERgp76 gene or the accumulation of its mRNA is increased in retinoic acid–treated F9 cells requires an ERgp76-specific cDNA probe (i.e., the DNA complementary to ERgp76 mRNA). This probe would be used in run-off assays to detect any differences in transcription rates in nuclear preparations from treated and untreated cells. The probe also would be used in Northern blot experiments to measure accumulation of ERgp76-specific transcripts in treated and untreated cells. An actin-specific cDNA would also be important in order to normalize the data collected with the ERgp76 probe. For example, use of the actin probe would permit correction for differences in RNA recovery in treated and untreated cells, since actin expression is unchanged with reinoic acid treatment.

70a. Removal of NFκB from the cytosol (lane 2) eliminates retardation of enhancer DNA by cytosol (lane 1). Thus NFκB retards enhancer DNA migration.

70b. The amount of NFκB in the cytosol preparation appears to be insufficient to increase the quantity of enhancer DNA retarded by the added purified NFκB (lanes 3–6).

70c. Addition of depleted cytosol to the incubation mixtures containing purified NFκB results in a decrease in the quantity of enhancer DNA that is retarded (lanes 3, 7–9). This observation suggests that depleted cytosol contains an inhibitor of NFκB activity.

71a. The data in Table 11-1 show that insertions in four

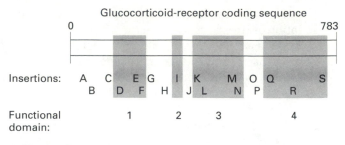

▲ Figure 11-6

different regions of the receptor coding sequence cause decreased induction. This finding suggests that the glucocorticoid receptor has four separate functional domains. These four domains correspond to the following insertions: domain 1 = insertion D, E, +F; domain 2 = insertion I; domain 3 = insertion K, L, M, +N; and domain 4 = insertion Q, R, +S (see Figure 11-6).

71b. Only insertions at Q, R, and S produce receptor proteins with decreased steroid-binding ability. Therefore, the region of the protein corresponding to these insertions is the steroid-binding domain.

71c. The data indicate that insertions in three domains block induction without affecting steroid binding. An EMSA-type assay (see problem 70) could be used to determine whether changes in any of these three functional domains affect DNA binding of dexamethasone-complexed receptor. Alternatively, the DNA-binding domain could be inferred by comparing the sequence of glucocorticoid receptor with that of known DNA-binding domains in other steroid hormone receptors.

72a. Incubation of poly U, EF2, and radioactive phenylalanine with the EF2-depleted translational system will result in poly U-dependent incorporation of labeled phenylalanine into polyphenylalanine. Addition of EF2 preparations containing progressively higher proportions of phosphorylated EF2 (pEF2), as established by isoelectric focussing, would cause progressive changes in polyphenylalanine synthesis if the effect of pEF2 on translation differs from that of EF2. If any changes observed are caused by EF2 phosphorylation and not some extraneous component of the pEF2 preparations, the addition of alkaline phosphatase, which removes phosphate groups from pEF2, should return polyphenylalanine synthesis to the levels observed in the absence of pEF2.

72b. Since EF2 is used as an elongation factor for the

translation of all eukaryotic mRNAs, the effect of EF2 phosphorylation should show no specificity with respect to mRNA species. In fact, phosphorylation of EF2 inhibits eukaryotic mRNA translation.

73a. Estrogen treatment stabilizes vitellogenin mRNA, as indicated by the longer half-life of vitellogenin mRNA in estrogen-treated than in untreated cells (curve a versus curve c).

73b. The estrogen effect on mRNA stability appears to be specific for vitellogenin mRNA, as the half-life of poly A – containing mRNA is the same in treated and un-

treated cells (curve d). However, the half-life value for total poly A – containing mRNA is an average for many different mRNA species, some of which are present in small amounts. For this reason, the presence of minor mRNA species that are stabilized by estrogen treatment cannot be proved or disproved from these data.

73c. The stabilization of vitellogenin mRNA by estrogen is reversible when estrogen is removed (curve b). Because the effect is reversible, it more likely results from reversible protein – nucleic acid interactions than from formation of stable covalent modifications of the mRNA.

C H A P T E R *12*

1. $5' \rightarrow 3'$
2. primer, 3'
3. deoxyribonucleoside triphosphates and ribonucleoside triphosphates
4. leading strand, lagging strand
5. gamma (λ)
6. semiconservative
7. replication origin
8. S
9. Okazaki fragments
10. ligase
11. primase
12. growing fork
13. aphidicolin
14. replisome
15. PCNA
16. replicon
17. topoisomerase II
18. catenanes
19. proofreading, $3 \rightarrow 5'$
20. helicase
21. processivity
22. linking number

23. SOS response
24. twist
25. Holliday
26. a b
27. a b c d
28. a b c d e
29. a c
30. b d e
31. a b c d e
32. a b c
33. a b d e
34. b c
35. a b c d
36. b
37. a
38. d
39. a
40. e

41. c
42. b

43. c d

44. b and e; DNA polymerase α and δ together have the same function as *E. coli* polymerase III.

45. c; both remove thymine dimers.

46. f; both have nuclease activity.

47. a; both catalyze synthesis of RNA primers. Primase acts in lagging-strand synthesis catalyzed by DNA polymerase α, and PCNA acts in leading-strand synthesis catalyzed by DNA polymerase δ.

48. d; both require ATP for activity.

49. g; both are DNA-repair enzymes.

50. f; both bind single-stranded DNA.

51. The nucleotide sequence at the 3' terminus of single-stranded parvoviral DNA can fold into hairpins, so that the end is double-stranded with a free 3' OH group. This end can then act as a self-primer for DNA replication, as diagrammed in Figure 12-8.

▲ **Figure 12-8**

52. It is hypothesized that the replisome is dimeric and contains two catalytic sites, one dedicated to leading-strand synthesis and one to lagging-strand synthesis. To insure that synthesis of both strands is in the same direction as the movement of the growing fork, the lagging strand is thought to contain a 180° loop, so that the $5' \rightarrow 3'$ synthesis on this strand is in the same direction relative to the growing fork as is the $5' \rightarrow 3'$ synthesis of the leading strand. See Figure 12–15 in MCB (p. 461).

53. Temperature-sensitive mutants are often used to study the enzymes carrying out the most critical steps in DNA replication because other types of mutations in these proteins generally are lethal.

54. It is generally accepted that the enzymes that replicate RNA viral genomes do not have the proofreading activity associated with DNA polymerases.

55. Both eukaryotic and prokaryotic DNA polymerases are thought to exist as dimers. Prokaryotic polymerase III is the structural and functional equivalent of a complex of eukaryotic polymerases α and δ. Polymerase III, together with the primosome, can carry out leading- and lagging-strand synthesis. In the eukaryotic complex, polymerase α plus primase synthesizes the lagging strand, and polymerase δ plus the associated PCNA (which acts as a primase) synthesizes the leading strand.

56. Viruses of eukaryotic cells replicate many times within a cell during the infectious cycle and therefore are not subject to cell-cycle control as is chromosomal DNA. A common way of studying viral DNA replication involves use of soluble in vitro systems containing cell extracts and biochemicals. These systems lack the structural features typical of chromosomes, which may place significant constraints on DNA replication. Nevertheless, studies of the requirements for replication of these small genomes have been crucial in obtaining the current information about eukaryotic DNA replication and are the necessary forerunners for studies of replication of the more complex chromosome.

57. See Figure 12-9.

58a. It is of considerable interest that *Alu* sequences can act as replication origins in this system; however, if T antigen is required for this activity, the universality of the phenomenon is diminished because only virally infected cells contain this protein.

58b. *Alu* sequences might be used as probes with extracts of cells to see if polymerases and/or other origin-binding proteins interact with them. To confirm the hypothesis conclusively, an *Alu* sequence at a specific location within a chromosome must be shown to act as an origin and deletion of this *Alu* sequence must be shown to eliminate the origin function.

59. It is not possible to distinguish unidirectional and bidirectional replication in electron-microscope bubble analysis because the bubble would appear the same regardless of which mechanism was used. Observation of bubbles in replicating viral DNAs that have been cut with a restriction endonuclease with a unique recognition site within the genome has shown that the center of the bubble remains a constant distance from the ends produced by the restriction enzyme. This finding is consistent with bidirectional replication but not with unidirectional replication. If unidirectional replication were occurring, one end of the bubble would remain at a constant distance from the restriction site. See Figure 12-9 in MCB (p. 456).

60. First, *E. coli* mutants that are sensitive to ultraviolet light (i.e., mutants that are defective in DNA repair) also

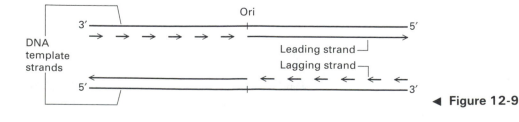

◄ **Figure 12-9**

show decreased recombination rates. Second, the presence in a cell of lesions caused by ultraviolet irradiation increases recombination. These findings suggest that in the first case, lack of repair enzymes concomitantly affects recombination, whereas in the second case, the presence of enzymes that repair damaged DNA increases recombination.

61. The first step in recombination is the formation of two single-strand breaks in strands of the same polarity in two homologous chromosomes. After strand exchange and rejoining, the crossover point can move down the helix, shifting a strand from one helix to the other, in a process called branch migration. If the chromosomes are heterozygous for an allele through which branch migration occurs, this region is now a heteroduplex. The recombination alleles occur because a helix is free to rotate 180° through the junction of the Holliday structure. (This junction is the point at which the crossover occurs, either before or after branch migration.) After rotation, nicking and rejoining of two strands will cause recombination of markers on either side of the branch point. See Figure 12-36 in MCB (p. 479).

62. The net effect of repeated sister chromatid exchange would be to produce individual DNA strands containing both heavy and light segments. Provided that the DNA can be extracted from the cells as long continuous molecules, the intermixed strands resulting from sister chromatid exchange would have an intermediate density that was neither light nor heavy. Thus the experimental data would appear to show that the "old" strand had not been conserved.

63. The result with the DNA polymerase III reaction mixture indicates that the enzyme continues to synthesize T7 DNA even after a large amount of an alternative template, T3 DNA, has been added. This result indicates that the enzyme has a high degree of processivity, a property that is to be expected for a replicative DNA polymerase such as DNA polymerase III. The result with the DNA polymerase I reaction mixture indicates that the enzyme switches templates when the T3 DNA is added, since most of the DNA synthesized corresponds to T3, not T7, DNA. This result, indicating that the enzyme is not highly processive, is expected for a DNA polymerase that is primarily involved in gap filling and repair.

64. The [³H]thymidine pulse will randomly label cells that are in different portions of the S phase. The cells labeled at the end of the S phase will be the first labeled cells to reach mitosis. Hence, if chromosome spreads are prepared at various times after labeling, the earliest chromosome spreads to show radioactive incorporation into the chromosomes will correspond to the cells that first traversed the G_2 phase of the cell cycle. Thus the time interval between the addition of the drug and the first appearance of labeled mitotic chromosomes is the minimum G_2 period of the HeLa cell cycle.

65. 8102 forks. To obtain this result, first calculate the number of base pairs synthesized per fork in one 12-h S phase:

$$(100 \text{ bp/s/fork}) \times 12 \text{ h} = 4.32 \times 10^6 \text{ bp/fork}$$

Dividing this value into the number of base pairs in the genome gives the minimum number of bidirectional forks:

$$\frac{3.5 \times 10^{10} \text{ bp/genome}}{4.32 \times 10^6 \text{ bp/fork}} = 8102 \text{ forks/genome}$$

66a. The rifampicin sensitivity of M13 DNA replication suggests that the RNA priming is catalyzed by the host E. coli RNA polymerase. The insensitivity of ongoing E. coli cellular DNA synthesis to rifampicin suggests that synthesis of RNA primers for lagging-strand replication, the repeated priming process during ongoing DNA synthesis, is catalyzed by primase rather than RNA polymerase.

66b. A mutation in the $5' \rightarrow 3'$ exonuclease activity of DNA polymerase I of E. coli would result in defective excision of RNA primers from Okazaki fragments formed during lagging-strand synthesis. Thus the Okazaki fragments would not be able to join together and defective DNA replication would occur. Such a mutation ought to be lethal.

67. Topoisomerase II activity is necessary for decatenation of newly replicated, closed circular DNA molecules (see Figure 12-28 in MCB, p. 471). The abundance of catenated mitochondrial DNA molecules in some mammalian cell lines suggests that these cells are deficient in mitochondrial topoisomerase II activity. Since these cell lines grow normally, they must not have a deficiency in nuclear topoisomerase II activity. For this to be true, there must be at least two different genes that code for topoisomerase II in mammalian cells, one that codes for a mitochondria-associated enzyme and one that codes for a nucleus-associated enzyme.

68a. In this experiment, the effect of pH on the sedimentation velocity of oak tree virus DNA is studied. The striking result is that as the pH is raised the sedimentation velocity of the DNA at first decreases and then increases greatly. The initial decrease occurs as a few base pairs begin to separate at the elevated pH. For a nicked or linear molecule, this would begin to lead to strand separation and the single-stranded portions would be more compact than the comparatively rigid duplex structure, leading to an increase in S value. However, for a closed circular DNA duplex, the initial strand separation, as denaturation begins, would be accompanied by loss in supercoiling. Overall the molecule would become less compact and have a decreased S value. For a linear DNA duplex or a nicked circular DNA, further pH increases should lead to complete DNA denaturation and strand separation. The separated strands would assume a random coil configuration and would not be very compact. In contrast, at high pH the S value of a closed circular DNA increases dramatically because the strands cannot separate and become tangled tightly upon each other. The observed pH profile of oak tree virus DNA thus suggests that is is a closed circular molecule. See Figure 12-23 in MCB (p. 468).

68b. At neutral pH, the oak tree virus DNA should appear as a form I, supercoiled, closed circular molecule by electron microscopy (see Figure 12-23 in MCB, p. 468). In electrophoretic experiments, the migration of this DNA in a gel should be sensitive to nicking and topoisomerase I incubation (see Figure 12-26 in MCB, p. 470).

69. Chemically deaminated cytosine is uracil. If uracil were a normal base in DNA, then deamination of cytosine would never be recognized as an error and be repaired.

70. Since the lengths of the arms in this chi form are equal, it must have resulted from a homologous recombination event. The allelic structure of homologous and nonhomologous chi forms are diagrammed in Figure 12-10.

71. Because replication functions are essential for cell multiplication, mutations in replication are lethal. Experimentally, lethal mutations can be maintained as conditional lethal mutations (e.g., a temperature-sensitive mutation) or as heterozygotes. Although repair functions are necessary for the long-term maintenance of genome constancy, mutations in repair functions are seldom lethal because the unrepaired DNA generally is not located in an essential region of the genome.

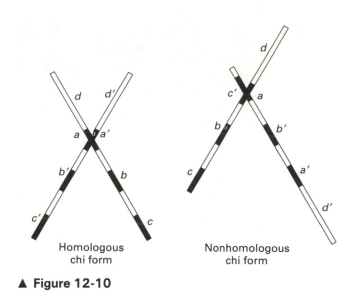

Homologous   Nonhomologous
chi form   chi form

▲ Figure 12-10

72a. The silver grains indicating [³H]thymidine incorporation are concentrated towards the centromere. Since these are the earliest labeled mitotic chromosomes, the labeled region must have been replicated during late S phase (see answer 64 for further explanation).

72b. The staining pattern indicates that two regions of the chromosome replicated late in the S phase. The first and major region is the centromere, which shows as a brightly staining region, corresponding to the region labeled by [³H]thymidine. The second and less bright region is about midway on each of the chromatids. In this experiment, only newly synthesized DNA that contains thymine instead of BrdU is stained.

72c. The dye staining method, which can resolve two late-replicating regions, appears to have a higher resolution than the autoradiographic method. The grains in the autoradiograph are more scattered and some are distal from the chromosome, so that only one late-replicating region (the centromere) is defined.

72d. The telomeric regions of the chromosome are certainly not late replicating, since neither silver grains nor dye is present near the ends of the chromosome. Thus the telomeres must be replicated either during mid or early S phase.

73a. For an organism, for example, the ΦX174 virus, DNA replication occurs if the gap is filled, but otherwise does not. In many cases, the error made will not result in a lethal mutation and "life will go on."

73b. Mechanistically, error-prone repair could be the outcome of increased error frequency by the normal repair DNA polymerase I as a result of the action of the products of the *umuC* and *umuD* genes. For example, UmuC and UmuD proteins might inhibit the proofreading $3' \rightarrow 5'$ exonuclease activity of the DNA polymerase. Alternatively, *umuC* and *umuD* might encode an entirely new DNA polymerase that is error prone, but nevertheless can produce gap filling.

73c. The Ames assay aims to detect all agents that can cause defects in DNA and hence mutations. Mutations arising from error-prone repair would increase the frequency of bacteria showing mutant phenotypes. Therefore, inclusion of the umuC and umuD genes in the test bacteria results in a more sensitive assay.

74a. The direction of replication at each of the probe sites is indicated in Figure 12-11. At the position between two replicons, the growing forks of the adjacent replicons are moving towards each other ($\rightarrow \leftarrow$). Hence three replicons are involved in replication of this DNA segment, as shown in Figure 12-11. Replicon I on the left in the figure must extend further in the left-hand direction, as only the rightward moving segment of the bidirectional fork is revealed in this experiment. Also, replicon III on the right, revealed by probes G and H, may extend further rightward; the right-hand end of this bidirectional fork is not delineated by these data.

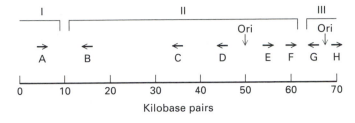
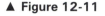

▲ **Figure 12-11**

At the origin, the two growing forks within a replicon are moving away from each other ($\leftarrow \longrightarrow$). In this segment of DNA only the origins of the middle and right-hand replicons (II and III) can be mapped, as shown in Figure 12-11. Of course, use of additional probes would locate these origins more precisely.

74b. The origin of replicon II clearly is to the right of the center of the replicon. This noncentral position presumably can occur because replicons are bounded by termination sequences, whose position may differ at the two ends. Although the origin of replicon III appears to be in the center, the right-hand terminus of this replicon is not mapped in this experiment. Thus this replicon also might have a noncentral origin.

75a. Holliday structure A is preferentially resolved at site I, as lane 2 shows more radioactivity in the POP′ band

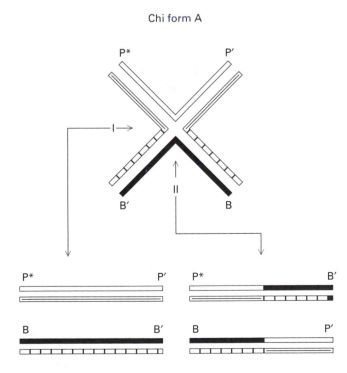

Chi form A

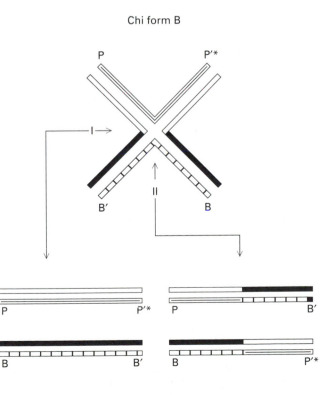

Chi form B

▲ **Figure 12-12**

than in the POB′ band. The presence of the chi band in lane 2 indicates that some of the Holliday structure A molecules have not been resolved.

Holliday structure B is preferentially resolved at site II, as lane 5 shows more radioactivity in the BOP′ band than in the POP′ band. Again some of the Holliday structure B molecules have not been resolved as indicated by the chi band.

75b. The nonradioactive resolution products from Holliday structure A are BOB′ and BOP′. The nonradioactive resolution products from Holliday structure B are BOB′ and POP′.

75c. Derivation of all four resolution products from chi form A and B is shown in Figure 12-12.

CHAPTER 13

1. phospholipid, leaflets, membrane proteins
2. α-helical
3. lipid anchors
4. cholesterol, fluidity
5. exoplasmic face, plasma membrane
6. chloroplast
7. cardiolipin
8. glycophorin, N-linked
9. lactoperoxidase
10. peripheral
11. Gorter and Grendel
12. cytoskeletal
13. electron spin resonance (ESR), phase transition
14. plasmadesmata
15. polarized cells, tight junctions, apical and basolateral
16. gap junctions
17. erythrocytes
18. a d
19. a b c d e
20. a b c
21. e
22. c d e
23. a c e
24. a
25. b c e
26. b c d e
27. c
28. a b c d
29. a b c d e
30. a b. See Figure 13-9.
31. a g
32. b
33. d
34. e
35. c f
36. h

37. At concentrations below its CMC, a detergent can still effectively solubilize integral membrane proteins. At such low concentrations, the detergent will be associated with the hydrophobic portion of the membrane proteins, but the proteins will not be dissolved in detergent micelles. At detergent concentrations above the CMC, integral membrane proteins will be dissolved in detergent micelles. See Figure 13-20, MCB (p. 505).

38. In water-soluble proteins and protein domains, the hydrophobic residues face toward the interior of the molecule. In the membrane-embedded segments of integral membrane proteins, the hydrophobic residues point outward toward the hydrophobic lipid environment and the hydrophilic residues point inward toward other α-helical chains in the transmembrane segments. Hence the description of membrane proteins as being "inside-out" is an accurate description of their membrane-embedded domains.

39. In the Gorter and Grendel experiment, the investigators measured the surface area of the erythrocyte and the surface area occupied by the extracted lipids when floated on the surface of a water solution. On a water surface, the lipids float as a monolayer. Their results showed that the area occupied by the lipids in a monolayer was twice that of the erythrocyte surface and led to the conclusion that the lipids must be arranged as a bilayer consisting of two lipid leaflets within the erythrocyte membrane.

40. The resealing of fragmented membranes occurs because exposure of the hydrophobic lipid domains at the "broken" edge of a bilayer to a hydrophilic environment is energetically unfavorable. For example, when a mammalian red blood cell is subjected to hypotonic solutions, hemoglobin is released from the cell through holes in the membrane; under appropriate conditions, the broken membrane reseals to give a sealed ghost.

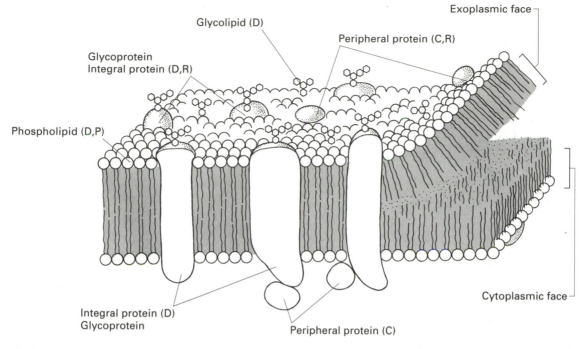

▲ Figure 13-9

41. The hydrophobicity of a fatty acid and the number of hydrophobic interactions between the chains increases with the length of the fatty acyl chains (i.e., with the number of $-CH_2$-groups). These hydrophobic interactions contribute to elevating the phase-transition (melting) temperature and to stabilizing a gel-like state. Thus, the shorter the fatty acyl chains, the fewer the interactions and the lower the phase-transition temperature. Double bonds (unsaturation) introduce kinks into a fatty acyl chain, which decrease the van der Waals interactions between chains. Thus the more double bonds, the lower the phase-transition temperature.

42. The mobility of lipids is much more restricted in biological membranes than in pure phospholipid liposomes because of the more complex structure of naturally occurring biological membranes. The presence of integral and peripheral membrane proteins in natural membranes permits lipid-protein interactions to occur. These interactions reduce the mobility of the lipids because the membrane proteins are comparatively large molecules and also may be associated directly or indirectly with cytoskeletal elements. In addition, biological membranes generally contain junctions (e.g., tight junctions), which may restrict movement of lipids to particular portions of the total plasma membrane.

43. Membrane anchoring probably is not important to "turning on" the enzymatic activity of transforming

proteins. Instead, the major effect of membrane anchoring is likely to be the placement of the transforming protein close to essential substrates for transformation. As a result, membrane anchoring probably increases the rate at which the substrate modifications essential for transformation occur.

44. Because ionic detergents are strongly denaturing, they not only solubilize membrane proteins but also cause their denaturation and concomitant loss of enzymatic activity. In contrast, nonionic detergents solubilize membrane proteins without denaturing them. Thus, if the goal is to preserve enzymatic activity, a nondenaturing, nonionic detergent is preferable to an ionic detergent.

45. The intestinal epithelial cell has a highly convoluted apical surface with a large surface area because of extensive microvilli. The function of this surface which is rich in glycocalyx, is absorption of nutrients from the intestinal lumen. This absorption requires transport systems, degradative enzymes to reduce the nutrients to transportable size, and protective separation of the cell surface from the lumen. The extensive glycocalyx is composed of molecules that confer the needed functional properties on the microvillar surface. For example, sucrase-isomaltase is a glycosylated degradative protein present in the glycocalyx of intestinal epithelial cells. Because these enzymes protrude from the cell surface,

they help to physically separate the cell surface from the intestinal lumen, perhaps serving a protective function as well as a degradative one.

46. Band 3 and connexin are transmembrane proteins that both traverse the membrane several times and associate as multimeric proteins. Multiple transmembrane α-helical segments combined with multimeric association of the protein can result in the exposure of hydrophobic residues on the lipid facing side of the α helices and of hydrophilic residues on the inward or protein-directed faces of the α helices. This arrangement allows for the creation of a local aqueous environment within a phospholipid bilayer. See Figures 13-50 and 13-51 in MCB (p. 525).

47. In animals, a sealed epithelial layer separates the lumen of an organ compartment from the rest of the body. For example, the lumen of the intestine is defined as a separate organ compartment by the sealed intestinal epithelium. This separation allows organ-specific processes to occur in a localized environment.

48. Metabolic coupling between cells is the result of gap junctions connecting the cells together. The open versus closed state of gap junctions is sensitive to intracellular levels of Ca^{2+}. Thus if intracellular Ca^{2+} goes up because Ca^{2+} leaks into a damaged cell from the extracellular environment, where Ca^{2+} is high, then the gap junctions connecting that cell to its neighbors close and the effect of the cell injury is limited to the immediately damaged cell.

49a. The minimum distance over which diffusion must occur during the 40-min post-fusion period to give complete intermixing is equal to one-fourth of the maximum circumference of the binucleate cell, as diagrammed in Figure 13-10. This distance can be calculated from the predicted radius r of the binucleate cell and standard geometric formulas describing a sphere.

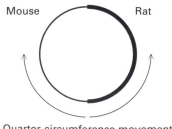

Quarter-circumference movement

▲ **Figure 13-10**

Assuming conservation of volume, the volume of the binucleate fused cell will be equal to the summed volume of the two individual cells. The volume of the individual cells is equal to

$$(4/3)\pi r^3 = (4/3)\pi(10\ \mu m)^3 = 4188\ \mu m^3.$$

The volume of the binucleate fused cell is twice this value, or 8376 μm³.

The radius of the spherical binucleate cell can then be calculated from its volume V:

$$r^3 = (V/(4/3\pi)) = (8376\ \mu m^3)/(4/3\pi) = 1999\ \mu m$$

$$r = 12.6\ \mu m$$

The circumference C of a circle of radius r is given by $C = 2\pi r$; thus the maximum circumference of the binucleate cell is equal to $2\pi(12.6\ \mu m) = 79.2\ \mu m$. One-fourth of this circumference is $79.2/4 = 19.8\ \mu m$.

The minimum net rate of diffusion required to give complete intermixing in 40 min thus equals 19.8 μm/40 min = **0.50 μm/min**.

49b. Diffusion rates for lipids in liposomes are about 1 μm/s. Thus the minimum rate of protein diffusion in the fused cells is more than 100-fold slower:

$$\frac{1\ \mu m/s}{0.50\ \mu m/min} = \frac{60\ \mu m/min}{0.50\ \mu m/min} = 120$$

50. Band 3, an integral erythrocyte plasma membrane protein, is immobile because it is linked by ankyrin to the cytoskeleton. As can be seen in Figure 13-2b, most of the cytoskeletal molecules such as spectrin and actin are retained in the shell after Triton X-100 treatment. Although ankyrin (band 2.1) is also retained, under the salt conditions used for the Triton X-100 treatment, band 3 is not retained. This finding indicates that the coupling of band 3 to ankyrin is by weak interactions that are not stable under the conditions used.

51a. The sequence indicates that this protein is amphipathic with hydrophilic segments at the amino terminus (amino acids 1–13) and the carboxyl terminus (amino acids 34–52) separated by an internal hydrophobic segment (amino acids 14–33). The hydrophobic segment is long enough (20 amino acids) to constitute a transmembrane domain. All these sequence features are typical of an integral membrane protein.

51b. See Figure 13-11. The exoplasmic and cytoplasmic face cannot be distinguished based on the sequence data alone.

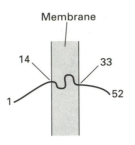

Membrane

14   33

1   52

▲ **Figure 13-11**

52. An algorithm based on amino acid hydrophilicity, particularly one that looked for hydrophobic transmembrane segments, would misassign lipid-anchored proteins to the soluble class because these integral membrane proteins have no obvious hydrophobic amino acid sequence feature.

53. As shown in Figure 13-3, the melting temperature first goes down and then goes up as the position of the double bond is shifted away from the polar head group. This occurs because the maximum effect of the kink caused by the double bond on van der Waals interactions between lipids is greatest when the kink is in the middle of the fatty acyl chain (i.e., in the middle of the hydrophobic portion). Disorder will be greatest and hence the phase-transition temperature lowest then.

54a. The observation that the proportion of spectrin existing as tetramers and higher oligomers is much less in HPP than in normal erythrocytes (60 percent versus 95 percent) indicates that HPP cells are deficient in the ability of spectrin to polymerize. This deficiency probably results from a mutation in the spectrin domains that interact during polymerization.

54b. Since oligomeric spectrin is a major component of the cytoskeleton in erythrocytes, HPP cells would be expected to have a defective cytoskeleton. A well-formed cytoskeleton, however, is necessary for the maintenance of erythrocyte shape and pliability. Thus HPP cells, which have a defective cytoskeleton, lack this flexibility and are trapped in the capillaries of the spleen and then degraded. Rapid erythrocyte turnover is the immediate cause of the anemia found in HPP patients.

55a. For a very high-conductance ESCK line, the seal between cells, which is the same seal as that between apical and basolateral surfaces, is leaky. This leakiness would be expected to minimize the differences in lipid and protein distribution between the apical and basolateral surfaces.

55b. Since low conductance results from the tightness of the tight-junction seal, the structural difference between low- and high-conductance ESCK cells must be in their tight junctions. Tight junctions can be observed morphologically in freeze-fracture micrographs. Thus morphological examination might reveal differences in the appearance of tight junctions in different ESCK cell lines.

56. The appearance of gap junctions as large clusters of connexons suggests that junctional clusters are held together by protein-protein interactions. If this is true, such interactions should not be disrupted by nonionic detergents. Treatment of cells with nonionic detergents should dissolve the rest of the cellular membranes and leave the gap junctions intact as protein-protein complexes. Such large protein-protein complexes could be isolated based on their density or sedimentation velocity.

57. Tight junctions appear as continuously thin strips that surround a cell like a thin ribbon. During cell homogenization, these ribbons should be sheared into shortish segments. Whether these ribbons would then be stable after detergent treatment to dissolve the rest of the membrane is questionable because the ribbons are not held together by multiple protein-protein interactions in the way that gap-junction clusters appear to be held together. The appearance of the tight junction suggests that it is held together at most by head-to-tail contacts between junctional "proteins." To date, the molecules involved in the formation of tight junctions are unknown.

58a. Gap junctions provide regulated openings for exchange of small molecules between cells. To maintain the common appearance and function of gap junctions among species, the features of connexin that probably are most conserved are (1) overall shape of the molecule, necessary for the same overall appearance; (2) its ability to interact with adjacent connexin molecules, which is essential for opening and closing of the "pore"; and (3) its Ca^{2+} responsiveness.

58b. The overall shape of connexin could be maintained despite considerable variation in amino acids among connexins from different species. In contrast, specific amino acid residues probably must be conserved to maintain the ability to form connexin-connexin interactions and to respond to Ca^{2+} because these properties are apt to depend on specific amino acids.

59a. As shown in Table 13-1, the half-time for transfer of sphingomyelin from a liposome to another bilayer is 1.2×10^5 min; this is equivalent to 83 days. Thus about 3 months is required for 50 percent of the sphingomyelin to move from one bilayer to another in the absence of carrier proteins. This is a long time relative to the division time (≤ 24 h) of many mammalian cells in culture.

59b. The naturally occurring phospholipids and N-rhodamine-phosphatidylethanolamine all have fatty acyl side chains containing more than 12 CH_2 groups, whereas the C_6 analogs have at least one short fatty acyl side chain. The C_6 analogs have more limited hydrophobic interactions with other lipids because the short chains contain fewer groups to interact; therefore, the analogs are not held tightly in liposomes and can transfer rapidly. In comparison, the naturally occurring lipids and N-rhodamine-phosphatidylethanolamine are held tightly in liposomes and transfer only slowly.

59c. At neutral pH, all of the lipids tested, with the exception of the ceramide C_6 analog, have bulky head groups, which are frequently charged. The ceramide analog has a small head group (an OH group), which is neutral in charge. For this reason, only the ceramide analog can flip-flop.

60a. The monoclonal antibody must react with a structural feature common to all these molecules. The most likely feature is a saccharide. Glycophorin is a glycoprotein with both N-linked and O-linked oligosaccharides. Many fibroblast membrane proteins also are glycoproteins, containing either N-linked or O-linked saccharides, or both. Plant thylakoids, a chloroplast component, contain galactolipids. Hence reactivity with the saccharide portion of glycophorin could explain the reactivity pattern of this antibody.

60b. Any monoclonal antibody that reacts with the saccharide portion(s) of membrane proteins might display widespread crossreactivity.

61a. Because of their hydrophobic domains, integral membrane proteins partition into the detergent phase, whereas peripheral membrane proteins partition into the aqueous phase. Inspection of the gel patterns in Figure 13-5a shows that band 3, band 5, and a species migrating faster than band 6 partition preferentially into the detergent phase; thus these bands correspond to integral membrane proteins. Conversely, bands 1, 2, 4.1, and 6 partition preferentially into the aqueous phase; thus these bands correspond to peripheral membrane proteins. Note that band 4.2 is found equally in both phases and cannot from these data be assigned to either class of membrane protein.

61b. All of the proteins revealed in Figure 13-5b are glycosylated, since they contain galactose. The exact correspondence of the galactose-staining species to the protein-staining species is not clear from these gels, perhaps because the glycoproteins do not stain readily or because they are minor constituents of these membranes. In fact, erythrocyte glycoproteins stain poorly with protein stains.

61c. Lipid-anchored proteins, because the lipid is hydrophobic, will partition into the detergent phase and be classified as integral membrane proteins.

62. The chief advantage of the phase-partition approach is its gentleness, as Triton X-114, a nonionic detergent, does not denature proteins. In contrast, the very alkaline pH 11 treatment can denature many proteins, leading to loss of enzymatic activity and possible loss of reactivity with specific antibody. Since both of these properties are commonly used to follow a protein during isolation, the pH 11 treatment is unsuitable for preparative purposes.

63a. Heterogeneity in the isoelectric focusing dimension indicates charge heterogeneity in the protein. If this were caused by differences in primary structure—that is, in the frequency of charged amino acids—then at least 16 amino acids would have to vary, as the two-dimensional gel pattern shows at least 16 mLAMP-1 spots. Alternatively, the heterogeneity could result from variation in secondary modifications—for example, the sialic acid content and extent of phosphorylation—both of which affect the isoelectric point.

63b. Experimentally, the source of heterogeneity could be established by determining either the secondary modification state or the primary sequence of mLAMP-1. Secondary modification can be investigated by enzymatic digestion with enzymes that would remove oligosaccharides, (i.e., glycosidases) or phosphates (i.e., phosphatases). The primary sequence of the protein and heterogeneity of the primary sequence could be investigated by normal DNA molecular sequencing approaches.

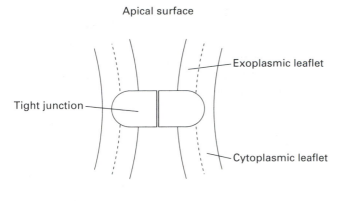

Apical surface

Exoplasmic leaflet

Tight junction

Cytoplasmic leaflet

Lateral surface

▲ **Figure 13-12**

64a. *N*-rh-PE introduced into the apical surface does not have equal access from both leaflets to the lateral surface. No labeling of the lateral surface occurs when *N*-rh-PE is introduced into the exoplasmic leaflet, whereas labeling of the lateral surface occurs when *N*-rh-PE is introduced into the cytoplasmic surface.

64b. Removal of Ca^{2+} would result in the tight junctions being disrupted. With tight junctions no longer present between the cells, free diffusion of lipid in either leaflet should occur.

64c. See Figure 13-12.

C H A P T E R *1 4*

1. partition coefficient   2. Fick's law

3. passive transport and facilitated diffusion; facilitated diffusion, passive transport

4. K_m 5. Nernst equation

6. ΔG, ion concentration gradient

7. active ion transport, ATP

8. ion pumps, Na^+-K^+ ATPase, Ca^{2+} ATPase

9. H^+ ATPases

10. antiport and symport; symport, antiport

11. group translocation   12. osmotic, turgor

13. endocytosis, phagocytosis

14. pinocytosis, receptor-mediated endocytosis

15. transcytosis   16. lysosome

17. coated pits, clathrin, endosomes, CURL

18. asialoglycoprotein

19. prelysosomal, pH-dependent

20. transferrin, ferrotransferrin, apotransferrin

21. human immunodeficiency virus (HIV), enveloped viruses

22. b   25. a b c d e

23. b c d   26. c e

24. a c   27. b d e

28. c d e   37d. P

29. a c   37e. P

30. a b d   38a. S

31. b e   38b. A

32. c d   38c. A

33. a b c e   38d. A

34. b c d   39a. R

35. b d   39b. T

36. b c d   39c. D

37a. F   39d. D

37b. V   39e. T

37c. F

40. The known structure of membrane transport proteins is inconsistent with carrier models in which transport protein molecules shuttle from one membrane face to the other. Transport protein molecules are large and bulky, and movement of a large ligand-binding hydrophilic domain into the hydrophobic membrane bilayer would be highly unfavorable energetically. We now know that these proteins are amphipathic and have multiple transmembrane α-helical segments, which are hydrophobic.

41. The uptake of glucose at the apical surface of an intestinal epithelial cell is mediated by a Na^+-glucose sym-

port. Entry of glucose against its concentration gradient is energetically favorable because it is coupled to the movement of Na$^+$ down a concentration gradient and with the membrane electric potential. Transport of glucose from the basolateral or serosal (blood-facing) membrane of an epithelial cell is mediated by a glucose permease and occurs down a concentration gradient. See Figure 14-14b in MCB (p. 546).

42. Transport ATPases are found not only in the plasma membrane and sarcoplasmic reticulum but also in mitochondria and chloroplasts. The Ca^{2+} ATPase a major sarcoplasmic reticulum protein) and the Na$^+$-K$^+$ ATPase (a plasma-membrane protein) have been shown to function reversibly. For example, if ADP and P$_i$ are placed outside of sarcoplasmic reticulum vesicles containing a high concentration of Ca^{2+}, ATP will be generated by a reverse pump action. Such results suggest that the chloroplast and mitochondrial H$^+$ ATPases generate ATP by the flow of H$^+$ ions down an electrochemical gradient.

43. The gene that confers multidrug resistance appears to encode an ATP-driven drug-transporter protein. This protein can pump lipid-soluble drugs out of tumor cells, thus effectively decreasing their intracellular concentration and protecting the tumor cell from the cytotoxic effect of the drug. Resistance correlates with amplification of the multidrug-resistance gene. See Chapter 10 (p. 387) in MCB for another example of gene amplification associated with drug resistance.

44. The movement of ions, sugars, and amino acids by cotransport is coupled indirectly to ATP consumption. The movement of a sugar, for example, against a concentration gradient, is linked to the cotransport of an ion with a favorable electrochemical gradient. Typically, the cotransported ion is Na$^+$. The Na$^+$-K$^+$ ATPase pumps Na$^+$ out of the cell, establishing a favorable gradient for Na$^+$ cotransport to drive glucose transport into the cell by a Na$^+$-glucose symport.

45. pH = $-\log$ [H$^+$]. Thus the stomach lumen has a pH of 1, corresponding to a H$^+$ concentration of 1×10^{-1} M. The pH of the cytosol of the adjacent cells is about 7.0, corresponding to a H$^+$ concentration of 1×10^{-7} M. The H$^+$ concentration in the stomach lumen is 10^6 times higher than that of the cytosol of the adjacent cells: $(10^{-1}$ M$)/(10^{-7}$ M$) = 10^6$.

46. The high osmotic pressure inside plant cells results in an inward flow of water and outward turgor pressure against the plant cell plasma membrane. If the same osmotic pressure could be experimentally induced inside mammalian cells, they would lyse. However, plant cells are surrounded by a cell wall that is rigid and able to withstand the high turgor pressure from within the cell. Consequently, the plasma membrane does not lyse. If the cell wall is removed, then the extracellular medium must be adjusted to balance osmotic conditions inside and outside of the cell to prevent cell lysis.

47. Phagocytosis involves the binding of a particle to receptors that trigger the phagocytic process. The particle is progressively enveloped by a "zippering" process to give a large phagocytic vesicle > 1 μm in diameter. Microfilaments are active participants in this engulfment process. In contrast, endocytosis is the invagination of small segments of the plasma membrane, which are typically coated with clathrin. In the case of receptor-mediated endocytosis, ligand-receptor complexes are engaged, in most instances, with the clathrin coat during the endocytic process. Endocytosis results in the invagination of small vesicles ≤ 0.1 μm in diameter. Microfilaments are not involved in the process. See Figure 14-26 in MCB (p. 557).

48. For a receptor to promote effectively the endocytosis of a circulating ligand from the mammalian bloodstream, the K_m of the receptor-ligand complex must be within the range of the normal bloodstream concentration of the ligand. If the ligand concentration was significantly less than the K_m of the ligand-receptor complex, the receptor would have little function under normal circumstances. Hence the expected concentration of asialoglycoproteins in the mammalian bloodstream is ~1×10^{-8} M.

49. Fusion of many enveloped viruses into mammalian cells requires an acidic pH because of the properties of the fusogenic protein present in the viral envelope. For example, the HA protein of influenza virus becomes fusogenic only when its conformation is altered by exposure to an acidic pH. HA then promotes the fusion of the viral envelope into the membrane of endosomes and the release of the nucleocapsid into the cytosol where it can be replicated.

50. Several types of evidence indicate that ligand binding and internalization are distinct steps in receptor-mediated endocytosis. First, ligand binding to endocytic receptors can occur at 4°C, whereas internalization only occurs as the cells are warmed to 37°C. Second,

mutant receptors that support binding but not internalization have been identified for low-density lipoprotein (LDL) and genetically engineered for other ligands. Finally, ligand binding to receptors does not require metabolic energy, but the internalization of ligand-receptor complexes does.

51. Familial hypercholesterolemia is characterized by high cholesterol levels in the circulation and the deposition of insoluble cholesterol in the bloodstream. The condition can be the result of defects in the internalization of low-density lipoprotein, a major cholesterol carrier in the mammalian bloodstream. Reported defects include deficiencies in the amount of receptor synthesized or in the structure of the synthesized receptor; mislocalization of the receptor; and deficiencies in pH-dependent recycling. These defects are caused by various mutations, which may be related to codon changes, promoter changes, etc.

  Phenotypically, all these defects result in very high cholesterol levels in the bloodstream and the development of atherosclerotic plaques in blood vessels. Many individuals with this disease suffer an early death, indicating that very high cholesterol levels are detrimental to human health.

52. Most ligands internalized by receptor-mediated endocytosis are delivered to lysosomes where they are degraded.

53. The use of receptor-mediated endocytosis to target pharmacological agents is based on the fact that receptors for some ligands are present only in certain cell types. One example of such targeting is the engineering of fusion proteins containing a bacterial toxin and an endocytic ligand (e.g., a hormone) whose receptor is present in a specific cell type. The toxin would then be internalized only in that cell type. An example of this approach is shown in Figure 14-46 in MCB (p. 572). Similarly, toxic drugs could be coupled to ligands for delivery to specific target cells. Practical limitations to this approach include the cell specificity of the target receptor, the possibility that modification of a ligand may affect its binding to receptor, and the possible decrease in a drug's effectiveness as the result of coupling to the ligand.

54a. This problem can be solved using the Michaelis equation

$$V = \frac{V_{max}}{1 + (K_m/C)}$$

by substituting the values for V_{max}, K_m, and the three glucose concentrations C. The calculated velocities in μmol glucose/ml packed cells/h are as follows: at 3 mM gluscose, $V = 333$; at 5 mM glucose, $V = 384$; and at 7 mM glucose, $V = 412$.

54b. The glucose concentration varies in the ratio $3:5:7 = 1:1.67:2.33$. The velocity of transport varies in the ratio $333:384:412 = 1:1.15:1.24$. Thus over this concentration varies by a factor of 2.33, whereas the glucose transport rate varies only by a factor of 1.24.

54c. The transport rates for a mutant permease with a K_m of 5 mM can be calculated as in answer 54a. The transport rates in μmol/ml packed cells/h for the mutant enzyme over this concentration range are as follows: at 3 mM glucose, $V = 187$; at 5 mM glucose $= 250$; and at 7 mM glucose $= 292$.

  Over these concentrations, which correspond to the range of physiological extremes in blood glucose levels, the high-K_m permease would be relatively inefficient. For example, under starvation conditions (i.e., 3 mM glucose), the mutant permease would transport glucose at about 37 percent of V_{max} (187/500), whereas the normal enzyme would transport glucose at 67 percent of V_{max} (333/500). During starvation it is particularly crucial for cells to be able to efficiently scavenge glucose from the bloodstream. As the glucose concentration is increased, the transport properties of the high-K_m permease and the normal permease become more and more similar.

55. Sugar and amino acid uptake from the intestine is mediated by a set of symports located in the apical membrane of intestinal epithelial cells. These symports use the cotransport of Na$^+$ down a concentration and electric potential gradient to drive sugar and amino acid movement up a concentration gradient. Impaired Na$^+$-K$^+$ ATPase function as a result of chlorea toxin produces a decreased Na$^+$ gradient because the Na$^+$-K$^+$ ATPase pumps Na$^+$ outward from the epithelial cell into the intestine. As a result of the decreased Na$^+$ gradient, uptake of sugars and amino acids into the epithelial cells is reduced.

56a. $E_K = +0.084$ or 84 mV. The electric potential across a membrane due to differences only in the K$^+$ concentration can be calculated from the Nernst equation

$$E_K = \frac{RT}{Z\mathscr{F}} \ln \frac{K_1}{K_r} \quad \text{(see MCB, p. 537)}$$

by substituting the following quantities: $R = 1.987$

cal/degree-mol; $T = 293°$; $Z = 1$; $\mathscr{F} = 23,062$ cal/mol-V; $K_1 = 140$ mM; and $K_r = 5$ mM.

56b.
$$\Delta G_{outward} = -326 \text{ cal/mol K}^+$$
$$\Delta G_{inward} = +326 \text{ cal/mol K}^+$$

ΔG for movement of K^+ is the sum of the free-energy change generated from the K^+ concentration gradient (ΔG_c) and the free-energy change generated from the membrane electric potential (ΔG_m).

$$\Delta G = \Delta G_c + \Delta G_m = RT \ln \frac{K_{in}}{K_{out}} + \mathscr{F} E_K \quad \text{(see MCB, p. 539)}$$

Substituting into the first term $K_{in} = 140$ mM, $K_{out} = 5$ mM, $R = 1.987$ cal/degree-mol, and $T = 293°$K and solving for ΔG_c gives the following:

$\Delta G_c = +1940$ cal/mol K^+ for movement inward

$\Delta G_c = -1940$ cal/mol K^+ for movement outward

Likewise, substituting into the second term $\mathscr{F} = 23,062$ cal/mol-V and $E = 0.07$ V and solving for ΔG_m gives the following:

$\Delta G_m = +1614$ cal/mol K^+ for movement outward

$\Delta G_m = -1614$ cal/mol K^+ for movement inward

Summing these values for ΔG_c and ΔG_m gives the following:

$\Delta G_{outward} = -1940$ cal/mol $+ 1614$ cal/mol $= -326$ cal/mol

$\Delta G_{inward} = +1940$ cal/mol $+ -1614$ cal/mol $= +326$ cal/mol.

Transport of K^+ outward is energetically favorable; it has a negative free-energy change. Transport of K^+ inward is energetically nonfavorable; it has a positive free energy change.

57a. The ouabain-binding site of the Na^+-K^+ ATPase must be located on the exoplasmic face of the plasma membrane, as microinjected ouabain does not inhibit the ATPase. The ATP-binding site must be on the cytoplasmic face of the plasma membrane, as ATP is made inside the cell and is present in the cell, not in the extracellular medium.

57b. The addition of a phosphate group to the enzyme (i.e., a phosphorylation event) would cause it to become more negatively charged. In the presence of extracellu-

lar ouabain, this negative charge, which normally is temporary, becomes permanent. The observation that ouabain treatment results in the α subunit becoming more negatively charged but has no effect on the charge of the β subunit suggests that the α subunit is phosphorylated during normal transport. This hypothesis could be tested by investigating the effect of a phosphatase on the charge properties of the α subunit following ouabain treatment. Most likely, ouabain treatment inhibits the Na^+-K^+ ATPase by blocking the reaction mechanism at the point which phosphate has been added to the enzyme. Thus the phosphorylated form of the enzyme would accumulate, blocking further ion transport.

57c. Since the removal of ouabain would probably let the transport reactions proceed, the phosphorylated α subunit would not accumulate; thus the enzyme would be less negatively charged.

58a. Since two Ca^{2+} cations are transported for each ATP molecule consumed, the rate of ATP consumption/enzyme molecule/s should be 15.

58b. Again by the same relationship, there would be 0.5 ATP molecules generated for each Ca^{2+} cation transported down a concentration gradient.

59a. Yes. The half-coated lymphocyte would bind to a macrophage via the F_c receptors on the macrophage. At 4°C, a temperature at which internalization cannot occur, the lymphocyte would sit, bound by F_c–F_c receptor interactions, on the surface of the macrophage as shown in Figure 14-7. At this temperature, the macrophage would not wrap its membrane around the surface of the lymphocyte.

59b. At 37°C, the capped lymphocyte would be only partially internalized by the macrophage because exposed F_c segments available for binding to macrophage F_c re-

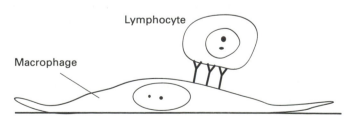

▲ **Figure 14-7**

ceptors are present only on a portion of the lymphocyte surface. Envelopment can occur only over the areas where there are exposed F_c segments.

60a. The normal LDL receptor is unusual in that it is found in coated pits in the absence of bound ligand. Presumably, the cytoplasmic domain of the receptor is interacting with assembly particles and this feature results in the unusual distribution. Therefore, the engineered hybrid receptor containing the AGR exoplasmic domain + the LDLR transmembrane and cytoplasmic domains would be expected to be present in coated pits in the absence of bound ligand. The LDLR cytoplasmic domain in this hybrid should interact with assembly particles, resulting in its being present in coated pits. The hybrid receptor containing the LDLR exoplasmic domain + the AGR transmembrane and cytoplasmic domains would be expected to be absent from coated pits and distributed randomly over the cell surface.

60b. In the presence of asialoglycoprotein, the hybrid receptor containing the AGR exoplasmic domain should bind asialoglycoprotein and be internalized by the transfected cells. Similarly, in the presence of LDL, the hybrid receptor containing the LDLR exoplasmic domain should bind LDL and be internalized. In both cases, the association of ligand with the receptor is pH sensitive. After internalization, the ligand-receptor complexes dissociate in CURL, and the hybrid receptors are recycled back to the cell surface, where they can support further rounds of internalization.

61. Assuming that an endosome is a sphere (which is an approximation), the volume of an endosome can be calculated from the formula

$$V_{sphere} = (4/3)\pi R^3, 1 \text{ cm}^3 = 1 \text{ ml} = 10^{-3}l$$
$$V_{endosome} = (4/3) \times 3.14 \times (0.1 \ \mu m)^3$$
$$= 4.2 \times 10^{-21} \text{ m}^3 = 4.2 \times 10^{-21} \text{ m}^3 \times (10^{-3} \text{ m/l})$$
$$= 4.2 \times 10^{-18}l$$
$$\text{pH } 7.0 = 1 \times 10^{-7} \ M \ H^+, \text{ pH } 6.0 = 1 \times 10^{-6} \ M \ H^+,$$
$$\text{pH } 5.0 = 1 \times 10^{-5}$$

Recall that pH equals the negative log of the hydrogen ion concentration. The number of free protons at pH 7.0 in the vesicle is

$1 \times 10^{-7} \text{ moles/}l \times 4.2 \times 10^{-18} l \times$ Avogadro's number
$= 1 \times 10^{-7} \text{ moles/}l \times 4.2 \times 10^{-18} l \times 6.023 \times 10^{23} \ H^+/\text{mole}$
$= 2.4 \times 10^{-1}$ hydrogen ions
$= 0.24$ hydrogen ions

Similarly, the number of free protons at pH 6.0 in the vesicle is

$$2.4 \times 10^0 \text{ hydrogen ions} = 2.4 \text{ hydrogen ions}$$

Similarly, the number of free protons at pH 5.0 in the vesicle is

$$2.4 \times 10^1 \text{ hydrogen ions} = 24 \text{ hydrogen ions}$$

The minimum number of protons that would need to be pumped into the vesicle to change the pH from 7.0 to 6.0 is

$$2.4 \text{ protons} - .24 \text{ protons} = 2.16 \text{ protons}$$

The minimum number of protons that would need to be pumped into the vesicle to change the pH from 6.0 to 5.0 is

$$24 \text{ protons} - 2.4 \text{ protons} = 21.6 \text{ protons}$$

Note that the movement of only a very small number of protons is sufficient to cause large and physiologically significant changes in vesicles pH. In the natural situation, endosomes have some buffering capacity. Hence a greater number of protons in reality would need to be moved. In the real physical world protons come in units of one. Statistically a pH of 7.0 in a vesicle could be achieved if on the average over time there were 0.24 free protons per unit time.

62a. Assuming that the HIV-binding site on CD4 corresponds to a simple linear sequence of amino acids, a synthetic peptide that spans this sequence should block HIV binding to T lymphocytes. Thus the binding site presumably could be identified by testing a series of synthetic peptides spanning different segments of the CD4 sequence for their ability to block binding. The validity of this approach depends on the assumption stated above; initial research indicates that it is a reasonable assumption.

62b. Any peptide that blocked HIV binding would also prevent infection. Thus blocking peptides might be introduced into the bloodstream of HIV-positive individuals to block further HIV binding to CD4 and perhaps reduce development of AIDS symptoms. Several problems might occur with this approach. The peptides might act as antigens and induce antibodies, which in turn could bind the peptides, making them ineffective. The peptides also might block the normal biological function of CD4.

63. Several different mutations can produce the phenotype of cellular resistance to an endocytosed toxin. These include mutations that are toxin specific (e.g., those affecting the production or structure of the receptor). In contrast, mutations that affect the fate of toxin-receptor complexes after their internalization may exhibit a wide specificity. For any given endocytosed toxin, most mutations will be unique for that toxin. However, a mutation affecting endosome acidification will affect the metabolism of any toxin that requires a pH-dependent step for toxicity. Thus selection for a mutation that gives simultaneous resistance to two toxins is likely to screen for mutations in a trait (e.g., endosome acidification) that is required for the toxicity of both toxins.

64a. The V_{max} and K_m values can be estimated by inspection of the kinetic curves in Figure 14-2. The values are as follows:

Subject	K_m	V_{max}
KD	1.5 mM	100 μmol glucose/ml packed cells/h
DZ	6.0 mM	500 μmol glucose/ml packed cells/h
Control	1.5 mM	500 μmol glucose/ml packed cells/h

64b. The defect in the erythrocytes of each of the two patients is different. In the erythrocytes from patient KD, the V_{max} is altered with respect to the control population; the V_{max} is five-fold lower than in the controls. In the erythrocytes from patient DZ, the K_m is altered; the K_m is four-fold higher than in the controls.

64c. For patient KD, the observed decrease in V_{max} could be accounted for if the number of permease molecules were decreased fivefold but the transport properties of each permease molecule were normal. Alternatively, the defect could be in the permease itself rather than in its numbers. For example, an alteration in permease amino acid sequence or in secondary modifications could lead to a decrease in V_{max}.

For patient DZ, the defect must affect the nature of the permease itself, since the K_m is increased compared with the control, indicating that the binding affinity between the permease and glucose must be lower.

65a. The temperature-dependence results suggest that valinomycin is a carrier and that gramicidin is a channel former. Valinomycin shows a sharp and very pronounced temperature dependence in its ability to conduct, (i.e., transfer) ions across the liposome membrane. This result is expected for a carrier that shuttles ions across a membrane because movement of carrier re-

quires a fluid lipid bilayer. Gramicidin conductance shows little temperature dependence. This result is expected for a channel former because the openness of the pores should show little dependence on temperature.

65b. Because an artificial liposome bilayer is typically composed of one phospholipid species and no cholesterol, it exhibits the narrowest possible melting profile for the membrane lipids. Naturally occurring membranes such as that of the erythrocyte ghost, contain several phospholipid and glycolipid species, proteins, and cholesterol. These more complex membranes exhibit a much broader melting profile under any condition.

66a. With addition of ATP, acidification of the tonoplast does occur under appropriate salt conditions. Since ATP is present in cells in the cytosol, the ATP-binding site must be on the cytoplasmic side of the tonoplast membrane.

66b. The data in Figure 14-4 indicate that Cl^- is required for tonoplast acidification. In the presence of $MgSO_4$ or $MgSO_4 + K_2SO_4$, little, if any, ATP-dependent acidification occurs. However, when Mg^{2+} is paired with Cl^- as the anion, acidification occurs at a rapid rate. If additional Cl^- is added as a salt with a monovalent cation, the acidification rate is increased further. There appears to be no specific requirement for the monovalent cation, as similar acidification rates are observed with several different monovalent cations paired with Cl^-. Since Mg^{2+} is generally required for ATP-dependent reactions, tonoplast acidification also requires Mg^{2+}, which forms a complex with ATP to give MgATP. The Mg^{2+} requirement is not actually demonstrated by the data presented.

66c. In vivo accumulation of lipid-soluble bases in tonoplasts might be driven by protonation of the bases. Uncharged base molecules would enter the tonoplast and then be protonated in the membrane-limited acidic compartment. Because the protonated base carries a positive charge, it would be membrane impermeable and be retained in the tonoplast. Continued inward diffusion of uncharged base molecules would lead to accumulation of the protonated base in the tonoplast.

Accumulation of lipid-soluble pigments in plant vacuoles might occur by a similar acidification mechanism.

67a. A key distinguishing trait of receptor-mediated endocytosis is saturation. That is, as the concentration of ligand is increased, the uptake rate reaches a maximal value, which is not increased at higher ligand concentrations. Uptake of dissolved solutes by pinocytosis does not exhibit saturation as the solute concentration in the medium is increased. The curves in Figure 14-5a indicate that LDL is internalized by receptor-mediated endocytosis, as its uptake rate saturates, and that sucrose is internalized by pinocytosis, as its uptake rate increases progressively (i.e., does not show saturation).

67b. The K_m for LDL uptake is a little greater than 1×10^{-9} M. By inspection of the graph for uptake versus concentration, a more precise value cannot be given. Because sucrose is internalized by pinocytosis rather than receptor-mediated endocytosis, a K_m value has no applicability in this case.

67c. As shown in Figure 14-5b, the uptake of both LDL and sucrose by cultured fibroblasts saturates with time (i.e., a maximal level is reached). Since LDL is degraded in lysosomes, the apparent plateau in LDL uptake results from the balance between the rate of uptake and the rate of degradation. The reason for saturation in the case of sucrose, which cannot be degraded by mammalian fibroblasts, is less obvious but may involve recycling of endocytic vesicles, which has been demonstrated in fibroblasts and other mammalian cells (see Figure 14-27 in MCB, p. 557). As part of this recycling, solute molecules can be recycled to the medium. As the internal sucrose concentration increases, the rate of sucrose efflux due to recycling would balance the rate of sucrose influx. Despite the occurrence of recycling, the primary reason for the saturation of LDL uptake is the balance between LDL uptake and LDL degradation.

68a. The gaseous air pollutant SO_2 will depress malate levels in guard cells because SO_2 inhibits phosphoenolpyruvate carboxylase activity. As malate is the major counterion for K^+ in guard cells, K^+ levels should be depressed by SO_2.

68b. Decreased K^+ levels should in the end result in decreased osmotic strength and an efflux of water from guard cells. Guard cells would then come together because of decreased turgor pressure and the stomates would close. Closure of stomates would prevent entry of CO_2, which is needed for photosynthesis; thus plant growth would be decreased.

C H A P T E R 15

1. glucose
2. ATP
3. sun
4. Embden-Myerhof
5. four (4)
6. exergonic
7. kinase
8. facultative anaerobes
9. ethanol
10. pyruvate
11. porin
12. cristae
13. pyruvate dehydrogenase
14. lipids, mitochondria
15. glycerol
16. acetyl CoA
17. oxygen
18. F_0F_1
19. membrane
20. proton gradient, electric gradient
21. anions
22. Nernst
23. F_0
24. antiport
25. proton
26. standard free energy ($G°'$)
27. spectroscopic
28. respiratory control
29. thermogenin
30. phosphofructokinase
31. b e
32. a b c
33. c
34. b d e
35. b c d
36. b e
37. e
38. a d
39. a b c e
40. c d
41. b e
42. a
43. a c d
44. a c d
45. d
46. b c d e

47. The energy released during hydrolysis of ATP ($\Delta G°'$) is used to drive most endergonic reactions that occur in the cell. Although other energy-related compounds that contain more energy than ATP (e.g., $FADH_2$) are

present in the cell, they are almost always converted to ATP.

48. The four protons and two of the four electrons produced in glycolysis are used to reduce the electron carrier NAD^+ (nicotinamide adenine dinucleotide) to NADH. NADH then shuttles its electrons to the electron transport chain of the inner mitochondrial membrane where their energy is ultimately converted into ATP. See Figure 15-21 in MCB (p. 603).

49. In the Embden-Myerhof pathway, endergonic reactions are coupled with exergonic reactions, the latter "pulling" glycolysis forward. As a result, the $\Delta G^{\circ \prime}$ for the entire pathway is negative.

50. Because yeast are facultative anaerobes, they will switch to aerobic metabolism if enough O_2 is available. The end products of fermentation, but not of aerobic respiration, are CO_2 and ethanol. Thus yeast in the presence of O_2 will not generate sufficient levels of either CO_2 or ehtanol to give acceptable wine or beer.

51. Both lactic acid and ethanol accumulation result from the lack of O_2 in the environment. In an aerobic muscle cells, glucose cannot be broken down completely and lactic acid accumulates. In an aerobic yeast cells, fermentation results in the production of carbon dioxide and ethanol.

52. The inner membrane has a several-fold higher concentration of proteins than does the outer membrane. Thus the number of "bumps" on the replica should be greater for the inner membrane than for the outer membrane.

53. These compounds gain access to the matrix by the action of permeases in the mitochondrial membrane. Many of these act as antiport systems and are powered by the proton-motive force.

54. The enzymes of the Krebs cycle exist in a multiprotein complex, which is embedded in a gel-like environment within the mitochondrial matrix. Even in dilute buffer the Krebs cycle enzymes maintain this large subunit organization, which guarantees the necessary proximity of one enzyme with the next, so the reactions can proceed.

55. The heme ring of the cytochromes has resonance forms that allow electrons to be delocalized and thus released to other intermediates. In addition, the cytochromes can contain copper or iron, both of which in the oxidized form can accept electrons to form the reduced ion.

56. The reduction potential gradient assures that electrons move in one direction only. This, in turn, results in the pumping of protons from the matrix to the intermembrane space.

57. The proton-motive force is dependent on both the transmembrane proton gradient and the electric potential. To demonstrate that each is sufficient for ATP production, it was necessary to show that the inner mitochondrial membrane can maintain a proton gradient. To test whether the pH gradient is, in part, responsible for the phosphorylation of ADP, one needs to add an uncoupler that diminishes only the pH gradient. However, most uncouplers (e.g., valinomycin, FCCP, DNP) affect both the pH and electric gradients; an alternative is to add ATP synthase (F_0F_1 particle) to synthetic liposomes. By monitoring ATP production and at the same time adjusting the pH gradient, it is possible to drive either the phosphorylation of ADP to ATP or the hydrolysis of ATP to ADP depending on the pH gradient.

58. Anaerobic metabolism does not generate nearly as much ATP as does glycolysis coupled with oxidative phosphorylation. Thus in eukaryotes that are rarely subjected to anaerobic environments, the presence of an anaerobic pathway capable of generating much less ATP than glycolysis would not be an advantage and would not be selected during evolution.

59. The various observed effects of compound X are consistent with its blocking the cytoplasm-to-mitochondria movement of ADP, NADH, etc. The compound probably acts by binding to one of the permeases in the inner mitochondrial membrane, which shuttle various metabolites from the cytoplasm to the mitochondrial matrix.

60a. To distinguish between cell types differing in metabolic activity, you could measure the amount of CO_2 evolved or O_2 used, two parameters that are related to the metabolic activity. The same number of cells should be used for these measurements. A convenient parameter would be to base measurements on the amount of protein. Although the amount of ATP produced also is related to metabolic activity, it is extremely difficult to distinguish ADP from ATP in a biochemical assay and to ensure that some of the ATP is not hydrolyzed during the collection and analysis of mitochondria.

60b. Growing cells in the presence of low O_2 would select for those cells that need less O_2 than normal to survive, presumably because they have a lower metabolic rate.

61. Although a number of factors might be contributing to the low ATP production, the pH probably is the major one. In order to generate ATP from inside-out submitochondrial vesicles, the external buffer must be more alkaline than the interior of the vesicles in order to establish the proton gradient necessary for ATP formation (see Figure 15-18 in MCB, p. 601). Thus increasing the pH of the external buffer should increase ATP production.

62. There could be several reasons for the different P-O ratios in the two preparations. The electron transport system seems to be operating equally well in both cases, since the same number of moles of O_2 is produced in each case. Thus the problem is not at this level. Possibly the membrane of the vesicles is leaky, thus decreasing the proton-motive force. Also, the permeases in the inner mitochondrial membrane may have been artifactually removed during preparation of the vesicles, thus allowing less ADP in and/or ATP out. Other possibilities include damage to the F_0F_1 particle and its subunit integrity.

63. The preferential binding of an antibody indicates that the electron transport carrier with which it interacts is exposed predominantly on either the cytoplasmic or exoplasmic face of the inner mitochondrial membrane. The asymmetric clustering of antibodies indicates that the electron transport carriers are nonuniformly arranged in the inner mitochondrial membrane. This arrangement may be critical for the vectorial transport of protons from the matrix to the intermembrane space. The clustering results from the bivalent nature of antibodies, which bind together, since the electron transport carriers are free to move within the inner mitochondrial membrane. This is analogous to capping and patching of plasma membranes by antibodies against cell-surface antigens. The observation that decreased ATP production is associated with antibody clustering indicates that the mobility of electron transport carriers is critical to their function. Once the antibody-carrier complexes become clustered, proton pumping is partially inhibited. The diminished rate of proton movement then leads to decreased ATP production.

64a. Since the effects of X and Y are not additive, and each has nearly the same inhibitory effect, they probably compromise the same parameter of mitochondrial function. However, the site affected by these compounds cannot be identified based on the data given. For example, both X and Y might block electron transport at the same point in the cascade. Alternatively, each might function as proton ionophores or act equally well to diminish the transmembrane electric gradient. However, since only a small portion of the ATP production in mitochondria results from the pH gradient, the former mechanism probably cannot account for the observed 50–60 percent decrease in ATP production.

64b. Compound Y probably acts as an inhibitor of ATP synthase, since addition of oligomycin, another inhibitor of ATP synthase, does not cause further inhibition. The observation that compound X and oligomycin have additive effects, whereas compound Y and oligomycin do not, suggests that X and Y affect two separate parameters that are involved in oxidative phosphorylation.

64c. The data in part (a) suggest that X and Y affect the same parameter, whereas the data in part (b) suggest that they affect different parameters. Although unusual, inhibitors affecting different parameters sometimes do not show additive effects; perhaps under the experimental conditions in part (a), this is what occurred. Thus it is most likely that X and Y affect different parameters.

65. Dinitrophenol dissipates the proton gradient across membranes, whereas valinomycin decreases the electric potential. Thus their relative effects suggest that the major component of the proton-motive force is the pH gradient in chloroplasts and the electric gradient in mitochondria.

66a. pH and electric potential

66b. Oligomycin is though to block the ability of protons to flow down their gradient through the F_0 particle. Thus oligomycin not only inhibits the phosphorylation of ADP to ATP but also slightly increases the proton gradient. Since the ability of mitochondria to retain rhodamine 123 depends partly on the proton gradient, the staining of mitochondria by rhodamine 123 would be expected to be slightly elevated in the presence of oligomycin.

66c. Nigericin increases the retention of rhodamine 123 in

mitochondria because it increases their transmembrane electric potential. Thus differences in inner membrane permeability cannot account for the observed variation in rhodamine 123 uptake among cells.

66d. Using the amount of O_2 utilized or CO_2 evolved as an indirect indicator of metabolic activity, you could measure the mitochondrial activity in transformed and non-transformed cells. By comparing the two sets of data (i.e., metabolic activity and dye retention), you could determine if a correlation existed between metabolic activity and retention. If this correlation held over a wide range of metabolic activities, induced by treatment with various drugs, then the evidence for a cause-effect relationship would be fairly conclusive.

67a. Temperature-sensitive (ts) mutants, which are genotypically identical, provide the best phenotypic matches. Thus any observed difference in the lethal effects of rhodamine on ts mutants at 39° and 37°C would probably be related to the difference in transformation state and not to other possible differences in the two cell types. Furthermore, the observation that neither the nonmutant parents nor the ts mutants are sensitive to rhodamine 123 at 39°C rules out the mutant gene as the causative factor for the enhanced lethality of rhodamine 123 on the mutant cells at 37°C.

67b. No. As noted in problem 66d, rhodamine 123 is retained for a much longer time in transformed cells than in normal cells. Thus the increased death rate observed with transformed cells might result from this increased retention rather than increased sensitivity to the dye.

67c. If the ts mutants were incubated *continuously* with dye over the course of the experiment, then both cell types would have rhodamine 123 in their mitochondria; that is, any possible effects of differential retention by the transformed and untransformed phenotypes would be negated. Under these conditions, if rhodamine 123 preferentially killed ts mutants at 37°C, you could conclude that mitochondria from transformed cells are more sensitive to the lethal effects of the dye than are mitochondria from untransformed cells.

68a. Since the electron transport system normally pumps protons out of the mitochondria, the pH of the matrix increases; thus the fluorescence of matrix-trapped BCECF should increase in intensity. The observed decrease in intensity suggests that the vesicles have an inverted orientation. In this case, protons would be pumped from the outside to the inside of the inside-out vesicles, leading to a decrease in matrix pH and consequently to a decrease in the emission intensity of trapped BCECF.

68b. The phosphorylation of ADP by ATP synthase on the inner mitochondrial membrane is coupled to and depends on the movement of protons down the pH gradient from the outside to the inside of the mitochondria. Exposure of the inside-out vesicles to dilute acid reverses the normal pH gradient; as a result, the ATP synthase hydrolyzes ATP to ADP.

68c. Dinitrophenol compromises the pH gradient and thus the resulting equilibration of protons results in a decrease in emission intensity. Valinomycin, a potassium ionophore, affects the electric potential more than the pH gradient. BCECF fluorescence does not reflect changes in the transmembrane electric potential.

69a. The effect of drug X on ATP production cannot be predicted from the data given. The responses of the fluorescent dyes suggest that Ca^{2+} ions are leaving the vesicles and that protons are entering them. Thus this system acts as a $Ca^{2+} - H^+$ exchanger, which could actually increase the transmembrane potential at the expense of the proton gradient (similar to the effect of nigericin; see Table 15-1). Since the electric potential is the more important contributor to the proton-motive force in mitochondria, ATP production might increase. However, it is not known drug X induces a one-for-one exchange, and therefore it is impossible to predict its effect on ATP production.

69b. The effect of the protease suggests that the vesicles have a transmembrane *protein* channel (e.g., a permease) that may act as an antiport system or exchanger pumping Ca^{2+} ions out and protons in and that is blocked by the drug. An alternative interpretation is that the Ca^{2+} and proton exchange operate independently through uniport systems, both of which are affected by binding of drug to the transmembrane protein channels.

70a. Chloroplasts generate ATP using photophosphorylation. Thus activation of the light-harvesting complexes are necessary to generate the transmembrane electric potential, which is being monitored in this experiment.

70b. SDS gel electrophoresis of the extracted material left in the supernatant after centrifugation of the remaining membranes should reveal all five subunits if CF$_1$ is removed by method D. The relative intensity of the stained gel bands also should correspond to the stoichiometry of the subunits.

70c. The data suggest that the loss of the δ subunit may result in a leaky thylakoid membrane containing a CF$_0$ particle.

70d. Phenol red could be used to monitor changes in the pH gradient across the membranes prepared by the various extraction methods under conditions similar to those for monitoring changes in the induced voltage (see Figure 15-6). In this way, it could be determined if method D, which removes the δ subunit, has a similar effect on the transmembrane proton gradient as it does on an induced electric potential.

70e. This experiment allows one to distinguish the dissipation of the voltage through CF$_0$ from the dissipation of the voltage through other intregral membrane proteins or parts of the (leaky) membrane. A full block (i.e., DCCD-dependent return to control levels) indicates that the dissipation of the voltage is probably all mediated through the CF$_0$ particle.

70f. This hypothesis could be tested by isolating the δ subunit by chromatography and then adding it to the membranes resulting from extraction methods A – D. If δ acts as a stopcock, then the voltage dissipation profile and pH gradient profile of membrane D + δ should be the same as those for membranes A, B, and C. Furthermore, the profiles for membranes A, B, and C should not be affected by the addition of δ.

71a. Oligomycin sensitivity indicates that the F$_0$F$_1$ particle is intact. ATP hydrolysis indicates that the F$_1$ particle is present in the preparation and operating efficiently. However, this assay is not critical for the experiments described in this problem.

71b. The results with samples 2 and 3 indicate that diamide alters oligomycin sensitivity but does not affect the F$_1$ particle (i.e., ATP hydrolysis). However, the order in which diamide is added is critical to its effect on oligomycin sensitivity.

71c. Sample 2 suggests that thiol groups may be important for the proper association between F$_0$ and F$_1$. Sample 4 suggests that these thiol groups are protected once the two units are coupled together.

71d. The ability of DTT to reverse the diamide effect indicates that diamide is probably exerting its effect by interacting with thiol groups and not in a nonspecific manner, which would not be reversible by DTT.

71e. These additional data also indicate that diamide inhibits the binding of F$_0$ and F$_1$ and that DTT can reverse this effect. There is a significant amount of F$_1$ in the supernatant of sample 2 but not in the supernatants of the other samples. In sample 2, F$_1$ was unable to bind to F$_0$ and sediment accordingly. However, all other preparations have most of the ATP hydrolysis activity in the pellet, implying that a complete F$_0$F$_1$ particle has been reconstituted.

C H A P T E R *16*

1. chloroplasts
2. sucrose
3. heme
4. water
5. water
6. NADP$^+$
7. chlorophyll *a*
8. plasma
9. cytochrome, quinone
10. Emerson
11. light-harvesting complex
12. manganese
13. 700, 680
14. triazine
15. Calvin
16. stroma
17. glyceraldehyde 3-phosphate
18. photorespiration
19. low, high
20. bundle sheath
21. b c

22. c

23. d e

24. a b c d e

25. c

26. e

27. a c

28. a b c d e

29. b d

30. b

31. b d

32. The chemical structure of porphyrins is highly conjugated, which facilitates delocalization of electrons.

33. Absorption of light causes electrons to move to the stromal surface, thus creating a transient positive charge on the luminal surface of the thylakoid. As a consequence, H_2O is split by enzymes located at the luminal surface, electrons are donated, and O_2 and protons produced.

34. In noncyclic photophosphorylation the sequence of electron flow is $PSII \rightarrow PSI \rightarrow NADP^+$. See Figure 16-16 in MCB (p. 630).

35. CF_0CF_1 (coupling factor) contains the ATP synthase in chloroplasts.

36. The reactions that result in carbon fixation can occur both in dark and light conditions, and they are indirectly dependent on light reactions.

37. The energy of a photon is inversely proportional to its wavelength. Because blue light has a shorter wavelength than red light, its absorption would deliver more energy to the photosynthetic apparatus than would absorption of red light.

38. The absorption spectrum of chlorophyll (i.e., the relative intensity of light of various wavelengths absorbed by it) is very similar to the action spectrum of photosynthesis (i.e., the relative rate of photosynthesis at various wavelengths of incident light). This similarity suggests that chlorophyll is critical to photosynthesis. See Figure 16-7 in MCB (p. 622).

39. During noncyclic electron flow in purple bacteria, hydrogen gas (H_2) or hydrogen sulfide (H_2S) rather than H_2O donates electrons to reduce oxidized cytochrome *c*. This oxidation produces sulfur gas from H_2S and H_2O from H_2. See MCB (p. 624) for details.

40. The differential solubility of the two photosystems suggests that PSII is more firmly embedded in the thylakoid membrane than is PSI or is surrounded by different lipids that are resistant to detergents.

41. Freeze-fracture has revealed that PSI and PSII particles are not uniformly distributed in thylakoid membranes. Mobile carriers are necessary to transfer electrons from PSII to PSI.

42. A protein kinase could phosphorylate the light-harvesting complex of PSII, thus decreasing the energy-transfer efficiency of PSII. Such a control mechanism does, in fact, operate in plants. See MCB (p. 630).

43. If the level of CO_2 falls below the K_M, then CO_2 cannot be fixed directly by ribulose 1,5-bisphosphate carboxylase in bundle sheaf cells. In such plants, called C_4 plants, CO_2 in mesophyll cells reacts with phosphoenolpyruvate to produce the four-carbon compound oxaloacetate, which is reduced to malate. Malate is then shuttled to the bundle sheaf cells, where it is decarboxylated, releasing CO_2 for the Calvin-cycle reactions. See Figure 16-20 in MCB (p. 635).

44. Because purple bacteria usually exhibit cyclic electron flow and do not evolve O_2 or reduce $NADP^+$ (see Figure 16-16 in MCB, p. 630), the absence of these reactions in the treated cells provides no information about the mechanism of compound X. Likewise, the observation that ATP is not synthesized is of little use in solving the problem because this could result from many different causes. Among the underlying causes of an inhibition of photosynthesis are (1) changes in the pH gradient, (2) blockage of f CF_0CF_1, and (3) blockage of the cytochrome bc_1 oxidoreductase complex or other component involved in electron flow. Not enough information is given in the problem to conclude which of these mechanisms applies to compound X.

45. Freeze-fracture electron microscopy is the only technique that would allow you to make these measurements. Freeze fracture reveals both PSI and PSII, which can be distinguished because of their difference in size. Thus the ratio of PSI to PSII can be determined by simple image analysis. Similarly, distances between the two can be accurately measured as well.

46. To generate a pH gradient, electrons must be transported from the lumen to the stroma. These electron

carriers must be associated closely enough with each other to effect this transfer but nonetheless must move the electrons from the lumen to the stroma. Thus their asymmetric arrangement is consistent with the lumen-to-stroma movement of electrons hypothesized in the current model of photophosphorylation. See Figure 16-16 in MCB (p. 629).

47. Our current understanding of noncyclic photophosphorylation suggests that electrons are moved through plastiquinone in PSII and ultimately to ferredoxin in the PSI phase. Thus an agent that blocks plastiquinone should block electron transfer to ferredoxin. Arnon's experiments suggest that PSI and PSII may not be coupled in linear electron flow.

48. Ammonium chloride is a base that can cross the thykaloid membrane into the lumen where it compromises the pH gradient across the membrane. This reduction in the pH gradient results in a drop in the photosynthetic efficiency of the membranes (i.e., decreased ATP production). However, electrons can still flow almost normally, so O_2 evolution is nearly the same in NH_4Cl-treated and untreated preparations.

49. In intact thylakoids, the oxygen-evolving complex (OC) is on the lumenal side of the membrane. When thylakoids are sonicated, some of them reseal right-side out, producing vesicles with the OC facing the lumen or cisterna; however, some reseal inside out, producing vesicles with the OC on the stromal side. Because antibodies cannot penetrate membranes, the anti-OC antibody used in the experiment would react only with inside-out vesicles. The right-side–out vesicles, which cannot react with the antibody, would remain in the supernatant in the experiment.

    Since protons accumulate in the lumen of intact thylakoids, they would accumulate on the outside of the inside-out OC vesicles, which in turn makes the medium acidic. Conversely, the right-side–out vesicles left in the supernatant accumulate protons within the lumen, thus causing the medium to become basic.

    Treatment of both vesicle preparations would make the membranes leaky, thus compromising the pH gradient and reducing the photophosphorylation of ADP to ATP.

50. The smaller vesicles probably contain only PSI, or they contain PSI and PSII but lack a component that functionally links the two photosystems. As a consequence, these vesicles can carry out cyclic, but not noncyclic, photophosphorylation. The larger vesicles contain functionally linked PSI and PSII and thus can perform noncyclic photophosphorylation, which is associated with formation of O_2 and NADPH.

51. The data provided indicate that DCMU does not affect the PSI cycle and does not interrupt electron flow before plastiquinone. Based on these data, DCMU seems to block photophosphorylation between plastiquinone and PSI. The actual site of action of this drug is between plastiquinone and cytochrome *f*, although this cannot be determined from the data given. See Figure 16-15 in MCB (p. 629).

52. To test this hypothesis, extract PSI from thylakoids by a relatively gentle treatment, which leaves PSII intact in the membranes. After separating chlorophyll *a* from the reaction center of PSI by acetone extraction followed by high-pressure liquid chromatography, determine the fluorescence profile of the extracted material in the presence and absence of compound Y. Next extract PSII from the membranes using more rigorous extraction procedures and perform the same spectrofluorometrical analysis. If compound Y can bind directly to the chlorophyll *a* in the reaction centers, then in the presence of Y a change in the excitation wavelength or the presence of an Emerson effect probably would occur.

53. The data in Figure 16-1a show that wild-type but not mutant chloroplasts can perform photophosphorylation (i.e., time- and light-dependent ATP production). However, the data in Figure 16-1b indicate that the defect is not in the CF_0CF_1 particles, since dephosphorylation of ATP to ADP occurs nearly as well in the wild-type and mutant chloroplasts. That is, ATP synthase probably is normal in the mutant plants. The defect might be in PSI, which would prevent both cyclic and noncyclic photophosphorylation.

54a. The most probable cause for the decrease in triose release is a poorly functioning antiport system, which brings CO_2 into the chloroplast stroma and transports triosephosphate out to the cytoplasm. The data indicate that PSI and PSII were operating normally and that the dark reactions also were normal, as ATP and NADPH levels remained relatively constant. The Triton X created a leak in the membranes, causing the accumulated triosephosphate to be released in a nonspecific manner. Thus compound Z appears to inhibit the transport of triosephosphate by the antiport system but does not affect the transport of CO_2. The time dependence of the decrease in release rate shown in Figure 16-2 reflects the time required for Z to enter the chloroplast membrane and begin to act.

54b. The accumulation of ATP and NADPH indicates that PSI and PSII were operating normally but suggests that the Calvin cycle was not. Since the dark reactions utilize ATP and NADPH, these metabolites should not accumulate in the stroma of chloroplasts if the Calvin cycle is operating normally. Thus compound Z appears to inhibit one or more steps in the Calvin cycle, thus reducing the production of triosephosphate. In this case, since the production not simply the release of triosephosphate is decreased, Triton X would not be expected to reverse the effect of compound Z.

55a. The fixation of CO_2 measured in this assay is indicative of carboxylase activity.

55b. The activase has no effect on the initial rate of CO_2 fixation by RBPase.

55c. No data from reversal experiments is given. In such an experiment, one would have to add activase at a point during the fallover (e.g., at the 30-min time point) and determine if the fixation rate increases after the addition of activase.

55d. RBPase and RuBP are associated via non covalent bonds, since they are separated by gel filtration. Also once RuDP is removed, RUBISCO is activated.

55e. First, the inhibitor is tightly bound to RBPase, since it cannot be separated by gel filtration. Second, phosphorylation is critical to the function of the inhibitor, since dephosphorylation reverses it.

55f. The precise point of inhibitor binding to RBPase could be determined with a photoaffinity analog.

56a. Both these agents fully extract the thylakoid membranes, thus releasing the pigments, which are more soluble in either methanol or acetone than in the membranes.

56b. Freezing and thawing might enhance the release of tenacious chlorophyll molecules. Freezing and thawing of proteins, however, often change their conformation

and, in turn, dramatically decrease their enzymatic activities.

56c. Rechromatograph the peak labeled chl c_1/c_2 under identical conditions. If a new chl c peak reappears, then there is a good possibility that it is an artifact resulting from alteration of chl c_1 and/or c_2 during chromatography.

56d. The differences in the spectra indicate that the three pigments are structurally dissimilar. Since structural differences often are associated with functional differences, the pigments also may have different functions.

56e. If a "new" pigment can be separated from closely related pigments using diverse techniques, then the argument that the "new" pigment is indeed novel and not merely an artifact of the extraction-separation scheme is further substantiated.

57a. The best approach would be to extract all the pigments and then separate them chromatographically, as described in problem 56. Spectral analysis of the separated peaks could be used to identify them. Compared with normal rice, mutants defective in chlorophyll b should either contain less of this pigment or an abnormal pigment with abnormal spectral properties.

57b. Because RBPase is a stromal enzyme, it should not be present in thylakoid membrane preparations. However, the great quantity of RBPase in the stroma makes it a common contaminant of thylakoid membrane preparations in which the enzyme sticks nonspecifically to the membranes.

57c. This subunit of RBPase probably is synthesized in the chloroplasts, not in the cytoplasm, since chloramphenicol blocks chloroplast (and mitochondrial) protein synthesis, whereas cyclohexamide blocks cytosolic protein synthesis. An alternative explanation would be that this RBPase subunit is synthesized in the cytoplasm but binds to a chloroplast-synthesized protein; however, current research not presented in the problem suggests that this is less likely than synthesis in chloroplasts.

57d. Chloramphenicol inhibits the synthesis of Qb, which comigrates with LHC-II, but does not affect the synthesis of the LHCs. In the absence of chloramphenicol, it is very difficult to discern LHC-II from Qb in these experiments.

57e. These data indicate that the synthesis rates are similar in all three strains.

57f. Two possible methods could be used to determine the amounts of each LHC synthesized quantitatively. First, densitometric tracings of the gels could be accomplished. Second, the LHC-I and LHC-II bands could be cut out of the gels and the amount of radioactivity in the separated proteins measured in a scintillation counter.

58a. No. Since the intensity of the Qb band decreased dramatically over time and the same number of counts were loaded on each lane, the increase in the intensity of the LHC-II band may be merely an artifact of this loading procedure and represent the LHC-II compensation of the decline of Qb.

58b. These data indicate that although the rates of synthesis of the LHC-I and LHC-II proteins are similar in the wild-type strain and mutant strains, the apoproteins from the wild type are more stable than those from the mutant.

58c. The results with chlorina 11 suggest that the effect of a mutation on LHC stability is correlated with the corresponding decrease in chlorophyll *b*. This finding supports the hypothesis that chlorophyll *b* stabilizes the apoproteins but is not definitive evidence.

C H A P T E R 17

1. ATP
2. amphipathic
3. endoplasmic reticulum
4. cytoplasmic
5. Golgi
6. free
7. bound ribosomes
8. cytoplasm
9. unfolded
10. endoplasmic reticulum (ER)
11. O-linked
12. conformation
13. trans
14. tunicamycin
15. degradation
16. NSF
17. phosphorylation
18. pH
19. propeptide
20. regulated
21. hormones
22. regulated
23. capacitance
24. b
25. d
26. b c e
27. c
28. b
29. a
30. c d e
31. e
32. a b c d
33. b
34. c d
35. a
36. c d
37. a b c

38. Bound and free ribosomes can be separated by disruption of cells followed by differential centrifugation. Studies of these fractions demonstrated that bound and free ribosomes are functionally similar. In addition, isolated bound ribosomes have been utilized to study the role of the signal recognition particle (SRP) and other components of cotranslational protein synthesis.

39. The best technique to follow a particular protein in a cell at this level is ultrastructural immunocytochemistry. Although movement of proteins can be detected using ultrastructural autoradiography, a specific protein can be tracked only by use of immunocytochemical methods.

40. Labeled leucine is preferable to other labeled amino acids (e.g., [^{32}S]methionine) because leucine is present in most cellular proteins and thus greater label incorporation occurs with this precursor than with others. For technical reasons, tritium is the best radioactive isotope to use in autoradiographic experiments such as these (see Chapter 6 in MCB).

41. The protein within zymogen granules is more than 100 times more concentrated than that within the endoplasmic reticulum and thus stains more intensely with heavy metals.

42. Studies with various artificial chimeric proteins containing a native signal peptide attached to the N-terminal end of a nonsecretory protein have indicated that the signal peptide generally is the only portion necessary for proper translocation.

43. Because proteases cannot pass across cell membranes, an exogenously added protease can digest only that portion of a membrane protein that is not embedded in the membrane; the rest of the protein, which is protected by the phospholipid bilayer, is not removed. By exposing both the inside and outside of vesicles to protease, stop-transfer anchor sequences (i.e., the parts of the protein not digested) can be characterized.

44. A signal sequence can be a topogenic sequence if it is embedded in the membrane and thus directs the topology of a protein within that membrane. One protein containing such a signal-anchor sequence is the glucose transporter (see Figure 17-19 in MCB, p. 658).

45. Mammalian cells can modify proteins post-translationally, whereas bacteria do not. Such modifications can be important for the proper activity of secreted proteins and often are necessary to inhibit their degradation by exogenous proteases.

46. A protein with Lys-Asp-Glu-Leu at the C-terminus is recognized as a resident ER protein and retained in the endoplasmic reticulum. Proteins that are normally secreted but have had this tetrapeptide sequence attached at the C-terminus are also retained in the ER.

47. Ultrastructural immunocytochemistry has demonstrated both a cis-to-medial and proximal-to-distal asymmetry in enzyme localization within the Golgi complex. Also, different types of modifications and sorting occur in different sacs. Thus the activities of the Golgi apparatus occur in an ordered sequence and exhibit spatial and functional specialization, like those of an assembly line.

48. The proper glycosylation of the LDL receptor is not necessary for its synthesis and insertion into the plasma membrane, but glycosylation seems to make the receptor resistant to proteolytic degradation by extracellular proteases.

49. Insulin associates with its receptor at pH 7.0 but not at pH 5.5, the pH of the vesicles in which the hormone is stored. Also, insulin is synthesized and stored as preproinsulin, which is less effective than the mature hormone in associating with the receptor.

50. The best method for examining fluorescently tagged phospholipids would be to use either a fluorescent or confocal microscope, the latter with fluorescent capabilities. The major problem in this approach is to ensure that the phospholipid molecules are indeed *in* the membrane and not *on* the membrane. This can be achieved by a vigorous washing and subsequent examination. About 30 min after adding a tagged phospholipid, it should be found in small vesicles (endocytotic vesicles) and later in the endoplasmic reticulum, usually the smooth endoplasmic reticulum.

51. One experimental approach is to pulse-label cells for various time periods with a radioactive fatty acid (e.g., [^{3}H]palmitate), which will be incorporated into newly synthesized phospholipids, then fractionate the cells, and measure the radioactivity in both the smooth ER and plasma-membrane fractions. The time between the first appearance of label in the smooth ER fraction and in the cell-membrane fraction corresponds to the time it takes newly synthesized phospholipid to be incorporated into the plasma membrane.

52. The results of a continuous pulse experiment will indicate the cellular compartments in which the labeled entities—in this case newly synthesized protein—are localized. If cells are examined after various labeling periods, this type of experiment also can demonstrate the sequence of compartments through which the labeled protein moves. From the results of pulse-chase experiments utilizing a series of chase times, one can also determine the time the labeled protein spends in each compartment and also whether a labeled protein retraces its movement back to a previous compartment. (This latter phenomenon has never been observed, but it should not be disregarded a priori.) In other words, with a pulse-chase experiment, one can determine both the sequence and timing of movement of newly synthesized material within the cell.

53. The enzymatic activity of many secretory proteins depends on the addition of sugar residues and other modifications that occur in the Golgi complex after synthesis by bound ribosomes. An enzymatic assay thus would not necessarily detect the synthesis of such a protein in an ER preparation. In contrast, an antibody may be able to specifically recognize its antigenic protein even with-

out these modifications; in this case, an antibody assay would detect synthesis of the protein in an ER preparation.

54. First the mutants should be analyzed to determine in which compartments (e.g., the rough ER, Golgi complex, secretory vesicles) proteins accumulate at high temperature. This can be accomplished by a shift analysis in which the ultrastructural localization of labeled secretory protein is compared at low and high temperatures. Second, by doing various crosses between the different strains identified by shift analysis and analyzing the proteins that are "stalled" in the maturation pathway, a thorough understanding of the specific types of protein modifications and their consequences in protein trafficking could be achieved. See Figure 17-10 in MCB (p. 650).

55a. One explanation is that the mRNA used in this experiment encodes a secretory or membrane protein and that the high-speed supernatant contained rough microsomes (i.e., bound ribosomes) as well as polysomes. Translation of the mRNA on the polysomes would produce the larger-molecular-weight species revealed by gel filtration. Translocation of this species across the microsomal membrane and removal of the signal by a signal peptidase would result in the smaller-molecular-weight protein observed. Other explanations are possible, but protein degradation probably is not one of them because both peaks are quite narrow.

55b. A number of techniques that utilize monoclonal antibodies—for example, immunoprecipitation and Western blot analysis—can be used to determine whether two protein species are similar. If they are related, antibodies to one should react with the other in either of these assays.

56. When mRNA encoding preprolactin, a typical secretory protein, was incubated in a cell-free translational system in the absence of SRP and microsomes, the complete protein with its signal sequence was produced. The addition of SRP to the incubation mixtures caused protein elongation to cease after 70–100 amino acids had been incorporated. If microsomes were then added to the incubations, protein synthesis recommenced and the complete protein was extruded into the lumen of the microsomes. See Figure 17-15 in MCB (p. 654).

57. The data indicate that the protein of interest is being synthesized by the neuronal cells but is not being re-

leased into the lumen of secretory vesicles. Thus the portion of the cDNA to alter would be the stop-transfer anchor sequence. Changing a few amino acid residues in this sequence might create a defective stop-transfer anchor sequence, thus allowing the protein to be secreted into the lumen of the secretory vesicles and ultimately to be secreted. With some luck, this modified protein would still demonstrate as much function as its native, particulate counterpart.

58. Both initial and terminal glycosylation can be tracked using ultrastructural autoradiography. Radioactive tracers such as [³H]glucose, [³H]fucose, or [³H]mannose are added to the medium bathing cultured cells, and then cells are sectioned for autoradiography. With the proper type of fixative, the unincorporated, labeled sugars are removed during the dehydration steps. However, those that have been covalently bound to proteins will remain in the cell and can be localized to particular cellular compartments (e.g., endoplasmic reticulum; cis, medial, and trans Golgi), via autoradiography.

59a. Because proteinase K cannot penetrate the microsomal membrane, it can digest only proteins that are not translocated into the lumenal space of the ER. Thus the diffuse bands in Figure 17-2 represent the degradation products of newly synthesized protein that was not translocated during the incubations. The discrete bands represent newly synthesized protein that was translocated into the ER lumen during the incubations; because this protein was enclosed by a membrane, it was protected from the action of the protease.

59b. Comparison of lanes 1 and 6 indicates that ATP or GTP or both is necessary for translocation. However, since translocation did not occur in the presence of ATP alone (lane 7), GTP must be a necessary factor. Comparison of lanes 3 and 4 indicates that hydrolysis of ATP but not of GTP is necessary for translocation, since AMP-PNP in the presence of GTP prevented translocation, whereas GMP-PNP in the presence of ATP did not. Thus, although both ATP and GTP are required for translocation, only ATP is hydrolyzed.

Lane 2 indicates that EDTA inhibits translocation of the nascent protein, suggesting that it "uncouples" the SRP-ribosome complex or prevents ATP hydrolysis, perhaps by chelating Ca^{2+} ions.

Finally, comparison of lanes 1 and 5 suggests that a signal peptidase is present in the translational system. When the incubation time was extended, the peptidase cleaved the 45-kDa nascent protein (lane 1) into the 40-kDa mature protein and a small signal peptide (lane 5).

60a. GTP binds to both the α and β subunit, since the covalently bound GTP photoaffinity analog is found in the gel at positions corresponding to both subunits in lane 1. ATP also may bind to the α subunit, since the ATP photoaffinity analog binds to this subunit (lane 6) and 20 μM ATP is able to displace some of the GTP photoaffinity analog from the α subunit (lane 4). Comparison of the intensity of the 68-kDa bands in lanes 1 and 7 indicates that GTP binds much more strongly than ATP to the α subunit.

60b. To respond to the reviewer's questions, you need to determine whether GTP binds to the separated α and β subunits. This can be accomplished by separating the subunits of the isolated SRP receptor on a SDS gel and then transferring the subunits to a nitrocellulose membrane (the Western technique). Now use a renaturing buffer to reconstitute the "native" form of the separated α and β subunits. Finally, add the GTP photoaffinity analog, irradiate, and do autoradiography to determine which of the two subunits has covalently bound the photolabel. Comparison of these data with those in Figure 17-3 (lane 1) should allow you to determine whether the photoaffinity analog of GTP binds to sites on both subunits, to a single site on one of the subunits, or to a single site shared by the subunits.

61a. Since glutaraldehyde cross-links proteins, the 400-kDa band in lane 2 represents a protein in which the four SP subunits have been cross-linked. The fact that the cross-linked protein is enzymatically active suggests that formation of the tetrameric protein occurs in the endoplasmic reticulum. The results of mild protease digestion suggest that the 180- and 150-kDa subunits, but not the 50- and 20-kDa subunits are on the "outside" of SP and thus are accessible to protease. Protease digestion results in a decrease in the molecular weight of the 150- and 180-kDa subunits, as evidenced by their slightly faster migration in lanes 3 and 4 compared with the migration of the native subunits in lane 1. The two lowest molecular weight species in lanes 3 and 4 represents the small portions of the 150- and 180-kDa subunits clipped off by the proteases.

61b. Modification (e.g., addition of sugars, proteolysis) of SP after it leaves the endoplasmic reticulum might alter its antigenic properties so that it no longer binds to the antibody. In this case, the modified protein located in Golgi vesicles would not be demonstrated by ultrastructural immunochemistry.

62. The appearance of a normal phenotype in B cells when

they are incubated with A cells suggests that mutant B cells can endocytose the normal lysosomal enzymes secreted by A cells; thus the lysosomal receptors probably are normal in B cells. In contrast, C cells probably have deficient receptors (or none at all) and thus cannot endocytose lysosomal enzymes secreted by A cells. The chromatographic data indicate that the lysosomal hydrolytic enzymes in A and C cells are identical, whereas the aberrant elution pattern of the B-cell enzymes suggest that these enzymes are defective. Thus B cells contain normal lysosomal receptors and abnormal lysosomal enzymes, whereas C cells contain abnormal receptors and normal lysosomal enzymes.

63a. Expressing these volumes as ratios corrects for any variation in cell size. Independent experiments in which all variables except cell size are held constant have indicated that the larger the cell, the larger the Golgi complex and TGN.

63b. After the cold-induced block is released, the TGN volume decreases in a time-dependent manner, whereas the Golgi volume increases. If the TGN : Golgi volume ratios are calculated for the various times and these values (0.76, 0.43, 0.28, and 0.15) are plotted against the time after release, a curved line results (Figure 17-6). This result indicates that the rates at which the two volumes change are different; thus the decrease in TGN volume is not exactly compensated by the increase in Golgi volume. If the changes balanced each other, then the plot of the ratio versus time would be a straight line.

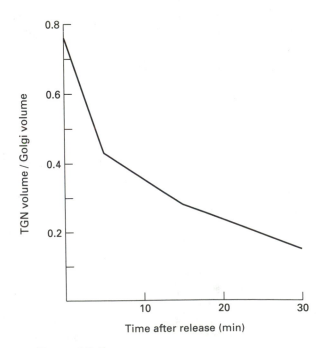

▲ **Figure 17-6**

CHAPTER 18 **341**

63c. Table 17-2 shows that there is an overall decrease in the Golgi surface area and an increase in the TGN surface area at 20°C compared with 32°C. The increase in the TGN, however, is much smaller than the decrease in the Golgi, suggesting that there is a loss of total membrane surface from these two organelles at the lower temperature. Addition of the values in the two rows confirms that the total surface area of the Golgi and TGN is smaller at 20° than 32°C.

63d. One possible explanation for the apparent loss of membrane surface at 20°C is that some of the "lost" Golgi membrane is incorporated into the rough endoplasmic reticulum. This hypothesis could be tested by determining the ratio of the rough ER surface area to the cell volume, using glucose 6-phosphatase as a cytochemical marker for this organelle.

64a. The Western blots indicate that there is a progressive decrease in the immunoreactivity of both MPR and LGP120R in the presence of H_2O_2 dependent on the exposure time to horseradish peroxidase. Since H_2O_2 inhibits antigenicity only in the presence of HPR, these data suggest that an active endocytic process is occurring to bring external HPR into vesicles containing MPR or LGP120R or both.

64b. Inspection of Figure 17-5 shows that the inhibition of immunoreactivity occurs more quickly in the case of LGP120R than of MPR. Thus the endocytic event that results in inhibition (see answer 64a) of LPG120R reactivity likewise occurs more quickly than the endocytic event that affects MPR reactivity. This suggests that there probably are two, but not necessarily, four separate species of vesicles. To determine if there are actually four types of vesicles, one would need to perform ultrastructural immunocytochemical studies using double labeling with antibodies to both receptors. Indeed, the results from such studies led to the hypothesis that four types of vesicles may exist.

C H A P T E R *18*

1. cytoplasm
2. pores
3. chimeric
4. intermembrane space
5. diameters
6. five
7. porin
8. ribulose 1,5-bisphosphate carboxylase
9. tail
10. matrix
11. conformation
12. mitosis
13. B
14. introns
15. uptake-targeting sequences
16. C and A
17. electron microscopy
18. folding
19. maternal
20. hybridization
21. kinase
22. decondensed chromatin
23. anaerobically
24. endosymbiosis
25. nucleus
26. fission
27. translocation of proteins
28. b c
29. c
30. a b c e
31. b c e
32. d
33. e
34. b c
35. b c d e
36. a b c d
37. c
38. b c d
39. a b c d

40. Phosphorylation of lamins precedes dissolution of the nuclear envelope.

41. No. It is thought that mitosis-promoting factor (MPF) triggers condensation of chromatin.

42. Because of the large pools of histones and lamins in the egg, the fertilized embryo can move more rapidly into the multicell stage than would be possible if these proteins had to be synthesized de novo. Experiments with inhibitors that block later, but not early, stages of embryogenesis have demonstrated that in the early embryo cell division in fact does occur with little or no synthesis of histones and lamins.

43. Because antibodies injected into the cytoplasm cannot pass into the interior of the nucleus where lamins are polymerized, they cannot block nuclear reassembly. To accomplish this, an intranuclear injecton or a similar in vitro approach is necessary to put the antibodies in the proper cellular compartment.

44. Nucleoplasmin tails act like a signal that facilitates entry of this protein into the nucleus, whereas the core facilitates retention of the tails in the nucleus. See Figure 18-3 in MCB (p. 684).

45. The ability of mild protease treatment to reduce nucleoplasmin uptake suggests that there is a nucleoplasmin receptor on the outer surface of the nucleus that is required for nucleoplasmin uptake. Protease degradation of this receptor would reduce nucleoplasmin-receptor interactions and hence uptake.

46. The existence of yeast mutants containing mitochondria but little mtDNA implies that biogenesis of mitochondria is dependent on nuclear-encoded genes.

47. Because RDPase is a chloroplast enzyme involved in the fixation of CO_2 in the Calvin cycle, genes encoding this protein would not be expected to be present in mtDNA. The presence of such genes in plant mtDNA suggests that DNA can be exchanged from chloroplasts to mitochondria.

48. Unfolding proteins bind to cytoplasmically synthesized mitochondrial proteins and are necessary for their translocation to mitochondria. Thus they "chaperone" mitochondrial proteins.

49. Proteinase K partially degrades the receptors for the matrix and stromal proteins, thus inhibiting uptake of these proteins by the organelle. Because these receptors are located in the outer membranes of chloroplasts and mitochondria, they are susceptible to protease degradation.

50. Because the precursor protein contains an N-terminal matrix-targeting sequence, the N-terminus of the precursor molecule is translocated first. An antibody to the C-terminal end would not prevent passage of the N-terminal end across the mitochondrial membrane. However, antibodies are too large to pass through a lipid

bilayer without the aid of a specific antibody receptor. Thus binding of an antibody to the C-terminal end of the precursor protein would inhibit passage of the C-terminus into and through the membrane, thus preventing completion of translocation. See Figure 18-15 in MCB (p. 698).

51. Energy is required for translocation of proteins to both chloroplasts and mitochondria. However, a membrane electric potential is necessary for translocation of matrix proteins to mitochondria, whereas ATP hydrolysis in the stroma of chloroplasts is necessary for translocation of proteins to the interior of this organelle.

52. One explanation of these data is that nuclear envelopes contain receptors only for lamin B; hence lamin A alone does not bind. However, the binding of lamin A in the presence of lamin B further suggests that A binds to B or to a site regulated by B. An alternative explanation is that A and B form a complex, which then binds to the envelope (i.e., the nuclear receptor is specific for the A-B complex). An additional experiment testing the ability of lamin B alone to bind could distinguish between these two possible explanations. In fact, the nuclear envelope binds only lamin B, and lamins A and C bind to lamin B.

53. Lamin B is associated with small vesicles during mitosis, whereas lamins A and C are free in the cytoplasm. Thus lamin B is not immunoprecipitated well in mitotic 3T3 cells because most of the lamin B is enclosed in vesicles of nuclear origin and thus is inaccessible to antibody. To test the specificity of the presumed anti-B antibody, the nucleus of nonmitotic should be used as the source of lamins; in this case, all three lamins can be detected with appropriate antibodies.

54. Since ^{32}P-labeled inorganic phosphate can pass into cells, the lack of lamin labeling results from dilution of the label by a large, nonradioactive intracellular pool of ATP and phosphate. This problem can be avoided by growing the cells in phosphate-free medium for 24 h to lower the pool size before adding the labeled phosphate. An alternative approach is to increase the labeling time or the specific activity of the ^{32}P, so as to increase the radioactivity of the ATP-phosphate pool.

55a. To control for the microinjection technique, you could use bulk injection with liposomes containing the mammalian DNA in their interior.

55b. To show that foreign material per se does not elicit the response, you could use the most closely related material that is known not to be linked to lamins in vivo; the best choice would thus be frog RNA.

56. To test whether small proteins can enter the nucleus, first radioactively label several purified small proteins. Then lyse cells with NP-40 or a similar agent and spin at 800 g to separate the nuclei. Resuspend the isolated nuclei and incubate them with each tagged protein for about 30 min. Finally, separate the nuclei from the incubation buffer and determine the percentage of the radioactivity in the nuclei and buffer.

   To examine whether a soluble factor(s) is involved, add cytosolic extracts to the incubations and determine if the percentage of radioactivity in the nuclei increases. In fact, most globular, non-nuclear proteins <9 nm in diameter can diffuse into the nucleus without the aid of a cytoplasmic factor.

57. Since protein synthesis in mitochondria and bacterial cells exhibit the same antibiotic sensitivities, mitochondria probably evolved from a prokaryote.

58. The finding that both types of mRNA can support in vitro protein synthesis suggests that their ribosomal-binding sites are similar.

59. The ability of cytoplasm, but not buffer alone, to effect translocation suggests that s soluble factor is involved in translocating this protein to mitochondria. This soluble factor may be a protein, since boiling of the cytoplasm destroys its transport ability. Boiling denatures proteins, causing loss of their enzymatic ability.

60. The nature of the energy inputs can be investigated in an in vitro mitochondrial translocation system (see Figure 18-11 in MCB, p. 694). Once successful translocation has been accomplished, add to the assay system an uncoupler of oxidative phosphorylation (e.g., valinomycin, FCCP, or DNP) that compromises the electrochemical gradient across mitochondrial membranes. As long as the ATP pools are sufficient in the buffer (~ 5 mM), such uncouplers should not affect the ability of unfolding proteins to maintain precursor proteins in the proper state for translocation. If the mitochondrial transmembrane potential is necessary for proper translocation, then there should be a rapid decrease in the translocation rate after addition of the uncoupler.

   To examine the role of ATP in maintaining precursor proteins in the proper state for translocation, AMPPNP, a nonhydrolyzable analog of ATP, is substituted for ATP in the assay system. If translocation proceeds normally with AMPPNP, then ATP is used to facilitate binding of precursor proteins and is not used as an energy source.

61. There probably are two receptors: one binds A and B, and the other binds C. Comparison of curves (1) and (3) confirms the conclusion stated in the problem that all three proteins require a receptor on the outer mitochondrial membrane to be translocated. When equal concentrations of A* + B or A + B* are incubated together, neither labeled protein achieves the same level of translocation as when each is incubated alone (curve 2 versus curve 3). These results suggest that A and B compete with each other for access to the same receptor. In contrast, translocation of C* is unaffected by the presence of A or B (curve 1), suggesting that C interacts with a different receptor than A and B do.

62. Translocation of proteins across the inner mitochondrial membrane is dependent on the membrane potential, but the rate of translocation is limited by the number of receptors on the outer membrane to which most, if not all, matrix proteins must bind in order to be transported. Thus a drug-induced increase in transmembrane potential causes a transient increase in the translocation rate, but the limited number of receptors prevents the rate from increasing beyond a certain maximal value, regardless of the drug concentration. In the example given, saturation is reached at about 3 μg/ml nigericin.

63a. These data suggest that tails are required for translocation but not necessarily a full complement of tails.

63b. The data do support the hypothesis, as the more tails present, the more nucleoplasmin is found in the nucleus. It is known, however, that tails prevent the degradation of the core (see MCB, p. 684); thus an alternative intepretation of the data is that all species translocate equally well, and the species with the fewer tails are degraded faster than those with more tails. The absence of low-molecular-weight bands at the bottom of the nuclear gel, however, makes this explanation unlikely. This alternative interpretation might also apply to the data in part (a).

63c. The fate of nucleoplasmin within the nucleus could be studied by microinjecting all the labeled species in equal molar concentrations into the nucleus. After incubation for a period of time, the nuclear protein is analyzed by SDS-PAGE and autoradiography. If one species is degraded more than the others, than the radioactivity associated with that protein should be decreased in the autoradiogram. If there is a direct relationship between the number of tails and stability of nucleoplasmin in the nucleus, then the autoradiographic gel profile resulting from this experiment should be similar to that shown for the nuclear fraction in Figure 18-5.

63d. In whole-cell experiments, the intracellular concentration of an inhibitor, not its external concentration, is important. Because egg cells are very large and have a small surface to volume ratio, the intracellular concentration of the ATPase inhibitor probably was not high enough to significantly affect translocation of nucleoplasmin to oocyte nuclei. However, most cells other than oocytes have an extremely high surface to volume ratio, and thus the intracellular concentration of the ATPase inhibitor could increase to a level that inhibited translocation of nuclear proteins.

63e. As noted in answer 63d, the use of inhibitors to demonstrate the ATP dependency of nucleoplasmin translocation in whole oocytes is inconclusive because of the unusually large size of egg cells. An alternative approach is to cool the cells, which should inhibit catalytic reactions but affect ATP-independent membrane events such as protein movement only slightly. Thus the observation that nucleoplasmin translocation is decreased at low temperatures suggests that it is dependent on enzyme-catalyzed ATP hydrolysis.

64a. The probable targeting domain is the C-terminal tripeptide S-K-L, which is serine-alanine-cysteine. However, the data do not exclude the possibility that R-E-I is required along with S-K-L. This could be tested by changing R-E-I in one of the proteins that translocates.

64b. The absence of peroxisomal localization of these two proteins indicates that the S-K-L tripeptide must be at the C-terminus to act as a targeting signal.

64c. Peroxisomal targeting of luciferase depends on 3- to 6-aa domain in the carboxyl end. In contrast, the targeting domain in mitochondrial and chloroplast proteins is in the amino end and contains 20–60 amino acids.

64d. If synthesis, but not sorting, of luciferase occurred in these experiments, then the recipient cells should have

exhibited a rather diffuse, cytoplasmic fluorescence compared with cells in which both synthesis and sorting occurred.

64e. Experiments with heterologous proteins consisting of a cytoplasmic protein coupled with the S-K-L sorting tripeptide at the carboxyl end could demonstrate whether this sequence alone is sufficient for peroxisomal targeting.

64f. Targeting sequences are cleaved in mitochondria, chloroplasts, and the endoplasmic reticulum. By analogy, the targeting sequences in peroxisomal proteins would be expected to be cleaved.

65a. Comparison of lanes 2 and 4 shows that proteins that comigrate with turkey lamins A and B are present in the yeast nuclear envelopes. Despite this similarity in migration behavior, additional evidence is needed to conclude that the turkey and yeast lamins are structurally identical proteins.

65b. The turkey samples serve as a positive control to monitor the extraction of the nuclear envelopes by the urea treatment used in these experiments.

65c. Either an immunoblotting or immunoprecipitation experiment should be done to confirm the presence of lamins A and B in yeast cells. If the antibodies used in the experiment are polyclonal, a limited proteolytic digestion might allow an immunoblot map demonstrating similar or dissimilar peptides from yeast and turkey lamins A and B.

65d. The curves in Figure 18-7 suggest that yeast nuclear envelopes may contain a lamin B receptor. However, a competition experiment must be done to demonstrate that other proteins of yeast origin do not act equally as well as the nuclear envelope preparation in this assay. To conclusively demonstrate the specificity of the putative yeast nuclear receptor for lamin B, it is necessary to show in competition experiments that nonradioactive lamin B can successfully compete with labeled lamin B in the urea-extracted yeast preparation, whereas other proteins cannot compete.

65e. The binding to yeast plasma membranes at high lamin B concentrations probably represents nonspecific binding. A competition experiment with unlabeled lamin B would help determine if there are low-affinity lamin B receptors in the plasma membrane.

66a. Comparison of lanes 1 and 2 indicates that preLHCP is being processed normally. This is a critical part of the experiment because it is imperative to show that this in vitro system mimics the in vivo situation.

66b. These treatments were used to determine if the mature LHCP is in fact in the thylakoid membrane, where it is found in vivo, or is soluble in the stroma or is only partially integrated in the thylakoid membranes. Resistance to extraction with NaOH and to proteolytic digestion are characteristic of integral membrane proteins.

66c. The decrease in molecular weight resulting from protease treatment suggests that part of LHCP is exposed to the stroma and thus is susceptible to partial protease degradation.

66d. The data in Figure 18-9 indicate that both ATP and stroma are necessary for binding of preLHCP to isolated thylakoid membranes; this binding protects the bound protein from degradation by protease (lane 2). However, neither ATP or stroma alone can support binding (lanes 3 and 5); in the presence of either one alone, the unbound preLHCP is susceptible to protease degradation (lanes 4 and 6).

67a. Since protein C, which contains only TS, is localized in the matrix, this sequence is sufficient to direct sorting into the matrix.

67b. Figure 18-11 indicates that only protein E, which contains both the IS and TS, sorts to the intermembrane space. Protein B, which contains only IS does not migrate to either the matrix or intermembrane space, probably because it cannot recognize the receptor on the outer mitochondrial membrane. Thus IS alone cannot direct sorting to the intermembrane space.

67c. One cannot tell from the data presented whether the ST sequence from VSV-G protein is recognized by mitochondria. Protein E, which does not contain this ST sequence, is found in the intermembrane space, whereas protein A, which contains it, is not. Since the only difference between A and E is the presence of the ST sequence, it is possible that the ST stopped translocation of protein A from the matrix to the intermembrane space. To clarify this question, experiments examining the inner membrane would have to be performed; in particular, one would need to determine whether protein A localizes to the inner membrane when it is incubated with isolated mitochondria.

67d. Comparison of proteins E and F suggests that TS must be located at the N-terminal end to direct a protein to the matrix. Thus its position is critical to matrix targeting.

C H A P T E R   *19*

1. endocrine
2. paracrine
3. hydrophilic or lipophilic
4. cholesterol
5. proteases
6. follicle-stimulating hormone (FSH)
7. estrogen
8. lower
9. integral
10. affinity chromatography
11. amino
12. cloning
13. catechol
14. terbutaline
15. fluorescent
16. tumor promoters
17. insulin
18. down regulation
19. receptor-mediated endocytosis
20. lysosomes
21. phosphate
22. cAMP
23. cytokinins
24. gibberellins
25. a
26. a c
27. a c
28. a b c
29. c
30. b c
31. c
32. a b
33. b
34. c
35. a b c d
36. c
37. b c
38. c
39. b

40. The response time of cells to hormones varies depending on the nature of the hormonal receptors, their location, and the type of ultimate effect exerted by a hormone. For instance, neurotransmitters can directly activate ion channels in cell membranes, causing changes in the membrane electric potential of cells often within <1 s. In contrast, steroid hormones must travel across many cellular barriers to reach their receptors in the nucleus. Once there, their final effect is on transcription/translation and can take hours to days to be manifested. In summary, rapid-acting hormones generally are ones that induce modifications (e.g., phosphorylation) in a protein *already* present in cells, whereas slow-acting hormones usually affect the synthesis of new proteins.

41. Polypeptide hormones cannot pass across the plasma membrane without the aid of endocytosis. Thus most polypeptide hormones exert a local effect at the plasma membrane by stimulating a second-messenger system. Endocytosed hormones are encapsulated within a phospholipid bilayer, and without a specific transport mechanism, they cannot enter the nucleus because of solubility problems.

42. These lipophilic compounds are not soluble in blood plasma and would precipitate if not attached to a carrier protein.

43. Aspirin can inhibit the synthesis of prostaglandins, some of which affect the ability of blood platelets to function in blood clotting. Thus aspirin may compromise the blood-clotting ability of individuals with clotting deficiencies to a dangerous extent.

44. Because the unprocessed precursor hormones are more stable and less biologically active than the corresponding mature hormones, they are less likely to be degraded or to stimulate their own receptors in the same cell.

45. Assuming that the insulin receptors in the transfected and nontransfected cells have the same K_M for hormone binding, then at a given insulin concentration, more insulin should bind to each transfected cell than to each nontranfected cell. However, the maximal physiological response of a cell to a hormone may occur when only a fraction of the receptors are occupied by ligand (see Figure 19-7 in MCB, p. 720). For this reason, a saturation level of insulin could have the same effect on both transfected and nontransfected cells. In addition, if the physiological response is limited by a second-messenger system or target protein(s), which are similar in both cell types, then the transfected and nontransfected cells would probably exhibit similar responses to insulin.

46. Differentiated neurons (e.g., sympathetic neurons) are derived from the adrenal gland. In fact, chromaffin cells from the adrenal gland can be induced to differentiate to a neuronal-like cell with nerve growth factor. Thus these tissues are developmentally related, and both can synthesize norepinephrine.

47. Because the G_s protein binds tightly to both GTP and GDP, both of these compounds are effective as affinity ligands for this protein.

48. Activation of adenylate cyclase, and the resulting rise in cAMP, causes a variety of metabolic responses in different tissues (see Table 19-4 in MCB, p. 723). The use of this common mechanism allows the integration of multihormonal signals early in the signal transduction cascade.

49. The insulin receptor undergoes autophosphorylation of tyrosine residues in the presence of insulin. The receptor subsequently phosphorylates various cytosolic proteins. Studies with both phosphatases and site-directed mutagenesis have demonstrated that the protein kinase activity of the receptor must be present for many insulin-responsive reactions to occur. See MCB, p. 745.

50. In the presence of a cAMP phosphodiesterase inhibitor, the extracellular levels of cAMP would increase and the gradient, which stimulates aggregation, would be destroyed. Thus aggregation would not occur. See Figure 19-37 in MCB (p. 752).

51. That a single-gene mutation makes cells insensitive to ethylene suggests that the effects of ethylene are receptor mediated and that only one type of receptor is involved in ethylene's signal transduction. However, several other explanations are possible.

52. Two different experimental approaches could be used to determine if an autocrine factor from the nontransformed cells mediates "detransformation." In one approach, transformed cells are placed in "conditioned medium" obtained from cultures containing only nontransformed cells in normal medium. An alternative approach is to grow both cell types in the same culture and

separate them by a partition or filter that prevents cell-to-cell contact but permits diffusion of an autocrine factor.

53. One interpretation of these data is that there are two receptor subtypes, one with a K_M of about 10^{-9} M and the other with a K_M of about 10^{-8} M. An alternative interpretation is that a single receptor can exist in two different forms having different affinities for NGF.

54a. Incubate tissue slices with hormone X under conditions that permit mRNA synthesis. Then permeabilize the cells and add labeled cDNA corresponding to a mRNA whose synthesis is known to be stimulated by hormone X. Those cells that contain the specific mRNA, which serves as an indicator of hormonal stimulation, will be revealed by autoradiography and are the target cells for hormone X.

54b. Cells known to synthesize the specific mRNA in response to hormone X should be used as a positive control. In other words, these cells should have the receptor for hormone X and synthesize the mRNA corresponding to the labeled cDNA. In the negative control, cells should be used that are known not to contain the mRNA that is being probed. The positive control ensures that the assay is working properly, and the negative control ensures that nonspecific mRNAs do not give a positive response.

55. The only thing that can be concluded from these data are that the receptors for both NGF and aFGF probably act through the same adenylate cyclase system. This is deduced from the observation that addition of both hormones does not have an additive effect on cAMP accumulation. However, since many different receptors can activate the same adenylate cyclase, one cannot determine from these data whether NGF and aFGF bind to the same or different receptors. To answer this question, you would have to perform binding studies.

56. The G_s protein and its GTP-binding site are located on the cytoplasmic side of the cell membrane. Since GMPPNP can penetrate the plasma membrane reasonably well, it binds to the G_s protein in both the vesicular membrane preparation and in living cells. Toxin X, however, probably cannot penetrate the plasma membrane and thus produces little response in nonpermeabilized living cells. In the vesicular membrane preparation, half of the vesicles would be expected to be inside out and half right-side out. The toxin can interact with

the G_s protein in the former but not in the latter; thus its activity is half that of GMPPNP.

57. The curves in Figure 19-4 show that the K_M for binding of epinephrine is about 5×10^{-6} M, whereas the K_M for epinephrine-stimulated cAMP accumulation is 100-fold greater. This difference suggests that production of cAMP depends on stimulation of a receptor distinct from that assayed in the binding experiment. The relatively high concentration of epinephrine needed to stimulate cAMP production suggests either that this stimulation is nonspecific or that a breakdown product of epinephrine is stimulating some other type of receptor.

58. Since the G_s protein can bind to both the hormone receptor and/or the adenylate cyclase, the six possible protein species formed by cross-linking are as follows: receptor, G_s adenylate cyclase, G_s-receptor complex, G_s-adenylate cyclase complex, and receptor-G_s–adenylate cyclase complex.

59. One possible explanation is that membranes from the undifferentiated cells contain only the stimulatory G protein complex, whereas membranes from the differentiated cells contain both the stimulatory and inhibitory G protein complex. If the G_i complex has a significantly lower affinity for GMPPNP than does the G_s complex, the parabolic curve shown in Figure 19-5 would be a likely result. See MCB, p. 728.

60. Separate the cAMP-PKs on an ion exchange column. Next label each protein fraction with the radioactive 8-azido analog, separate on a gel, and perform autoradiography to detect the bands. If there is a fraction-by-fraction correlation between kinase catalytic activity and binding of 8-azido cAMP, then there is an excellent chance that the two kinase activities noted were indeed cAMP-PKI and cAMP-PKII. The possibility that other cAMP-PKs could be labeled does not exist because the problem states that the only two found in these cells are cAMP-PKI and cAMP-PKII.

61. The extracellular concentration of free Ca^{2+} is about 5 mM, whereas the intracellular concentration of free Ca^{2+} is in the low micromolar range. Thus there is a concentration gradient of 1000-fold favoring the entry of Ca^{2+} into the cytoplasm through various Ca^{2+} channels. The binding of Ca^{2+} to calmodulin is cooperative; that is, binding of each Ca^{2+} ion enhances the binding of subsequent Ca^{2+} ions. Thus the Ca^{2+}-dependent con-

formational change in calmodulin, which permits it to affect the activities of various enzymes, can be induced by a smaller change in the intracellular Ca^{2+} level than if no cooperativity existed.

62a. The insulin-stimulated thymidine incorporation in the control cells is mediated by the endogenous level of insulin receptors (~ 2000/cell). Expression of mutant receptors in transfected CHO cells (TM sample) does not affect the activity of these native insulin receptors. However, when the tyrosine kinase antibodies are introduced in the TW + AB sample, the antibodies depress the tyrosine kinase activity of both the endogenous receptors (C sample) and the transfected wild-type receptors (TW sample). As a result, insulin-stimulated thymidine incorporation in the TW + AB sample is below that of the control cells.

62b. No. Since virtually no endogenous, cytosolic proteins that are phosphorylated by insulin have been identified, the single phosphorylated protein on the gel is most probably the phosphorylated form of the insulin receptor itself. Although increasing the receptor number through transfection should increase the chances of detecting phosphorylated cytosolic proteins, which are in low abundance in the cell, the inability to do so has nothing to do with the validity of the data presented.

63a. The inability of the cAMP antagonist to block the effect of NGF [(Rp)-cAMPS + NGF curve] and the additive effect of the cAMP agonist and NGF [(Sp)-cAMPS + NGF curve] both suggest that NGF does not act through the cAMP pathway. In other words, two separate pathways exist to mediate the effects of NGF and cAMP.

63b. The data Figure 19-8 suggest that NGF and cAMP may have a synergistic effect. The forskolin-only curve indicates that at concentrations below 10^{-8} M, forskolin has little stimulatory effect on differentiation of PC12 cells; the maximum forskolin effect of 20 percent occurs at about 10^{-6} M. At low forskolin concentrations ($<10^{-8}$ M), the NGF + forskolin curve represents the maximal activity of NGF, which is present at saturating levels. If NGF and forskolin had merely additive effects, then the maximal combined effect should have been about 50 percent (30 percent from NGF + 20 percent from forskolin) at 10^{-6} M. Since the observed combined effect at this concentration was 70 percent, NGF and forskolin probably are acting synergistically.

64a. The data in Figure 19-9 suggest that ODC activtation is mediated by two separate pathways. Pretreatment with PMA, which causes down regulation of protein kinase C, has no effect on the basal level of ODC activity in buffer but eliminates the ability of PC12 cells to increase their ODC activity in the presence of PMA. In contrast, NGF induction of ODC activity was unaffected by pretreatment with PMA, which down-regulated protein kinase C.

64b. Since the mutant cells lack NGF receptors and do not exhibit NGF induction of ODC activity, the activity in the PMA^+/NGF^+ sample represents PMA induction only. Therefore, the PMA^+/NGF^- sample should be similar to the PMA^+/NGF^+; in both cases only PMA induction is occurring. The PMA^-/NGF^+ sample should be similar to the PMA^-/NGF^- sample, which exhibits only the basal uninduced ODC activity.

65a. Mutants ATP, CT, and TM do not exhibit autophosphorylation; thus they appear to be deficient in tyrosine kinase activity.

65b. Endocytic down regulation is not regulated autophosphorylation, as this activity proceeds normally in the autophosphorylation-deficient mutants.

65c. Since the NOKI mutant exhibits autophosphorylation but not conformational change, these activities clearly do not have to occur together.

65d. Protein kinase C is activated by Ca^{2+}. Since wild-type PDGF receptors can mediate phosphoinositol hydrolysis and increased Ca^{2+} flux, both of which increase intracellular Ca^{2+} levels, it is possible that PDGF stimulates protein kinase C. To confirm this, the activity of protein kinase C in the presence and absence of PDGF should be measured. In fact, PDGF has been shown to stimulate protein kinase C activity in most systems.

66a. The large difference in the time courses of cGMP and cAMP accumulation suggest that there is a single receptor for cAMP that when stimulated first activates synthesis of cGMP within seconds; the increased cGMP levels then induce synthesis of cAMP, which takes several minutes to accumulate, in a cascade pathway. To demonstrate that there are two different cAMP receptors, one activating an adenylate cyclase and the other activating a guanylate cyclase, you would have to do binding studies with labeled cAMP and look for two different K_M values.

66b. Mutants defective in cGMP production but normal in cAMP synthesis would have to be isolated and scored for their abilities to aggregate. If such mutants do aggregate, then presumably cGMP production is not vital to aggregation. Since the mutants in Figure 19-12 are defective in both cGMP and cAMP synthesis, these data do not resolve the question.

66c. The time courses in Figure 19-12 might be related to the dissociation kinetics, but the differences in cAMP and cGMP accumulation also might be the consequence of the activity of G proteins, phosphodiesterases, etc.

66d. Since adenylate cyclase is normal in the mutants, the G protein, the adenylate cyclase, and the receptors must be of the wild type (the latter is noted in the problem). Thus the defect must be *between* the cAMP cell-surface receptors and the G protein.

C H A P T E R 20

1. axon
2. myelinated
3. inhibitory
4. threshold
5. interneurons
6. ganglia
7. depolarized
8. hyperpolarized
9. milliseconds
10. ionic concentration gradients and selective permeability
11. Nernst
12. voltage-gated
13. Na^+
14. permeability, conductivity
15. faster
16. nodes of Ranvier
17. K^+
18. synaptophysin
19. α-bungarotoxin
20. desensitization
21. acetylcholinesterase
22. slow postsynaptic
23. morphine
24. habituation
25. voltage-gated Ca^{2+} channels
26. facilitator neurons
27. b c e
28. b
29. a
30. d
31. b
32. a
33. b c
34. e
35. a b c d e
36. b e
37. a
38. b c
39. c d e
40. b d e
41. c
42. a b c d

43. Electric synapses are connected via gap junctions, so that the electric impulse is transmitted directly from one cell to the next. These junctions are common in circuits where speed is of the highest priority. Impulse transmission at chemical synapses involves the release of a neurochemical at the presynaptic cell, its diffusion across the synaptic cleft, and its binding to a postsynaptic cell. Because chemical synapses can be modified more easily than electric synapses, they are prevalent in tissues where modulation and regulation of signals is more important than transmission speed.

44. Both parasympathetic and sympathetic neurons are autonomic. One type usually stimulates, whereas the other inhibits specific tissues or organs. See Figure 20-9 in MCB (p. 769).

45. Fibroblasts have no voltage-sensitive channels, a necessary prerequisite for the generation and maintenance of action potentials.

46. The resting potential is close to E_K at this time.

47. An increase in the permeability of K^+ or Cl^- ions or a decrease in the Na^+ leak current could lead to hyperpolarization.

48. Cyanide and CO inhibit ATP synthesis in mitochondria, leading to a decrease in cytosolic pools of ATP; this reduction, in turn, decreases the activity of the Na^+-K^+

ATPase. However, because very few Na$^+$-K$^+$ ions traverse the plasma membrane during the course of the action potential, it takes several hours for cyanide or CO to compromise the Na$^+$-K$^+$ gradient to the extent that neither the resting potential nor the action potential can be sustained.

49. Following depolarization of the neuronal membrane and induction of an action potential, the voltage-gated Na$^+$ channels close spontaneously and are inactive until they are repolarized to the resting potential. Since each region of the membrane is inactive for a few milliseconds after an action potential has passed, the action potential can be propagated only in the forward direction, and not in the retrograde direction. See Figure 20-15 (p. 775) and Figure 20-17 (p. 777) in MCB.

50. The Na$^+$ channel protein has been purified by affinity chromatography on columns containing bound neurotoxins such as tetrodotoxin that bind specifically and with high affinity to this channel protein.

51. Both Na$^+$ and K$^+$ ions are hydrated; that is, there is a shell of bound water surrounding these cations. The pore of the Na$^+$ channel must thus be large enough to permit passage of the hydrated cations, which are considerably larger than the unhydrated cations.

52. Inhibitors of protein synthesis block long-term memory but not short-term memory in *Aplysia*, suggesting that certain newly synthesized proteins are required for the former but not for the latter to occur.

53. In the dark, the rod is depolarized and secretes neurotransmitter. When "stimulated" with light, it hyperpolarizes and secretion of neurotransmitter is inhibited.

54. Cyclic GMP can open both Na$^+$ and Ca^{2+} channels in rod cells. Light elicits a reduction in cGMP levels, which leads to a closing of both Na$^+$ and Ca^{2+} channels. The resulting drop in intracellular Ca^{2+} stimulates the synthesis of more cGMP, thus "resetting" the cells so that they are less sensitive to small changes in light levels. See Figure 20-51 in MCB (p. 807).

55. An action potential in any postsynaptic cell depends on the timing of the arrival of electric impulses at the axon hillock of the postsynaptic cells. Since the action potential of neuron C is not in synchrony with those of neurons A and B, it does not contribute to generation of the postsynaptic action potential; thus inhibition of neuron C does not prevent firing of the postsynaptic cell. The data also indicate that stimulation of both neurons A and B is necessary for firing of the postsynaptic cell.

56. The slow depolarization might indicate that the cell is dying, which would, in turn, cause the ATP levels to decrease, so that the resting potential would not be able to be maintained. This process, however, usually takes a much longer time than is indicated in the graph (see answer 48). In actual practice, the slow depolarization most probably results from a leaky seal between the microelectrode and the plasma membrane of the axon. Often this will correct itself with time.

57. The most probable reason for the hyperpolarized resting potential and lack of an action potential is that there is too little or no Na$^+$ in the external medium. The resulting decrease in the Na$^+$ "leak current" would account for the hyperpolarized resting potential. Furthermore, in the absence of a Na$^+$ gradient, no action potential could be generated even when the Na$^+$ channels open normally. Although an increased permeability to K$^+$ ions also could account for the hyperpolarized resting potential, only a defect in E_{Na} could result in no action potential. The lack of Na$^+$ ions in the form of NaCl would ultimately cause lysis of the cells due to an osmotic imbalance.

58. One possibility is that the drug maintains the voltage-gated Na$^+$ channels in an open position once they have opened, although the drug is not capable of opening voltage-gated channels by itself. Another possibility is that it prevents the activation of the voltage-gated K$^+$ channels that are responsible for the quick return to the resting potential.

59. The effect of the neurotoxin could be mimicked by increasing the K$^+$ concentration and decreasing the Na$^+$ outside the neuron. The former would cause a decrease in the resting potential, and the latter would cause a decrease in the action potential.

60. A defective Na$^+$ channel that cannot be inactivated in the wild-type manner also could contribute to delayed repolarization of neurons. The best techniques for demonstrating such a defect are voltage clamping and patch clamping.

61. Molecules of known charge and size can be covalently linked to the fluorescent probe, and the ability of the modified dye to pass through gap junctions under various conditions determined.

62a. The slight hyperpolarization exhibited by cell 2 indicates that this cell has an inhibitory response to the neurotransmitter.

62b. To determine if cell 1 and cell 2 are connected by a synapse, depolarize cell 1 and look for a response in cell 2; then depolarize cell 2 and look for a response in cell 1. A response will occur in the nondepolarized cell only if the cells are synaptically connected.

62c. A neurotransmitter is not inhibitory or excitatory by nature. Rather, its effect depends on the response of the postsynaptic receptor to the neurotransmitter and the receptor's specificity to ion species. Thus a single neurotransmitter can act both as an excitatory transmitter, as in the case of cell 1, and as an inhibtory transmitter, as in the case of cell 2.

63. Impulse transmission across electric synapses occurs more rapidly than across chemical synapses. Thus the two types of synapse can be distinguished by stimulating the presynaptic cell and recording the changes in membrane potential with time in both the pre- and postsynaptic cells. The time course of these change is characteristic for each type of synapse (see Figure 20-28, p. 785, in MCB).

    Most chemical synapses, but not electric synapses, are inhibited by various toxins. For example, the addition of cobalt or cadmium ions usually blocks most chemical synapses because these metals inhibit the uptake of Ca^{2+} ions at the presynaptic neuron, thus preventing the release of the neurotransmitter into the synaptic cleft. In addition, chemical synapses with specific postsynaptic receptors will respond to specific stimulators and toxins; for example, the nicotinic acetylcholine receptor is stimulated by nicotine and inhibited by α-bungarotoxin. Transmission across electric synapses is not affected by such modulatory substances.

64. (a) You could determine if the mutant cells contain less neurotransmitter than wild-type cells by quantitative ultrastructural immunochemistry using an antibody directed against acetylcholine. If the defect is located here, then mutant cells would contain fewer gold particles in transmitter vesicles than wild-type cells. (b) Ra-

dioimmunoassay techniques could be used to compare the ability of mutant and wild-type cells to release acetylcholine into the synaptic cleft. In this case, 50 mM K$^+$ is added to the extracellular medium to depolarize all the neurons in the culture and then the extracellular medium is collected and analyzed for acetylcholine. (c) A defect in the postsynaptic receptors could be detected by measuring the changes in membrane potential in mutant postsynaptic cells with a microelectrode following addition of a suprathreshold level of acetylcholine to the extracellular medium. Similar postsynaptic responses in both mutant and wild-type cells would indicate that the postsynaptic receptors are functional in both cases.

65. The cells should contain voltage-sensitive channels, one of the characteristic features of an "excitable" cell. In addition, they should be able to synthesize, store, and release neurotransmitters.

66. Since half-maximal stimulation of the receptor occurs at about 5 μM dopamine, the lower-affinity subtype B of the dopamine receptor probably is present in the retina.

67. The toxin most probably inhibits cGMP phosphodiesterase. This results in the inability of the retina to break down 3',5'-cGMP to 5'-GMP. The resulting accumulation of cGMP causes the Na$^+$ channels to remain in the open mode, and the light-induced activation of cGMP phosphodiesterase would thus be prevented.

68a. The hypothesis is supported by the observation that cell death is much greater in the presence than in the absence of NMDA (+NMDA curve versus control curve). In addition, the specificity of this effect is indicated by the ability of MK801, a blocker of NMDA, to inhibit it; that is, the survival curve in the presence of MK801 + NMDA is similar to the control curve.

68b. The absence of any effect of NMDA on cell survival during the first 7 days in culture suggests that either no NMDA receptors or no functional receptors were present in the embryonic hippocampal cells during this period. Additional studies would be needed to confirm this conclusion, however (see part d).

68c. MK801 alone increased the survival rate from day 1 in culture compared with control cultures, whereas MK801 + NMDA blocked the NMDA-elicited rapid cell death beginning at day 7. These findings suggest that MK801 can inhibit cell death by two different mechanisms, one of which prevents activation of NMDA receptors by NMDA and the other of which does not involve these receptors.

68d. The best way to distinguish these explanations would be binding studies with [^{3}H] glutamate, using NMDA to displace the radioactive glutamate. If receptors are present, then binding should be observed; in this case, presumably, the receptors, though present, do not allow an influx of Ca^{2+} ions early in the culture period.

68e. One way to demonstrate whether NMDA-elicited cell death is related to the influx of Ca^{2+} ions would be to compare the NMDA effect at high and low external Ca^{2+} concentrations. One problem with this experimental approach is that cells need some extracellular Ca^{2+} to remain attached to the substratum; thus depleting external Ca^{2+} might cause some effects unrelated to the NMDA effect.

68f. The data in Figure 20-8 indicate that 2-day cells are less sensitive to the toxic effects of Ca^{2+} than 10-day cells, presumably because they are less permeable to Ca^{2+} ions. That is, an approximately 100-fold higher concentration of Ca^{2+} ionophore is required to cause 50-percent cell death in 2-day cells than in 10-day cells. These findings suggest that the inability of NMDA elicit cell death during the first 7 days of culture, as shown in Figure 20-7, results from the inability of the stimulated receptor to permit an influx of Ca^{2+} ions from the extracellular medium.

68g. This experiment provides no useful information because the glutamate binding observed may reflect nonspecific binding to various cell structures and/or binding to the other two glutamate receptor subtypes. In order to demonstrate the presence of the NMDA receptor, competition experiments using NMDA to displace the radioactive glutamate are necessary.

69a. Assuming that the fast depolarization component of the action potential is due to conventional channels, TTX would block the inward movement of Na^+ ions.

69b. Assuming that the fast depolarization component of the action potential is due to conventional channels, TEA would block the outward movement of K^+ ions.

69c. Since the slow depolarization occurs in the presence of TEA + TTX, it probably is associated with channels other than Na^+ and K^+ channels. This experiment also shows that the slow depolarization occurs in the absence of the previous Na^+ and K^+ ion fluxes associated with the action potential; thus the slow depolarization does not depend on passage of Na^+ or K^+ ions.

69d. Since both cadmium and cobalt block voltage-sensitive Ca^{2+} channels, these results suggest that the slow depolarization results from an inward flux of Ca^{2+} ions.

69e. If Ca^{2+} were removed from the extracellular bathing medium, which typically contains 5 mM $CaCl_2$, the latent depolarization would be abolished. Under these conditions, even though Ca^{2+} channels may open, no inward current would be exhibited because no Ca^{2+} gradient would be present.

69f. The only suitable technique is the use of an intracellular fluorescent dye that would indicate the relative levels of free intracellular Ca^{2+}. One such dye is Fluo-3, whose emission intensity increases as the intracellular concentration of free Ca^{2+} increases.

70a. When wild-type DLM fibers are stimulated, their neurotransmitter evokes EJPs but then is recycled back to the synapse where it is taken up and reincorporated into vesicles.

70b. There is a correlation between the number of vesicles per section and the EJP magnitude: the fewer the number of vesicles, the lower the EJP.

70c. Addition of cobalt or cadmium would probably block the release of vesicles in the wild-type synapses.

70d. If the defect in shi mutants involved a habituation of the postsynaptic receptor, then bathing the synapse in exogenous neurotransmitter should have little effect on the reduced EJP exhibited at 29°C. In contrast, if the defect involves an inability to release vesicles, then the EJP should increase in the presence of exogenous neurotransmitter.

70e. A defect in the ability of DLM fibers to reincorporate neurotransmitter into vesicles could be demonstrated by adding tritium-labeled neurotransmitter and measuring its uptake using ultrastructural autoradiography. The presence of grains in the synaptic process but not in vesicles would suggest that the defect is in reincorporation into vesicles.

70f. The possible role of this protein in endocytosis by non-neuronal cells might be demonstrated by comparing the in vitro uptake of an electron-dense material by various types of cells isolated from wild-type flies and shi mutants. Interpretation of the results from this type of experiment might be complicated because bulk-phase endocytosis (pinocytosis) may have to be distinguished from receptor-mediated endocytosis.

70g. The correlation between the number of vesicles per synapse and the EJP is compelling but not conclusive evidence in favor of the hypothesis.

71a. These data indicate that free synapsin I can bind to stripped synaptic vesicles with half-maximal binding occurring at a synapsin I concentration of about 50 nM in 40 mM NaCl. High salt concentrations, however, reduce the binding of synapsin I to the vesicles.

71b. The reduced binding of synapsin I and high salt concentrations and to protease-treated vesicles suggests that the stripped vesicles contain a receptor protein on their surface that facilitates binding of synapsin I.

71c. These results suggest that the tail domain binds to a protein receptor on the vesicle but that the head merely interacts nonspecifically with phospholipids of the vesicle.

71d. This experiment suggests that added synapsin I probably is incorporated into stripped vesicles in a manner similar to that which occurs in vivo.

71e. The most definitive experiment would be to subject the isolated proteins to Western blot analysis using anti–synapsin I antibodies.

C H A P T E R 21

1. polymerize
2. multigene
3. alpha (α) and beta (β)
4. microtubule-organizing center (MTOC)
5. chromosomes
6. discontinuous
7. colchicine
8. microtubule-associated proteins (MAPs)
9. centrioles
10. plus
11. synthesis
12. GTP
13. tau
14. basal bodies
15. isotypes
16. depolymerize
17. minus
18. acetylation
19. fast
20. genetic, mutants
21. flagella and cilia
22. tyrosine
23. minus, plus
24. triplet
25. radial spokes and central pair
26. shortening, elongation
27. polarizing
28. b c d e
29. b d
30. c
31. a b d
32. a d e
33. a e
34. c e
35. b
36. a b c d e
37. c
38. d
39. b c d
40. c
41. a
42. b

43. The most plentiful and stable source of both actin and myosin filaments is muscle.

44. Each of the 13 protofilaments that make up a microtubule has two distinct ends because the repeated tubulin dimers within the protofilaments are all oriented in the

same direction (e.g., $\alpha\beta \rightarrow \alpha\beta \rightarrow \alpha\beta$, etc.). See Figure 21-1 in MCB (p. 816).

45. Microtubules must disassemble and reassemble in response to the changing structural requirements of the cell. This is especially true in the case of spindle microtubules, which are assembled and then disassembled during the cell cycle. The microtubules in flagella, in contrast, are quite stable. Recent experiments with colchicine and cold-induced depolymerization suggest that the differences in the stability of various microtubules may be related to the presence of different tubulin isotypes.

46. Slow cooling or heating can depolymerize microtubules; thus quick freezing not only preserves these structures by preventing general degradation but also keeps them in the polymerized state. Deep etching is a rotary application technique in which a thin layer of metal is sprayed on several planes of a structure, so that the relationship of one type of structure with another (e.g., microtubules and filaments) is revealed. Standard thin sections have very little depth of field and thus cannot reveal such relationships.

47. If axonal microtubules disassemble, then the axon will retract and synaptic contact between neurons would be lost. Thus the stability of axonal microtubules, which is mirrored in their long half-life, is critical for the maintenance of the structural integrity of neuronal networks.

48. At the critical concentration of tubulin dimers, a steady-state condition exists in which the addition of tubulin dimers to microtubules is balanced by their loss, so that the average length of microtubules does not increase or decrease. However, because there is a turnover of tubulin subunits, labeled tubulin dimers would be released from previously labeled microtubules.

49. The free primers have both ends available for polymerization and depolymerization. In the presence of excess tubulin, polymerization will occur at both ends, thus lengthening the microtubules. Microtubules that are growing from centrioles, however, have only the (+) ends available for polymerization because the (−) ends are capped by the centriole. In the presence of excess amounts of tubulin, therefore, polymerization could occur only from the (+) ends.

50. Autoradiography allows one to monitor the polymerization of newly synthesized radiolabeled tubulin dimers into structures such as flagella as illustrated in Figure 21-15 in MCB (p. 825). Fluorescent immunocytochemistry using anti-tubulin antibodies cannot as clearly demonstrate newly added tubulin subunits.

51. Tyrosine-modified α-tubulin is present in most microtubules, whereas untyrosinated α-tubulin is found preferentially in cytoplasmic microtubules. When injected into cells, antibodies against tyrosine-modified α-tubulin will bind to the tyrosinated form of tubulin, making it unavailable for incorporation into cycling microtubules. The inhibitory effect of antibody injected into cells will depend on the relative incorporation of tyrosinated and untyrosinated α-tubulin in various microtubular classes.

52. With video-enhancement Nomarski optics, one can examine a living axon and determine whether a particular structure is moving in a particular direction. However, because of the limited resolution of light microscopy, transmission electron microscopy is necessary to visualize the relationship between the transported structure and microtubules. See Figure 21-21 in MCB (p. 832).

53. These observations indicate that kinesin binds to microtubules in the presence of ATP but that hydrolysis of ATP to ADP is necessary for kinesin-dependent movement of vesicles along microtubules. See Figure 21-25 in MCB (p. 835).

54. Both kinesin and MAP1C cause vesicles to move along microtubular networks. However, kinesin causes vesicles to move from the (−) to the (+) end, whereas MAP1C causes vesicles to move from the (+) to the (−) end.

55. Individuals with defective dynein produce flagella and cilia that are nonmotile. Since cilia in the trachea are responsible for removing debris, which otherwise could enter the lung, defective cilia can result in coughing. Similarly, movement of sperm cells and of eggs require active flagella and cilia, respectively. Individuals with defective dynein will have nonmotile sperm or egg cells and thus are sterile.

56. Animal cells contain centrioles at the center of the MTOC, whereas plant cells do not. It is not known what structure(s) in plants acts as a MTOC-like structure to organize the microtubular array during cell division.

57. This experiment demonstrates that the forces pushing and pulling chromosomes during mitosis are equal in magnitude but opposite in direction.

58. The method chosen must allow the investigator to label α- and β-tubulin monomers during interphase and then to determine whether this label is present in spindle microtubules during mitosis. For example, cells can be labeled with [^{35}S]methionine for a few hours during interphase and then chased with unlabeled methionine. These labeled cells are synchronized in mitosis and the spindle tubulin immunoprecipitated, run on a SDS gel, and autoradiographed. The presence of a highly radioactive tubulin band would indicate that the mitotic spindle is formed from tubulin dimers synthesized in interphase. In fact, this has been demonstrated. If a similar experiment is performed with cells close to mitosis being labeled (rather than interphase cells), the tubulin band will contain minimal radioactivity, indicating that newly synthesized tubulin is not used in spindle formation. The use of drugs that block protein synthesis would not be appropriate in this case because this treatment probably would stop the cell cycle in G$_2$ and prevent mitosis.

59. The most probable reason for these results is that the primer preparations were opposite in polarity. The A primers probably had their (−) ends covalently linked to the Sephadex beads, thus exposing the (+) ends to the buffer. Because tubulin adds much more rapidly to the (+) end than to the (−) end, rapid elongation occured with primer A. In contrast, the B primers probably had their (+) ends linked to the Sephadex beads, so that the slower-polymerizing (−) ends were exposed to the buffer.

60. During the lag phase, small microtubule primers form slowly. During the elongation phase, tubulin dimers are added to the primers; this addition reaction occurs much more rapidly than does the initial generation of short primers and leads to the rapid lengthening of the microtubules. Polymerization enters the plateau phase when the rates of addition and loss of dimers become equal. This steady state occurs at the critical concentration of free tubulin dimers (see Figure 21-8 in MCB, p. 821).

61. Even though the length of "stable" microtubules in cells remains constant, these microtubules are turning over; that is, tubulin dimers are being added and lost at about the same rates, so that there is no net change in microtubular length. Binding of colchicine to free tubulin dimers inhibits their addition to microtubules. Thus in the presence of colchicine, the rate of addition of tubulin to stable microtubules will be decreased, so that it no longer equals the rate of loss. As a result, there will be a time- and concentration-dependent colchicine-induced decrease in the length of stable microtubules.

62. The turnover rate is 1 μm/min. To determine this value, first find two points on the curve that when added together equal 100 percent. For instance, the time difference between when half of the labeled tubulin is incorporated into microtubules (1 min) and when half is released at (6 min) can be compared. Likewise, the time at which the labeled tubulin just begins to be incorporated (0 min) can be compared with the time at which the tubulin begins to be released (5 min). In either case, the time difference is 5 min. Since the length of the microtubules in question is 5 μm, the turnover rate is 1 μm/min.

63. In the native gel, the 400,000-MW species in lane 1 is the $\alpha\beta$-tubulin dimer complexed with microtubule-associated proteins. As the purification proceeds, the microtubule-associated proteins are lost, as indicated by the descending molecular weight of tubulin. By step number 4, most, if not all, of the microtubule-associated proteins have been dissociated from tubulin, leaving the tubulin dimer with a molecular weight of about 100,000. SDS gel electrophoresis disrupts noncovalent bonds and disulfide bridges in proteins, thus splitting the tubulin dimer into the α- and β-tubulin monomers. These appear as separate, closely spaced bands in the SDS gel, each with a molecular weight of 50,000. The fact that both of these bands are present and remain constant throughout the purification suggests that many microtubule-associated proteins are linked to tubulin by weak bonds that are disrupted during the SDS gel electrophoresis.

64. The autoradiographs should reveal that after 1 min only the tips of microtubules from the MTOC are labeled, whereas after 30 min nearly all the microtubules in the cell are labeled. Overlayering of the autoradiographs on photomicrographs of the same cells previously prepared for anti-tubulin fluorescent immunocytochemistry can help determine where new polymerization is occurring relative to the entire preexisting microtubule.

65. The data indicate that pigment granules move away from the cell center in the presence of agent Z, but the radial symmetry evident in the light control (− Z) is not evident. The fact that there is movement at all suggests

that Z probably does not affect ATP synthesis, binding, or kinesin. The change in the direction of movement suggests that Z may alter the cytoarchitecture of microtubules in cells. An in vitro method of testing this hypothesis is to add Z to isolated microtubules. A disruption in the linearity of the microtubules following this treatment would support the theory that Z causes a perturbation of the microtubular cytoarchitecture. An additional experiment is to analyze cells in the presence and absence of Z by immunocytochemistry with antitubulin antibodies. If Z affects microtubular architecture, a different staining pattern would be likely.

66. Preparation B probably had its dynein removed or compromised, so that movement cannot occur at all. Preparation C probably had its nexin removed, thus allowing the doublets to slide relative to each other. See Figure 21-29 in MCB (p. 837).

67. Compound D probably affects the ability of the kinetochore fibers to find the kinetochores, which serve to cap the (+) ends of kinetochore fibers and thus inhibit their depolymerization. As discussed in MCB (p. 849–850), if this capping is experimentally disrupted, kinetochore fibers will depolymerize. Since the astral and polar fibers are not similarly affected by compound D, this drug probably does not interact in a general way with all microtubules as does colchicine, which binds to free tubulin and after incorporation into microtubules prevents the addition of more tubulin.

68. If the concentration of free tubulin at the (+) ends is different from that at the (−) ends of the microtubules and both are at the critical level at which no net loss or addition of tubulin occurs, then a microtubule can maintain the same length but not turn over. This demands different, local, soluble pools of free tubulin in cells.

69a. The gel filtration profiles indicate that high salt (profile B) inhibits the binding of CKII to itself or to unidentified cytosolic proteins; as a result, a sharp peak of CKII activity is observed. In contrast, at low salt (profile A), CKII activity appears diffusely throughout the column, suggesting that the enzyme associates with various protein species of differing molecular weights.

69b. Western blot analysis using anti-tubulin antibodies could verify that the 50-kDa protein is indeed tubulin.

69c. Lane A serves as the control to ensure that the SDS gel electrophoresis/autoradiography technique used in this experiment can detect phosphorylated proteins. The fact that casein is phosphorylated as revealed in lane A indicates that the system is operating properly.

69d. These results indicate that there is a CKII-like activity in the rat tubulin preparation, suggesting that CKII and tubulin are associated with each other and that CKII might be a tubulin kinase.

69e. Each $\alpha\beta$-tubulin dimer binds to two molecules of GTP, one of which is hydrolyzed during the addition of a dimer to a microtubule. Cytosolic proteins may play a role in this process; one that could utilize GTP, as CKII can, would be a good candidate.

69f. The data in Figure 21-9 indicate that CKII associates with tubulin but can be released after several assembly-disassembly cycles.

69g. A double-label fluorescent immunocytochemistry experiment would be able to determine whether tubulin and CKII associate in vivo. Cells are stained with fluorescent anti-tubulin and anti-CKII antibodies prepared with two different dyes (e.g., FITC, which is green, and rhodamine, which is red). Colocalization of these probes along the microtubular network in doubly stained cells would indicate that tubulin and CKII are associated in intact cells. Avila's group has in fact obtained such evidence.

70a. NGF appears to cause an increase in the length of neurites with time. The day 6-1 sample shows that removal of NGF leads to a rapid decrease in neurite length, suggesting that NGF is necessary for the maintenance of these processes.

70b. Figure 21-11 shows that the amounts of isotypes I, IV, and V was almost the same in all samples, whereas the amounts of isotypes II and III increased progressively at days 2, 4, and 6. The increase in isotypes II and III correlates well with the development of long neurites; in addition withdrawal of NGF resulted in the decrease of isotypes II and III to non-NGF levels. These data suggest that NGF induces the synthesis of isotypes II and III and that the presence of NGF is necessary to maintain the increased levels of these tubulin isotypes.

70c. To determine whether the synthesis of tau and MAP2 are causally related to the prior synthesis of isotypes II and III, selective isotype blockers would have to be used. Cells would be incubated with these blockers and the amounts of tubulin isotypes II and III, tau, and MAP2 determined. Comparison of the results would reveal whether a corresponding preferential decrease in the various proteins occurs, a finding that would support the hypothesis.

70d. NGF causes an increase in tubulin synthesis, as evidenced by the increased cytoskeletal and soluble pools from day 2 to day 6. The much greater increase in the cytoskeletal pool than in the soluble pool indicates that most of the newly synthesized tubulin is polymerized. In addition, withdrawal of NGF (sample 6-1) leads to an immediate decrease in the polymerized cytoskeletal pool, indicating that NGF is necessary for the *maintenance* of polymerized tubulin.

70e. The anti-tubulin antibody that recognizes all isotypes serves as an internal standard by which the amounts of each of the five isotypes can be quantified.

71a. These data indicate that tyrosinated α-tubulin becomes localized in the apical cytoplasm of cells, whereas acetylated α-tubulin becomes localized in the cortex near the point of contact between cells during this time period. See Figure 21-13 ($-$ nocodazole) for a diagram of this polarization.

71b. Since nocodazole prevents the assembly of microtubules, the staining patterns in the presence of this drug reflect the stability of already formed microtubules. The data thus suggest that microtubules containing tyrosinated α-tubulin are more stable than those containing acetylated α-tubulin.

71c. This finding that all microtubules are uniformly staining with both antibodies in the presence of taxol suggests that microtubules consisting of both tyrosinated and acetylated tubulin might form in these cells.

71d. If the difference in distribution of the tyrosinated and acetylated tubulin subtypes was due to a preferential localization of the enzyme(s) that catalyzes the post-translational modification, then it might be expected to be preferentially located in regions where the acetylated tubulin microtubules are concentrated. A uni-

form distribution of acetyltransferase is consistent with the suggestion stated in the answer to 71c.

72a. ATP is necessary for binding of kinesin, and the dephosphorylation of ATP to ADP results in movement. Thus if ATP is used in this step, kinesin will bind and then immediately release when ATP is dephosphorylated. Use of AMPPNP results in stable kinesin binding to the microtubular affinity matrix.

72b. If gel electrophoresis could be accomplished on a preparative scale, the gel slices could be cut out corresponding to this putative kinesin, eluted in buffer, and then added to a preparation of isolated microtubules on a glass slide. If the isolated protein is kinesin, then addition of ATP should result in movement detectable by polarizing light microscopy.

72c. The lamellar-like extensions seen in Figure 21-14 would not be expected in typical kinesin-microtubule preparations.

72d. After obtaining a dark-field micrograph, fix the cells and analyze them by fluorescent immunocytochemistry using anti-tubulin antibodies. Examine the two sets of micrographs for areas that appear bright in the dark-field but that do not stain with anti-tubulin antibodies; such areas would contain membrane networks not based on tubulin microtubules.

72e. The ability of Triton-X to disperse portions of the network indicates that it is probably partially composed of lipids, not tubulin.

72f. This finding is further evidence that the network is partly composed of lipid, since lipid aggregates can be easily disrupted by altering the osmolarity of the surrounding medium. The same is not true of microtubules.

72g. This observation indicates that formation of the membrane network depends on hydrolysis of ATP, not simply its presence, and suggests that microtubular movement may be required for network formation.

73a. Because X-rhodamine tubulin labels all microtubules in cells, some method was needed to eliminate all microtubules except the kinetochore fibers, which are the focus of the research.

73b. Kinetochore microtubules are more stable than other microtubules in the cell, as evidenced by their persistence in a buffer that causes disassembly of other microtubules.

73c. Cold depolymerization or colchicine/nocodazole treatment might induce disassembly of most microtubules but leave the kinetochore tubules intact.

73d. Anti-tubulin immunocytochemical analysis of the cell would reveal antibody labeling in the photobleached zone as well as across the rest of the spindle.

73e. Anaphase microtubules move toward the poles, but exhibit no turnover of the microtubules.

73f. These data indicate that metaphase kinetochore tubules turn over, whereas anaphase kinetochore tubules do not.

CHAPTER 22

1. actin
2. viscosity
3. polymerization and depolymerization
4. ATP
5. plus, minus
6. heavy and light
7. thick filament
8. S1
9. minus
10. involuntary
11. rigor
12. creatine phosphate
13. Ca^{2+} ions
14. Ca^{2+} ATPase
15. troponin C
16. tropomyosin, myosin heads
17. myosin light-chain kinase
18. isoforms
19. α-actinin
20. titin
21. lower
22. villin
23. uvomorulin
24. sol, gel, gel solin
25. filamin
26. filipodia
27. stress fibers
28. intermediate filaments
29. vimentin
30. a c d
31. a c
32. d
33. c d e
34. a c d
35. c
36. a c d
37. e
38. a b
39. b d e
40. b c d e
41. a b c d e
42. e
43. a b c d e

44. This finding indicates that actins from different species are so similar in structure that they recognize and interact with each other, suggesting that actin arose early in evolution and was highly conserved.

45. First, the observation that all bound S1 fragments point in one direction, toward the so-called (−) end, demonstrates the polarity of actin filaments and provides a method for distinguishing the two ends. Second, the relative rates of growth at the (+) and (−) ends of actin filaments can be demonstrated by reacting actin monomers with decorated primers and examining the elongated products in the electron microscope. See Figure 22-2 in MCB (p. 862).

46. Under these conditions, no significant change in the average length of actin filaments would be expected because the higher concentration of actin at the (−) ends would compensate for the slower rate of addition at this end.

47. Myosin tails consist of two protein α helices, each of which is a rigid coil. These helices wrap around each other, forming a coiled-coil molecule, which has considerable structural rigidity.

48. The arrowheads would point away from the Z disks because the (+) ends of thin actin filaments are attached to the Z disk and bound S1 fragments point toward the (−) ends of actin filaments.

49. The indicated structures are (a) myosin thick filaments, (b) actin thin filaments, and (c) cross-bridges formed by the association of myosin heads with actin filaments. If the muscle had been prepared in the presence of ATP, the myosin heads would not be associated with the actin filaments; thus the cross bridges would not be visible.

50. In the rigor complex, myosin heads bind at a 45° angle to the actin filaments and cannot be released; thus the thick and thin filaments cannot slide past each other. Rigor occurs when muscle is depleted of its high-energy stores.

51. The mechanisms by which a rise in Ca^{2+} triggers contractions differs in smooth and striated muscle. In striated muscle, binding of Ca^{2+} to troponin C leads to muscle contraction (see Figure 22-18 in MCB, p. 872). Contraction of smooth muscle is triggered by activation of myosin light-chain kinase by Ca^{2+}-calmodulin (see Figure 22-19a in MCB, p. 874). Thus only smooth muscle is inhibited by microinjection of antibodies to myosin light-chain kinase.

52. These data suggest that creatine phosphate serves as the major energy source in muscle by transfering a high-energy phosphate to ADP to form ATP in a reaction catalyzed by creatine kinase.

53. By shearing microvilli from intestinal epithelial cells and pooling them for analysis, researchers have obtained rather pure preparations of microvilli containing actin.

54. Because actin is a key part of the contractile ring, which contracts during cell division, actin inhibitors can affect this process.

55. Movement of the ER in *Nitella* results from myosin-actin interactions, whose orientation is dictated by the polarity of the actin filaments. Because all the actin filaments run in one direction, the cytoplasm is propelled in one direction only. See Figure 22-34 in MCB (p. 885).

56. Profilin has been characterized as a protein that can bind 1:1 with monomeric actin; this binding inhibits actin polymerization. Thus cells containing profilin are likely to have a large cytoplasmic pool of monomeric actin. See Figure 22-35 in MCB (p. 886).

57. The protease probably binds to or partially proteolytically digests the S1 fragment of myosin, the site of the actin-stimulated ATPase activity. By pretreating myosin with papain, one can produce S1 fragments with full ATPase activity. If the coelenterate protease inhibits the binding of S1 to actin, treatment of functional S1 fragments with the protease would prevent "arrowhead" binding as visualized in the electron microscope. Thus one could determine whether the protease affects the ATPase activity itself and/or the binding of myosin to actin.

58. Preparation A exhibits the behavior typical of properly isolated S1, HMM, or myosin; that is, in the presence of the nonhydrolyzable analog it binds to actin, and in the presence of ATP it exhibits movement. Preparation B probably is a defective form of S1, HMM, or myosin that can bind to actin but has lost its ATPase activity. Preparation C has lost the part of the S1 fragment that is necessary for binding to actin.

59a. 23 sections would be needed in a sarcomere 2 μm (2000 nm) long (2000 nm ÷ 90 nm/section = 23 sections).

59b. The indicated structures are (a) myosin thick filaments and (b) actin thin filaments. Since the section contains very few actin filaments surrounding the myosin filaments, it probably comes from the AH zone.

60a. Because actin is highly conserved among species and thus is not very immunogenic, little or no production of antibodies occurs when actin from one species is injected into another species. If actin is modified by reacting it with an aldehyde that cross-links the protein, the modified actin is immunologically different from "native" actin and is likely to elicit antibodies. Other types of covalently modified actins also can stimulate better production of antibodies than the native protein.

60b. One approach is to extract the basal fibers from MDCK cells and react them with S1 myosin fragments. The appearance of the typical arrowheads would indicate that the fibers probably are actin. Another approach is to react the fibers with a fluorescent derivative of phalloidin, which binds to F actin. Successful labeling with this compound would indicate that the fibers are actin.

61. This question could be investigated by treating MDCK cells with cytochalasin D (which inhibits the assembly of

actin filaments) during the time period that the basement membrane normally forms in untreated cells. The absence of basement membrane deposition in the presence of cytochalasin D would suggest that this process depends on the formation of stress fibers. However, this drug disrupts the polarity of cells, causes microvilli to contract, and has other effects on membrane protein localization. Thus this experiment could provide only suggestive, not definitive, evidence.

62. The most plausible reason for the differences in preparations A, B, and C is that they differ in the degree of overlap between the myosin and actin filaments. In other words, the actin-myosin overlap in preparation A may be similar to that depicted in Figure 22-13 (no. 2 and 3) MCB (p. 868), which permits maximum interaction between actin and myosin and hence maximum tension generation. In contrast, in preparations B and C, the overlap may be too little or too much to generate maximum tension.

63a. To demonstrate that compound X does not directly inhibit the ability of chymotrypsin to act as a protease, a control experiment should be performed with a different substrate known not to be affected by compound X. Observation of the same proteolytic rate and product profile in the presence and absence of compound X in this control experiment would indicate that the poison does not act directly on chymotrypsin.

63b. The resistance of myosin to chymotrypsin degradation in the presence of X suggests that the poison makes the "hinge" in the intact myosin molecule less flexible, thus producing a state of rigor. The results of the motility assay indicate that X does not affect the S1 fragment.

64a. An alternative approach is to microinject an antimyosin antibody into a wild-type nonmuscle cell about to undergo cytokinesis. Inhibition of cytokinesis by this treatment would support the notion that myosin plays an important role in cytokinesis.

64b. In situ decoration with myosin S1 fragments would indicate the polarity of the actin filaments in the mutant cells but would not provide any information about the presence or function of myosin in these cells.

65. Extract A causes a net depolymerization of actin filaments, as evidenced by the steady increase in the amount of soluble labeled actin over time. In the presence of

extract B, the actin filaments maintain a constant length; however, they must be loosing and adding actin filaments at identical rates, otherwise there would be no soluble actin monomers in solution. The filaments in extract C are likewise maintaining a constant length; the low basal level of labeled monomeric actin in this case suggests that C contains an unidentified capping entity that binds to the actin filaments and inhibits their assembly and disassembly.

66. The most probable protein component in the extract is gelsolin, which in the presence of 1 μM calcium can cut actin filaments thus converting it to a more sol-like form. See Figure 22-39 in MCB (p. 889).

67a. Ultrastructural immunocytochemistry employing the commercially available antibodies to cytokeratins would demonstrate that these fibers contain keratin, an intermediate-filament subunit protein.

67b. If the cells are gently extracted, intermediate filaments should remain intact with the desmosomes. It is critical that the buffer contain Ca^{2+} to ensure that desmin still connects the desmosomes. Although both microtubules and actin filaments will depolymerize considerably during this extraction, intermediate filaments should remain intact. Their continuity could be demonstrated by preparing whole mounts of a monolayer and viewing it with a standard 100-KV transmission electron microscope.

68a. It is *not* possible to conclude from these data that the number of gp135 molecules per cell increased for several reasons. First, because the cell number increased during the study period, the increase in total binding of anti-gp135 shown in Figure 22-7 may be due to the increase in cell number alone. Second, the experimental protocol only labels the apical surface, not the basal surface; thus the observed binding reflects only this portion of the cell.

68b. The ratio is 24:1 apical to basal. This can be determined by counting the gold particles as an indirect indicator of gp135 distribution. Although fluorescence microscopy would reveal the same immunocytochemical staining, this technique is not nearly as amenable to quantification as ultrastructural immunocytochemistry.

68c. One assumption is that gp135 on the apical and basal surfaces has the same portion of the protein exposed to the antibody and thus has the same binding affinity for exogenously added antibody.

68d. One possible control is to repeat the experiment with a membrane protein that does not have a preference for apical or basal domains and thus presumably is not influenced by actin. If the distribution of this protein is not affected by cytochalasin treatment, then the drug probably is acting on the actin filaments and not directly on the membrane proteins, including gp135. An alternative control is to clone the gene encoding gp135, express it in a cell type that does not exhibit polarity, and examine the effect of cytochalasin on the distribution of gp135 in these cells. Again, if no drug-induced change in distribution is observed, then one can conclude that cytochalasin does not act directly on gp135.

68e. Double labeling of cells with fluorescent stains for gp135 and actin could provide evidence to support or refute the hypothesis. For example, actin filaments could be stained with phalloidin-rhodamine and gp135 could be stained with FITC-labeled anti-gp135 antibodies. Examination of such doubly stained cells by fluorescence microscopy would reveal if actin filaments and gp135 colocalize, a result that would support the hypothesis.

68f. The permeability and transepithelial resistance data indicate that the low-calcium medium disrupts the tight junctions in these cells. Thus the proposed experiment would indicate whether the cell-to-cell integrity of tight junctions (and probably to a lesser extent of desmosomes) is, in part, directing the distribution of gp135 molecules.

69a. The data presented in Figure 22-9 do not justify these conclusions. Because only the basal cells divide, they are the only ones that can incorporate [³H]thymidine; however, much of the protein measured in these experiments comes from the terminally differentiated cornified layer. Thus, if this cornified layer is more extensive in one preparation than in the other, the data would reflect this difference, which is not indicative of a difference in the rate of cell division. Likewise, the results with retinoic acid could indicate either that cell division has been stimulated or that the ability of the two epidermal systems to terminally differentiate has decreased.

69b. Light microscopy autoradiography can demonstrate the location of the proliferative cells, over which the autoradiographic grains would be concentrated.

69c. Retinoic acid inhibits expression of K1, K6/K10, and K16 in the SCC-13 cells but does not inhibit expression of any of the indicated keratins in NHEK cells.

69d. Two-dimensional gel electrophoresis would probably separate the comigrating keratins.

69e. A double-labeling experiment in which cells labeled with [³H]thymidine are subjected to immunohistochemical analysis with antibodies against K6 and K16 followed by autoradiography of the same sections would reveal whether only proliferating cells express these keratins. In fact, nonproliferating cells contain K6 and K16.

70a. There are two ways to calculate the change in granule speed resulting from CG1 injection. The slight decrease in granule speed that occurs with CH291 probably reflects the effect of injection itself, since the data in Figure 22-11 shows that this antibody does not bind to any of the tropomyosin isoforms. In all likelihood, the decreased speeds observed with CG3 and CGβ6, which do bind tropomyosin isoforms, also result from an injection effect. Thus averaging the values with CH291, CG3, and CGβ6 and subtracting the value with CG1 would eliminate this injection effect. The calculated change in speed due to CG1 is 12.6 μm/min using this calculation method. Alternatively, the value with CG1 can simply be subtracted from the uninjected value to give a change in speed of 15.9 μm/min; this calculated value, however, includes some experimental artifact.

70b. It is not possible to tell from these data which isoform is necessary for granule movement. Although antibody CG1, which recognizes isoforms 1 and 3, inhibits movement of granules, antibody CG3, which recognizes isoforms 1 and 3, does not inhibit movement of granules.

70c. Absorb CG1 on isoforms 1 and 3 before injecting it into the cells. The absence of a CG1 effect in this case would indicate that CG1 must interact with isoforms 1 and 3 to effect a change in granule speed.

70d. The data in Figure 22-12 suggest that the inhibitory effect of antibody CG1 is reversible with time. Alternatively, protein synthesis subsequent to antibody injection may overcome the effect of CG1.

70e. The purpose of this experiment is to determine if CG1 can strip tropomyosin from actin filaments. If it does, then the tropomyosin isoforms would appear in the soluble supernatant fraction. In fact, as the data in Figure 22-13 show, tropomyosin does not appear in the supernatant, indicating that CG1 cannot cause tropomyosin to dissociate from actin filaments. Since CH291 does not recognize any of the tropomyosin isoforms, it should have no ability to strip tropomyosin from actin filaments. Thus the appearance of soluble tropomyosin in the CH291 sample would indicate that spontaneous dissociation, unrelated to the presence of specific antibodies, occurred under the conditions of this experiment.

71a. Form 1 myosin is a monomer with a bent tail. Form 2 is a monomer with a straight tail. Form 3 can be described as a parallel dimer with intertwined rod sections.

71b. The absence of form 1 and form 3 at high salt concentrations suggests that these forms are stabilized by intramolecular and intermolecular interactions, respectively, which are known to be disrupted at high salt. An alternative explanation is that high salt induces a conformational change in myosin that precludes formation of forms 1 and 3. Little is known about myosin organization in nonmuscle cells, although it is known that myosin molecules overlap to form thick filaments in skeletal muscle. Thus any evidence for regulation of intra-and intermolecular interactions in nonmuscle myosin may provide clues about how myosin filaments are assembled in nonmuscle cells.

71c. React isolated myosin with tagged MY1 antibodies at low salt and then by electron microscope autoradiography determine the distance between the bound antibody on the two myosin monomers in a form 3 molecule. Since the antibody binds at the same site on both monomers, this distance is a precise measure of the amount of stagger in form 3 myosin.

71d. Although there is a correlation between myosin HC phosphorylation and the percentage of form 1 molecules (bent monomers), this is not conclusive evidence because it is not clear from the data that all the incorporated ^{32}P is in bent monomers.

71e. Two controls would be useful: one in which a kinase that does not phosphorylate myosin is used and one in which the nonhydrolyzable analog AMPPNP is substituted for ATP. In both cases, there should be no incorporation of ^{32}P.

71f. After phosphorylation of *Dictyostelium* myosin in vitro, separate the three myosin forms from each other by differential centrifugation and measure the amount of incorporated ^{32}P in each form. Incorporation of ^{32}P only in form 1 (bent monomers) would be more definitive evidence for the hypothesis than the data shown in Figure 22-15. However, this type of experiment might be technically difficult to accomplish.

72a. A double-label immunocytochemical experiment using antibodies to myosin I and to actin filaments could demonstrate any possible colocalization of these components in the isolated plasma membrane preparations.

72b. Possibly some myosin I is attached to the plasma membrane and some is present in the cytoplasm. In this case, the observations in parts (a) and (b) would be consistent.

72c. This finding suggests that most of the myosin I probably is located on the plasma membrane and that the soluble myosin I noted in part (b) is an artifact of the fractionation and separation procedures.

72d. The data do not support this hypothesis. If it were true, the decrease in the amount of membrane-bound myosin I should parallel the decrease in membrane-bound actin in both salt-extracted samples. However, because some decrease in membrane-bound myosin I was observed in both cases, the data are somewhat inconclusive.

72e. Since the data suggest that myosin I is not linked to the plasma membrane via actin filaments, there should be a concentration-dependent increase in the association of myosin I with the KI-extracted plasma membranes.

72f. Preincubation of plasma membranes with actin filaments would not increase binding of myosin I because this binding is not mediated by actin filaments (see answers 72d and 72e).

CHAPTER 23

1. homotypic
2. collagen
3. proteoglycan
4. duplication
5. propeptides
6. C-terminal
7. rough endoplasmic reticulum
8. lysyl oxidase
9. II
10. VI
11. basal lamina
12. hyaluronic acid
13. glycoprotein
14. integrin
15. Ca^{2+}
16. uvomorulin
17. N-cadherin
18. splicing and glycosylation
19. agrin
20. meristem
21. pectins
22. extensin
23. lignin
24. plasmodesmata
25. c d
26. a b c d e
27. b
28. c
29. d
30. a d e
31. b d
32. c d e
33. a
34. b e
35. a d
36. b e

37. The presence of glycine is critical because the side chain of this amino acid is a hydrogen atom, the only R group that is small enough to fit into the space in the center of the three-stranded helical structure characteristic of the collagens.

38. Proline and hydroxyproline cross-link adjacent chains of the collagen triple helix.

39. Fibrous collagen consists of triple-helical collagen I molecules packed together side by side in a staggered array. Adjacent molcules are cross-linked by covalent bonds between two lysine or hydroxylysine residues at the C-terminus of one chain and similar residues at the N-terminus of the adjacent chain. See Figures 23-7 (p. 908) and 23-8 (p. 909) in MCB.

40. This finding suggests that the disulfide bonds are important in stabilizing the association of the three chains in procollagen before formation of the triple helix.

41. Hyaluronic acid has many hydrophilic residues, which can bind water forming a highly hydrated gel-like matrix. In addition, binding of cations by the COO^- groups increases the osmotic pressure of the gel, causing more water to be taken up into the gel. This results in high turgor pressure.

42. Hyaluronic acid is the central component of cartilage proteoglycan aggregates.

43. Because type IV collagen is the major collagen present in the basal lamina, this structure is sometimes referred to as the type IV matrix.

44. Laminin can bind to both cell-membrane receptors (i.e., integrin receptors) and to type IV collagen in the basal lamina, one type of extracellular matrix.

45. This finding suggests that the fibronectin receptor may be a transmembrane receptor that binds actin-related proteins in the cytosol and fibronectin in the extracellular matrix. See Figure 23-27 in MCB (p. 923).

46. This type of experiment provides information about induction. During normal development, induction occurs when two types of cells that originate in different embryonic layers interact, leading to synthesis of tissue-specific proteins in one or both partners of the pair. See p. 929 in MCB for discussion of this phenomenon.

47. Binding of FGFs to heparan sulfate in the basal lamina protects these growth factors from proteolytic degradation, thus ehancing their ability to induce cell proliferation and differentiation.

48. Variations in the spatial configuration of a limited number of cell-adhesion molecules on the surface of neurons could produce a large number of surface recognition markers.

49. Both cellulose and collagen are extracellular matrix components that confer tensile strength on their respective cells or tissues.

50. Certain oligosaccharide fragments, called elicitors, from cell walls of fungi cause the induction of the enzymes for

phytoalexin synthesis in plants. At least one elicitor also induces production of β-glucosidase by infected plants; this enzyme clears the fungal cell wall further, producing more elicitor and thus enhancing the production of β-glucosidase and phytoalexin by a positive-feedback mechanism.

51. Cross-linking of pectins to hemicelluloses forms a complex intercellular network between the cell walls of adjacent plant cells.

52. This organization of the fiber layers imparts more rigidity to these structures than would a random organization.

53. The orientation of cellulose microfibrils is disrupted when plant cells are treated with microtubule-disrupting drugs. Also, there is a correlation between the direction of microtubules and the direction of deposited cellulose microfibrils.

54. Because type I collagen from adults is cross-linked via covalent bonds, it is not soluble in water. Acid can hydrolyze these bonds and thus type I collagen can be solubilized in dilute acid and can be applied to the surface of the dish in a dilute (usually ethanolic) acid solution.

55. No. Although some collagen diseases are thought to result from an inability to assemble triple-helical collagen in the extracellular matrix, all collagens need N- and C-terminal propeptides to aid in the formation of the triple helix. These propeptides are cleaved and not present in the mature, extracellular matrix collagen. Thus denatured type I collagen from normal individuals does not renature to form the native triple helix.

56a. The indicated bands correspond to the α chains recognized by the antibody that are derived from (a) triple-helical procollagen, (b) triple-helical mature collagen, and (c) procollagen. The absence of any extracellular triple-helical procollagen or mature collagen bands in the presence of the drug suggests that the drug inhibits one or more of the intracellular reactions leading to the formation and subsequent secretion of procollagen triple helices (see Figure 23-10 in MCB, p. 911). For example, the effect of the drug might be similar to that of vitamin C deficiency; however, unlike vitamin C deficiency, no degradation occurs, as evidenced by the lack of low-molecular-weight species in the + drug intracellular profile.

Another possibility is that a direct drug-collagen interaction physically prevents the release of procollagen triple helices. Whatever the precise mechanism, the drug permits formation of some type of procollagen species, which is not released and thus accumulates intracellularly. Since these profiles were obtained on denaturing SDS gels, it is not possible to tell whether this species is single stranded or triple stranded.

56b. The indicated bands correspond to (a) triple-helical procollagen, (b) triple-helical mature collagen, and (c) single-stranded procollagen that is recognized by the antibody. These profiles, coupled with those in Figure 23-1a, indicate that the drug inhibits formation of the interchain disulfile bonds among the three procollagen chains, thus preventing formation of the procollagen triple helix. This inhibition leads to an internal accumulation of monomeric procollagen chains, which migrate more slowly on the gel than triple-stranded procollagen.

57. There is probably a defect in the extracellular proteases that clip off the N- and/or C-terminal propeptides. Alternatively, the recognition sequence for these proteases on the collagen molecule could be mutated so that the proteases, although present in wild-type amount and activity, are unable to remove the propeptides.

58. To determine whether collagen synthesis is similar in cells growing on both surfaces, first label both cell forms with [^{35}S]methionine for 30 min. Then immunoprecipitate type IV collagen using an anti-type IV antibody and determine the radioactivity present in each immunoprecipitant. If the same number of counts per minute are incorporated in the cells grown on microporous membranes as in those grown on plastic, then their rates of collagen synthesis are similar. If cells are radioactively labeled for longer periods of time (many hours) and a similar immunoprecipitation experiment accomplished, the possible accumulation of type IV collagen could be similarly quantitated.

59. Experimentaly distinguishing between surface-bound laminin and soluble laminin in this system is difficult. For example, addition of anti-laminin antibodies would reduce the effectiveness of both surface-attached and soluble laminin to stimulate neuronal development. Indeed, no relatively simple method is available by which you could conclusively demonstrate that when laminin is added, only soluble laminin and not surface-attached laminin is present in this system.

60. This proteolytic fragment probably contains the sequence Arg-Gly-Asp-Ser, which is required for binding of fibronectin by cells. CL5 cells pretreated with this fragment before plating on a fibronectin surface would most probably adhere less well to the surface than would untreated control CL5 cells.

61. Epithelial cells are considered morphologically differentiated if they exhibit polarity. In other words, they should have microvilli on the apical surface and the nucleus should be located nearer to the basolateral area. Epithelial cells are considered biochemically differentiated if they secrete or synthesize identifiable proteins similar to those seen in vivo. Thus in the case of mammary cells, the more casein they produce, the more differentiated they would be. This protein could be used as an appropriate biochemical marker for differentiation of mammary cells.

62. Calcium chelators remove calcium from E-cadherin, thus facilitating dissociation, and trypsin degrades cadherin, thus preventing reassociation. Pectinase or cellulase would be the most appropriate enzymes for dissociating plant cells.

63a. No. Any label that would give an indication of cell number could be used in these experiments. Based on its cost, half-life, and other characteristics, ^{35}S is a cost-effective probe.

63b. PMA increases attachment of CHO cells to a fibronectin surface as evidenced by the increase in [^{35}S]methionine on plates treated with PMA compared with control untreated plates.

63c. PMA probably exerts its effect through a specific interaction, since the dose-response curve exhibits saturation.

63d. Additional support for this hypothesis might be obtained from several types of experiments. First, the effects on cell adhesion of other compounds known to increase protein kinase C activity could be monitored. Second, the PMA-dependent increase in protein kinase C could be correlated with the increase in adhesion of CHO cells. Third, the effect of PMA on adhesion of mutant cells that are refractory to a PMA-induced increase in protein kinase C could be determined.

63e. If PMA increased receptor number, then treated cells would exhibit higher maximal adhesion than control cells. This is not indicated by the data. Rather, the PMA appears to increase the affinity of receptors for fibronectin, as evidenced by the lower fibronectin concentration required for 50% half-maximal adhesion by treated cells.

63f. The GRGDSP peptide can interact with the RGD-binding site on fibronectin receptors, thus blocking their subsequent binding to fibronectin.

63g. PB1 would have the same effect as GRGDSP if it recognizes and blocks the RGD-binding site on the receptor. However, not all antibodies to the fibronectin receptor would necessarily block the RGD-binding site.

63h. This finding indicates that most probably PMA specifically modulates the fibronectin receptor or receptor number and is not exerting a nonspecific effect on cell adhesion.

63i. One possible hypothesis consistent with the information given in the problem is that PMA stimulates protein kinase C, which catalyzes formation of a phosphorylated form of the receptor that has a higher affinity for fibronectin than the unphosphorylated form. One approach to testing this hypothesis would be to demonstrate the phosphorylation of the fibronectin receptor in cell membranes in the presence of PMA. Also, the PMA-induced increase in cell adhesion could be correlated with the PMA-induced increase in intracellular protein kinase C. An alternative hypothesis is that PMA modulation of the fibronectin receptor is mediated through actin and/or actin-binding proteins (e.g., vinculin and talin), which are known to be enriched in the regions where cells adhere to the substratum (see Figure 23-27 in MCB, p. 923).

64a. Because mature type I collagen is present only in the extracellular matrix, an antibody that recognizes only type I collagen would not show any intracellular immunofluorescent staining even if procollagen were present in cells. To demonstrate intracellular collagen, it is necessary to use an antibody that recognizes only type I procollagen; to demonstrate extracellular mature collagen, an antibody that recognizes only type I collagen should be used.

64b. The most appropriate positive control would be to make cryosections of dog tissue containing true mesenchymal cells. If these cells showed a positive staining pattern using the same fixatives and similar protocol as with the MDCK cells, then the lack of staining in part (a) would indicate the absence of procollagen was not an experimental artifact.

64c. Type I collagen is not present in the basal lamina.

64d. This finding suggests that there are no tight junctions in fusiform cells. However, a circular, plasma membrane-associated staining pattern probably would appear when the polarized cells are exposed to anti-Z01.

64e. MDCK cells grown on a microporous membrane have a type IV-containing basal lamina, whereas fusiform cells do not. Because procollagen is secreted in vesicles, the extracellular type IV staining exhibits a punctate pattern.

64f. The data presented do not allow you to answer this question.

64g. The cells grown on a microporous membrane have a basement membrane, which probably contains laminin. This bound laminin may exert a negative feedback on the cells that turns off intracellular laminin synthesis. However, fusiform cells, which have no basement membrane, probably release laminin to the surrounding medium; this soluble laminin would become diluted and thus be capable of exerting less negative feedback on intracellular laminin production.

65a. Because cell-adhesion molecules lie on the surfaces of the cells, it would be necessary to obtain a cross-section of a thin section of the cell membrane in order to express the data in terms of square micrometers. With routine thin sectioning, this is not feasible, and the linear measure is more practical.

65b. The fibroblast-containing coculture served as a negative control. If expression of cell-adhesion molecules had been decreased in this case, it would indicate that the coculture-induced decrease observed with Schwann cells and dorsal root neurons was not a neuronal-specific effect.

65c. Add the "conditioned medium" resulting from growth of pure cultures of each cell type to pure cultures of the other cell type. If the coculture effect depends on direct cell contact, then L1 and N-CAM expression in these "conditioned medium" cultures should not be decreased.

65d. The results of this experiment would indicate whether the decrease in L1 and N-CAM in Schwann cells shown in Table 23-1 is reversible. If this decrease is reversible, then the amount of these cell-adhesion molecules in the Schwann cells in cocultures should increase after immunocytolysis of the dorsal root neurons.

65e. The N-CAM antibody probably inhibits cell-to-cell attachment of these embryonic neurons.

65f. A major difference between N-CAMs in embryonic and adult tissue is their PSA content, which is much higher in embryonic neurons than in adult neurons. The high PSA content of embryonic N-CAMs may hinder their ability to bind anti–N-CAM antibodies.

65g. Endoneurominidase N can cleave PSA from embryonic N-CAMs, thus enhancing their adhesive properties and making the cell-to-cell attachment of embryonic neurons similar to that of adult neurons.

66a. These blots indicate that expression of the laminin receptor mRNA peaks before birth (lane 1 in LMR blot), whereas the laminin B1 chain mRNA peaks at about 1 week postnatal (lane 3 in LMB1 blot). The LMA blot shows that no laminin A chain mRNA is expressed in developing mice kidneys during the times sampled.

66b. The F9 cells, which are known to contain LMR, LMB1, and LMA, serve as a positive control. This is particularly important for the LMA blot, which show no bands in any kidney lane (1–5).

66c. No predictions can be made about the protein bands detected in Western blots based on the Northern blots of the corresponding mRNAs. For example, an mRNA may be present, but its protein not yet synthesized. Conversely, a protein may be detected in a Western blot, but if its mRNA is rapidly degraded, it may not be detected in a Northern blot.

66d. The laminin A chain is 400 kDa and the laminin B1 and B2 chains are both ~200 kDa. In this gel system, the latter two migrate very closely together and cannot be easily distinguished from each other. To determine if both B chains are present, a gel system that can separate the B1 and B2 chains must be used.

66e. The doublet may correspond to two receptor subtypes that have similar molecular weights. Another possibility is that partial degradation of the laminin receptor occurred during immunoprecipitation. Still another possibility is that one band is an immature form of the other (e.g., an underglycosylated or uncleaved form).

66f. Because laminin has a binding site for type IV collagen, it may have been copurified with this collagen. During subsequent production of anti–type IV antibodies, some anti-laminin antibodies also would be elicited. Thus the antibody preparation used in the immunoprecipitation experiment analyzed in lane 4 probably was contaminated with anti-laminin antibodies, so that laminin, as well as type IV collagen, was detected.

66g. The immunoprecipitate obtained with the anti–type IV collagen antibodies could be subjected to a Western blot analysis using the anti-laminin antibody. If the contaminating band in lane 4 of Figure 23-6 is a laminin chain, a band should be observed in this Western blot.

CHAPTER 24

1. benign, malignant
2. basal lamina
3. line, strain
4. transformation
5. transforming growth factor
6. plasminogen activator
7. fibronectin
8. oncogene
9. retrovirus
10. long terminal repeat (LTR)
11. promoter insertion
12. Epstein-Barr
13. HIV
14. electrophiles
15. ultimate carcinogens, cytochrome P-450s
16. Ames
17. x-rays and UV radiation
18. tyrosine
19. epigenetic
20. promoters
21. antioncogene
22. b d e
23. d e
24. b c
25. b c e
26. c d e
27. a b c e
28. a e
29. a b d e
30. b c e
31. a c e
32. d e

33.

	Product	Location
a. *jun, myc, fos, ski*	NTF	N
b. *mos, raf(mil)*	PK	C
c. *sis*	GF	S
d. Ha-*ras*, N-*ras*	GTPase	PM
e. *erbA*	THR	N
f. *src, abl, met, fps*	PTK	C
g. *crk*	PLC	C
h. *ros, erbB, fms*	RPK	PM

34. G_0 is the part of the cell cycle in which cells are quiescent, or at rest. While in G_0, the cells have unduplicated DNA. Indeed G_0 is more or less a prolonged part of G_1 — sort of a detour off the G_1 phase — although some cell biologists do not distinguish G_0 from G_1. Rapidly growing cancer cells skip the G_0 phase and proceed through the G_1 phase quickly. Normal cells vary from type to type in the length of time, if any, spent in G_0, but, in general, they spend more time in $G_1 + G_0$ than cancer cells. See Figure 24-4 in MCB (p. 960).

35. In permissive cells, the virus first directs synthesis of early proteins, which induce the cell to shift from the G_1 or G_0 phase to the S phase. Induction of the S phase leads to the late phase of viral infection during which synthesis of viral coat protein, replication of viral DNA, and

production of mature virions occur. This massive formation of virions leads to cell death. In nonpermissive cells, the induction of the S phase by viral early proteins does not lead to replication of viral DNA, probably because the cellular enzymes are not compatible with some viral sequence or viral protein (perhaps the viral origin of replication?). Thus in infected nonpermissive cells, production of virions and cell death do not occur. However, as long as early proteins are produced, infected nonpermissive cells continue to move through the cell cycle without resting in the G_1 phase. In most such cells, the virus is lost or degraded during cell proliferation, and the cells return to normal. Occasionally, the viral DNA is integrated into the genome; in this case, the continued production of viral early proteins in a nonpermissive cell causes the cell to become permanently transformed. See Figure 24-10 in MCB (p. 968).

36. Most retroviral infections in somatic cells are asymptomatic. The retroviral reverse transcript is integrated into the host cell DNA, is transcribed, and can direct synthesis of new virions. However, this uses only a small fraction of the host cell's structures, and since most retroviruses do not carry transforming genes, the host cell can grow relatively normally.

37. First, infection by a *transducing retrovirus* can lead to cell transformation by introducing an oncogene into the host-cell DNA. Second, integration of a *slow-acting retrovirus* into the host-cell DNA can activate a cellular proto-oncogene, leading to cell transformation.

38. Usually when there is one double-strand break in the DNA, there are other breaks. Without a template for repair, repair enzymes sometimes join fragments together in other-than-the-original manner. Joining of fragments from different chromosomes produces chromosomal translocations. These types of rearrangements may lead to cellular transformation by activating a proto-oncogene.

39. Activation of a proto-oncogene to an oncogene may occur when (1) transcription of the gene is increased, (2) a deletion in the gene occurs, or (3) another alteration, for example, a point mutation, occurs in the gene. Alternately, a proto-oncogene product may be "activated" to oncogene-product status by loss of an antioncogene that is producing a functional product.

40. The "two-step" model of transformation proposes that a non-nuclear oncogene product (e.g., a *ras* gene prod-

uct) and a nuclear oncogene product, (e.g., a *myc* gene product) are required to achieve complete transformation. In primary rat embryo cells, for example, transfection with a *ras* oncogene produces morphological changes in the cell but not immortality. In contrast, transformation with *myc* in addition to *ras* leads to immortality and the morphological changes characteristic of transformation. A similar division of activity between gene products has been demonstrated for the papovavirus early-gene products. Evidence demonstrating that this model is probably too simple has been obtained by placing either *ras* or *myc* under the control of very strong promoter and enhancer sequences; in these cases, either oncogene alone can produce both immortality and the morphological changes associated with transformation of a primary cell.

41. Although teratocarcinoma cells are genetically identical to normal cells, their behavior depends on the presence or absence of some unidentified factor(s) in their environment. Epigenetic alterations may underlie formation of malignant teratocarcinomas.

42. The most important factors determining whether a cigarette smoker gets cancer are probably (1) the exposure of the smoker to other carcinogens (initiators) and (2) the level of exposure and length of exposure to cigarette smoke. Another factor, which is probably less important, is the smoker's genetic makeup.

43. The *RB* gene encodes a nuclear protein that may suppress transcription by complexing with transcriptional activators, including those encoded by oncogenes. Loss of the *RB* gene causes transformation and induction of retinoblastoma, an inherited tumor found in children. Because the absence of the *RB* gene product may allow certain oncogene products to function, this gene is referred to as an antioncogene.

44. Retrovirus Q is a transducing retrovirus, which lacks all or part of the genes necessary to make new virions (i.e., the genes encoding the reverse transcriptase, the viral structural proteins, and the envelope protein). The chicken cells used in this study apparently carry the genes necessary to make these proteins, whereas the rat cells do not. In other words, the chicken cells were previously infected with the nontransforming virus from which retrovirus Q is derived and probably carry the viral genes integrated into their chromosomal DNA. The host-cell DNA thus acts as a "helper virus" and provides the viral proteins necessary for production of retrovirus Q virions. If rat cells were coinfected with

retrovirus Q and a wild-type retrovirus encoding its missing or defective proteins, retrovirus Q virions would be produced. See Figure 24-15 in MCB (p. 975).

45a. Most carcinogens cannot act unless they are converted to electrophilic *ultimate carcinogens* by liver enzymes called mixed function oxidases, which include the cytochrome P-450s. The rat liver extract in the Ames test contains enzymes for converting suspected carcinogens to compounds that would be physiologically relevant cancer-causing agents in a mammal.

45b. If the strain of *Salmonella* used in the Ames test had a defective *recA* gene, few revertant colonies would be seen because most changes in the DNA sequence that lead to reversion of the *his* mutation are not caused by the chemical damage to the DNA alone but by mistakes in DNA repair performed by the inducible RecA protein. See pp. 481–482 in MCB for discussion of RecA protein.

45c. Yes. UV radiation, as well as chemical carcinogens, can cause DNA damage that is repaired by the error-prone RecA protein.

46a. Either the early embryonic environment provides some sort of factor (a cell-surface ligand, perhaps) that stimulates the cells to develop normally or the adult environment provides a factor that stimulates malignant development. The observation that teratocarcinoma cells form tumors in a tissue culture, which is not likely to contain any type of factor in appreciable amounts, suggests that the putative factor stimulates normal development in the early embryo and that its absence, in tissue culture and adult animals, leads to tumor formation.

46b. Identifying this factor obviously is not an easy task or it would have been accomplished. It seems unlikely that the factor is secreted by early embryo cells, because explants of early embryo cells into tissue culture also produce tumors. Thus the factor probably is produced by something in the uterine environment other than the embryo cells. Fractionation of mouse uterus or mouse serum would be one reasonable approach. The factor could theoretically be detected by determining whether addition of a fraction to teratocarcinoma cells in tissue culture caused the cells to develop normally. More sophisticated approaches to identification of the factor are certainly possible.

47a. Together these proteases may allow tumor cells to penetrate the basal lamina, capillary walls, and intersti-

tial connective tissue, so that the tumor can spread and establish metastases. There is also evidence that the proteases act on cells to help them maintain some of the morphological characteristics of tumor cells.

47b. Because of the receptors, plasminogen may be activated to plasmin on the surface of tumor cells. The resulting bound plasmin may be unaffected by high levels of serum plasmin inhibitors. Thus plasminogen/plasmin localized to tumor-cell surfaces, rather than the soluble enzymes, may be responsible for tumor invasiveness.

48a. The wild-type A431 cells, which have increased numbers of EGF receptors and decreased ability to undergo differentiation, probably would be more tumorigenic than the EGF-resistant cells. Studies on breast and bladder cancers have indeed indicated that overexpression of EGF receptors is correlated with malignant potential.

48b. It seems likely that the tumor cells are exposed to less than 0.1 n*M* EGF, as the data in Figure 24-1 show that these concentrations stimulate growth and higher concentrations inhibit growth. In fact, serum EGF levels are known to be less than 1 n*M*; the concentration of EGF in extracellular fluid at the site of a tumor is probably even less.

49. One possibility is that this protein disrupts actin polymerization, which is necessary for microfilament formation, resulting in the loss of microfilaments. Another possibility is that the *fgr* protein acts to phosphorylate vinculin, perhaps by associating with it specifically. Since adhesion plaques are composed largely of vinculin, which helps to bind microfilaments to the membrane, it would be reasonable to hypothesize that either of these mechanisms might be related to the losses of adherence and anchorage dependence that are characteristic of transformation.

50a. These data indicate that diethylnitrosamine acts as a tumor initiator, a single dose of which can induce cancer without the addition of other compounds. Phenobarbitol acts as a tumor promoter, whose continued presence can increase the cancer-forming potential of diethylnitrosamine.

50b. If the phenobarbitol diet alone is given, no lesions would be found because a tumor promoter cannot induce a tumor on its own.

51a. A reasonable hypothesis is that the *tat* gene product acts to increase the expression of a cellular gene whose product is, in turn, responsible for the cellular changes leading to Kaposi's sarcoma. This mechanism has not, however, been experimentally confirmed as yet.

51b. These data suggest that the *tat* gene product may be responsible, or at least important, in the development of Kaposi's sarcoma lesions. The findings thus suggest the HIV may cause cancer as well as acquired immunodeficiency syndrome. Interestingly, Kaposi's sarcoma is also more common in males than in females in the human population; even varieties of Kaposi's sarcoma unrelated to AIDS are much more common in males than in females. The reason for the prevalence in males is unknown.

52a. The data suggest that 3T3 cells under "normal" growth control contain a factor in their membranes that regulates the growth of adjacent 3T3 cells. Presumably, 3T3 cells have cell-surface receptors for this factor, and interaction of the factor on one cell with its receptor on an adjacent cell causes the adjacent cell to stop synthesizing DNA (and stop growing). When fraction S_4 is added to sparse 3T3 cells, the cells are "fooled into thinking" that they are touching another cell and reduce their synthesis of DNA; thus this fraction contains the inhibitory factor. In fact, the same effect is seen if 3T3 cell-surface membranes are added rather than the purified factor.

52b. The sensitivity of fraction S_4 to low pH and to heat suggest that the growth-inhibitory factor is a protein.

52c. The data indicate that SV40-transformed cells have lost the normal mechanism of growth control mediated by the S_4 factor. Perhaps transformed cells lack the cell-surface receptor that recognizes the growth-inhibitory factor. Alternatively, the transformed cells may be deficient in some component of the pathway by which recognition of the factor leads to inhibition of DNA synthesis.

53. In order to study the activity of the transmembrane precursors of these growth factors, one must block their proteolytic cleavage. This might be accomplished by addition of a protease inhibitor. However, many cellular events depend on proteolysis, so it would be difficult at best to find a protease inhibitor specific enough for this cleavage that it would not inhibit other events affecting cell growth. Certainly, it would be helpful if an inhibitor were hydrophilic enough that it could not cross the plasma membrane. Another approach, which has been used successfully by Wong and coworkers and Brachmann and coworkers involves blocking proteolytic cleavage by mutating the nucleotide sequence encoding the TGF-α precursor protein to eliminate the proteolytic cleavage site. Cells transfected with plasmids containing such mutated genes expressed a mutant TGF-α precursor on their surfaces but did not release any of the growth factor into the media. Such cells, nonetheless, were capable of stimulating the intrinsic tyrosine kinase activity of the EGF receptor in other cells. Addition of a solubilized form of the mutant TGF-α precursor also promoted anchorage-independent growth of cells in soft agar. In both cases, however, approximately 100-fold more precursor TGF-α was required to produce the same effect as mature TGF-α.

54. The fact that the woman had antibodies to the P and P_1 antigens before the blood transfusion suggested that she had been previously exposed to these antigens. Since expression of altered cell-surface molecules is common in malignant tumors, one possibility is that the tumor cells were expressing P and P_1 antigen on their surface, which would account for the presence of antibodies against these foreign antigens in the woman's serum. In this case, even a small trial transfusion of blood containing P and P_1 antigens would cause a dramatic rise in the level of serum antibodies directed against these tumor antigens, triggering a series of reactions that would destroy the tumor cells displaying the incompatible P and P_1 antigens. Philip Levine and coworkers in Sen-itiroh Hakomori's laboratory, in fact, demonstrated that the antibodies in the woman's serum were induced by the tumor, thus substantiating this explanation.

55a. Comparison of experiments 5 and 6 suggests that mevalonate is required for activity of the Ras protein. When the endogenous synthesis of mevalonate was blocked by compactin (experiment 5), the Ras protein did not promote cell division as it did in the absence of compactin (experiment 2). Addition of mevalonate circumvented the negative effect of compactin, allowing the Ras protein to function to promote cell division (experiment 6).

55b. The addition of a lipophilic, mevalonate-derived group to the Ras protein may provide a membrane anchor for the Ras protein. Support for this hypothesis comes from studies on the localization of a yeast Ras protein in cells with a mutation in post-translational processing that is known to affect the addition of a farnesyl group (a lipophilic mevalonate derivative) to another yeast protein. In the presence of this mutation, the normally membrane-bound yeast Ras protein is cytoplasmic.

55c. If the Ras protein requires mevalonate or a derivative for function, then compounds that inhibit the synthesis of the required compound may be useful in controlling the growth of tumors in which the *ras* gene is active. Of course, the treatment would have to inhibit isoprenoid synthesis in tumor cells enough to reduce Ras protein activation but not severely affect the ability of normal cells to synthesize products necessary for cell viability and growth.

56a. Probes U, V, and W reveal that the tumor DNA from some patients has only one restriction fragment that hybridizes with markers on chromosome 22, whereas the leukocyte DNA from the same patients has two different restriction fragments hybridizing with these markers. This finding suggests that the tumor cells of these individuals have lost at least some of the DNA from one of their copies of chromosome 22.

56b. Retinoblastoma involves a small deletion on chromosome 13.

56c. These data suggest that a deletion on chromosome 22 brings about tumorigenesis. The deletion may occur by a relatively rare somatic-cell mutation. Although a deletion could bring about activation of an oncogene directly by increasing the transcription of a gene or altering a gene, it is not obvious how this mechanism could be inherited. Instead, acoustic neuroma, like retinoblastoma, appears to be brought about by the loss of a gene or genes in a cell. The loss of a normal gene presumably unmasks a recessive, defective, homologous gene that was previously masked by the presence of the normal gene. (It is this defective recessive gene that is the inherited defect.) The fact that the *loss* of this gene causes the tumor suggests that the (normal) gene functions as an antioncogene. Antioncogenes produce products that control the function of another gene or gene product, and their absence allows cells to proliferate, forming a tumor.

56d. Possible explanations include the following:

- The tumors in these patients arose by some mechanism that does not involve the loss of genes on chromosome 22.

- These tumor cells may have the same defective allele as those of other patients, but the change that occurred in the DNA of the normal allele of the tumor-associated gene chromosome 22 may not have altered the size of the restriction fragment produced. For example, a point mutation that made the gene nonfunctional may have occurred.

- A deletion may have occurred affecting the gene associated with the tumor, but in these cases, it did not include the DNA in the restriction fragments that hybridized to the probes.

57a. Transformation apparently requires both the *ras* gene, HPV-16 DNA, and the presence of another effector; both dexamethasone and progesterone are effective. On the basis of the data shown here, however, it cannot be excluded that progesterone could cause transformation in the absence of transfection with one or both DNAs.

57b. A reasonable hypothesis is that the viral DNA contains an oncogene with a nuclear product. This idea is supported by the finding that HPV-16 can transform primary cells in cooperation with an activated *ras* gene, but not in conjunction with activated *c-myc*.

57a. Since steroid hormones can affect gene expression, a reasonable hypothesis is that these hormones affect the transcription of one or both of the oncogenes. Indeed, there are data indicating that the noncoding region of the HPV-16 genome contains a "glucocorticoid-reactive element"; this region is activated by dexamethasone and produces increased transcription of viral genes.

58. Pertussis toxin sensitivity of the *ras*-transfected cells would have suggested that the Ras protein acts as an analog of the G_i protein in the thrombin pathway described. Since this sensitivity was not observed, the Ras protein probably does not act in this way. However, the Ras protein could act, like the G_i protein, to stimulate phospholipase C, if the Ras protein is not ADP-ribosylated or affected by ADP-ribosylation. Alternatively, the Ras protein may act at a later step in the thrombin pathway, or the Ras protein may act in a pathway other than the one described involving the G_i protein.

59. One reasonable approach would be to examine other cell types undergoing differentiation for the presence of the N-*myc* oncogenes at various stages. This could be done either in whole organisms or in cultured cells, such as erythroleukemic cells, that can be induced to differentiate in culture. Another approach might be to investigate the effect of an antisense RNA probe for N-*myc* mRNA on cellular differentiation. The antisense RNA could be introduced into a large cell type by microinjection or could be introduced into cultured cells by transfecting them with an expression vector (plasmid) that contains a gene encoding the N-*myc* antisense RNA. Inhibition of differentiation would imply that the N-*myc* gene product plays a critical role in differentiation.

CHAPTER 25

1. immunoglobulins

2. determinant

3. effector and binding

4. T-cell receptor

5. memory cells or plasma cells

6. somatic mutation

7. tolerance

8. pathogens

9. clonal selection

10. heavy

11. CTLs, T_H cells

12. class switching

13. macrophages and B cells

14. MHC

15. myelomas

16. allelic exclusion

17. antigen-independent

18. complement

19. macrophages and B cells

20. lymphokines

21. a b c

22. a b c d e

23. a b c d e

24. a c e (also b, since the lymphokine known as LT can kill cells)

25. a b c d e

26. a b c d e

27. a c d e

28. b c

29. a b c d

30. a b d e

31a. IgG is secreted by B cells (plasma cells). It recognizes and binds foreign antigens in blood and extracellular fluid; acts as an activator of complement; and crosses the placenta, conferring passive immunity to the fetus.

31b. IL-2 is secreted by T_H cells. It stimulates growth of these cells.

31c. Class II MHC proteins are associated with all vertebrate cells. These proteins act as determinants of "self."

31d. CD4 is located on T_H cells and some other cells such as macrophages and neurons. It functions in antigen presentation and activation of T_H cells and is a receptor for HIV.

31e. IgA is secreted by B cells (plasma cells). It is associated with epithelial cells and is found in many secretions. It functions to bind foreign antigens before they can enter the body proper.

31f. IgM is located on B cells in the blood and extracellular fluid. It activates macrophages, initiates complement action, and recognizes and binds foreign antigens in the blood and extracellular fluid.

31g. IgD is located on the surfaces of many cells, including perhaps B memory cells, primarily in tissues. Its function is unknown.

31h. IgE is secreted by B cells (plasma cells) and is found primarily in tissues. It recognizes and binds to antigens on the surface of mast cells, stimulating release of histamine, which is thought to be an important defense against parasitic infection.

32a. SCID results from lack of the enzyme adenosine deaminase.

32b. Myeloma results from an excess number of B cells.

32c. AIDS results from lack of functional T_H cells.

32d. Allergies result from an excess of IgE leading to excess histamine release by mast cells.

33.

Immunoglobin class	Heavy-chain type	Relative serum concentration
IgA	α	4
IgM	μ	3
IgG	γ	5
IgD	δ	2
IgE	ϵ	1

34a. ID

34b. ID

34c. D

34d. D

34e. D

34f. D

34g. D

34h. ID

34i. I

34j. D

35. The following observations support a selective theory and are inconsistent with an instructive theory of antibody generation:

- Antibodies can be denatured and renatured in the absence of antigen and retain their specificity.

- Antigen-specific antibodies can be detected on the surface of lymphocytes before the specific antigenic stimulus is presented to an animal.

- Clones of B cells and B-cell derivatives (hybridomas) produce a single antibody with a single specificity.

DNA, RNA, and protein sequencing have also provided evidence supporting a selective theory.

36. Macrophages function in antigen presentation to T cells and in phagocytosis of bacterial cells.

37. The diversity of antibodies results from DNA breakage and recombination, use of alternative polyadenylation sites in processing of pre-mRNAs, somatic mutation, and addition of random bases by terminal transferase.

38. When children are first exposed to a particular pathogen, they can mount only a weak primary immune response, which generally cannot prevent infection. In contrast, adults have many memory T cells and B cells resulting from previous exposure to many different pathogens. Because these cells can respond quickly to a pathogen, adults generally are less susceptible to infection than children. However, adults are just as susceptible as children to infection by pathogens to which they have not been previously exposed.

39. T cells contain the T-cell receptor on their cell surface; interact with macrophages, which present antigen to them; produce lymphokines, which stimulate other immune system cells; and can kill cells directly, acting as the effectors of the cellular arm of the immune system. B cells have immunoglobulin on their cell surface, do not interact directly with macrophages, and produce immunoglobulins, which carry out the effector functions of the humoral arm of the immune system.

40. A foreign antigen that has been introduced into a newborn mouse will not elicit an immune response in that mouse when reintroduced after the mouse is mature. However, when the same antigen is administered to an immunologically naive mature mouse, it will produce a normal immune response.

41. A shift in the class of immunoglobulin produced by a B cell and its progeny — called class switching — occurs by recombination between the VDJ region and the downstream constant regions specific for each class in heavy-chain DNA. The constant region for the μ chain (C_μ) lies upstream of the constant regions for all the other heavy chains, including C_γ. Thus an IgM-producing B cell can switch to production of any of the other immunoglobulin classes. However, a cell producing IgG cannot switch to IgM production because its DNA has previously undergone a switch recombination event that eliminated the C_μ region. See Figure 25-40 in MCB (p. 1043).

42. During the recombinations joining the V_κ and J_κ regions in light-chain DNA, a few nucleotides are randomly lost. Because of this random loss of nucleotides, the resulting V_κ-J_κ joint may maintain the original reading frame (i.e., be in phase) or it may change the reading frame (i.e., be out of phase) and generate stop codons downstream in the $V_\kappa J_\kappa$ unit. Since the DNA coding sequence is read in groups of three nucleotides, these random recombinations result in two nonproductive, out-of-phase joints for each productive, in-phase joint. See Figure 25-25 in MCB (p. 1025).

43. A virgin B cell may never encounter antigen, in which case it will die after a few days in circulation. If antigen is encountered, the cell will divide and generate two types of progeny. One is the immunoglobulin-producing cell called the plasma cell, which persists for several weeks in circulation. The other is the long-lived memory B cell.

44. CTLs from a mouse that has been infected with a virus will kill virally infected cells from an identical (syngeneic) mouse, but will not kill virally infected cells from an otherwise identical mouse that differs only in the MHC region of the genome.

45. In the first place, cancers tend to be more common in cells with a high proliferative potential, such as skin cells, gastrointestinal cells, mammary epithelial cells, and red and white blood cells. Secondly, the occurrence of DNA breakage and recombination during differentiaton in T and B cell lineages increases the probability that cellular DNA will either break or rejoin inappropriately. In nearly all cases examined so far, leukemias result from translocation of an immunoglobulin regulatory sequence (including enhancers) to a chromosomal site near a proto-oncogene. As discussed in Chapter 24 of MCB, inappropriate transcription and translation of oncogene products can result in loss of cellular growth control and cancer. In hematopoietic cells, activation of proto-oncogenes would probably not occur in the absence of DNA strand breakage and recombination.

46a. Maternal antibodies against the Rh surface antigen cross the placenta and enter the fetal bloodstream. In the presence of complement, binding of these antibodies to red blood cells, which have bound Rh antigen on their surface, leads to cell lysis. Thus, Rh⁺ newborns whose mothers produce antibodies to the Rh antigen during their gestation will have low red blood cell counts.

46b. IgG is responsible for erythroblastosis fetalis because it is the only class of immunoglobulin that can cross the placenta.

46c. Isohemagglutinins are IgM molecules, not IgG molecules. Because IgM antibodies cannot cross the placenta, they do not react with fetal and newborn red blood cells.

47a. The heavy-chain composition of the immunoglobulins produced by different cell types can be determined by isolating the DNA from each type of immune cell and subjecting it to Southern blot analysis with probes specific for all the possible heavy-chain constant regions.

47b. The correct order is H-J-L-K-I. Since the earliest expressed immunoglobulin is IgH and the cells at this stage contain all the heavy-chain constant regions, the H region must be closest to the variable region in the heavy-chain DNA. The H and J regions are both lost in cells that secrete IgL; thus the H and J regions must be closely linked and must precede the other heavy-chain constant regions. Cells secreting IgI have lost the K and L regions, as well as the H and J regions; thus the I region must be last in the chromosomal order. See Figure 25-40 in MCB (p. 1043).

48a. The variable region of each κ light-chain gene contains one V_κ and one J_κ segment. Random recombination within a genomic library containing 400 V_κ regions and 10 J_κ regions would generate $400 \times 10 = 4 \times 10^3$ functional light-chain genes. See Figure 25-23 in MCB (p. 1024).

The variable region of heavy-chain gene contains one V, one D, and one J segment. Random recombination within a genomic library containing 400 V, 10 D, and 10 J regions would generate $400 \times 10 \times 10 = 4 \times 10^4$ functional heavy-chain genes. See Figure 25-26 in MCB (p. 1027).

48b. Each immunoglobulin molecule contains two identical copies of the light chain and two of the heavy chain.

Therefore, this organism theoretically could generate $4000 \times 40000 = 1.6 \times 10^8$ different immunoglobulin molecules.

48c. If all three reading frames can be used, then the number of possible immunoglobulins would be increased by a factor of three. $1.6 \times 10^8 \times 3 = 4.8 \times 10^8$.

49a. Normal fibroblasts from strain A mice would not be killed because they lack the viral Z antigen.

49b. Z-infected fibroblasts from strain A mice would be killed because they have both the viral Z antigen and the appropriate class I MHC gene product expressed on their cell surfaces.

49c. Normal fibroblasts from strain B mice would not be killed because they lack both the viral Z antigen and the appropriate class I MHC determinant.

49d. Z-infected fibroblasts from strain B mice would not be killed because they lack the appropriate class I MHC determinant.

49e. Q-infected fibroblasts from strain A mice would not be killed because they lack the appropriate viral antigen. (For discussion of the role of MHC gene products in CTL killing, see MCB, pp. 1034–1035 and Figure 25-32.)

50. IgG is the most common immunoglobulin in the blood, and thus the easiest to purify in large quantities. In addition, it is very effective in neutralizing protein toxins such as those found in snake venom, and it has a long half-life in the circulation (approximately 3–4 weeks).

51. These observations imply that phagocytosis and/or metabolism of the antigen is necessary to induce an immune response. Monomeric proteins are not easily endocytosed by macrophages; thus peptides from these proteins will not be generated, nor will the peptides be presented to T cells. Even though nonmetabolizable substances may be endocytosed, the lysosomal enzymes in the macrophage cannot digest the substances; so again no fragments are produced for presentation to T cells at the cell surface. It seems to be generally true that forms of antigen that cannot be endocytosed or de-

graded by macrophages will be tolerogenic rather than immunogenic. Changing the form of a metabolizable substance (e.g., by heat-induced aggregation or by adding it to an easily endocytosed adjuvant) will make the substance immunogenic in most cases. However, the exact mechanism whereby tolerance to self-antigens is developed remains to be elucidated.

52. These observations are consistent with the hypothesis that the causative agent for ankylosing spondylitis is an unidentified pathogen, such as a bacterium or a virus, which may interact with alleles of MHC genes. For example, the pathogen might use the B27 gene product as a cell-surface receptor; only cells with that receptor can become infected by the pathogen. Alternatively, the antigenic determinants of the pathogen may serendipitously resemble the B27 gene product, so that the pathogen is seen as a "self"-antigen and is not attacked by the immune system. A variant of this mechanism is that the antigenic determinants of the pathogen, in combination with particular MHC gene products, are not recognized by the immune system cells; this is called the "hole in the T-cell repertoire" theory. Finally, it is possible that the B27 allele is not directly responsible for the disease but rather is closely linked to a defective gene that is the cause of the disease, or that another gene is also required.

53. IgM is particularly well suited for the early immune response because of its pentameric structure. Since antigen-antibody interactions depend upon a number of noncovalent bonds, the *avidity* of an antibody with 10 antigen-combining sites (IgM) will be much higher than the avidity of an antibody with only two (IgG) or four (IgA), even if the *affinity* of the antigen-combining sites is equal. Since antigens produced during the early phases of an immune response are usually of lower affinity than those produced later (after somatic mutation fine-tunes the antigen-combining region), IgM can function more effectively than the other immunoglobulins during the early response.

54. The polymorphic nature of MHC proteins is most probably the consequence of co-evolution of the vertebrate immune system and pathogens. Consider the case of a primitive organism with a single MHC protein, which enabled it to recognize its own cells as self. A pathogen whose antigens closely resembled this MHC protein could escape the immune system and probably seriously damage or kill the host organism. Mutant organisms with varient MHC alleles would be capable of responding to the presence of such a pathogen, however, and would be spared. Sequential mutations of pathogens

and their host organisms would result in the highly polymorphic MHC proteins that we see in modern vertebrates.

55a. The data indicate that the combination of gp120+ antiserum with anti-gp120 specificity can almost completely inhibit T_H activation. Neither antiserum nor gp120, which is known to bind to the CD4 protein on the surface of T_H cells, can inhibit this activaton if added separately. A hypothesis to explain these data is that gp120, when bound to the surface of T_H cell, can still react with anti-gp120 antibodies. Binding of these specific antibodies somehow blocks T_H activation, which is mediated *via* a cell-surface molecule that binds to the stimulatory agent. The observed inhibition of cell activation indicates that CD4 is necessary for T_H cell activation. Additionally, the data imply that, in the blood of patients with active HIV infection, the combination of viral gp120 and specific anti-gp120 antibody could further cripple the immune system by blocking the activation of T_H cells. This inhibition would be in addition to the known inhibition of HIV-infected T_H-cell function.

55b. The first experiment that should be performed would use purified anti-gp120 antibody (rather than crude antisera) in order to eliminate the possibility that some other component of the serum was participating in the observed inhibition. If purified antibodies (either polyclonal or monoclonal) were as effective as crude serum, then one could use other known activators of T_H cells (e.g., interleukin-1) to test for the effects of gp120, antiserum, and the combination on T_H cell activation by other mediators. Other indirect activators of Ca^{2+} influx could also be tested in order to eliminate the possibility that the combination of gp120 and anti-gp120 antiserum artifactually inhibited the Ca^{2+} transport protein.

56a. The first mechanism — deficiency of the normal products of the missing enzyme — can be eliminated because both inosine and deoxyinosine can be made de novo by mammalian cells. Thus SCID patients probably can produce sufficient quantities of these metabolites despite the absence of ADA.

The similarity in the symptoms associated with ADA and PNP deficiency suggest that both syndromes result from a common toxic effect. Although no common toxic metabolite is obvious from the pathways shown in Figure 25-2, it is possible that at least one of the substrates of both ADA and PNP are toxic to lymphocytes at high concentrations. In fact, high levels of deoxyadenosine and deoxyguanosine are known to be toxic to human lymphocytes. Thus the second mechanism —

accumulation of toxic levels of substrate — probably is involved in the pathogenesis of both of both SCID and the PNP-deficiency syndrome. Currently it is thought that accumulated deoxyadenosine and deoxyguanosine are converted to their respective nucleotides, both of which are potent inhibitors of ribonucleotide reductase; this inhibition accounts for the toxic effects of high levels of these compounds.

In addition, deoxyadenosine is a potent inhibitor of *S*-adenosylhomocysteine hydrolase. This inhibition results in accumulation of S-adenosylhomocysteine, which competes with *S*-adenosylmethionine in important methylation reactions of nucleotide biosynthesis. Thus the third mechanism — accumulation of a toxic metabolite from an alternative reaction — may also operate in pathogenesis of ADA-deficiency SCID.

Since neither ADA nor PNP normally detoxifies any common toxic metabolite, the fourth pathogenic mechanism (accumulation of a toxic metabolite which is normally detoxified by side-reactions of the missing enzymes) seems unlikely.

56b. Evidence for or against the two most likely mechanisms identified in part (a) could be obtained by chemical analysis of bodily fluids from patients for the metabolites that would be expected to accumulate. These are the deoxynucleotides dATP and dGTP, derived from excess deoxyadenosine and deoxyguanosine, and *S*-adenosylhomocysteine, resulting from direct deoxyadenosine-mediated inhibition of *S*-adenosyl-

homocysteine hydrolase. These interrelationships are diagrammed in Figure 25-6.

In fact, patients with ADA-deficiency SCID have high levels of dATP and *S*-adenosylhomocysteine in their lymphocytes. Patients with PNP deficiency have high levels of dGTP but do not accumulate *S*-adenosylhomocysteine in their lymphocytes. The peculiar sensitivity of lymphocytes to elevated levels of deoxyadenosine, deoxyguanosine, and *S*-adenosylhomocysteine must result from the high DNA synthetic rates of these cells. The apparent additional specificity for T cells over B cells observed in SCID patients remains to be elucidated. These data do reinforce the point, however, that lymphocytes have a high rate of DNA synthesis; leukemias derived from these cells can thus be quite susceptible to drugs that affect DNA synthesis.

57. The inability of investigators to find T_s-cell-specific surface markers cannot be construed as conclusive evidence of T_s cells; it is merely a lack of evidence for their existence. ("Absence of evidence is not evidence of absence.") One strategy for resolving the issue might be to undertake an extensive search for a T_s-cell marker, using monoclonal antibodies against populations thought to be enriched in T_s cells. This would entail screening large numbers of monoclonal antibodies for one or more that could inactivate or kill the T_s activity of a T-cell population.

The absence of the I-J gene within the MHC is also not damaging by itself. It is clear from classical genetic

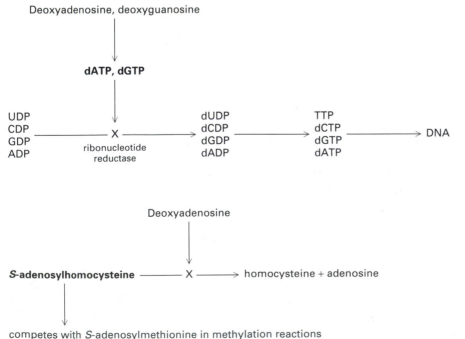

◀ **Figure 25-6**

studies that the polymorphism of the I-J determinants is influenced by the I region of the MHC. Perhaps a class II gene product encoded by this region interacts with an altered T-cell receptor to produce the polymorphic I-J determinants. The polymorphism thus would reside in the uniqueness of the T-cell receptor, rather than in the MHC itself. This hypothesis also would explain the T-cell receptor rearrangements observed in T_s hybridomas. The lack of a classical T-cell receptor rearrangement simply suggests that suppressor T cells might function in a manner not explained by the classical T_H and CTL paradigms.

Although none of the observations against the existence of T_s cells is irrefutable by itself, the sum of these observations do weigh strongly against their existence. Nonetheless, most of the evidence may simply indicate that T_s cells do not function exactly as predicted from studies of T_H cells and CTLs. This controversy should serve to remind us to question our assumptions as well as our data.

58a. Seven of the nonexpressed alleles (lanes e, f, h, i, l, m, and n) have been rearranged; six (lanes a, b, d, g, j, and k) have not. The rearranged alleles are characterized by shorter probe-protected nuclease fragments, which migrate further in the agarose gel. One allele (lane c) has been deleted entirely.

58b. The finding that translationally viable heavy-chain genes are not expressed argues against the feedback model proposed. Other mechanisms must be operating (perhaps in addition to feedback control) to establish allelic exclusion in these cells.

58c. Other mechanisms could include deletion or inactivation of enhancer sequences, inhibition of transcription by DNA methylation, defects of translation initiation, defects in heavy-chain structure that inhibit membrane insertion, or the inability of the expressed heavy chain to form functional dimers with the expressed light chain.

59a. These data argue that memory B cells persist for only a few weeks in the absence of antigen stimulation. The decay of the memory response, seen in Figure 25-4, is rapid and is consistent with a memory B cell half-life only 2–3 weeks in these animals.

59b. It is possible that memory B cells require specific T-cell help in order to be retained in the circulation. Since this adoptive transfer also included T cells, this explanation is unlikely, but it merits further investigation. An alternative explanation is that antigen stimulation is also present in animals for years after first encounter with the antigen. The persistence of antigen in lymph nodes or on dendritic cells is one possibility. This persistent antigen could enable the continued stimulation of B and T cells for many years, allowing the memory B cells to remain active and in circulation. If this hypothesis is true, the memory immune response can be viewed as an ongoing, low-level primary response. Antigen persistence, if proven, would have profound implications for vaccine development as well.

C H A P T E R 26

1. oxygen

2. RNA

3. proteins

4. RNA, protein

5. self-splicing, guanosine

6. transplicing

7. reverse transcriptases

8. stromatolites

9. archaebacteria, eubacteria, and eukaryotes

10. progenote

11. archaebacteria, eocytes

12. endosymbiont, mitochondria, chloroplasts

13. stable

14. exon shuffling

15. a b e

16. b c d

17. c d e

18. d e

19. a b d e

20. e

21. a c e

22. a c d

23. First, you would want to examine the inner part of the meteorite rather than the surface, which is more likely to be contaminated. In an abiotically synthesized sample, nonbiological amino acids, as well as both D- and L-enantiomers of biological amino acids would be present. In organisms on earth, the L-isomers of the twenty plus biological amino acids predominate.

24. Useful properties for the first biological macromolecule to have had certainly include the ability to replicate itself, the ability to catalyze other reactions involving itself, and the ability to carry out as many other complex functions as possible. Because it is unlikely that proteins could have had the ability to replicate themselves, they do not qualify as the primordial molecule of life. Like DNA, RNA can be replicated by complementary strand copying. However, RNA can also function as a catalyst for self-cutting, self-splicing, and self-elongation. The ability of RNA to carry out other complex functions is demonstrated by its present-day ability to act as mRNA, tRNA, and rRNA; as a primer in DNA synthesis; as a part of the protein secretion apparatus; and in other diverse functions. Present-day DNA does not exhibit such a diverse array of functions. The facts that deoxyribonucleotides can be synthesized by all cells from ribonucleotides, that deoxythymidylate is synthesized from deoxyuridylate, and that RNA can be copied into DNA by the enzyme reverse transcriptase, which has now been found in many diverse organisms, all are consistent with the idea that DNA followed RNA in the evolutionary sequence. In all likelihood, RNA was the primordial molecule of life.

25. Three lines of evidence have led to the suggestion that mitochondrial maturases do not act catalytically but simply promote self-splicing. First, there is a homology between the sequence of these mitochondrial introns and the sequence of other self-splicing group I introns. Second, the mechanism of the maturase reactions is analogous to that of self-splicing reactions of other group I introns. For example, both reactions require guanosine as a cofactor, and the guanosine is incorporated into the spliced product. Third, one of maturases has a sequence similarity to a tRNA synthetase, suggesting that it might function similarly; that is, it might bind to RNA in a specific conformation and stabilize that conformation.

26. Transplicing is a mechanism by which useful parts of separate RNA molecules might have been united, leading to RNAs with progressively more complex catalytic and/or coding abilities.

27a. The existence of RNA editing is supported by the finding that mitochondrial cDNA from these protozoans contains (1) some rare cDNAs that match the mitochondrial DNA, (2) a few cDNAs containing some sequences that match the final protein-coding mRNAs and some sequences that match the mitochondrial DNAs, and (3) many cDNAs that match the protein-coding mRNAs but not the mitochondrial DNA. This repertoire of mRNA species strongly suggests that mitochondrial DNA is transcribed in the usual fashion and that the primary transcripts undergo progressive editing proceeding from one end (as evidenced by the partially edited intermediates).

27b. During RNA editing, uridylate residues are introduced into or deleted from a primary RNA transcript. In this way, translatable RNAs are created from untranslatable RNAs by introduction of U to make the AUG start codon and/or by changes in the 3′ end. See Figure 26-15 in MCB (p. 1062).

28. Mitochondria are thought to have been acquired by eukaryotes about 1.5 billion years ago. Perhaps *Giardia lamblia* diverged from other eukaryotes before the fusion that produced mitochondria occurred.

29. Present-day proteins may have been built during evolution by the fusion of separate elements of DNA, each coding for a particular protein structural domain. The fusions may have occurred within the "extra" DNA of the introns. According to this concept, each exon in a present-day protein-coding gene would be expected to code for a particular function or structural domain.

30. In all likelihood, eubacteria (and archaebacteria) have lost most of the introns that were carried by the progenote, thus "streamlining" their genomes and allowing faster replication.

31. DNA is preferable to RNA as the genomic material because it is less chemically reactive than RNA. The deoxyribose in DNA contains fewer hydroxyl groups than ribose; thus fewer derivatives can be formed. The presence of deoxyribose also makes DNA more stable than RNA to alkaline hydrolysis. In addition, the presence of thymine in DNA promotes stability because deamination of cytosine to uracil in DNA can be recognized and excised by cellular enzymes; such a change in RNA could not be recognized and could produce a change in the inherited sequence if RNA were the genomic material.

32. A possible scenario for the origin of the first cells relies on the ability of amphipathic molecules to self-associate and form organized semipermeable membranes. If a sufficient concentration of amphipathic molecules was

isolated in a tide pool at high tide (along with the self-replicating polymers), evaporation of the water would have concentrated the polymers and dried the amphipathic molecules in a film at the bottom of the dried-up pool. When the high tide returned, the action of the waves would have served to disperse the amphipathic molecules, trapping some of the self-replicating polymers inside these structures. This scenario is entirely analogous to the formation of liposomes by vigorous dispersion of dried lipid films in buffer, a method used extensively today to form structures that can encapsulate polymers like DNA, RNA, and protein. This scenario also assumes that the concentration of amphipathic molecules in the primordial sea was sufficiently high that the amphipathic molecules would remain associated rather than monomeric. Thus the origin of cellular life depends not only upon self-replicating polymers and amphipathic molecules, but perhaps also upon the presence of a moon to give rise to tidal rhythms.

33a. The presence of redundant genes for a mitochondrial protein is consistent with two different hypotheses regarding the origin of nuclear genes for mitochondrial proteins. One hypothesis is that cDNA copies of mitochondrial genes, synthesized by reverse transcriptase, became incorporated into the nuclear genome. The other hypothesis is that genes for these proteins were present in both nuclear and endosymbiont genomes originally. The primitive eukaryotic cell probably had genes for electron transport chain and other respiratory enzymes, as did the cyanobacterial precursor of mitochondria. The mitochondrial genes may have been inactivated in many cases, and later lost if the nuclear genes were able to compensate for this loss by producing the appropriate gene product. Such a mechanism would have allowed time for the evolution of the complex machinery and signals necessary to incorporate cytoplasmically synthesized polypeptides into the various mitochondrial compartments.

33b. Both hypotheses are compatible with the available data. However, the first hypothesis would require several more assumptions (e.g., reverse transcriptase in the mitochondria, transfer of nucleic acid through two bilayer membranes); the second hypothesis requires only the reasonable assumption that the precursors of eukaryotic cells had genes for electron transport and oxidative phosphorylation. Experimental evidence for the presence of an intron(s) in the nuclear gene for a mitochondrial protein, especially if the introns are absent from the redundant mitochondrial gene, would argue against the reverse transcriptase model because cDNA copies should not have introns.

34. One possible integrated approach would be to assume that both theories are essentially correct and that two complementary processes operated in prebiotic times. That is, thermal polycondensation of amino acids to yield proteinlike structures and formation of self-replicating RNA-like polymers could both have occurred in the prebiotic phase. Association of these different types of polymers could have eventually led to integration of their divergent properties to yield early self-replicating protein/RNA systems. The addition of a semipermeable membrane (see problem 32) would have added the final component necessary for cellular life.

35a. Scheme III is most consistent with the available data, which imply that mitochondria are very primitive and are more closely related to each other than to prokaryotes. According to this theory, animal, plant, and fungal mitochondria are considered to be endosymbionts derived from primitive free-living cells. These endosymbionts retain some ancient features such as codon usage.

35b. According to scheme III, mitochondria represent the closest living relatives to primitive living systems; indeed, animal mitochondria are considered to be the most primitive and simplest genetic systems known. By the same token, plant cells, containing nuclei, mitochondria, and chloroplasts, represent the pinnacle of cellular evolution.

36a. These data are consistent with the separation of the archaebacteria from the eubacteria at an early time in evolutionary history. In other words, they are consistent with the present thinking that archaebacteria are as unrelated to eubacteria as are eukaryotes, in contrast to former theories that grouped archaebacteria and eubacteria.

36b. The pressure of extreme environments, including very hot and very acidic places, may have selected for ether linkages because these linkages are more stable in such environments than ester linkages, which can be hydrolyzed with relative ease.

37. The evolutionary tree showing the relative times at which these four species diverged is presented in Figure 26-7. To derive the evolutionary relationships among the four species, first determine from the curves in Figure 26-2 the difference in the $T_{\frac{1}{2}}$ values for the A* + A and B* + B homoduplexes and their corresponding labeled heteroduplexes. For example, the difference in

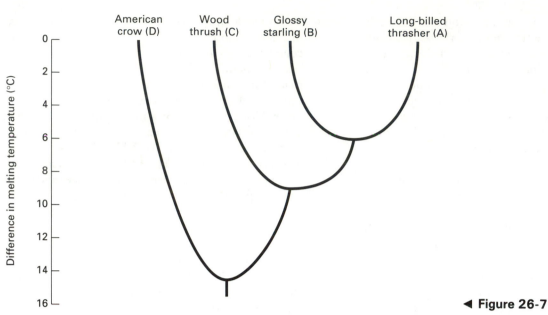

American crow (D)   Wood thrush (C)   Glossy starling (B)   Long-billed thrasher (A)

◀ **Figure 26-7**

the $T_{\frac{1}{2}}$ values for A* + A and A* + D equals 14.5°C (87.0° − 72.5°). From these values, the evolutionary tree shown in Figure 26-7 can be constructed. These data indicate that the American crow (D) diverged first, then the wood thrush (C), and finally the long-billed thrasher (A) and the glossy starling (B).

38a. Based on the sequence data in Figure 26-3a, the total number of base differences between each pair of sequences is as follows:

	A	B	C	D	E
A	—	16	8	17	9
B	16	—	20	15	17
C	8	20	—	17	9
D	17	15	17	—	15
E	9	17	9	15	—

The greater the number of base differences between two DNA sequences, the earlier in evolutionary time they diverged. Thus these pairwise comparisons clearly indicate that species A, C, and E diverged most recently. With eight differences between A and C and nine differences between A and E and between C and E, it seems that each of these species must have diverged from the other two at about the same point in time. Species B and D differ from each other in 15 bases and from species A, C, and E in 15–20 bases; thus B and D clearly diverged both from each other and from the other three species at an earlier, similar point in time.

In fact, the gene shown here is part of the sequence of NAD dehydrogenase, and the species are A = chimpanzee; B = orangutan; C = gorilla; D = gibbon; and E = human.

38b. Tree II best fits the data.

39a. HGS17-1 is probably not the functional gene for this protein because it contains only the exons and has lost all the introns. It represents a **processed pseudogene,** which is thought to arise from integration of a cDNA copy (made by reverse transcriptase) into the genome of the human organism. This probably occurred some time ago in evolutionary terms, since the sequence has diverged somewhat from the present, functional coding sequence.

39b. One way to determine this would be to use these λ clones to transfect cultured cells (e.g., hamster cells) and ascertain if the human DNA could be transcribed (look for complementary mRNA) and translated (look for human ribosomal protein S17). When this was done, only the HGS17-6 clone was transcriptionally active in transfected hamster cells. Another option would be to sequence the ribosomal protein and compare the predicted protein sequences based on DNA sequences with the actual sequence.

40a. Transfer RNAs could evolve simply from the RNA species that were charged with the amino acids. This initially homogeneous population of pre-tRNAs could diverge, accompanied by variation and divergence of the variant tRNA "synthetases." At some later time, the aminoacylating activity of the RNA replicase would be replaced by a protein (or ribonucleoprotein such as RNase P) species, leaving only the aminoacyl-NMP in-

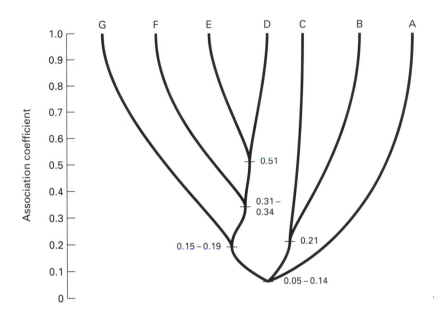

▲ **Figure 26-8**

termediate as a molecular fossil of the original tRNA synthetase.

40b. The model presented in part (a) does not explain the association of specific anticodons with specific amino acids. The original association of amino acids with particular codons would merely reflect a fortuitous affinity of that amino acid for a particular RNA species.

40c. Before these events, there was only one type of RNA, which had template and other functions. After these events, there were two types of RNA: genomic (template) RNA and functional (nontemplate) RNA. This specialization of structure and function was the important breakthrough required for evolution of protein synthesis and living cells.

40d. This hypothesis casts a different light on the following observations among others:

• tRNAs serve as primers for retroviral reverse transcription.

• tRNA genes punctuate the primary transcripts of bacterial and chloroplast operons.

• tRNA coding regions in DNA serve as promoters for transcription by eukaryotic DNA polymerase III.

• $Q\beta$ replicase and other viral RNA replicases require a 3′-terminal CCA for initiation of replication.

41a. The greater the association coefficient S_{AB} between the 16S rRNAs from two species, the more related the two species are. Conversely, the lower the value of S_{AB}, the further back in evolutionary time the two species diverged. For example, the highest association coefficient in Figure 26-6 is $S_{DE} = 0.51$; thus species D and E diverged most recently. The next highest values, $S_{DF} = 0.34$ and $S_{EF} = 0.31$, indicate that species F diverged from both species D and E at a somewhat earlier, approximately identical time. From similar analysis of the remaining data, the evolutionary tree shown in Figure 26-8 can be constructed.

41b. The association coefficients between *Sulfolobus acidocaldarius* (species G) and the other three archaebacteria (species D, E, and F) range from 0.15 to 0.19, whereas its association coefficients with the eubacteria and eukaryotes (species A, B, and C) are 0.06–0.07. Thus *Sulfolobus acidocaldarius* is more closely related to other archaebacteria than to eubacteria or eukaryotes. However, this species certainly diverged from the other archaebacteria at an early evolutionary time, as indicated in Figure 26-8.

Index

DNA primase, 21
cDNA probe, 92
DNA virus, mutation in, 98
domain(s)
    alpha-helical transmembrane, 103
    amino acid, in DNA-binding proteins, 90
    antibody, 221
    functional, in hormone receptors, 90
    plasma membrane, 103
donor atom, 1
dopamine receptor, 173
dorsal root ganglion cell, cell adhesion
        molecule in, 208–209
down regulation, 158
Drosophila melanogaster
    amino acid sequences in, 90–91
    gene control in, 90
    gene library in, 47
    mutation in, 91–92
    P element in, 81, 82
drug resistance, 82–83, 117
Ds element, 83
dunce mutant, 170
duplex DNA, 72, 73–74, 84–85
dyneine, 179, 180

E-cadherin, 206
*Eco*RI, in molecular cell biology research, 45
Edman degradation, 43, 47
EDTA, 206
effector, 51, 54
egg cell, size of, 28
EGTA, 206
elastin, 202
electric gradient, 122
electric potential
    membrane. *See* membrane potential
    in oxidation-reduction reaction, 6
electrical synapse, vs. chemical synapse, 170,
        172
electron, in photosynthesis, 133
electron carrier, 126
    in photosynthesis, 133, 134–135
electron flow
    cyclic, 133
    noncyclic, 133
electron microscope bubble analysis, of DNA
        replication, 98
electron microscopy, 26
    applications of, 27
    freeze-fracture, 134
    transmission, of microtubules, 180
electron spin resonance, 104
electron transport, 125
electrophile, 212
electrophoresis, 44, 45, 46, 47, 74
electrophoretic mobility shift analysis, 92–93
elicitor, 204
elongation factor, 17
Embden-Meyerhoff pathway, 121
Emerson effect, 132
end-plate potential, miniature, 169
endergonic reaction, 2
endocrine signaling, 157. *See also* signaling
endocytic vesicle, 146
endocytosed receptors, recycling of, 114
endocytosed toxins, resistance to, 118
endocytosis, 114, 115
    antigenic, 225
    receptor-mediated, 114, 116, 117, 158
        ligand binding in, 118
        saturation in, 120
    vs. phagocytosis, 117

endonuclease
    assay for, 84
    restriction, 43
        in gene mapping, 46, 47, 48
        in genetic mapping, 46–47
endoplasmic reticulum, 139, 140
    movement of, 193
    protein transport across, 139–146
    rough, 26, 139
        in anterior pituitary cells, 29
        density of, 29
        marker molecule for, 28–29
        in Schwann cells, 29
    smooth, 26
        in ceruminous gland cells, 29
        density of, 29
        in Leydig cells, 29
        marker molecule for, 28–29
        in Schwann cells, 29
endosome, 114, 118
endosymbiont hyposthesis, 232, 234
endosymbiosis, 148
energy
    activation, 2
    forms of, 2
    free. *See* free energy
    potential, 4
energy pump, 6
enhancer, transcriptional, 61, 64
enkephalins, 168
entactin, 202
enthalpy, in hydrophobic interaction, 5
entropy, 2, 3, 5
    in hydrophobic interaction, 5
envelope
    nuclear, 148–149
        creation of, 151
        lamins in, 150
    viral, 114, 117
enzyme(s), 2. *See also* specific enzymes
    active sites on, 8, 9
    basal activity of, 12
    binding sites on, 8, 11
        catalytic activity of, 9, 11, 12
    in DNA replication, 98
    genetic, 74
    hydrolysis sites on, 12
    Krebs cycle, 124–125
    lysosomal, 145
        activation of, 141
        in osteoclasts, 29
    in metabolic pathways, 11
    in molecular cell biology research, 45
    in nucleotide salvage pathway, 34
    phosphorylation of, 8. *See also*
        phosphorylation
    restriction, in gene mapping, 43, 46, 47, 48
    substrate affinity of. *See* K_M
    substrate binding by, 8, 12
eocyte, 232
epidermal growth factor, 216
    cancer and, 215
epigenetic condition, 212
epinephrine, cAMP and, 158, 161
epithelial cell
    culture of, 39
    differentiation of, 205–206
    dissociation in, 206
    intestinal, size of, 28
Epstein-Barr virus, 212
equilibrium, chemical, 2
equilibrium constant, 5–6
equilibrium density–gradient centrifugation,
        vs. differential velocity centrifugation, 28

erythroblastosis fetalis, 224
erythrocyte, glucose uptake by, 119
erythrocyte membrane, 104, 107, 110
erythroleukemia, 33
Escherichia coli
    β-galactosidase synthesis and, 53 culture of,
        39–40
    lysogenic, 56
    in molecular cell biology research, 35
Escherichia coli DNA polymerase I, 97
Escherichia coli DNA polymerase III, 97
Escherichia coli dnaA, 97
ESCK epithelial cell layer, 108
estrogen, 157
    vitellogenin mRNA and, 93–94
ethanol, 122, 124
ethylene, 160
eubacteria, 232
    separation of from archebacteria, 235
euchromatin, 71
eukaryote, 232. *See also* bacteria
    cell-type specific proteins in, 88
    characteristics of, 27, 63
    constituents of, 17
    cytoskeleton of, 26
    development in, molecular genetics of,
        87–94
    DNA integration into, 35, 37
    DNA replication in, 97
    DNA synthesis in, vs. prokaryote, 37
    gene control in, 87–94
    gene expression in, 64
    mitochondria of, evolution of, 232, 233
    processes occurring in, 27
    transformed cells in, 34
    vs. prokaryote, 26
eukaryotic chromosome, 71–77
    characteristics of, 72
    molecular anatomy of, 79–85
eukaryotic elongation factor, phosphorylation
        of, 93
evolution
    cell, 231–239
    efficiency of, 83
    mobile elements in, 81
evolutionary trees
    avian, 235–236
    construction of, 239
    DNA sequencing and, 236
    molecular, 234–235
excitatory junctional potentials, synaptic
        vesicles and, 175
exocytosis, 25
exon, 62
    in protein evolution, 233
    stability of, 232
exon-intron mapping, 82
exon shuffling, 232
exonuclease, 44
exothermic reaction, 2
extensin, 202
extracellular matrix, 201–210
    cell differentiation and, 207–208
    functions of, 202
    spatial and temporal expression of, 209–210

facilitated diffusion, 113, 114–115
facilitator neuron, 168
facultative anaerobes, 122, 124
familial hypercholesterolemia, 117
fat molecules, hydrophobic interaction of, 5
fatty acids, 12–13
    in epithelial cell culture, 39